Moss Flora of China

English Version

Hookeriaceae—Thuidiaceae

6

Authors' Addresses

Wu Peng-cheng (see above)

Lin Bang-juan
South China Institute of Botany, Academia Sinica
Guangzhou 510650, Guangdong, China

Gao Chien (see above)

Cao Tong
Life and Environment College
Shanghai Normal University
Shanghai 200234, China

Li Zhi-hua (see above)

Benito C. Tan (see above)

Si He (see above)

Jia Yu (see above)

Wang Mei-zhi
Institute of Botany, Academia Sinica
Beijing 100093, China

Fu Xing
Institute of Applied Ecology, Academia Sinica
Shenyang 110015, Liaoning, China

Sun Jun
Institute of Applied Ecology, Academia Sinica
Shenyang 110015, Liaoning, China

Zhong Ben-gu
Department of Biological Sciences & Biotechnology
Guizhou Normal University, Guiyang, 550001, China

The national boundaries of China on this map are drawn after the 1:4M "Relief Map of People's Republic of China" published by China Cartographic Publishing House in 1989

CHINA'S ADMINISTRATIVE DIVISIONS

ANHUI – Southeast
BEIJING – Northeast
FUJIAN – Southeast
GANSU – North Central
GUANGDONG – Southeast
GUANGXI – South Central
GUIZHOU – South Central
HAINAN – South Central
HEBEI – Northeast
HEILONGJIANG – Northeast
HENAN – Southeast
HONG KONG – Southeast
HUBEI – Southeast
HUNAN – Southeast
JIANGSU – Southeast
JIANGXI – Southeast
JILIN – Northeast
LIAONING – Northeast
MACAO – Southeast
NEI MONGOL – North Central
NINGXIA – North Central
QINGHAI – North Central
SHAANXI – North Central
SHANDONG – Northeast
SHANGHAI – Southeast
SHANXI – Northeast
SICHUAN – North Central
SOUTH CHINA SEA ISLANDS – Southeast
TAIWAN – Southeast
TIANJIN – Northeast
XINJIANG – Northwest
XIZANG – West Central
YUNNAN – South Central
ZHEJIANG – Southeast

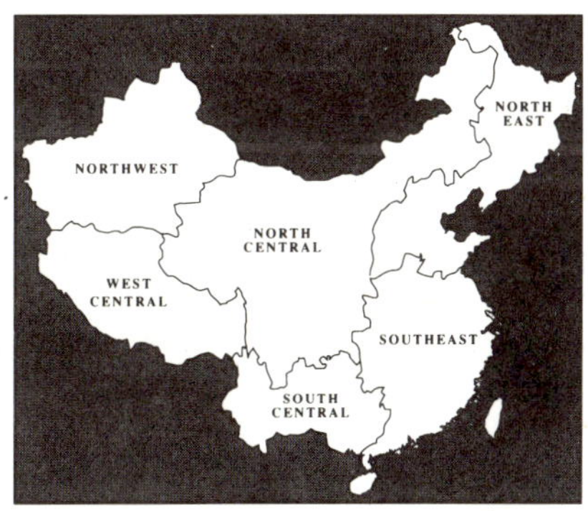

Moss Flora of China

English version
Volume 6. Hookeriaceae–Thuidiaceae

中国藓类植物志

英文版

第六卷： 油藓科—羽藓科

Wu Peng-cheng
Marshall R. Crosby
Editors-in-Chief

Si He
Editor

With Contributions from
Wu Peng-cheng, Lin Bang-juan, Gao Chien, Cao Tong, Li Zhi-hua, Benito C. Tan, Si He, Jia Yu, Wang Mei-zhi, Fu Xing, Sun Jun, and Zhong Ben-gu

Science Press (Beijing, New York)
&
Missouri Botanical Garden Press (St. Louis)

Moss Flora of China
English Version
Volume 6

ISBN 7-03-010289-4/Q·1141
ISBN 1-930723-12-1 (V.6)
ISBN 0-915279-72-X (entire work)

Science Press (Beijing)
16 Donghuangchenggen North Street, Beijing 100717, China

Missouri Botanical Garden Press (St. Louis)
MBG Press, P.O. Box 299, St. Louis, MO 63166-0299, U.S.A.

Cover illustration: *Abietinella abietina* (Hedw.) Fleisch., drawn by Guo Mu-sen.

Contents

Foreword

China contains one of the tectonically most important and complicated terrains in the world. It has three terraces from the eastern coast to the Qinghai-Xizang (Tibet) plateau in the west, with the highest altitude of 8,848 meters at the peak of Zhumulangma (Mount Everest), the so-called "roof of the world." The complicated topography and diverse climate of China, coupled with its great longitudinal (74°–134° east) and latitudinal (18°–54° north) ranges, have resulted in tremendous vegetational and floristic diversity. China is the only country in the world with an unbroken landmass extending from tropical to boreal regions, represented by continuous transitions of plant communities from tropical, subtropical, temperate to boreal forests, and from deserts and steppes to alpine tundras or to permanent glaciers. The well-known diversity of vascular plants found in China is paralleled by the diversity of bryophytes.

China has about 2,500 species of mosses and has the richest and the most diverse moss flora in the Northern Temperate zone. This number is about 20% of all the mosses currently recognized in the world. Knowledge of the Chinese moss flora, like that of the vascular plant flora, is of importance to the understanding of the underlying dynamics of past plant migrations, vegetational history, and the significance of local refugia in Asia. In some instances, bryophytes are better species indicators for past vegetational changes or directions of plant migration than vascular plants, because they survive in sheltered microhabitats after the macro-environments have been altered or destroyed. The Chinese moss flora is also significant to the understanding of the floristic composition and geographical distribution of the bryophytes of North America, Europe, and temperate Asia, since many Chinese taxa have evolved disjunctive ranges either between eastern Asia and eastern North America or between eastern Asia and western Europe. It can be said that not much progress in the bryophyte geography of temperate Asia can be counted upon without adequate knowledge of Chinese bryophytes.

Until the early 20th century, very little was known about Chinese bryophytes either in China or elsewhere. Between the late 19th and the early 20th centuries, European missionaries and explorers collected bryophytes while traveling in China. The mosses were studied and, in many cases, described as new species by western bryologists, such as V. F. Brotherus, H. Dixon, W. Mitten, and C. Müller. The study of Chinese bryology began to grow in China in the late 1930s. Greater advances in our knowledge of Chinese mosses were achieved during the 1950s and 1960s due to the work of the late professor Chen Pan-chieh and his students. Unfortunately, for nearly fifteen years from the late 1960s to the early1980s bryological research in China progressed very little when compared with that in North America, western Europe, and Japan.

Two important developments in the study of Chinese bryophytes have occurred recently. First, numerous Chinese bryologists have traveled abroad for training and studying in North America, Europe, and Japan, resulting in revisions of numerous families and genera. At the same time, western bryologists have visited China in increasing numbers to work with Chinese colleagues. These activities have greatly increased the number of Chinese specimens available to western researchers. In addition, several major expeditions have been organized by the Chinese Academy of Sciences together with several U.S., Canadian, and Finnish academic institutions to locations in remote regions of Guizhou, Hainan, Heilongjiang, Henan, Hubei, Hunan, Liaoning, Qinghai, Sichuan, Yunnan, and Xinjiang.

Second, the Chinese Academy of Sciences (CAS) established the Editorial Committee of the *Cryptogamic Flora of China* in 1973 and in the early 1980s initiated an important project, *Flora Bryophytorum Sinicorum*, which parallels *Flora Reipublicae Popularis Sinicae* (FRPS). Joint revisions of the FRPS are now being carried out by U.S., European, and Chinese botanists, and the results are being published. For mosses, two volumes of the *Flora Bryophytorum Sinicorum* have been published, and the third is in press. These works are largely the results of a series of expeditions to Qinghai-Xizang (Tibet) and Hengduan Mts. (Southwest China) supported by the Chinese Academy of Sciences between 1980 and 1986. Future volumes of the Chinese mosses will continue to be published in the next few years. However, even when the entire *Flora* is published, much of the information on Chinese bryophytes will still be inaccessible to western bryologists, because it is written in Chinese.

The desirability of producing a *Bryophyte Flora of China* in English is evident. The project was conceived at a U.S.-China bryological workshop held in Beijing in March 1990. As a result of this meeting an agreement was reached to jointly produce the English version of the *Bryophyte Flora of China*, and at the same time to continue to promote fieldwork in remote areas of China. A formal agreement was signed by Zeng Cheng-kui and Peter H. Raven on July 2, 1992, and renewed on July 29, 1998. Since its inception, the project has received full support from the Missouri Botanical Garden and the CAS, which considers the *Bryophyte Flora of China* to be one of the important projects in the Academy's basic science research. The project has also received support from Harvard University (U.S.A.), the University of Alberta (Canada), the University of Helsinki (Finland), and the Hattori Botanical Laboratory (Japan).

To mobilize the cooperation and support for this project from the worldwide bryological community, an international editorial committee was established. The committee members include five Chinese bryologists (Gao Chien, Hu Ren-liang, Li Xing-jiang, Li Zhi-hua, and Wu Peng-cheng) and five non-Chinese bryologists (Marshall Crosby, Zen Iwatsuki, Timo Koponen, Benito Tan, and Dale Vitt). The Institute of Botany, CAS, Beijing, and the Missouri Botanical Garden, St. Louis, are the coordinating and managerial centers of the project. Si He and Jia Yu are the project coordinators responsible for the management of the project.

The *Bryophyte Flora of China*, English version, is an offshoot, but not a direct translation of the ongoing project, *Flora Bryophytorum Sinicorum*. The current project is a separate program with different emphases on the format and taxonomic treatments. The treatments in the English version are based on revisions resulting from an international collaboration between Chinese bryologists and their western colleagues. Rather than fully illustrating each species, the *Moss Flora of China*, English version, includes only illustrations of East Asiatic and Chinese endemic species. All data dealing with nomenclature, Chinese types, morphological description, distribution, and habitat information will be widely disseminated through Internet access or other computerized interactive communication systems.

Over the past few years, the editorial committee members have obtained funds from the Chinese, Japanese, Canadian, and U.S. governments to carry out revisionary works and manuscript preparation. The activities carried out under this program have promoted botanical exchanges among the international bryological communities. The present publication has greatly enhanced our understanding of the classification of Chinese mosses and will lead to a greater access to Chinese moss collections by bryologists in the West. The printed *Flora* and the computerized database will become indispensable and authoritative references for bryologists, botanists, ecologists, foresters, conservation biologists, horticulturists, students, and others interested in the rich diversity of Chinese mosses. We are very grateful and much indebted to all bryologists and people involved in the project and to those institutions and governmental agencies for providing financial support to the project.

Zeng Cheng-kui
Editor-in-chief
Editorial Committee
Cryptogamic Flora of China
Chinese Academy of Sciences

Peter H. Raven
Director
Missouri Botanical Garden

Introduction

The *Moss Flora of China*, English version, is a series of eight volumes dealing with all mosses known to be native or naturalized in China. It contains descriptions, identification keys, illustrations of Chinese endemic and East Asian taxa, habitat information, distributional ranges outside China, specimen citations, and distribution maps of all accepted taxa. The *Flora* is arranged phylogenetically and basically follows the same sequence of the families treated in the *Genera Muscorum Sinicorum* (P.-C. Chen et al. 1963, 1978), which adopts a modified classification system of Brotherus (1924, 1925). Nevertheless, the circumscription of some families will reflect our present understanding of the groups. Genera and species are arranged alphabetically for ease of use.

The present volume (volume 6) is the third to be published in the series. It treats 54 genera, 189 specific and infraspecific taxa. The following families are treated: Hookeriaceae (9 gen., 24 spp., 8 subspp.), Symphyodontaceae (1 gen., 5 spp.), Leucomiaceae (1 gen., 1 sp.), Hypopterygiaceae (4 gen., 19 spp.), Theliaceae (2 gen., 6 spp.), Fabroniaceae (10 gen., 29 spp.), Leskeaceae (9 gen., 26 spp., 2 subspp.), Anomodontaceae (4 gen., 19 spp.), and Thuidiaceae (14 gen., 50 spp.). Volume 3 of the *Flora* will follow afterwards. It will treat the Grimmiaceae, Ephemeraceae, Funariaceae, Splachnaceae, and Tetraphidaceae. Future volumes of the series will be published more or less in accordance with the publishing schedule of the *Flora Bryophytorum Sinicorum*.

Taxonomic treatments are presented from the level of families to genera to species and infraspecies. Since the taxonomic accounts presented in this *Flora* are not intended to be monographic, intermediate ranks above species level, such as subsections, sections, subgenera, and subfamilies are not employed. Familial names are provided without citations of authors and bibliographic references. For genera and recognized species, authors and bibliographic citations are given. Each taxon has a description, illustration references, if not illustrated in the *Flora*, and information about its distribution. Either subspecies or varieties are recognized. Forms are not treated. Generic and species descriptions are condensed, giving emphasis only to features critical for determination. All keys are dichotomous, indented, and in general artificial.

All morphological descriptions and keys are based on Chinese specimens examined, except where no material was available to the authors, information from literature is used. When certain important structures are not observed in Chinese material, descriptions of the missing structures may be based on non-Chinese material or cited from published references. Different opinions among authors on the circumscription of certain taxa are discussed briefly, if necessary. Descriptions of families are followed by the number of genera and species occurring in China and other comments. If a species represented in China has more than one infraspecific taxon, a full description is provided for the species and only diagnostic features are presented under each infraspecific taxon. If an infraspecific taxon of a species occurring in China is not the autonym, there will be no description under the species, but a full description will be given under the infraspecific taxon. All descriptions follow a conventional order from gross morphology to cellular details, and they are in botanical English. The terminology basically follows Magill's (1990) *Glossarium Polyglottum Bryologiae*.

The synonymy is generally limited to Asiatic taxa and those that were reported from China by Redfearn, Tan and He (1996). The synonyms given were based on examination of accepted names or type specimens in consultation with opinions presented in other major publications dealing with East Asian mosses. Type localities of all accepted taxa are given. Type information of all Chinese taxa, including that of the Chinese synonyms, is provided with detailed data. At the generic level, synonyms used in the literature are included. At the specific and infraspecific levels, only main synonyms, especially those used in major floras of neighboring countries, are included. Names included in the literature due to misidentifications are not considered to be synonyms. Synonyms of taxa occurring outside of China and those of uncertain taxa are not included. *Nomina nuda* are generally not listed. The bibliographical citations of all recognized taxa and their basionyms cited in this treatment have been verified.

All measurements are metric. If only one measurement of a character state is given, it stands for length. For dimensional measurements, both length and width are used and they are presented as length × width. A range of measurements is separated by a dash (–). In case of discontinuous states found in a measurement, such character states are given in parentheses ( ).

The information about distribution of a taxon in China is generally given in the section of representative Chinese specimens examined. Geographic names are arranged in an order of decreasing political magnitude from provinces or autonomous regions to county municipalities, which are the smallest units cited in the flora. The word "province" is omitted after a provincial name. The word "City," however, is given after a city name. County is abbreviated as "Co." following the county names. The official spellings of Chinese provinces and municipalities are as follows: Anhui, Beijing, Chongqing, Fujian, Gansu, Guangdong, Guangxi, Guizhou, Hainan, Hebei, Heilongjiang,

Henan, Hubei, Hunan, Jiangsu, Jiangxi, Jilin, Liaoning, Nei Mongol, Ningxia, Qinghai, Shaanxi, Shanghai, Shanxi, Shangdong, Sichuan, Taiwan, Tianjin, Xinjiang, Xizang, Yunnan, and Zhejiang. For distribution ranges outside China, generally only those countries that are adjacent to China are listed. For a widely distributed taxon, more general descriptions, such as circumboreal, Eurasia, the Pacific Islands, Old World Tropics, Europe, North America, and South America, are given. Each specimen citation is presented in the following order: province or autonomous region, county, collector with given names abbreviated followed by family names, collection number or the date of collection, and herbarium acronyms, indicating where the examined specimen is deposited.

Each recognized taxon in the *Flora* has a Chinese name. It is usually the same as the one that was used in the *Flora Bryophytorum Sinicorum*. The Chinese names are followed by their corresponding transliteration in *Hanyu Pinyin*. A new Chinese name is given for the taxon that was not named previously in Chinese literature. Two indexes are included at the end of each volume: one for botanical Latin names and another for Chinese moss names.

Abbreviations for western authors' names follow mainly Sayer et al. (1964). The alphabetized spellings of modern Chinese authors' names are based on *Hanyu Pinyin* when the author's name is not romanized. The Chinese authors of botanical names are spelled out exactly the same as those appearing after the scientific names in the original publications. Generally, Chinese authors names in reference citations are given as follows: family names are spelled out completely preceded by given names, which are abbreviated using a capital letter followed by a period and a hyphen to connect each syllable. Chinese authors' names are spelled out completely in the title page and family treatments with a hyphen to indicate the breaking of two syllables.

Abbreviations of scientific journals follow Bridson's *Botanico-Periodicum-Huntianum/ Supplementum* (BPH/S) (Pittsburgh, Hunt Institute, 1991). The book abbreviations follow Stafleu & Cowan's (1976–1988) *Taxonomic Literature*, ed. 2 (TL-2), but the convention for capitalization is consistent with BPH.

Acknowledgments

The U.S. National Science Foundation grants (DEB-9525938, DEB-0072657) to the Missouri Botanical Garden and the grants from the China National Natural Science Foundation, Chinese Academy of Sciences, and the Ministry of Sciences and Technology to the Editorial Committee of the *Cryptogamic Flora of China* financially supported this work. Preparation of the work was also financially supported by authors' home institutions from various countries, particularly from the Institute of Botany (Beijing), Institute of Applied Ecology (Shenyang), Kunming Institute of Botany (Kunming), South China Institute of Botany (Guangzhou), Zhongshan University (Guangzhou), East China Normal University (Shanghai), and Guizhou Normal University as well as from the Missouri Botanical Garden (St. Louis) and Harvard University (Boston). Funding provided by the governments of Canada, Japan, Finland, and the Finnish-Chinese Botanical Foundation in supporting the visits of Chinese authors is very much appreciated.

The Farlow Cryptogamic herbarium of Harvard University, the Hattori Botanical Laboratory, the National Natural Science Museum of Tokyo, the University of Alberta (Edmonton), and the University of Helsinki are particularly thanked for the support in making their facilities available for the authors. The scientific exchanges with the following bryologists are greatly appreciated: Drs. Timo Koponen, Pekka Isoviita, Sinikka Piippo, Dale Vitt, Benito C. Tan, William R. Buck, Marshall Crosby, Si He, Andries Touw, Zen Iwatsuki, P. K. Holmgren, and G. Moggi. The Curators of the Botanische Staatssammlung München (M), British Museum (BM), Missouri Botanical Garden (MO), Muséum National d'Histoire Naturelle (PC), and New York Botanical Garden (NY) are thanked for the arrangements of loans. Hu Ren-liang, Li Zhi-hua, Wang You-fang, and Marshall R. Crosby are gratefully acknowledged for their reviews of the manuscripts. Peter H. Raven is deeply appreciated for his unconditional support and commitment to this project, as well as to other bryological research programs in China. Numerous people have played significant roles at various stages of the project. They include Zeng Cheng-kui, Wei Jiang-chun, An Jian-ji, Tian Jin-xiu, Han Xing-guo, Fu De-zhi, Qin Hai-ning, Robert E. Magill, Bruce Allen, Barbara Mack, and Paul L. Redfearn, Jr. Publications staff, Zeng Jian-fei, Peng Ke-li, and Huo Chun-yan of the Science Press and Amy S. McPherson, Diana Gunter, and Victoria Hollowell of the Missouri Botanical Garden Press, are deeply appreciated for their editorial assistance and publication support.

Illustrations in this volume are largely adopted from volume 6 of the Chinese edition of *Flora Bryophytorum Sinicorum* with some modifications and rearrangements. Therefore, illustrators of that publication are gratefully acknowledged. They include Guo Mu-sen, Lin Bang-juan, Gao Chien, Cao Tong, Wang Wen-he, Zheng Pei-zhong, Li Zhi-hua, and Wu Peng-cheng. The illustration legends, organization of the drawings, and the distribution maps are all prepared by Si He.

HOOKERIACEAE

油藓科 you xian ke

by Lin Bang-juan and Benito C. Tan

Plants small to medium-sized, often soft and flaccid. *Stems* rarely branched, often flattened, loosely to densely foliate, 4–8 ranked. *Leaves* variable in outline, often bordered; costa single, double or absent, usually ending below the apex; leaf cells smooth or papillose, often rounded or hexagonal, lax; alar cells not differentiated. *Setae* elongate, smooth or scabrous; capsules oblong-ovoid to cylindrical, inclined or horizontal; peristome double; exostome teeth often with a wide median furrow line, papillose or striolate; opercula conic to rostrate. *Calyptrae* mitriform, lobed or ciliate at base, smooth, scabrous or pilose.

The family has 27 genera worldwide, of which 9 genera, consisting of 23 species, 2 subspecies and 7 varieties, are treated in China.

1. Costa single 2
1. Costae double, short or absent 3
   2. Leaves complanate, heteromorphic; upper laminal cells rounded hexagonal or polygonal; exostome teeth striate-lamellate .... 7. *Distichophyllum*
   2. Leaves not complanate, isomorphic; upper laminal cells rhomboidal to oblong; exostome teeth papillose .... 6. *Daltonia*
3. Costae elongate, strong, with at least one branch reaching half the leaf length 4
3. Costae short (never more than half the leaf) or absent 7
   4. Leaf borders strongly differentiated; upper laminal cells rounded hexagonal, more than 30 μm in diameter .... 5. *Cyclodictyon*
   4. Leaf borders not or weakly differentiated; upper laminal cells oblong, rhomboidal to elongate, if rounded hexagonal, less than 30 μm in diameter 5
5. Stems erect; leaf borders weakly differentiated; margins entire; exostome teeth papillose .... 1. *Actinodontium*
5. Stems creeping; leaf borders not differentiated; margins denticulate or serrate; exostome teeth cross-striate 6
   6. Plants green to yellowish green; costae reaching near the apex, strongly ridged, toothed or lamellate abaxially; laminal cells hexagonal to short rhomboidal .... 2. *Callicostella*
   6. Plants often purplish green; costae reaching about the leaf middle, not protruding abaxially; laminal cells oblong to rhomboidal .... 9. *Hookeriopsis*
7. Leaves with distinct borders .... 3. *Calyptrochaeta*
7. Leaves without borders 8
   8. Margins entire; leaf cells broadly oval, large, lax, more than 18 μm wide .... 8. *Hookeria*
   8. Margins serrate; leaf cells irregularly rhomboidal to shortly linear, projecting or papillose, less than 18 μm wide 9
9. Costae double, short, with two unequal forks (longer one never more than half the leaf); leaf cells only projecting at upper ends; operculum short-rostrate .... 4. *Chaetomitriopsis*
9. Costae very short, mostly less than 1/4 the leaf length; leaf cells dorsally papillose; operculum long-rostrate .... *Chaetomitrium* (reported from Taiwan, not treated here)

1. ***Actinodontium*** Schwaegr. 假黄藓属 jia-huang xian shu

Sp. Musc. Frond., Suppl. 1, 2(2): 75. 1826.

Plants small, flaccid, tuft-forming, yellow to yellowish green, often glossy. *Stems* flattened, up to 1.5 cm long, simple or sparsely branched; central strand not differentiated. *Leaves* crowded, shrunken when dry, not complanate, erect-spreading when wet, oblong-lanceolate, gradually acuminate; leaf margins entire, without differentiated border, involute at lower half; costae double, extending beyond midleaf; laminal cells elongate or rhomboidal, becoming oblong-hexagonal apically and rectangular basally. *Autoicous*. *Setae* erect, smooth, short; capsules erect, ovoid-cylindrical; opercula conic-rostrate; exothecial cells non-collenchymatous; exostome teeth narrowly lanceolate, keeled, papillose. *Calyptrae* mitrate, laciniate at base, covering only the apex of capsule.

This is a small genus with about seven species worldwide. Two species are known from tropical Asia and one from China.

1. ***Actinodontium rhaphidostegum*** (C. Müll.) Bosch & Sande-Lac. 皱叶假黄藓 zhou-ye jia-huang xian Bryol. Jav. 2: 37. 1862. *Hookeria rhaphidostega* C. Müll., Syn. Musc. Frond. 2: 677. 1851. Type. Indonesia: Java. *Blume s.n.*

Plants small, flaccid, tuft-forming, yellowish green. *Stems* up to 1.5 cm long. *Leaves* crowded, shrunken when dry, not complanate, erect-spreading when wet, oblong-lanceolate, gradually acuminate, 3–5 mm × 0.5–1.0 mm; leaf margins entire, without differentiated border, involute at lower half; costae double, extending beyond midleaf; laminal cells elongate or rhomboidal, 80–125 µm × 15–21 µm, becoming oblong-hexagonal apically and rectangular basally. *Setae* smooth, 4–6 mm long; capsules ovoid-cylindrical, 1.0–1.5 mm long; exothecial cells non-collenchymatous; exostome teeth papillose. *Calyptrae* mitrate, laciniate at base, covering only the apex of capsule.

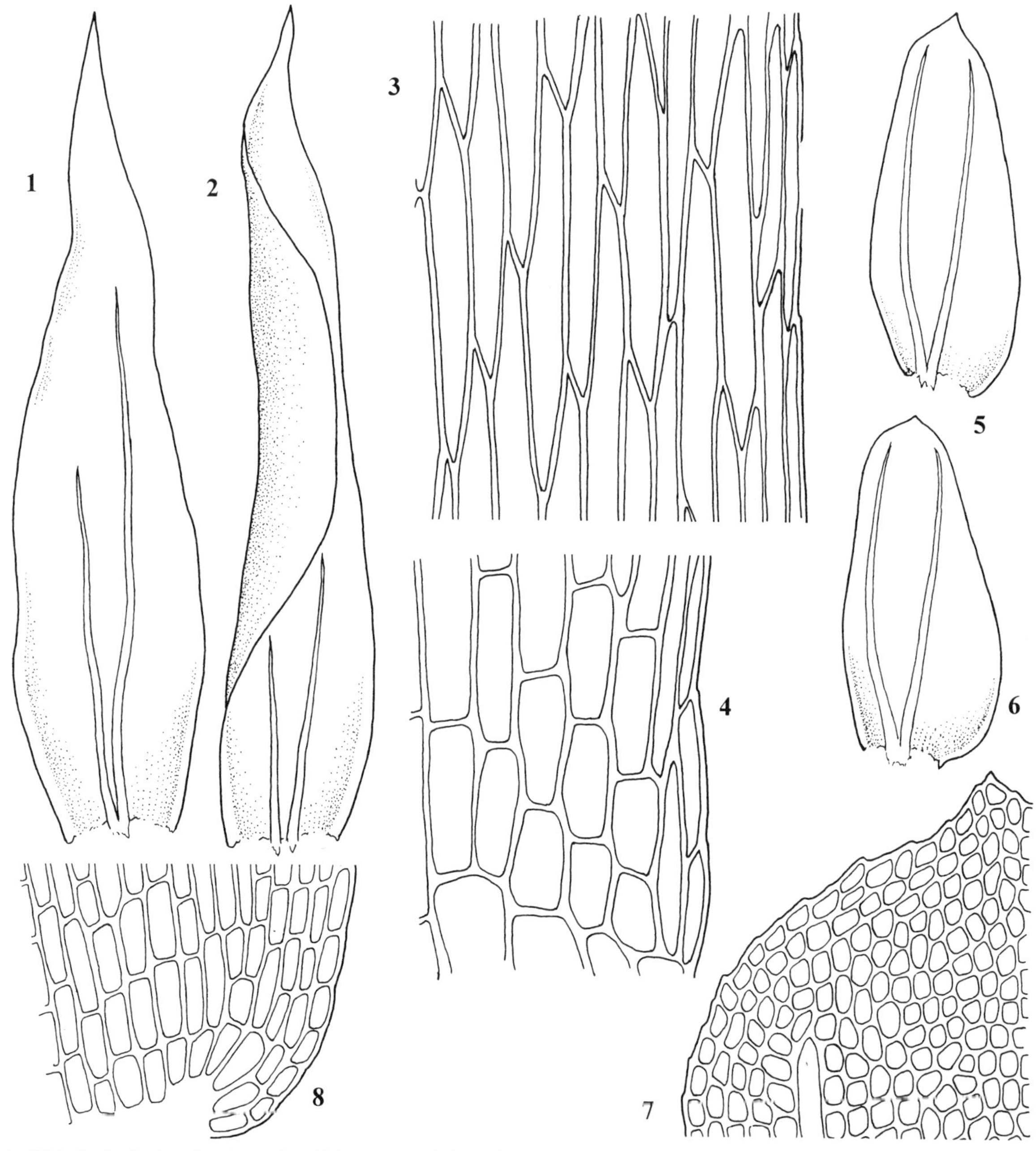

Plate 396. 1–4, *Actinodontium rhaphidostegum*. 1, leaf (× 53). 2, perichaetial leaf (× 53). 3, upper leaf cells (× 530). 4, basal leaf cells (× 530). 5–8, *Callicostella prabaktiana*. 5–6, leaves (× 42). 7, apical leaf cells (× 430). 8, basal leaf cells (× 430). (1–4, drawn from *C.-K. Wang 203*, NICH; 5–8, from *P.-Z. Zheng 303*, IBSC.)

Illustrations: Plate 396, figs. 1–4.
Habitat: on humus over rocks.
Distribution: China, Indonesia, Malaysia, and the Philippines.

Chinese specimen examined: TAIWAN: Nan-tou Co., *C.-K. Wang 203* (NICH).

This species is easily recognized by its ascending growth pattern, oblong-lanceolate, bicostate leaves with gradually acuminate apices and undifferentiated leaf borders, rhomboidal to elongate laminal cells, and papillose exostome teeth.

2. ***Callicostella*** (C. Müll.) Mitt. 强肋藓属 qiang-le xian shu
J. Proc. Linn. Soc., Bot., Suppl. 1: 136. 1859. *Hookeria* sect. *Callicostella* C. Müll., Syn. Musc. Frond. 2: 216. 1851.

Plants often slender, medium-sized, mat-forming, light yellow, dark green to dark brown in color. *Stems* long creeping, subpinnately branched, rhizoids dense, secondary stems 3–5 cm long, in cross section with several layers of colored, slightly thickened co[illegible]cal cells surrounding a parenchymatous core, central strand absent. *Leaves* complanate, forming dorsal, lateral, and ventral rows, flexuous and plicate when dry, ovate to oblong-lanceolate, rounded or abruptly short-acuminate, toothed above; costae double, strong, reaching leaf apex; leaf margins plane, asymmetrically folded on one side at leaf base, without differentiated borders, irregularly serrulate above, entire near base; leaf cells oval-hexagonal, mostly papillose, elongate, smooth below. *Setae* lateral, smooth or weakly papillose distally; capsules ovoid-cylindrical, horizontal to inclined, strongly constricted at mouth; exostome teeth cross-striate, with wide median furrows; opercula conic-rostrate. *Calyptrae* mitrate, laciniate at base. *Spores* spherical.

This primarily South American and African genus has about 26 well-established species and nearly 100 published species names worldwide. In tropical Asia, five species were reported, and two species occur in China.

1. Cells of dorsal and lateral leaves mostly papillose .... 1. *C. papillata*
1. Cells of dorsal and lateral leaves smooth .... 2. *C. prabaktiana*

1. ***Callicostella papillata*** (Mont.) Mitt. 强肋藓 qiang-le xian
J. Proc. Linn. Soc., Bot., Suppl. 1: 136. 1859. *Schizomitrium papillatum* (Mont.) Sull., U. S. Expl. Exped., Musci 23. 1859. *Hookeria papillata* Mont., London J. Bot. 3: 632. 1844. Type. Indonesia: Java.

Plants mat-forming, light yellow, dark green to dark brown. *Stems* 3–5 cm long. *Leaves* ovate to oblong-lanceolate, 1.5–2.0 mm × 0.4–0.7 mm, broadly acute to abruptly acuminate, at times mucronate; costae double, strong, reaching leaf apex, protruding abaxially forming a spine or a lamella distally; laminal cells oval-hexagonal, 13–22 μm wide, moderately thick-walled, becoming smaller apically, rectangular and oblong basally, 22–35(–45) μm × 13–16 μm, unipapillose in dorsal and lateral leaves, often smooth in ventral leaves. *Setae* 1–2 cm long; capsules ovoid-cylindrical, 1.0–1.5 mm long. *Spores* 15–18 μm in diameter.

Illustrations: P.-J. Lin & B. C. Tan 1995 (Fig. 22, A–F).
Habitat: on shaded and wet root stumps, soil, rocks, decaying logs, and tree trunks; alt. 50–970 m.
Distribution: China and paleotropical.

Chinese specimens examined: GUANGDONG: Bo-lou Co., *Crosby 15309* (MO), *P.-Z. Zheng 1033* (FH, IBSC), *P.-Z. Zheng 1034* (MO); Mt. Ding-hu, *D.-K. Chen 1* (IBSC, MO). HAINAN: Chang-jiang Co., *P.-J. Lin et al. 254* (FH, IBSC); Mt. Diaoluo, *P.-J. Lin 3545* (IBSC, MO); Mt. Jian-feng, *P.-C. Chen et al. 318a* (FH, H, IBSC, MO, PE); Le-dong Co., *Reese 17805* (MO); Ling-shui Co., *P.-C. Chen 403* (FH, IBSC), *Reese 17815* (MO); Qiong-zhong Co., *P.-J. Lin et al. 2539* (FH, IBSC). HONG KONG: Tai Po Kau, *P.-J. Lin et al. 1046, 1063, 1071, 1090, 1882, 1886* (IBSC); Mui TSZ Lam, *P.-J. Lin et al. 1543, 1569* (IBSC); Mt. Tai-mo, *P.-J. Lin et al. 2050, 2056, 2066* (IBSC); Shing Mun, Lead Mine Pass, *P.-J. Lin et al. 1245* (IBSC). TAIWAN: Tai-pei Co., *S.-H. Lin s.n.* [Bryophytes of Taiwan Exsiccatae ser. 3, no.106] (MO, NICH); Tao-yuan Co., *M.-J. Lai 0276* (MO). XIZANG: Mo-tuo Co., *Y.-G. Su 4608* (MO). YUNNAN: Jing-hong Co., *Redfearn et al. 33725a* (MO); Meng-yang, *W.-X. Xu 6767* (MO).

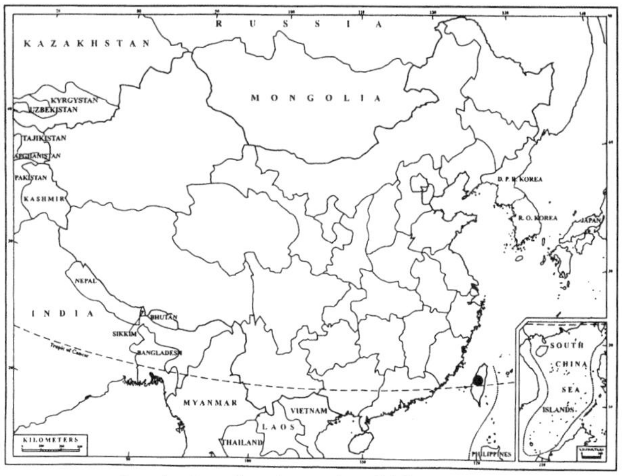

1. *Actinodontium rhaphidostegum*

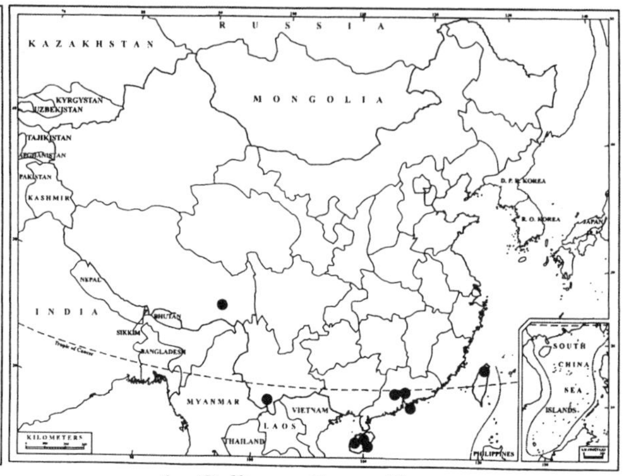

1. *Callicostella papillata*

The papillae of leaf cells, the marginal serration, and the leaf apices of this species are highly variable. Sometimes, the leaf cells, especially those of the ventral leaves, become smooth when this species grows in wet or very shady habitat.

2. ***Callicostella prabaktiana*** (C. Müll.) Bosch & Sande-Lac. 平滑强肋藓 ping-hua qiang-le xian
Bryol. Jav. 2: 40. 1862. *Schizomitrium prabaktianum* (C. Müll.) H. Miller, H. Whittier & B. Whittier, Bryophyt. Biblioth. 16: 251. 1978. *Hookeria prabaktiana* C. Müll., Syn. Musc. Frond. 2: 678. 1851. Type. Indonesia: Java.

Plants resembling *C. papillata* in morphological details, except that the leaf cells of this species are smooth and the spores are smaller (11–13 μm in diameter). The rhizoids are variable from smooth to weakly papillose.

Illustrations: Plate 396, figs. 5–8.
Habitat: on rocks, sandy soil or tree trunks by streamlets; alt. 175–350 m.
Distribution: China and Malaysian regions.

Chinese specimens examined: HAINAN: Le-dong Co., Mt. Jian-feng, *P.-Z. Zheng 303* (FH, IBSC); Ling-shui Co., *P.-J. Lin et al. 3306* (FH, IBSC).

3. ***Calyptrochaeta*** Desv. 毛蒴藓属 mao-shuo xian shu
Mém. Soc. Linn. Paris 3: 226. 1825.

Plants brownish green, in lax tufts. *Stems* simple or branched, with complanate, flexuous leaves when dry, flat, laxly foliate, usually with conspicuous clusters of brown, filiform propagula in upper leaf axils. *Leaves* dimorphic, lateral leaves spreading, leaves of dorsal and ventral rows smaller and erect, ovate, shortly acuminate, bordered, serrate above; costa short, forked; laminal cells lax, smooth, rhomboidal or fusiform in the upper half of the leaf, becoming narrowly rectangular at base. *Perichaetial leaves* smaller. *Setae* lateral, spinose; capsules ovoid or oblong-ovoid, horizontal; exostome teeth striolate below, becoming papillose above; opercula conic, beaked. *Calyptrae* campanulate, lobed at base.

This genus was known in the earlier literature as *Eriopus* Brid., an illegitimate name. It is characterized by ascending stems with complanate foliation, often bearing attenuate branches with highly branched, filiform propagula in the leaf axils. The very short double costae separate it from the genera with strong double costae, such as *Actinodontium*, *Callicostella*, *Cyclodictyon*, and *Hookeriopsis*.

Calyptrochaeta has about 30 species worldwide, and 2 species are recognized in China.

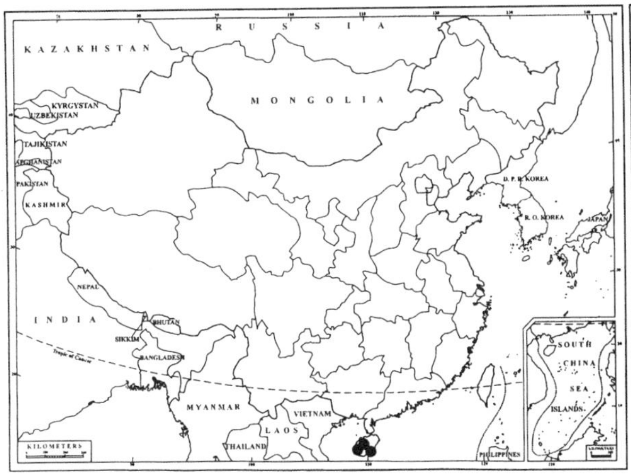

2. *Callicostella prabaktiana*

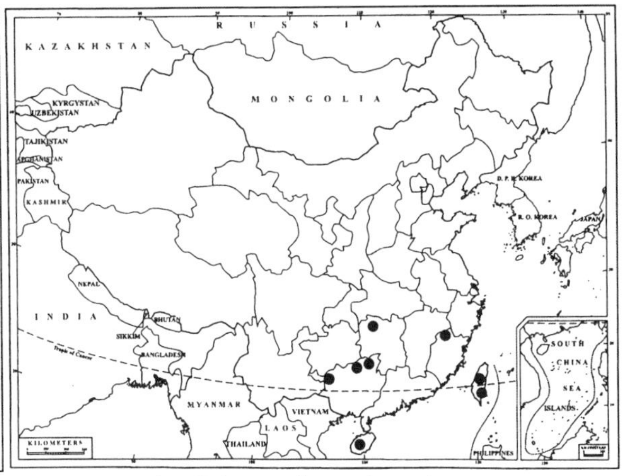

1. *Calyptrochaeta japonica*

1. Leaf margins weakly toothed; upper leaf cells shortly oval-hexagonal (1.5:1), thin-walled .... 1. *C. japonica*
1. Leaf margins strongly toothed; upper leaf cells oblong-rhomboidal to fusiform .... 2. *C. ramosa* subsp. *spinosa*

1. ***Calyptrochaeta japonica*** (Card. & Thér.) Iwats. & Nog. 柔叶毛蒴藓 rou-ye mao-shuo xian

J. Hattori Bot. Lab. 46: 236. 1979. *Eriopus japonicus* Card. & Thér., Bull. Acad. Int. Géogr. Bot. 17: 2. 1907. Type. Japan: Naze, *Ferrié s.n.*

Eriopus mollis Card., Bull. Soc. Bot. Genève, ser. 2, 3: 278. 1911.

Plants tufted, yellowish green when dry, simple or branched, with complanate, flexuous leaves. *Stems* up to 3 cm long, lacking central strand in cross section. *Leaves* dimorphic, lateral leaves larger, obovate, 2–3 mm × 1.0–1.8 mm, median leaves broadly ovate; leaf apices acute to mucronate; laminal cells smooth, shortly oval to rhomboidal, 34–56 µm × 27–34 µm (about 1.0–1.5:1), thin-walled, becoming longer at the lower half of the leaf; leaf margins plane, weakly toothed, with 1–2 rows of differentiated rectangular to elongate cells, thin-walled. *Dioicous* or autoicous. *Perichaetial leaves* much smaller in size, ovate, about 1.0 mm long, cuspidate. *Sporophytes* not seen.

Illustrations: Plate 397, figs. 1–2.
Habitat: on shaded, humicolous soil, rocks and boulders, wet walls and trunk bases; alt. 580–2300 m.
Distribution: China and Japan.

Chinese specimens examined: FUJIAN: Mt. Wuyi, *P.-J. Lin 2587* (FH, IBSC). GUANGXI: Long-sheng Co., *P.-C. Wu et al. 1047* (IBSC, MO, PE); Rong-shui Co., *M.-Z. Wang 46332a* (MO, PE); Tian-lin Co., *P.-J. Lin et al. 2507* (FH, IBSC). GUIZHOU: Mt. Fan-jing, *Crosby 15726* (MO), *B.-C. Tan 91-1183* (FH); Sui-yang Co., B.-C. *Tan 91-977* (FH). HAINAN: Bai-sha Co., *Redfearn 35784* (MO). HUNAN: Zhang-ja-jie City, *D.-K. Li 17919* (IBSC). TAIWAN: Nan-tou Co., *M.-J. Lai 2552* (as *Eriopus mollis*, MO), *S.-H. Lin 35* (FH, NY); Tai-tung Co., *S.-H. Lin s.n.* [Bryophytes of Taiwan Exsiccatae ser. 3, no.115, as *Eriopus parviretus*] (FH, MO).

2. ***Calyptrochaeta ramosa*** (Fleisch.) Tan & Robinson subsp. ***spinosa*** (Nog.) P.-J. Lin & Tan 叉枝毛蒴藓刺边变种 cha-zhi mao-shuo xian ci-bian bian-zhong

Harvard Pap. Bot. 7: 29–30. 1995. *Calyptrochaeta spinosa* (Nog.) Ninh, Acta Bot. Acad. Sci. Hung. 27: 159. 1981. *Eriopus spinosus* Nog., J. Sci. Hiroshima Univ. Ser. B, Div. 2 (Bot.), 3: 51. 1937. Type. China: Taiwan, Mt. Taihei, *Noguchi 6601a.*

Plants tufted, yellowish green to dark green. *Stems* simple or branched, 1.5–4.0 cm high, with complanate, flexuous leaves when dry, without a central strand in cross section. *Leaves* dimorphic, lateral leaves oblong-obovate, 1.5–5.0 mm × 1–3 mm, median leaves ovate to oblong-ovate, acute to long mucronate; laminal cells mostly oblong-fusiform in the upper half of the leaf, (50–)72–101(–112) µm × 22–33 µm, moderately thick-walled, becoming narrowly rectangular at base; leaf margins plane, strongly toothed, at times bi-geminate, with 1–2 rows of differentiated linear cells, thick-walled, not obviously pitted. *Perichaetial leaves* small, ovate, 2 mm long, abruptly acuminate to cuspidate. *Perigonial leaves* small, ovate to lanceolate. *Setae* lateral, about 5–6 mm long, spinose; *capsules* ovoid; exostome teeth as *C. japonica*; opercula beaked. *Calyptrae* not seen.

Illustrations: Plate 397, figs. 3–6.
Habitat: on rocks, tree trunks, trunk bases, and decaying logs in dense forests; alt. 400–2200 m.
Distribution: China, Nepal, and Vietnam.

Chinese specimens examined: GUANGDONG: Ru-yuan Co., *P.-J. Lin 880* (FH, IBSC). GUANGXI: Rong-shui Co., *M.-Z. Wang 46690a* (MO, PE); Tian-lin Co., *P.-J. Lin et al. 2448* (FH, IBSC); Xing-an Co., *P.-J. Lin et al.*

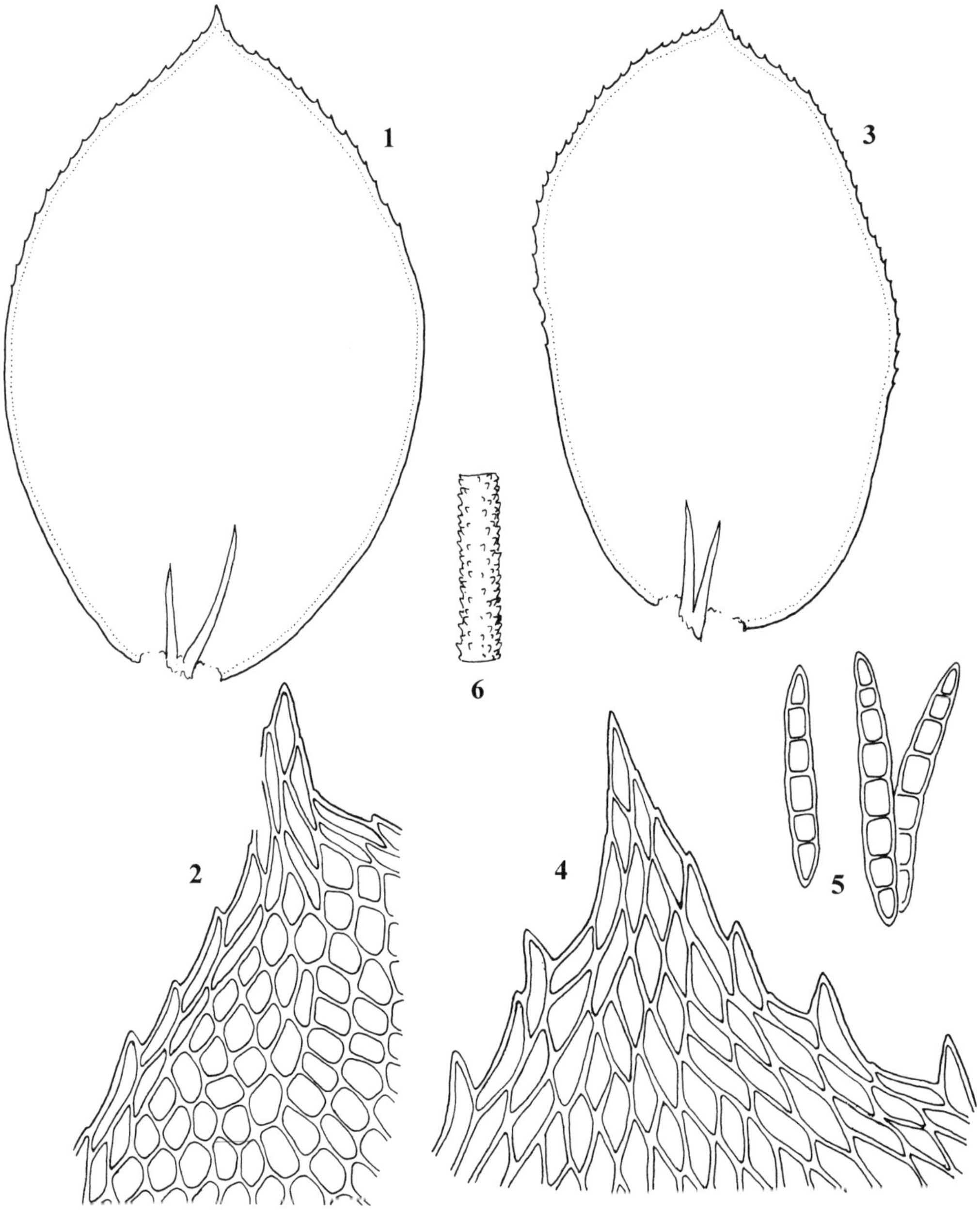

Plate 397. 1–2, *Calyptrochaeta japonica*. 1, leaf (× 45). 2, apical leaf cells (× 220). 3–6, *Calyptrochaeta ramosa* subsp. *spinosa*. 3, leaf (× 26). 4, apical leaf cells (× 200). 5, gemmae (× 200). 6, a portion of seta (× 26). (1–2, drawn from *P.-J. Lin 2587*, IBSC; 3–6, from *P.-J. Lin 367*, IBSC.)

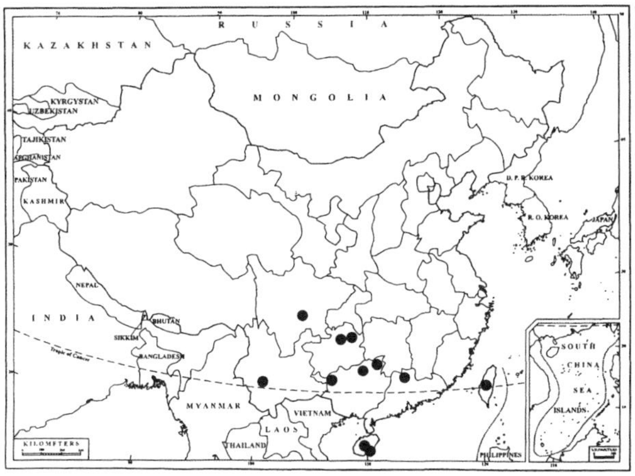

2. *Calyptrochaeta ramosa* subsp. *spinosa*

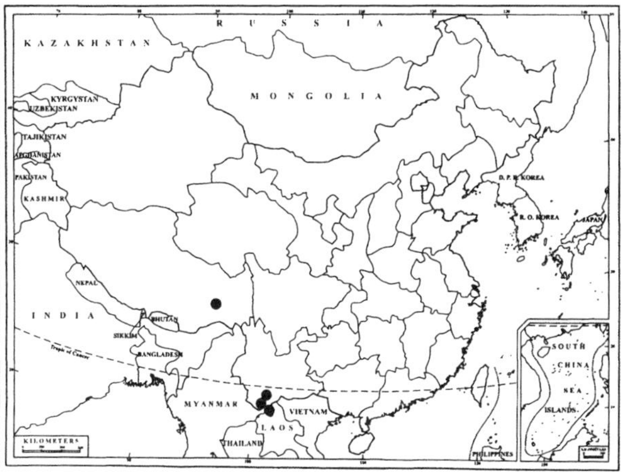

1. *Chaetomitriopsis glaucocarpa*

1294 (FH, IBSC). GUIZHOU: Fanjing Nature Reserve, *Crosby 15838* (MO); Sui-yang Co., *Crosby 15977* (MO), *Tan 91-977* (FH). HAINAN: Mt. Bawangling, *Reese 17586* (IBSC, LAF, MO), *P.-J. Lin s.n.* (FH, IBSC); Mt. Diaoluo, *Y.-X. Lin 397* (IBSC); Mt. Limu, *P.-J. Lin et al. 131* (FH, IBSC), *P.-J. Lin et al. 86* (MO); Ling-shui Co., *Redfearn 36179a* (MO, SMS, as *C. japonica*); Mt. Wuzhi, *P.-J. Lin et al. 2009* (FH, IBSC). SICHUAN: Mt. Omei, *P.-J. Lin 3282* (IBSC). TAIWAN: Mt. Ali, *Iwatsuki 2155* (FH, IBSC, NICH); Mt. Daxue, *Iwatsuki 2855b* (FH, IBSC, NICH). YUNNAN: Jing-dong Co., *J.-K. Zhang 1094-2* (FH, IBSC).

4. ***Chaetomitriopsis*** Fleisch. 灰果藓属 hui-guo xian shu
Musci Buitenzorg 4: 1371. 1923.

The generic description is the same as for the species. The genus is monotypic.

1. ***Chaetomitriopsis glaucocarpa*** (Reinw. ex Schwaegr.) Fleisch. 灰果藓 hui-guo xian
Musci Buitenzorg 4: 1372. 1923. *Campylophyllum glaucocarpum* (Reinw. ex Schwaegr.) Fleisch., Nova Guinea, Bot. 12(2): 123. 1914. *Campylium glaucocarpum* (Reinw. ex Schwaegr.) Broth., Nat. Pflanzenfam. I(3): 1042. 1908. *Stereodon glaucocarpus* (Reinw. ex Schwaegr.) Mitt., J. Proc. Linn. Soc., Bot., Suppl. 1: 115. 1859. *Hypnum glaucocarpon* Reinw. ex Schwaegr., Sp. Musc. Frond., Suppl. 3, 1(2): 228a. 1828.

Plants slender and elongate, up to 10 cm or longer, yellowish green, glossy, in dense tufts. *Stems* prostrate, pinnately to bipinnately branched, rhizoids clustered here and there. *Stem leaves* sparsely arranged, branch leaves similar, 0.7–0.8 mm × ca. 0.4 mm, broadly cordate-ovate, long slender to shortly acuminate, squarrose; margins serrulate nearly throughout to base; costae double, unequally forked (longer one never more than half the leaf length) or sometimes short to indistinct; leaf cells narrowly rhomboidal, 28–32 μm × 4–6 μm, projecting at upper ends. *Autoicous*. *Inner perichaetial leaves* plicate, slenderly elongate at apex. *Setae* slender, 2–3 cm long; capsules cylindrical, horizontal to pendulous; opercula conic, short-rostrate; annuli consisting of inflated cells, deciduous; peristome double; exostome teeth lanceolate, cross-striolate, with a median zigzag line; endostome segments narrowly lanceolate, pale, papillose, perforate; basal membrane high; cilia rudimentary. *Calyptrae* cucullate, sparsely pilose. *Spores* spherical, medium-sized.

Illustrations: Bartram 1939 (Pl. 20, fig. 348); Gangulee 1977 (Fig. 752).
Habitat: on trunks or rotten logs.
Distribution: China, India, Vietnam, Laos, Thailand, Indonesia, the Philippines, and New Guinea.

Chinese specimens examined: XIZANG: Mo-tuo Co., *Y.-G. Su 2839*, *2981a* (HKAS, MO). YUNNAN: Jing-hong Co., *Redfearn 33699a*, *33731a* (MO), *W.-X. Xu 6612*, *6778*, *6815* (MO, PE); Meng-hai Co., *W.-X. Xu 6089* (MO, PE); Meng-la Co., *M.-Z. Wang 4054* (MO, PE); Si-mao Co., *W.-X. Xu 6949*, *11419* (MO, PE).

The species was recorded from Taiwan (Redfearn et al. 1996), but no Taiwanese specimen was seen.

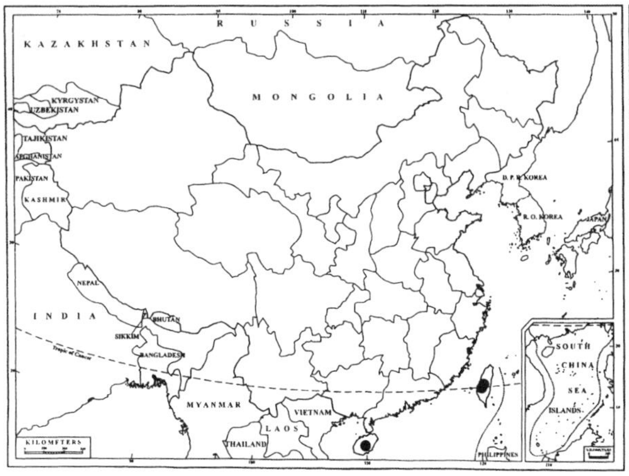

1. *Cyclodictyon blumeanum*

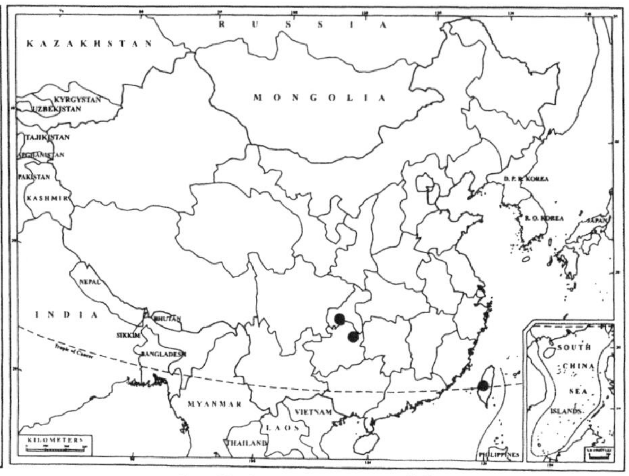

1. *Daltonia angustifolia*

5. ***Cyclodictyon*** Mitt. 圆网藓属 yuan-wang xian shu
J. Proc. Linn. Soc., Bot. 7: 163. 1863.

Plants slender, medium-sized, soft, delicate, yellowish green, not glossy, in flat mats. *Stems* prostrate, irregularly branched, flat, rhizoids sparse. *Leaves* laxly areolate, in 5–8 rows, complanately arranged, dimorphic and asymmetric, ventral and dorsal leaves appressed on stems, lateral leaves larger, ovate to oblong-lanceolate, concave, acute, acuminate to mucronate; costae double, extending beyond the midleaf; margins plane, entire, toothed apically, with a strongly differentiated border; leaf cells large, rounded, oval-hexagonal to shortly rhomboidal, smooth. *Autoicous* or dioicous. *Perichaetial leaves* small. *Setae* elongate, smooth; capsules ovoid to oblong-ovoid with a short neck; opercula conic, long-rostrate; exostome teeth slenderly lanceolate, transversely striolate; endostome segments papillose; cilia present. *Calyptrae* mitrate, laciniate at base.

The genus is primarily Neotropical and African in distribution. It consists of some 100 species in need of a critical taxonomic revision. One species occurs in China and in Asia.

1. ***Cyclodictyon blumeanum*** (C. Müll.) O. Kuntze 南亚圆网藓 nan-ya yuan-wang xian
Revis. Gen. Pl. 2: 835. 1891. *Hookeria blumeana* C. Müll., Syn. Musc. Frond. 2: 676. 1851. Type. Indonesia: Java.

Plants mat-forming, light green when dry. *Stems* irregularly branched, primary branches 15–17 mm long, 2.0–2.5 mm wide with leaves, in cross section with transparent, parenchymatous cells, central strand absent. *Leaves* complanately arranged, wrinkled when dry, ovate to oblong-lanceolate, 1.5–1.7 mm × 0.5–0.8 mm, acute, acuminate to mucronate; costae double, long, reaching 3/5 of the laminal length; margins plane, entire, toothed apically, with a strongly differentiated border, consisting of mostly two rows of linear cells; leaf cells large, rounded, oval-hexagonal to shortly rhomboidal, 45–67(–78) µm × 27–31 µm, smooth, thin-walled, becoming smaller at apex, rectangular at leaf base. *Autoicous*. *Perichaetial leaves* much smaller in size, ovate, acuminate, ecostate. *Sporophytes* not seen in Chinese material. Based on Philippine specimens: *Setae* about 2 cm long; capsules oblong-ovoid, about 1.0–1.3 mm long, horizontal; exostome teeth transversely striolate, with a wide median gap. *Calyptrae* naked, mitrate, with laciniate basal margin. *Spores* not seen.

Illustrations: Plate 398, figs. 1–6.
Habitat: on rotten logs and rocks covered with soil beside streamlets; alt. 820 m.
Distribution: This species is widespread and common in Malaysia, reaching as far as the Pacific Islands and Australasia. It appears to be rare in China, being known only from the offshore Islands, Taiwan and Hainan.

Chinese specimens examined: HAINAN: Qiong-zhong Co., Mt. Wu-zhi, *P.-J. Lin et al. 2544* (FH, IBSC). TAIWAN: Chia-yi Co., *S.-H. Lin s.n.* [Bryophytes of Taiwan Exsiccatae ser. 3, no. 62] (MO); Nan-tou Co., Chitou, *M.-J. Lai 11538* (NICH).

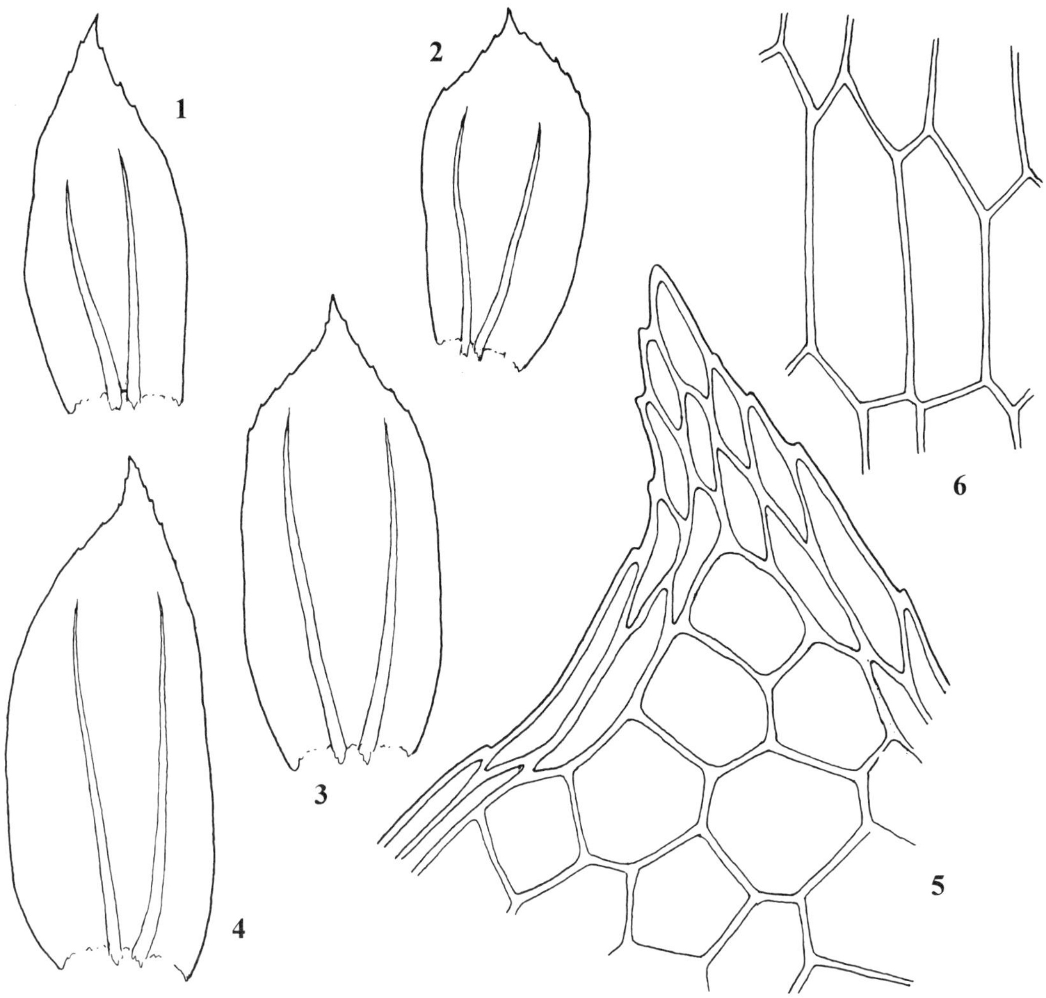

Plate 398. *Cyclodictyon blumeanum*. 1–4, leaves (× 53). 5, apical leaf cells (× 530). 6, basal leaf cells (× 530). (1–6, drawn from *P.-J. Lin et al. 2544*, IBSC.)

6. ***Daltonia*** Hook. & Tayl. 小黄藓属 xiao-huang xian shu
Muscol. Brit. 80: 1818.

Plants slender, small to medium-sized, green to yellowish brown, more or less glossy, in dense tufts. *Stems* prostrate or suberect, simple or forked, densely foliate, not or scarcely flattened. *Leaves* ovate or lanceolate, acuminate, subentire, with a well differentiated border; costa single, strong, somewhat carinate, ending some distance below the apex; leaf cells smooth, fusiform to narrowly rhomboidal, linear at margins. *Autoicous*. *Perichaetial leaves* small, costa none or indistinct. *Setae* elongate, lateral, scabrous above, twisted when dry, reddish; *capsules* ovoid to oblong-ovoid, suberect; exostome slenderly lanceolate, densely papillose, keeled; endostome segments linear, densely papillose, perforated; basal membrane low; opercula conic, slenderly rostrate. *Calyptrae* mitrate, deeply fringed at base. *Spores* papillose.

This is a tropical and subtropical genus of about 50 species concentrated in East Asia and northern South America. Three species are recognized in China.

1. Leaf apices aristate or abruptly long cuspidate .... 2. *D. aristifolia*
1. Leaf apices gradually acute or acuminate; if abruptly constricted, the acumens mucronate to shortly cuspidate 2
   2. Leaves plane along costa .... 1. *D. angustifolia*
   2. Leaves carinate along costa .... 3. *D. semitorta*

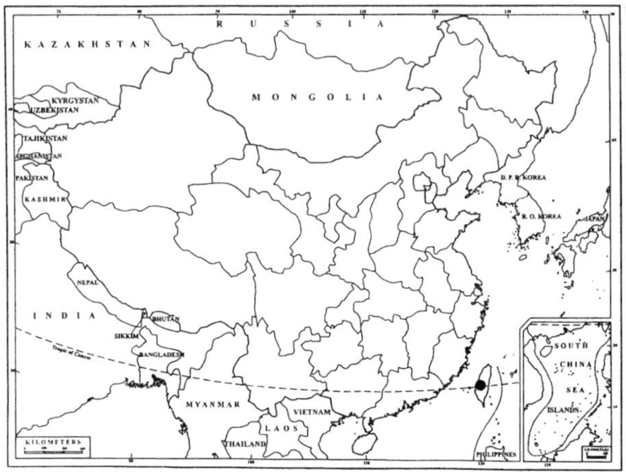

2. *Daltonia aristifolia*

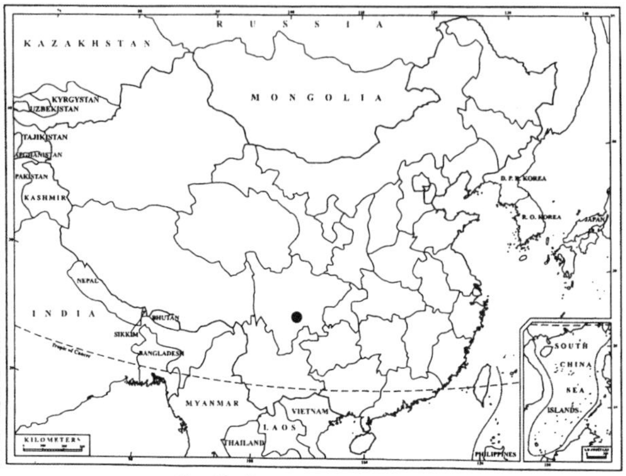

3. *Daltonia semitorta*

1. ***Daltonia angustifolia*** Dozy & Molk. 狭叶小黄藓 xia-ye xiao-huang xian
Ann. Sci. Nat. Bot., ser. 3, 2: 302. 1844. Type. Indonesia: Java.
Daltonia angustifolia Dozy & Molk. var. *strictifolia* (Mitt.) Fleisch., Musci Buitenzorg 3: 959. 1908. *Daltonia strictifolia* Mitt., J. Proc. Linn. Soc., Bot., Suppl. 1: 146. 1859.

Plants small, brownish when dry, tuft-forming, often few in a cluster. *Stems* simple, erect or branched at base, 4–6 mm long, in cross section with an outermost layer of colored, moderately thick- to thin-walled cells enclosing a core of colorless, parenchyma cells, central strand absent. *Leaves* spirally crowded, flexuous when dry, erect-patent when wet, narrowly lanceolate to oblong-lanceolate, 1.5–2.0 mm × 0.25–0.50 mm, gradually long acuminate, with a well differentiated border of 2–3(–4) rows of linear cells; margins entire, recurved on both sides; costa single, strong, somewhat carinate, ending some distance below the apex; leaf cells smooth, irregular in shape, oblong, fusiform to narrowly rhomboidal, 22–33 µm × 7–10 µm, moderately thick-walled, becoming shorter near apex, longer and laxer at leaf base. *Synoicous*. *Perichaetial leaves* smaller, ovate, 0.2 mm long, ecostate. *Setae* lateral, 3–5 mm, reddish, smooth below, scabrous above; capsules 1.0–1.3 mm long, ovoid, inclined to erect; exostome teeth and endostome segments nearly the same length, ca. 2/3 the length of the theca, papillose. *Calyptrae* mitrate, fringed with a basal row of long hairs, covering only the long rostrate operculum. *Spores* 10–11 µm in diameter, papillose.

Illustrations: Plate 399, figs. 1–5.
Habitat: on tree trunks and branches, rarely on rocks; alt. 1900 m.
Distribution: China and widely distributed in East Asia and Australasia.

Chinese specimens examined: CHONGQING: Nan-chuan Co., Mt. Jin-fo, *P.-C. Chen 1493d* (PE). GUIZHOU: Fanjing Nature Reserve, *Crosby 15800* (MO). TAIWAN: Chia-yi Co., *Iwatsuki, Sharp & Sharp 376 p.p.* (mixed with *D. aristifolia*, NICH); Nan-tou Co., *Shevock 17901a* (MO); Mt. Tahsuch, *Iwatsuki, Sharp & Sharp 2989* (NICH).

T.-Y. Chiang and C.-M. Guo (1989) reported *D. angustifolia* var. *gemmiphylla* Fleisch. from Taiwan, but from the description and illustration it appears to be close to *D. contorta* C. Müll. The occurrence of this species in China awaits the confirmation of the specimens.

2. ***Daltonia aristifolia*** Ren. & Card. 芒叶小黄藓 mang-ye xiao-huang xian
Rev. Bryol. 23: 105. 1896. Type. Indonesia: Java, *Massart 1395*.

The present species differs from *D. angustfolia* in having long aristate leaf apices and relatively smaller upper and apical leaf cells (10–17 µm × 5–6 µm). The upper leaf width (ca. 0.5 mm) of *D. aristifolia* is also broader than that of *D. angustifolia*, which measures less than 0.25 mm. Without sporophytes, *D. aristifolia* can be confused with *Distichophyllum cuspidatum*, which has similar abruptly cuspidate to aristate leaf apices.

Illustrations: Plate 399, figs. 6–10.
Habitat: on small tree trunks and branches.
Distribution: China, and widely scattered in tropical Asia, reaching north to the Himalayan region.

Chinese specimens examined: TAIWAN: Taihoku, Mt. Taihei, *Noguchi 6316* (NICH); Chia-yi Co., *Iwatsuki & Sharp 376 p.p.* (mixed with *D. angustifolia*, NICH).

3. ***Daltonia semitorta*** Mitt. 折叶小黄藓 zhe-ye xiao-huang xian
J. Proc. Linn. Soc., Bot., Suppl. 1: 146. 1859. Type. India: Sikkim, *Hooker s.n.*

Plants small, corticolous, less than 5 mm high. *Leaves* to 2 mm × 0.50–0.75 mm, broadly lanceolate to lanceolate, strongly carinate along the costa, flexuous when dry, oblong-lanceolate, acuminate or shortly cuspidate with a slightly twisted acumen; margins narrowly recurved on one or both sides of the leaf, bordered by 2 rows of cells; costa single, reaching upper 2/3 the lamina length; upper laminal cells irregularly rhomboidal-hexagonal to shortly rectangular, 11–14 µm × 6–8 µm, thin-walled, becoming rectangular in lower 1/3. *Perichaetia*, perigonia and sporophytes not seen.

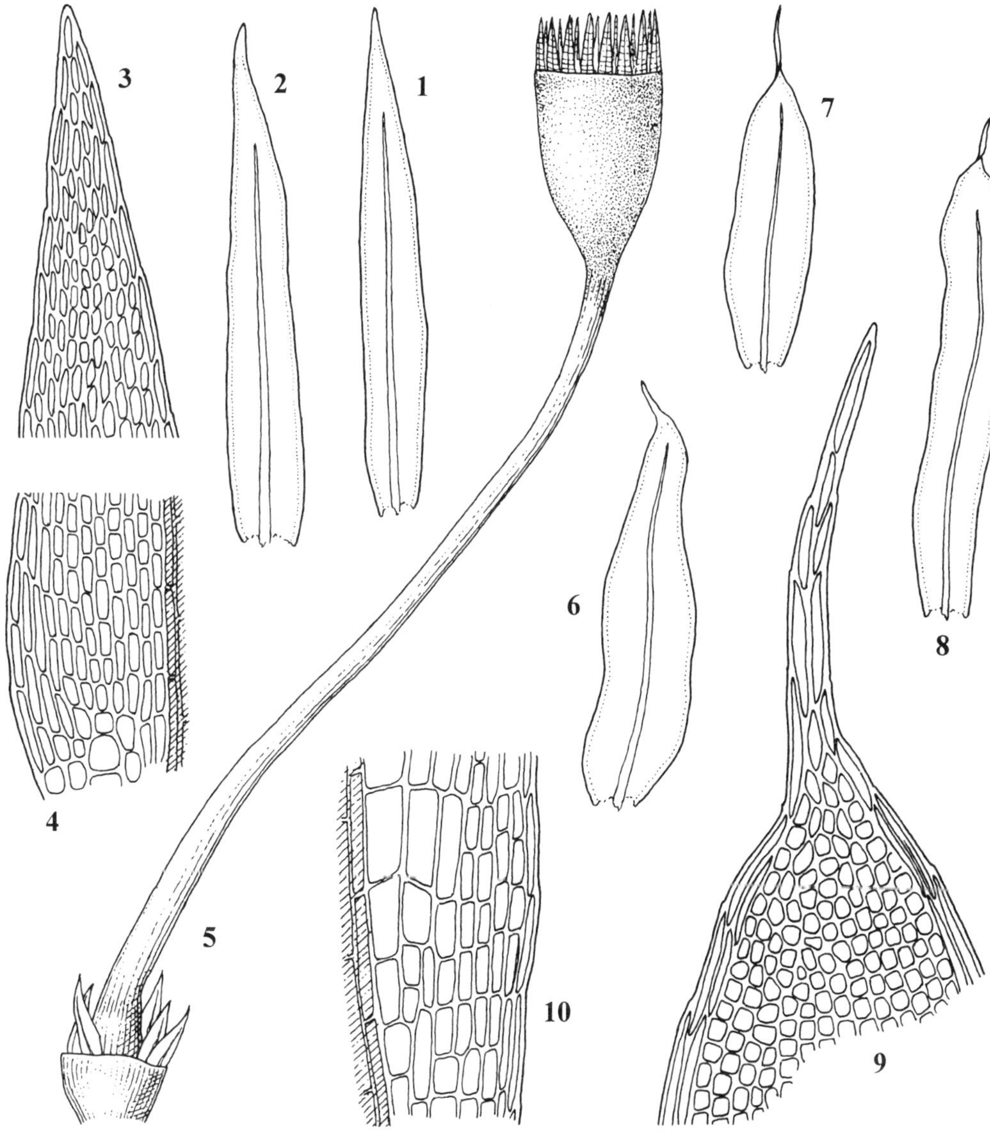

Plate 399. 1–5, *Daltonia angustifolia*. 1–2, leaves (× 43). 3, apical leaf cells (× 130). 4, basal leaf cells (× 130). 5, capsule (× 43). 6–10, *Daltonia aristifolia*. 6–8, leaves (× 45). 9, apical leaf cells (× 220). 10, basal leaf cells (× 220). (1–5, drawn from *Crosby 15800*, MO; 6–10, from *Noguchi 6316*, NICH.)

Illustrations: Gangulee 1977 (Fig. 734).
Habitat: unknown.
Distribution: China, Sikkim, and Nepal.

Chinese specimen examined: SICHUAN: Mt. Omei, *L.-Y. Tai 201a* (PE).

The examined specimen deposited at PE bore a herbarium name, "*Daltonia taiana* Chen, n. sp." Only a few plants of the present species were found mixed with specimens of *Zygodon* Hook. & Tayl.

7. ***Distichophyllum*** Dozy & Molk. 黄藓属 huang xian shu
Musci Frond. Ined. Archip. Ind. 99. 1846.

Plants usually soft, yellowish green, somewhat glossy, in dense flat mats. *Stems* flat, sparingly branched; central strand absent. *Leaves* in 6–8 rows, heterophyllous; dorsal leaves usually smaller and different in shape from the lateral and terminal ones, crowded, complanate, lingulate, oblanceolate, obovate to spathulate, pointed or rounded at apex; margins plane, entire, a narrow border of elongated cells differentiated; costa single, ending near or above midleaf; upper leaf cells rounded to hexagonal, smooth, laxer at base. *Autoicous*. *Perichaetial leaves* small, lingulate or ovate-lanceolate, border not differentiated, usually ecostate. *Setae* slender, smooth or papillose, reddish brown; capsules erect or inclined, ovoid to oblong-ovoid with a distinct neck; exothecial cells moderately to strongly collenchymatous; exostome teeth narrowly lanceolate, hookerioid with numerous cross-lamellae and a deep median groove; endostome segments lanceolate, papillose, perforate; basal membrane high; cilia absent; opercula conic, long-rostrate. *Calyptrae* horn-shaped, finely fringed at base. *Spores* small, weakly to moderately papillose.

In stem cross sections, all Chinese species of *Distichophyllum* lack a central strand, while most species outside China exhibit a well differentiated hyalodermis and an outer cortical layer consisting of cells smaller than the cells of the inner cortex and central cylinder; the outermost row of cortical cells next to the hyalodermal cells is dark brownish in sharp contrast to the transparent inner cortical cells.

Distichophyllum, consisting of more than 100 species worldwide, is the largest genus of the Hookeriaceae in China. The genus has a pantropical distribution range with many centers of species diversity in the Old and New Worlds. The southwestern region of China (Guizhou, Sichuan, and Yunnan), together with the regions of Malesia and Australasia, are three important centers of species diversification. There are 13 species and 7 varieties in China.

1. Leaves strongly dimorphic; gemmiferous leaves broadly and roundly ovate with a large, constricted hood-like apical lobe .... 6. *D. meizhii*
1. Leaves not strongly dimorphic; specialized gemmiferous leaves absent 2
 2. Leaves distinctly carinate .... 1. *D. carinatum*
 2. Leaves not carinate 3
3. Leaf apices long-cuspidate or aristate, acumens more than 100 µm long .... 4. *D. cuspidatum*
3. Leaf apices round, apiculate, acute, acuminate or mucronate, acumens less than 100 µm long 4
 4. Plants small; leaves less than 2.5 mm long 5
 4. Plants large; leaves more than 2.5 mm long 17
5. Leaf apices round to obtusely muticous or apiculate 6
5. Leaf apices acute, acuminate or stoutly mucronate 12
 6. Leaf apices without a differentiated border, denticulate near apex .... *D. obtusifolium* (reported from China)
 6. Leaf border reaching apices, entire throughout 7
7. Submarginal cells distinctly smaller than the paracostal cells in upper one third of lamina 8
7. Submarginal cells not smaller than the paracostal cells in upper one third of lamina 9
 8. Leaf borders thin and narrow .... 7. *D. mittenii* in part
 8. Leaf borders stout and broad .... 2b. *D. cirratum* var. *elmeri*
9. Leaf cells large, lax, more than 18 µm wide 10
9. Leaf cells small, dense, less than 18 µm wide 11
 10. Leaves broadly obovate to nearly orbicular; laminal cells more than 30 µm wide .... 10a. *D. subnigricaule* var. *subnigricaule*
 10. Leaves obovate to oblanceolate; laminal cells smaller, less than 30 µm wide .... 10b. *D. subnigricaule* var. *hainanense*

11. Leaves mostly 2 mm long, broadly spathulate .... 12. *D. wanianum*
11. Leaves mostly l mm long, oblanceolate to obovate .... 8b. *D. oblongum* var. *fanjingensis*
12. Leaf borders strong and broad, 20–50 µm wide, consisting of 2–4 rows of thick-walled, elongate to linear cells 13
12. Leaf borders not strong, narrow, 10–20 µm wide, consisting of 1–2(–3) rows of thin to moderately thick-walled, elongate to linear cells 16
13. At least 3–5 rows of submarginal cells distinctly smaller than the paracostal cells in upper 1/3 of lamina .... 2a. *D. cirratum* var. *cirratum*
13. Only 1–2 rows of submarginal cells slightly smaller than or the same size as the paracostal cells in upper 1/3 of lamina 14
14. Leaves more than 2 mm long; upper laminal cells large, more than 20 µm in diameter .... 3c. *D. collenchymatosum* var. *pseudosinense*
14. Leaves less than 2 mm long; upper laminal cells small, less than 20 µm in diameter 15
15. Leaves oblong to oblanceolate .... 8a. *D. oblongum* var. *oblongum*
15. Leaves obovate to spathulate .... *D. jungermannioides* (reported from China)
16. Upper laminal cells large, more than 20 µm wide; leaf apical acumens stout, more than 50 µm long (except var. *brevirostratum*), often twisted .... 3a. *D. collenchymatosum* var. *collenchymatosum*
16. Upper laminal cells small, less than 20 µm wide; leaf apical acumens weak, less than 50 µm long, not twisted, at times not well differentiated .... 5. *D. maibarae*
17. Leaves incompletely bordered 18
17. Leaves bordered all around 19
18. Only apical portion of the leaves lacking a border........ *D. obtusifolium* (reported from China)
18. Upper 1/3 of lamina lacking a border .... 9. *D. osterwaldii*
19. Leaf apices rounded or apiculate, with a tiny mucro 20
19. Leaf apices acute to mucronate; acumens mostly long and stout, often twisted 21
20. Upper laminal cells homogeneous, large .... 11. *D. tortile* in part
20. Upper laminal cells not homogeneous, submarginal cells distinctly smaller than the paracostal cells .... 7. *D. mittenii* in part
21. Upper leaf cells thin-walled, not collenchymatous; setae papillose .... 11. *D. tortile* in part
21. Upper leaf cells thin- to moderately thick-walled, slightly collenchymatous; setae smooth .... 3b. *D. collenchymatosum* var. *brevirostratum*

1. ***Distichophyllum carinatum*** Dix. ex Nichols. 折叶黄蘚 zhe-ye huang xian
Rev. Bryol. 36: 24. 1909. Type. Austria: Salzburg, *Dixon & Nicholson s.n.*

Plants small, light yellowish green, densely tufted. *Stem leaves* crowded, helically crisped when dry, ovate-lanceolate when wet, 1.0–1.5 mm × 1 mm, acuminate to shortly cuspidate, strongly carinate in upper half of the lamina, often keeled; leaf margins reflexed below, bordered with 1–2 rows of thin-walled, linear cells; costa single, not reaching the leaf apex; laminal cells large, roundly rhomboidal to hexagonal, nearly homogeneous, 36–45 µm × 15–18(–22) µm, thin-walled, becoming smaller near the leaf apex, rectangular near leaf base. Only immature sporophytes seen. *Setae* about 10 mm long, smooth below, slightly papillose above. *Calyptrae* mitriform. *Filiform propagules* present.

Illustrations: P.-J. Lin & B. C. Tan 1995 (Fig. 26, D–F).
Habitat: confined to wet limestone forests.
Distribution: China, Japan, and Western Europe.

Chinese specimens examined: CHONGQING: Nan-chuan Co., *X.-Y. Hu 0558* (as *D. cuspidatum*, FH, MO, PE); ibid, *X.-Y. Hu 493* (as *D. cuspidatum*, PE). SICHUAN: E-mei Co., *Allen 6514* (MO).

This is perhaps one of the most easily identified species by virtue of its carinate and conduplicately infolding leaves. The Chinese specimens were misidentified as *D. cuspidatum*. The latter has a much longer, cuspidate leaf apex and a plane and not infolding lamina.

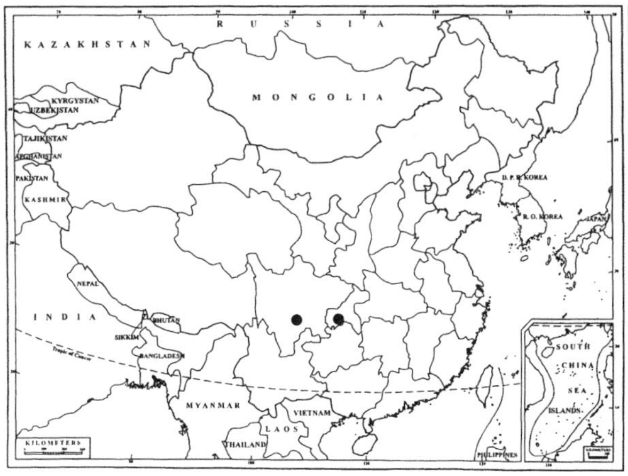

1. *Distichophyllum carinatum*

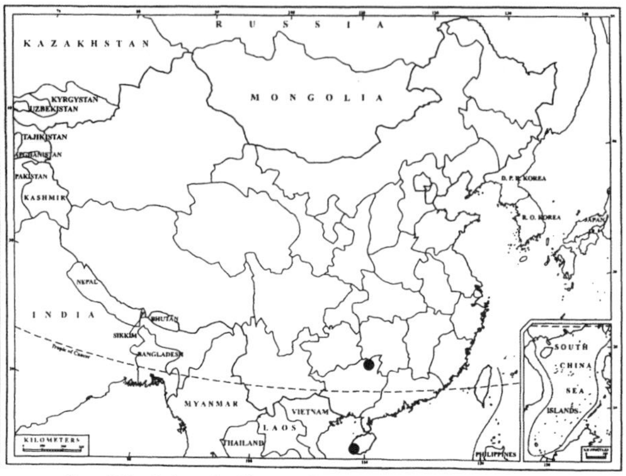

2a. *Distichophyllum cirratum* var. *cirratum*

2a. ***Distichophyllum cirratum*** Ren. & Card. var. ***cirratum*** 卷叶黄藓原变种 juan-ye huang xian yuan bian-zhong
Rev. Bryol. 23: 104. 1896. Type. Indonesia: Java, Tjibodas, *Massart 1397* (isotype FH).
Distichophyllum perundulatum Dix., J. Linn. Soc., Bot. 50: 106. 1935, fide Tan and Robinson (1990).

Plants small to medium-sized, mat-forming. *Stems* prostrate, 2–3(–6) cm tall, 2–5 mm wide with leaves when dry, irregularly branched, laxly foliate. *Leaves* strongly crisped, often twisted lengthwise when dry, variable in shape, ovate, oblong-elliptic to lingulate, 2.0–2.5 mm × 0.5–1.0 mm, the apices obtuse or with a stout mucro; leaf margins entire, often strongly undulate even when moist, borders strongly differentiated, consisting of 2–3(–4) rows of moderately thick-walled, linear cells and measuring 30–45 µm wide; costae strong, reaching near leaf apex, sometimes becoming continuous with the leaf mucro; laminal cells hexagonal to polygonal, moderately thick-walled, 18–22(–33) µm wide, with 3–4 rows of smaller submarginal cells, 11–18 µm wide. *Sporophytes* not seen in Chinese collections. *Setae* reportedly 7–8 mm long, smooth.

Illustrations: Plate 400, figs. 1–4.
Habitat: on soil in shady sites.
Distribution: China, Malaysia, and Thailand.

Chinese specimens examined: GUANGXI: Long-sheng Co., *P.-C. Wu et al. 127b*, *344a*, *419*, *1298* (FH, IBSC, PE). HAINAN: Mt. Jianfengling, *P.-J. Lin 4061*, *4241* (FH, IBSC).

Fleischer (1908) and Tan and Robinson (1990) considered this species to be conspecific with *Distichophyllum nigricaule* Mitt. ex Bosch & Sande Lac. Mohamed and Robinson (1991), however, concluded that the two species are separable. When treated separately, *D. cirratum* differs from *D. nigricaule* in having stronger and wider leaf borders and by the presence of numerous rows of smaller submarginal leaf cells. In contrast, laminal cells of *D. nigricaule* are more or less homogeneous and its leaf borders are narrower, measuring 15–25 µm wide. Because of having mucro leaf apices, *Distichophyllum cirratum*, sometimes, may be confused with *D. collenchymatosum,* which also has similar broad leaf borders. Nevertheless, the former can be identified by the presence of submarginal rows of cells that are smaller than the paracostal laminal cells. This difference in cell sizes between the peripheral and the costal areas of the leaf is best observed at the upper 1/3 of the leaf length. In addition, average leaf cell size of *D. cirratum* is smaller than that of *D. collenchymatosum*.

2b. ***Distichophyllum cirratum*** var. ***elmeri*** (Broth.) P.-J. Lin & B. C. Tan 卷叶黄藓南亚变种 juan-ye huang xian nan-ya bian-zhong
Harvard pap. Bot. 7: 36. 1995. *Distichophyllum nigricaule* Bosch & Sande Lac. var. *elmeri* (Broth.) Tan & Robinson, Smithsonian Contr. Bot. 75: 22. 1990. *Distichophyllum elmeri* Broth., Leafl. Philipp. Bot. 2: 656. 1909. Type. The Philippines: Luzon, Baguio, *Elmer 10453* (holotype H).

Plants are similar to var. *cirratum* in habit, size, and leaf shape. The submarginal leaf cells [14–18(–23) μm wide] of the present variety are smaller than the paracostal cells [23–36(–52) μm wide]. In addition, there are up to 8 rows of small, submarginal leaf cells in var. *elmeri*. The leaf borders of var. *elmeri* are as broad as in var. *cirratum*, consisting of 3–4 rows of linear cells. The leaf apices of var. *elmeri* are muticous or shortly mucronate (5–10 μm long) and the costae never reach the leaf apices. Setae are 10–12 mm long, yellowish brown, smooth, becoming coarsely wrinkled near the bases of capsules.

Illustrations: Plate 400, figs. 5–8.
Habitat: on rocks.
Distribution: China, Thailand, Malaysia, North Kalimantan, and the Philippines.

Chinese specimens examined: HAINAN: Mt. Diaoluo, *P.-J. Lin et al. 2834* (IBSC, MO), *P.-Z. Zheng 401* (FH, IBSC); Mt. Jianfengling, *P.-C. Chen et al. 593*, *814* (FH, IBSC, PE), *W. D. Reese 17792* (FH, IBSC, LAF, MO); Mt. Limuling, *P.-J. Lin et al. 137* (IBSC). TAIWAN: (see Yang and Lee 1964, as *D. nigricaule* sensu Bartram).

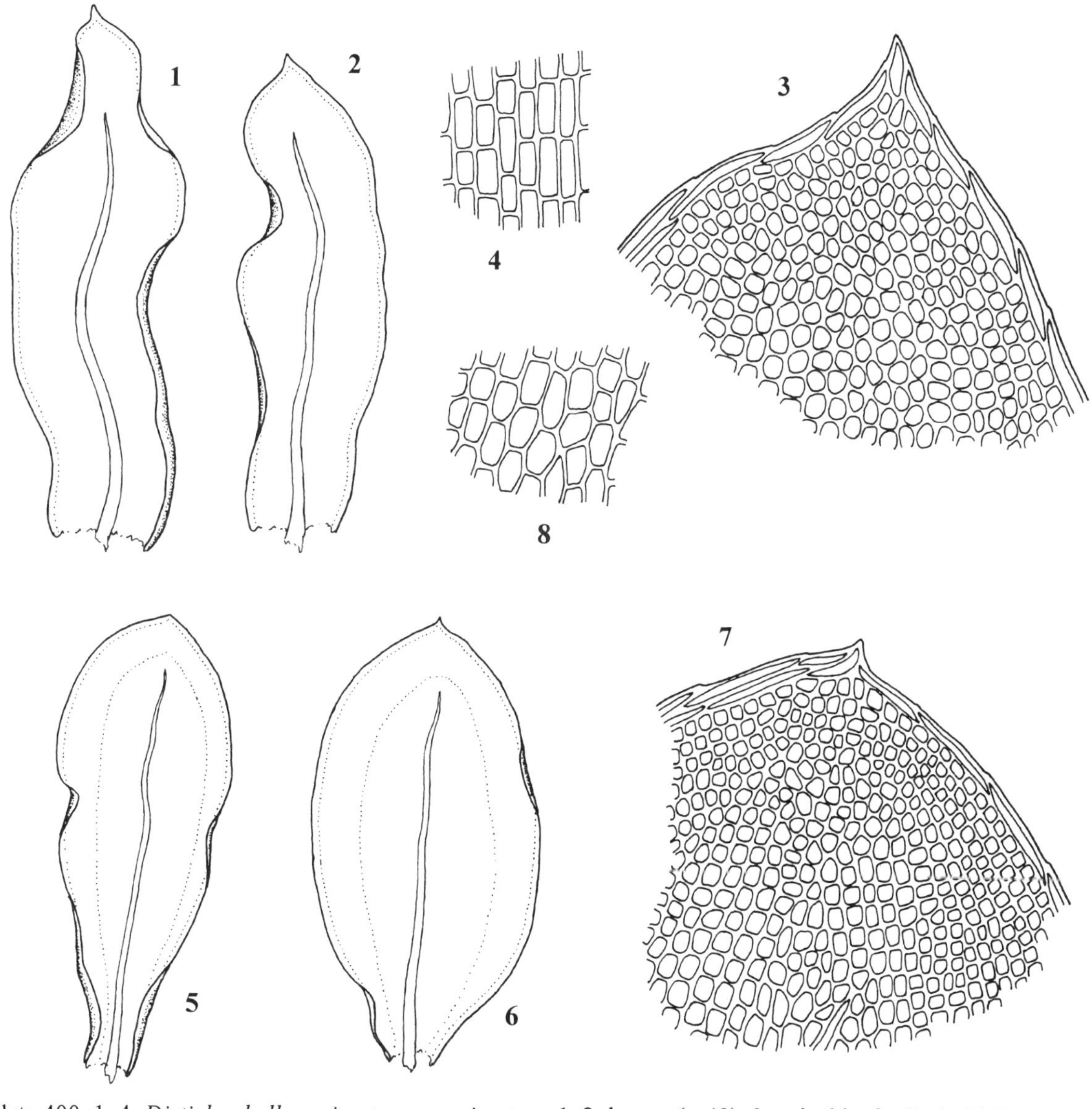

Plate 400. 1–4, *Distichophyllum cirratum* var. *cirratum*. 1–2, leaves (× 43). 3, apical leaf cells (× 200). 4, median leaf cells (× 200). 5–8, *Distichophyllum cirratum* var. *elmeri*. 5–6, leaves (× 40). 7, apical leaf cells (× 200). 8, median leaf cells (× 200). (1–4, drawn from *P.-C. Wu et al. 1298*, PE; 5–8, from *P.-J. Lin 137*, IBSC.)

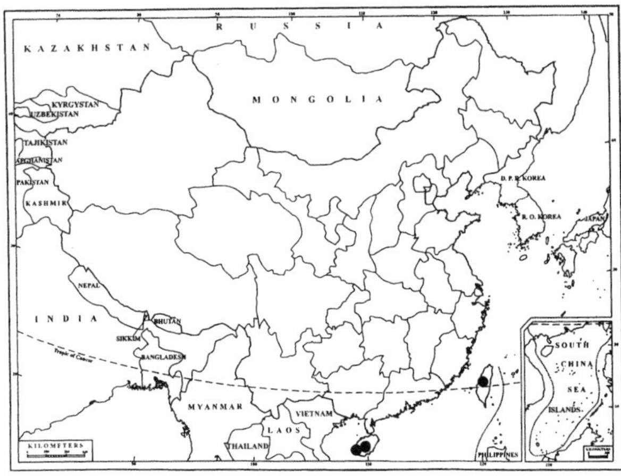

2b. *Distichophyllum cirratum* var. *elmeri*

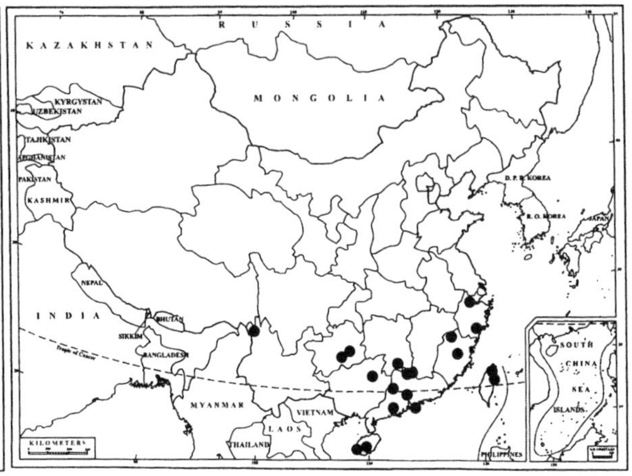

3a. *Distichophyllum collenchymatosum* var. *collenchymatosum*

Distichophyllum elmeri has been unjustifiably considered to be a synonym of *D. cirratum* by Mohamed and Robinson (1991). Our study of the Chinese specimens reveals that these two species are not alike morphologically and can be separated using microscopic characters. Sometimes, usual forms of this species might be difficult to distinguish from *D. cirratum*. Following Noguchi and Iwatsuki (1972), we recognize *D. elmeri*, but consider it to be a variety of *D. cirratum*.

3. ***Distichophyllum collenchymatosum*** Card. 厚角黄藓 hou-jiao huang xian
Bull. Soc. Bot. Genève, ser. 2, 3: 278. 1911. Type. Japan.

1. Leaves firm and plane; laminal cells more regularly hexagonal, thick-walled; marginal borders thick and broad, more than 20 µm wide .... var. *pseudosinense*
1. Leaves flaccid and wavy; laminal cells round-polygonal, thin-walled; marginal borders thin and narrow, less than 20 µm wide 2
   2. Leaf apices acuminate, acumens well developed, stout, more than 30 µm long.......... var. *collenchymatosum*
   2. Leaf apices mucronate or short apiculate, acumens small, slender, less than 30 µm long..... var. *brevirostratum*

3a. ***Distichophyllum collenchymatosum*** var. ***collenchymatosum*** 厚角黄藓原变种 hou-jiao huang xian yuan bian-zhong

Distichophyllum cavaleriei Thér., Monde Pl., ser. 2, 9: 22, 1907. Type. China: Guizhou, Tong Tcheou, 1904, *J. Cavalerie s.n.* (lectotype PC, designated by P.-J. Lin & B. C. Tan 1995).
Distichophyllum sinense Dix., Hong Kong Naturalist., Suppl. 2: 22. 1933. Type. China: Guangdong, White Cloud Mountain, *Herklots 302C*.
Distichophyllum cf. *perlimbatum* Broth., in Z.-H. Li & Piippo, Trop. Bryol. 9: 40. 1994, *nom. nud.*

Plants small to medium-sized, densely tufted. *Stems* often irregularly branched, complanately and densely foliate, 1–3(–6) cm tall, 3–5 mm wide with leaves. *Leaves* contracted and undulate when dry, ovate-oblong to broadly lingulate, 1.5–3.0 mm × 1.0 mm, with a distinct, well developed and stout apical acumen or mucro measuring 20 to 50 µm long; leaf margins entire, plane to wavy when wet, with a well developed border consisting of 2–3 rows of thin-walled, linear cells, measuring 12–18 µm wide; costa well developed, extending to leaf apex or ending well below the apex; laminal cells large, round-polygonal, (27–)34–56 µm wide, thin-walled, slightly collenchymatous, nearly uniform in size from the margin to costa, becoming slightly smaller only near the apex and extreme periphery, rectangular at leaf base. *Perichaetial leaves* narrowly oblong, small, less than l mm long, with distinct border. *Sporophytes* lateral. *Setae* elongate, 10–15 mm long, smooth; capsules ovoid, about 1.0–1.5 mm long. *Gemmae* filamentous, highly branched, multiseptate present on the abaxial, basal part of leaf costa.

Illustrations: Plate 401, figs. 1–2.
Habitat: on shady rocks, decaying logs, forest floor and road banks; alt. 200–1200 m.
Distribution: China, Korea, Japan, and the Philippines.

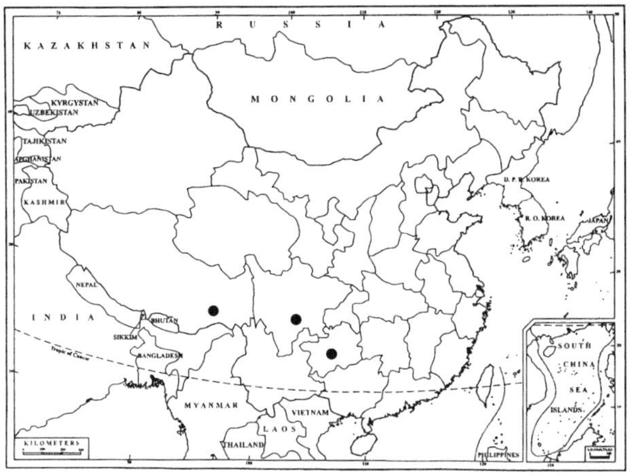

3b. *Distichophyllum collenchymatosum* var. *brevirostratum*

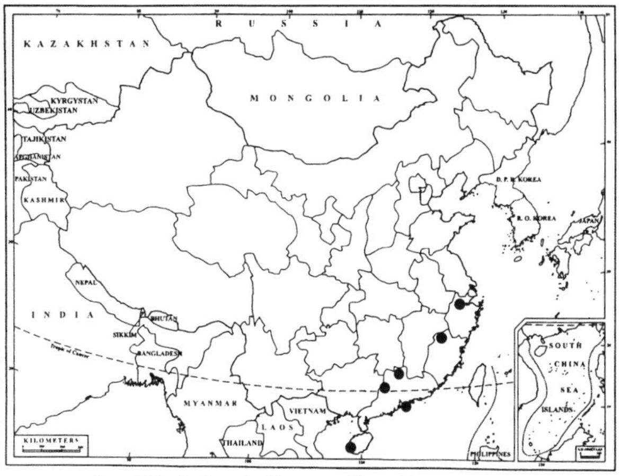

3c. *Distichophyllum collenchymatosum* var. *pseudosinense*

Chinese specimens examined: FUJIAN: Gu-tian Co., *D.-K. Li 20279* (FH, IBSC); Mt. Wuyi, *D.-K. Li 16303* (FH, IBSC). GUANGDONG: Feng-kai Co., *P.-J. Lin 36* (FH, IBSC); ibid, Heshiding Nature Reserve, *J.-L. Zhang & S.-H. Shi 121*, *141* (FH, SYS); Ru-yuan Co., *P.-J. Lin 758* (IBSC, MO), *L. Zhang 98* (FH, IBSC); Shi-xing Co., *P.-J. Lin 4491* (IBSC); Yang-chun Co., *B.-H. Liang 89785* (FH, IBSC). GUANGXI: Li-po Co., *Z.-H. Zhang 2534* (FH). GUIZHOU: Mt. Leigong, *B.-C. Tan 91-789* (FH). HAINAN: Mt. Diaoluo, *P.-J. Lin 3346* (FH, IBSC); Mt. Jianfeng, *P.-J. Lin et al. 577* (FH, IBSC). HONG KONG: (as *D. sinense*, see Dixon 1933). HUNAN: Mang Shan (Mt.), *P.-J. Lin et al. 1135* (FH, IBSC); Yi-zhang Co., *P.-J, Lin et al. 940* (IBSC). TAIWAN: Ilan Co., *W.-C. Lee 256* (FH, IBSC); Tai-pei Co., *C.-C. Chuang 5019* (FH, IBSC); Tao-yuen Co., *M.-J. Lai 9417* (H). YUNNAN: Gong shan (Mt.), *C.-W. Wang s.n.* (FH, IBSC). ZHEJIANG: Mt. Xitianmu, *P.-C. Wu 7911* (MO); Wen-zhou City, *D.-K. Li 8294* (FH, IBSC).

Dixon (1933) characterized this variety as having narrow, robust and shortly cuspidate leaves and pellucid, rather large leaf cells, which are slightly but not very markedly smaller toward the margins. The large, pellucid leaf cells separate this species from *Distichophyllum maibarae*, which has smaller leaf cells and a much shorter leaf apical mucro. The nearly uniform leaf cells across the blade from the margins to the costa separate *D. collenchymatosum* from *D. cirratum*, which has several submarginal rows of distinctively smaller leaf cells.

Thériot (1907) compared *Distichophyllum cavaleriei* with *D. nigricaule* and *D. heterophyllum* and concluded that it differed from the latter two species by the larger leaf cells and broader leaf borders. Our examination shows that these distinguishing features fall within the accepted range of characters of *D. collenchymatosum*.

3b. ***Distichophyllum collenchymatosum*** var. ***brevirostratum*** (Thér.) P.-J. Lin & B. C. Tan 厚角黄藓短喙变种 hou-jiao huang xian duan-hui bian-zhong

Harvard Pap. Bot. 7: 37. 1995. *Distichophyllum brevirostratum* Thér., Mondes Pl., ser. 2, 9: 22. 1907. Type. China: Guizhou, Pin-fa, *J. Cavalerie s.n.* (holotype PC).

Plants medium-sized. *Stems* to 2(–4) cm tall. *Leaves* similar to those of var. *collenchymatosum* except in having a shorter apical mucro measuring 18–24 µm long; laminal cells large, round-hexagonal, 33–56 µm wide. *Setae* 1.5 cm long, smooth; capsules about 1.5 mm long, inclined.

Illustrations: Plate 401, figs. 3–4.
Habitat: on soil and rocks.
Distribution: endemic to China.

Chinese specimens examined: SICHUAN: Mt. Omei, *P.-J. Lin 3286* (FH, IBSC). XIZANG: Mo-tuo Co., *Y.-G. Su 3294* (HKAS, IBSC).

Thériot (1907) characterized this variety as having a short opercular beak. However, the type specimen was found to contain broken beaks. When compared to var. *collenchymatosum*, the leaves of var. *brevirostratum* have a much shorter apical mucro and a slightly narrower and less distinctive leaf border. Intermediate specimens between the two varieties sometimes exist. With its short leaf apical mucros, specimens of var. *brevirostratum* can be mistaken for *D. maibarae*. The latter species, however, has smaller leaf cells.

3c. ***Distichophyllum collenchymatosum*** var. ***pseudosinense*** P.-J. Lin & B. C. Tan 厚角黄藓沿海变种 hou-jiao huang-xian yan-hai bian-zhong
Harvard Pap. Bot. 7: 38. 1995. Type. China: Hainan, Mt. Jianfeng, *P.-J. Lin 3938* (holotype IBSC; isotype FH).

Plants often appear coarser and more stiff-looking than the other two varieties. Stems are about 2–3 cm long and 4 mm wide with the leaves. Because of the thick and stout leaf borders, which are ca. 50 µm wide, leaves of var. *pseudosinense* are stiff or less crisped than those of var. *collenchymatosum* when dry; the cell walls of the leaf border are thick and pitted; laminal cell walls are equally thick and firm, making the cell outlines more regularly hexagonal; the leaf apical mucros of the present variety are stout and large, ca. 270 µm long.

Illustrations: Plate 401, figs. 5–6.
Habitat: on rocks and soil.
Distribution: This variety seems to be confined to the wetter, oceanic southern part of China, which includes Guangdong, Hainan, Fujian, and Zhejiang provinces. Its presence in Taiwan can be safely predicted.

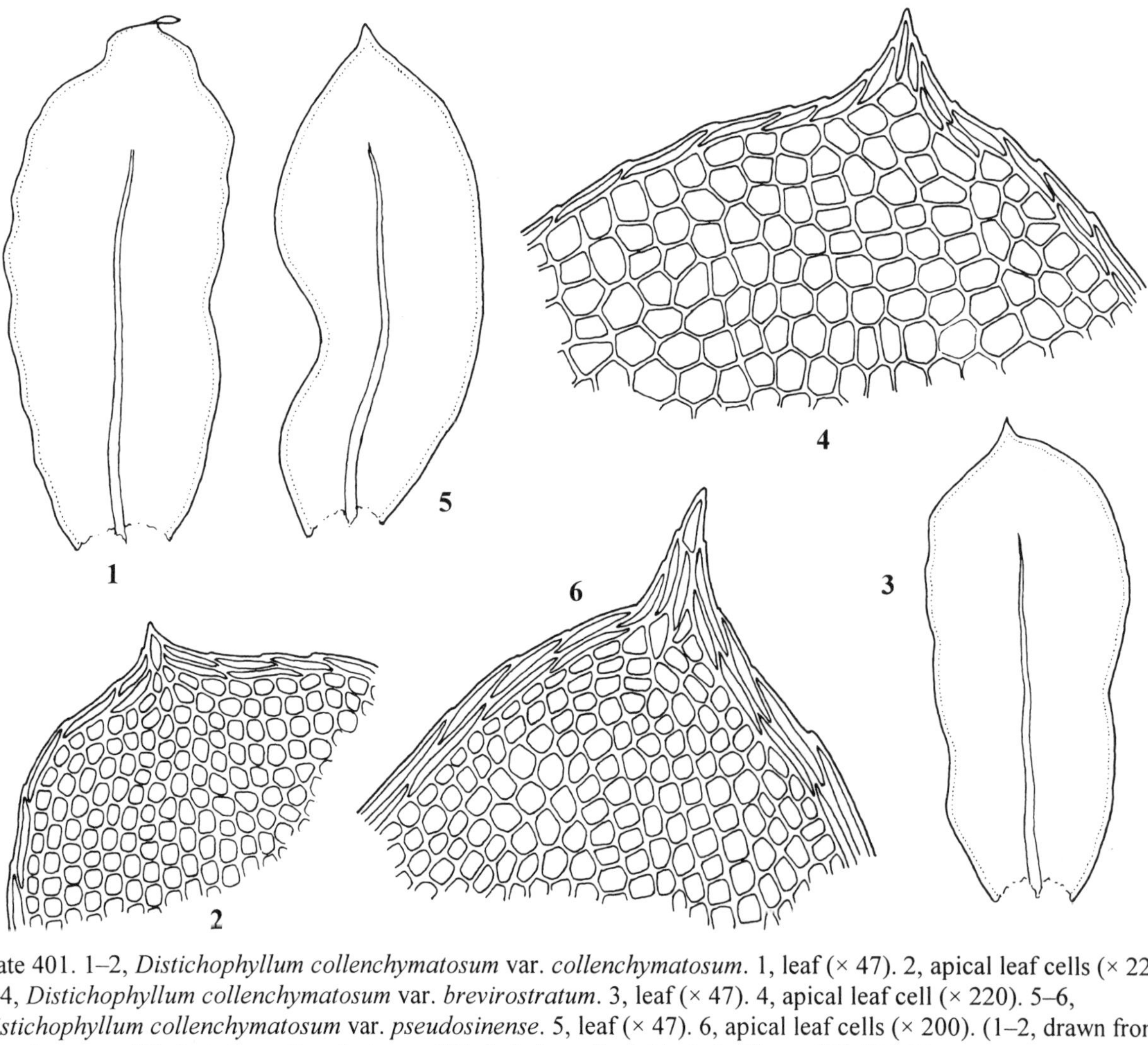

Plate 401. 1–2, *Distichophyllum collenchymatosum* var. *collenchymatosum*. 1, leaf (× 47). 2, apical leaf cells (× 220). 3–4, *Distichophyllum collenchymatosum* var. *brevirostratum*. 3, leaf (× 47). 4, apical leaf cell (× 220). 5–6, *Distichophyllum collenchymatosum* var. *pseudosinense*. 5, leaf (× 47). 6, apical leaf cells (× 200). (1–2, drawn from *Cavaleria s.n.*, PE; 3–4, from *Cavaleria s.n.*, PE; 5–6, from *P.-J. Lin & L. Zhang 577*, IBSC.)

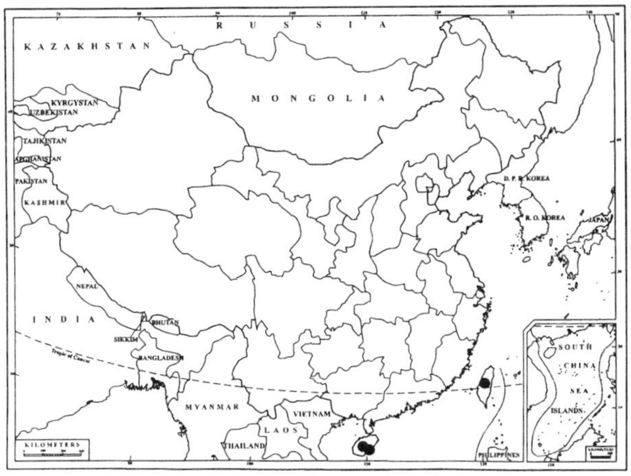

4. *Distichophyllum cuspidatum*

5. *Distichophyllum maibarae*

Chinese specimens examined: FUJIAN: Mt. Wuyi, *P.-C. Chen et al. 194* (PE). GUANGDONG: Feng-kai Co., *J.-L. Zhang & S.-H. Shi 73* (FH, SYS); Ru-yuan Co., *P.-J. Lin 866* (FH, IBSC). HAINAN: Mt. Jianfeng, *P.-J. Lin & L. Zhang 577* (IBSC). HONG KONG: Mui Tsz Lam, *P.-J. Lin et al. 1550* (IBSC); Lantau Island, *P.-J. Lin et al. 1737* (IBSC); Mt. Tai-Mo, *P.-J. Lin et al. 2027, 2041, 2046* (IBSC); Tai Po Kau, *P.-J. Lin et al. 1072, 1880* (IBSC); Yin Ngan, *P.-J. Lin et al. 900* (IBSC). ZHEJIANG: West Tianmu Shan (Mt.), *P.-C. Wu 7984* (PE).

4. ***Distichophyllum cuspidatum*** (Dozy & Molk.) Dozy & Molk. 尖叶黄藓 jian-ye huang xian
Musci Frond. Ined. Archip. Ind. 4: 101. 1846. *Hookeria cuspidata* Dozy & Molk., Ann. Sci. Nat., Bot., ser. 3, 2: 305. 1844. Type. Indonesia: Java.

Plants forming aggregate tufts. *Stems* ascending, 1.0–1.5 cm tall, 4 mm wide with leaves. *Leaves* erect, moderately crisped, often twisted when dry, not complanate, oblong to broadly lingulate, at times oblanceolate, 3–4 mm × 1 mm, concave above, the leaf apices more or less contracted into long-cuspidate acumens; leaf acumina variable in length, 150–300 µm long; leaf borders strong, consisting of 2–3(–5) rows of thick-walled, linear cells; leaf margins entire, plane above, revolute below, at times undulate; costae reaching more than 2/3 the leaf length; leaf cells nearly uniform above, round-hexagonal, 18–40 µm, often collenchymatous, becoming shortly rectangular at base, 34–56 µm × 14–20 µm. *Perichaetial leaves* very small, about 1 mm long. *Sporophytes* lateral; setae mostly smooth, becoming papillose distally, ca 5 mm long; capsules shortly ovoid, erect or inclined.

Illustrations: P.-J. Lin & B. C. Tan 1995 (Fig. 26, A–C).

Habitat: on decaying logs, trunk bases, but more often on tree trunks and branches; alt. 800–1300 m. Outside China, *D. cuspidatum* is a common epiphyte in mid-montane rainforests.

Distribution: China, India, Sri Lanka, Thailand, Vietnam, Malaysia, Indonesia, Kalimantan, the Philippines, and New Guinea.

Chinese specimens examined: HAINAN: Mt. Diaoluo, *P.-J. Lin et al. 949, 968* (FH, IBSC, MO); Mt. Limu, *P.-J. Lin 124* (FH, IBSC); Ling-shui Co., *Redfearn 36185, 36189* (MO), *Reese 17922* (MO). TAIWAN (see Yang and Lee 1964).

The long-cuspidate leaf apices of this species are distinctive in the field and provide an easily observed character. Nevertheless, confusion with *D. collenchymatosum* exists. The latter species, however, has a shorter leaf apical acumen, a weaker leaf border, and ironically, larger, but not distinctively collenchymatous leaf cells. Without sporophytes, *Distichophyllum cuspidatum* may be mistaken for *Daltonia aristifolia*, which has similar, long-cuspidate leaf apices. However, the leaf apices of *Daltonia aristifolia* are gradually narrowed into a long acumen, while those of *Distichophyllum cuspidatum* are contracted into a long acumen. The leaf cells of *Daltonia aristifolia* are irregularly rhomboidal to shortly polygonal, while those of *Distichophyllum cuspidatum* are uniformly round-hexagonal.

5. ***Distichophyllum maibarae*** Besch. 东亚黄藓 dong-ya huang xian

J. Bot. (Morot) 13: 40. 1899. Type. Japan: Maibara, *Faurie 11130* (isotype FH).

Distichophyllum duongii T. Ninh, Acta Bot. Acad. Sci. Hung. 27: 154. 1981.

Distichophyllum stillicidiorum Broth., Akad. Wiss. Wien Sitzungsber., Math.-Naturwiss. Kl., Abt. 1, 131: 217. 1923. Type. China: Hunan, Yolu-shan, near Tschangscha, *Handel-Mazzetti 11489* (holotype H).

Plants variable in size, in dense tufts. *Stem leaves* slightly contracted to moderately undulate when dry, broadly ovate-oblong, 1.0–2.0(–2.6) mm × 0.8–1.0(–3.5) mm; lateral and terminal leaves larger, oblong to shortly spathulate; leaf apices roundly apiculate, acute to shortly acuminate, entire, with very obscure teeth at leaf tips owing to the end walls of border cells, the acumens small, mostly less than 50 µm long; leaf margins plane, with narrowly differentiated borders consisting of 1–3 rows of linear cells; median laminal cells homogeneously quadrate to round-hexagonal, 16–34 µm × 11–16(–23) µm, thin-walled, slightly or not collenchymatous, becoming rectangular at leaf base; apical leaf cells more uniformly hexagonal, 11–18(–23) µm × 11–13(–18) µm. *Perichaetial leaves* much smaller, 1.5 mm long, broadly ovate, somewhat long acuminate with indistinct borders. *Sporophytes* lateral; setae 8–9(–15) mm long, smooth, except at the base of capsule; capsules oblong-ovoid, about 1 mm long, with strongly wrinkled apophysis; exostome teeth hookerioid, with cross-lamellae. *Calyptrae* mitriform, with basal long hairs. *Spores* 8–10 µm, smooth to weakly papillose.

Illustrations: Plate 402, figs. 1–4.
Habitat: on wet soil, shaded cliffs and rocks by streamlets, and decaying logs in forests; alt. 400– 2500 m.
Distribution: China, India, Malaysia, and Indo-China.

Chinese specimens examined: CHONGQING: Chong-qing City, *P.-C. Chen 1400* (PE). FUJIAN: Mt. Wuyi, *D.-K. Li 1867* (FH, IBSC). GUANGDONG: Mt. Dinghu, *Crosby 15228* (MO), *P.-Z. Zheng 954* (FH, IBSC, MO); Le-chang Co., *L. Zhang 493* (FH, IBSC, MO); Ru-yang Co., *L. Zhang 233* (FH, IBSC). GUANGXI: Long-sheng Co., *P.-C. Wu et al. 1279d* (IBSC, PE); Rong-shui Co., *M.-Z. Wang 46029* (PE); Xing-an Co., *P.-J. Lin et al. 1376* (FH, IBSC). GUIZHOU: Chi-shui Co., *Crosby 16151* (MO), *B.-C. Tan 91-1059* (FH); Mt. Fanjing, *Crosby 15746* (MO), *B. C. Tan 91-1175* (FH); Mt. Leigong, *Crosby 15622, 15629* (MO), *B. C. Tan 91-783* (FH). HAINAN: Ling-shui Co., *Redfearn 36222* (MO); Mt. Bawang, *P.-J. Lin et al. 3286* (FH, IBSC); Mt. Jianfengling, *P.-C. Chen et al. 519* (IBSC, H); Mt. Hutou, *P.-J. Lin 251* (IBSC); Mt. Limu, *P.-J. Lin et al. 188* (IBSC). HONG KONG: Mt. Tai-Mo, *P.-J. Lin et al. 314, 1365, 2051* (IBSC). HUNAN: Zhangjia-jie, *T. Cao 42626* (IFSBH). JIANGXI: Lu shan (Mt.), *Z.-L. Wan 101* (PE). TAIWAN: Mt. Ali, *M.-J. Lai 2888* (as *Daltonia contorta*, NICH); Ping-tung Co., *C.-C. Chuang 1467* (FH, UBC).YUNNAN: Da-guan Co., *X.-J. Li 4241* (IBSC); Jiang-dong Co., *J.-K. Zhang 1834* (IBSC); Gong Shan (Mt.), *Q.-W. Wang 7011* (IBSC); Xishuangbanna, *W.-X. Xu 6733* (FH, IBSC). ZHEJIANG: Hang-zhou City, *C.-H. Gao 7652* (FH, IBSC); West Tianmu Shan (Mt.), *P.-C. Wu 7911* (PE).

This is the most commonly encountered species of *Distichophyllum* in China. It is highly variable with plants measuring from less than 1 cm tall to 2 cm tall and setae from 8 mm to 1.5 cm long. In spite of its variable size, *D. maibare* can be identified by its nearly uniform and small upper laminal cells and a small leaf mucro, which often appears to be incompletely differentiated from the adjacent laminal cells.

6. ***Distichophyllum meizhii*** B.-C. Tan & P.-J. Lin 贡山黄藓 gong-shan huang xian

Trop. Bryol. 10: 55. 1995. Type. China: Yunnan, Gong-shan Co., Du-long-jiang (river), alt. 1300 m, *M.-Z. Wang 10040* (holotype PE).

Plants large, in lax tufts, flaccid. *Stems* tall, slender, sparsely branched, ascending, up to 4.5 cm long, 3 mm wide with leaves. *Leaves* erect-spreading to complanate, strongly dimorphic, vegetative leaves wrinkled when dry, broadly obovate to lingulate with a short narrow leaf base, 2.0–2.5 mm × 1.25–1.50 mm, the leaf apices broadly rounded with a short mucro measuring 40–50 µm long, the lower half of the lamina carinate; leaf margins entire below the apex, irregularly denticulate at the leaf mucro, wavy even when wet, with a narrow border consisting of 2–3 rows of thin to moderately thick-walled, linear cells; costae in vegetative leaves slender, ending well below the apex, often split distally into two branches, but becoming coarse and wide in gemmiferous leaves; laminal cells irregularly quadrate-polygonal, 22–33 µm wide, thin-walled, not collenchymatous, gradually becoming slightly smaller at the margins and apex, 13–22 µm wide, rectangular at leaf base, 56–101 µm × 18–27 µm; gemmiferous leaves broadly ovate, the upper portion of the lamina constricted into a rounded hood with numerous basally branched, filiform gemmae borne on the adaxial surface of the much expanded leaf costa. *Autoicous*? *Perichaetial leaves* small, about 0.3 mm long, ovate-lanceolate, acuminate, ecostate, with narrow border; archegonia numerous per bract, about the length of the perichaetial leaf. *Sporophytes* not seen.

Illustrations: Plate 402, figs. 5–9.
Habitat: on boulder; alt. 1300 m.
Distribution: endemic to China.
Specimen examined: see the type cited above.

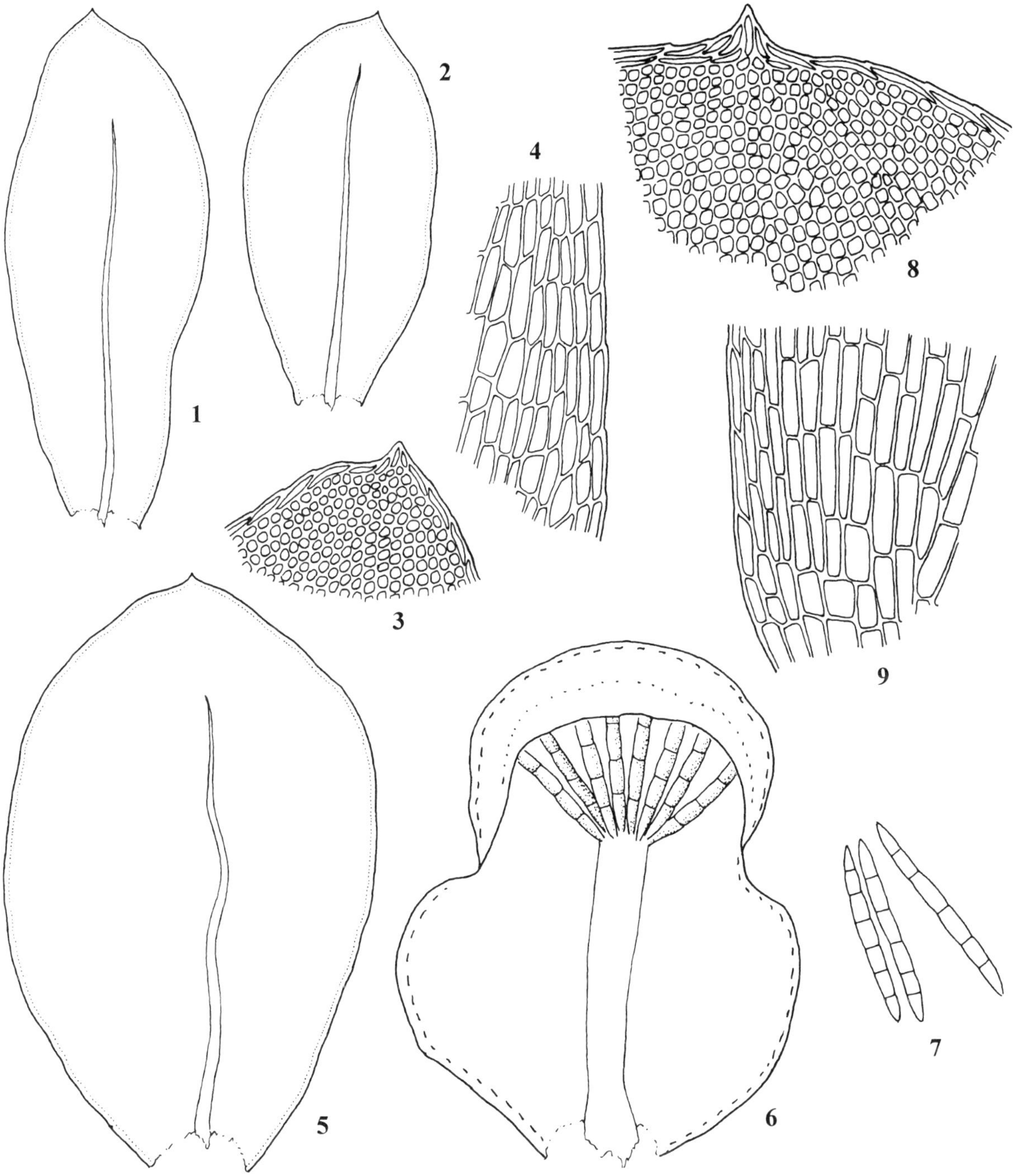

Plate 402. 1–4, *Distichophyllum maibarae*. 1–2, leaves (× 50). 3, apical leaf cells (× 250). 4, basal leaf cells (× 250). 5–9, *Distichophyllum meizhii*. 5, leaf (× 42). 6, gemmiferous leaf (× 42). 7, gemmae (× 42). 8, apical leaf cell (× 200). 9, basal leaf cell (× 200). (1–4, drawn from *P.-C. Wu 1279*, PE; 5–9, from *M.-Z. Wang 10040*, PE.)

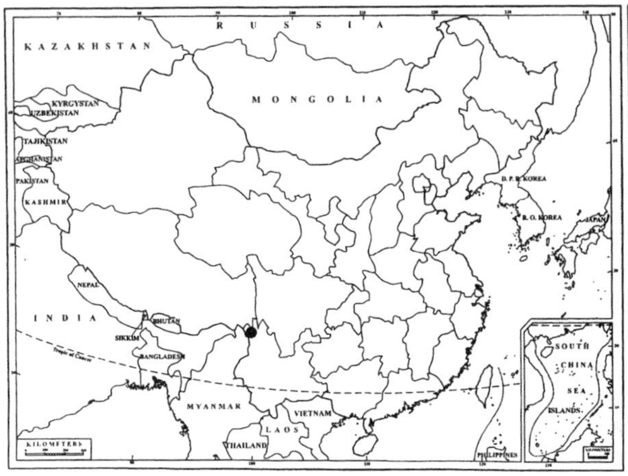

6. *Distichophyllum meizhii*

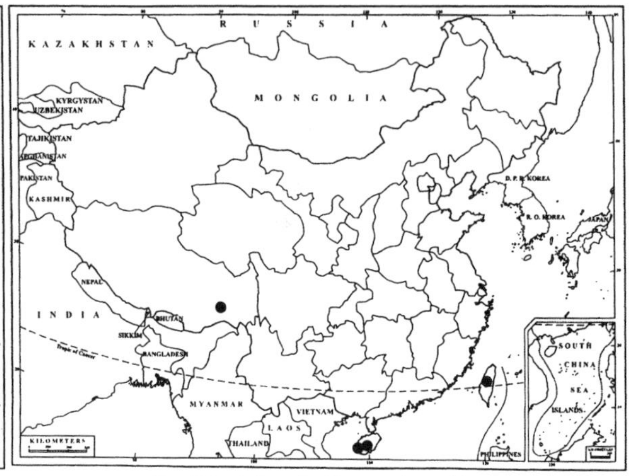

7. *Distichophyllum mittenii*

The strong dimorphism shown between the vegetative and gemmiferous leaves in this species is very striking. It is closely related to *D. heterophyllum* (Mitt.) Par. from Darjeeling, India. Both species show an evolutionary trend toward the development of morphologically dissimilar gemmiferous leaves. Unfortunately, both taxa are known from only gametophytic materials, and perhaps maintain their natural populations by asexual reproduction.

7. ***Distichophyllum mittenii*** Bosch & Sande Lac. 钝叶黄藓 dun-ye huang xian
Bryol. Jav. 2: 25. 1861. Type. Indonesia: Java, *Amann s.n.*

Plants medium-sized to large, mat-forming, laxly foliate, complanate. *Stems* prostrate or ascending, to 2 cm tall, 4 mm wide with leaves. *Leaves* slightly undulate to moderately crisped when dry, shiny, spathulate to obovate from a narrow leaf base, 3–4 mm × 2 mm, the apices broadly rounded with a tiny apiculus or mucro measuring less than 15 µm long; costae strong, ending well below apex; leaf margins entire, slightly undulate when wet, with a narrow border of 1–2 rows of thin-walled, linear cells throughout; leaf cells hexagonal, 15–22 µm wide, thin- to thick-walled, slightly collenchymatous, the submarginal cells in upper half of the leaf forming several rows, of much smaller cells, 9–15 µm; basal laminal cells rectangular, 67–90(–110) µm × 27–33 µm. *Sporophytes* lateral; setae to 10 mm long, scabrous; capsules ovoid, the exothecial cells strongly collenchymatous.

Illustrations: P.-J. Lin & B. C. Tan 1995 (Fig. 31, A–E).

Habitat: on rocks and decaying logs in shady places; alt. 1000–1650 m.

Distribution: China, Japan, Sri Lanka, Thailand, Kampuchea, Vietnam, Malaysia, Indonesia, Kalimantan, the Philippines, and New Guinea.

Chinese specimens examined: HAINAN: Mt. Diaolu, *P.-J. Lin et al. 2710, 2898, 3516* (FH, IBSC); Mt. Jianfengling, *L. Zhang 1546* (IBSC), *P.-C. Chen et al. 215* (FH, IBSC, MO). TAIWAN: Nan-tou Co., *M.-J. Lai 2692* (IBSC, MO). XIZANG: Mo-tuo Co., *Y.-G. Su 3167, 3189* (FH, IBSC).

Many Chinese specimens of *Distichophyllum mittenii*, especially those from Hainan, were mistaken for *D. obtusifolium*. Diagnostically, leaf apices of *D. obtusifolium* are devoid of a clearly differentiated leaf border and are broadly rounded without an apiculus or mucro as is the case in *D. mittenii*. Furthermore, the submarginal leaf cells of *D. obtusifolium* are only slightly smaller than the inner paracostal cells, quite unlike the markedly reduced submarginal leaf cells seen in *D. mittenii*. Thus far, we have not seen any authentic Chinese collections of *D. obtusifolium*, although its presence in China is to be expected.

8a. ***Distichophyllum oblongum*** B. C. Tan & P.-J. Lin var. ***oblongum*** 圆尖黄藓原变种 yuan-jian huang xian yuan bian-zhong
Bot. Bull. Acad. Sin., n.s. 32: 310. 1991. Type. China: Guangxi, Mt. Miao-er, *P.-J. Lin 1748* (holotype FH; isotype IBSC).

Plants small, in dense mats. *Stems* procumbent, branched, reaching 1 cm tall and less than 1 mm wide with leaves when dry. *Leaves* shrunken, twisted when dry, complanate, spreading when wet, mostly oblong, also oblanceolate to shortly spathulate, about 1 mm long, less than 0.5 mm wide, acute to short acuminate, the acumens 45–100 µm long; leaves bordered throughout, stout, 14–25 µm wide, entire, consisting of 1–2 rows of thick-walled, linear cells; costae ending well below apex; laminal cells irregular in shape, mostly polygonal, more hexagonal at mid-leaf portion, 11–17 µm long, thick-walled, not collenchymatous, homogeneous across the lamina, becoming rectangular at leaf base. *Perichaetial leaves* small, oblanceolate, acuminate, bordered, ecostate. *Sporophytes* lateral; setae 3–4 mm long, smooth; capsules shortly obovoid, less than 1 mm long, erect to nodding.

Illustrations: Plate 403, figs. 1–4.
Habitat: on soil crust over rocks and decaying logs; alt. 2000–2200 m.
Distribution: endemic to China.

Chinese specimens examined: GUANGXI: Mt. Miao-er, *P.-J. Lin 1302*, *1340* (FH, IBSC).

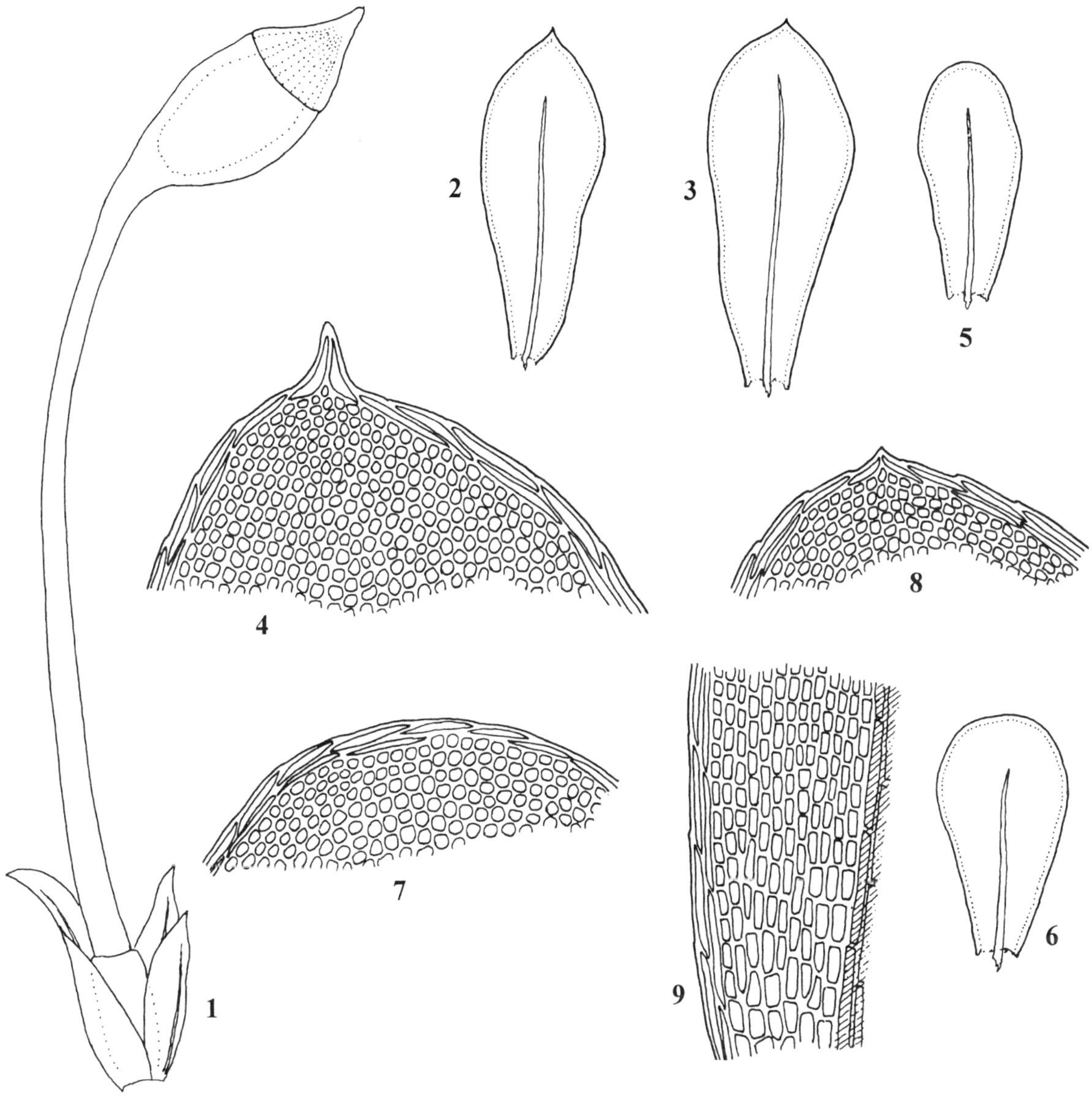

Plate 403. 1–4, *Distichophyllum oblongum* var. *oblongum*. 1, sporophyte (× 47). 2–3, leaves (× 47). 4, apical leaf cells (× 220). 5–9, *Distichophyllum oblongum* var. *fanjingense*. 5–6, leaves (× 42). 7–8, apical leaf cells (× 200). 9, basal leaf cells (× 200). (1–4, drawn from *P.-J. Lin et al. 1340*, IBSC; 5–9, from *B. C. Tan 91-1215*, FH.)

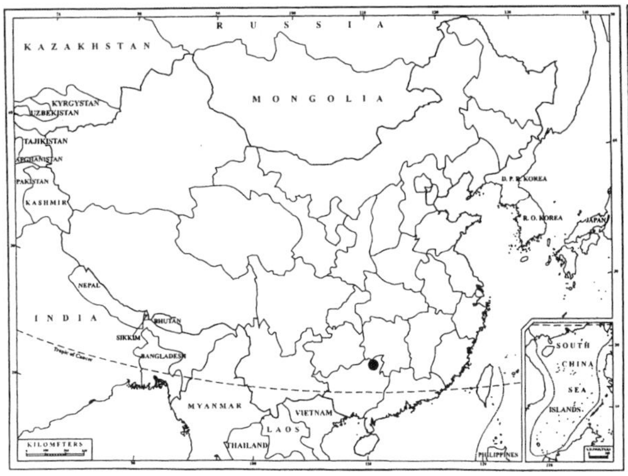

8a. *Distichophyllum oblongum* var. *oblongum*

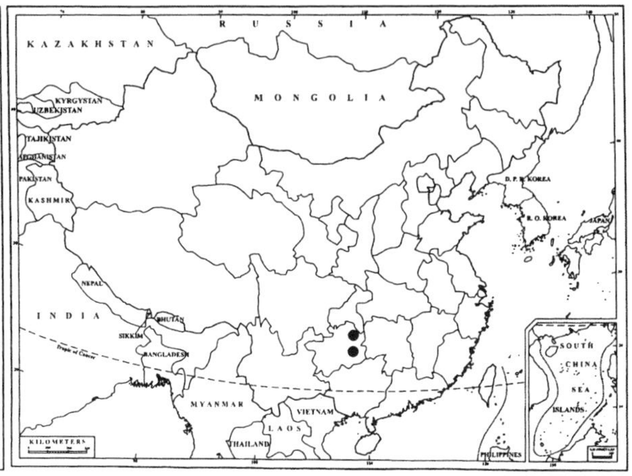

8b. *Distichophyllum oblongum* var. *fanjingense*

8b. ***Distichophyllum oblongum*** var. ***fanjingense*** P. J. Lin & B. C. Tan 圆尖黄藓贵州变种 yuan-jian huang xian gui-zhou bian-zhong

Harvard Pap. Bot. 7: 41, 1995. Types. China: Guizhou, Mt. Fanjing, on trunk of *Magnolia*, summit dwarf forest near Nine Dragon Pond beyond Jin-ding, *B. C. Tan 91-1215* (holotype FH; isotype IBSC); ibid, *Crosby 15830* (paratype MO); Lei-shan Co., summit vicinity of Mt. Leigong, *Crosby 15490* (paratype MO).

Plants forming dense mats. *Stems* small, less than 10 mm long. *Leaves* complanate, slightly wrinkled when dry, obovate to short spathulate, about 1.0–1.5 mm × 0.25–0.50 mm, round obtuse, at times with minute apiculus or mucro; leaf borders strong, entire, 13–20 µm wide; costa variable in length, often extending beyond 2/3 of the leaf length; laminal cells irregularly polygonal, 11–18 µm wide, becoming short rectangular near base, 22–34 µm × 11–18 µm. *Perichaetial leaves* small, oblong, with rounded and obtuse apices. *Sporophytes* lateral; setae 3–4 mm long, smooth, reddish brown; capsules small, ovoid, slightly nodding.

Illustrations: Plate 403, figs. 5–9.
Habitat: on trunks.
Distribution: endemic to China.

Chinese specimens examined: see the type information cited above.

This variety differs mainly from var. *oblongum* in having broadly spathulate to obovate and roundly obtuse leaves. This leaf character is best observed among terminal stem leaves, which produce a widely opened fan appearance. If the apiculus is present, it is much smaller than that seen in var. *oblongum*. The perichaetial leaves of this variety have round-obtuse leaf apices. *Distichophyllum oblongum* var. *fanjingense* appears to approach *D. obovatum* (Griff.) Par. from East India (Gangulee 1977) in many gametophyte details. However, the present variety has smooth setae while the setae of *D. obovatum* were reported to be scabrous along the entire length.

9. ***Distichophyllum osterwaldii*** Fleisch. 大型黄藓 da-xing huang xian

Musci Buitenzorg 3: 994. 1908. Type. Indonesia: Java.

Plants medium-sized to large, loosely tufted. *Stems* ascending, 1.5–3.0 cm tall, laxly foliate, complanate, 4 mm wide with leaves when dry. *Leaves* strongly undulate, twisted when dry, lingulate, obovate to spathulate from a narrow leaf base, 1.5–3.0 mm × 1.0–1.5 mm, broadly rounded, muticous, slightly denticulate at apex, otherwise entire; leaf border distinct only in the lower half of the leaf, narrow, consisting of 1–2 rows of linear cells; costa slender, reaching 3/4 the leaf length; paracostal laminal cells (18–)23–33 µm wide, rounded hexagonal, becoming slightly smaller submarginally and apically, shortly rectangular at base, 38–45(–67) µm × 11–23 µm. *Sporophytes* lateral; setae 5–10 mm long, strongly papillose throughout; capsules ovoid-oblong, 1 mm long.

Illustrations: Plate 404, figs. 1–4.

Habitat: on rocks; alt. ca. 1700 m.
Distribution: China, Japan, Indonesia, Malaysia, and the Philippines.

Chinese specimens examined: FUJIAN: Nan-jing Co., *M. Zang 190*, *200* (FH, IBSC). GUANGXI: Lin-gui Co., *P.-C. Wu et al. 1037* (IBSC); Rong-shui Co., *Y. Jia 1686* (PE); Xing-an Co., *P.-J. Lin 1327* (FH, IBSC). TAIWAN: (see B.-Y. Yang and W.-C. Lee 1964).

The Chinese specimens rarely attain the full size of their counterparts in other countries. Its leaf outlines and areolation, especially the incompletely developed leaf borders, may create confusion with *D. obtusifolium*. But the latter is a smaller plant and the leaf borders disappear only near the leaf apex and not half way below, as in *D. osterwaldii*. From the equally large plants of *D. tortile*, the present species can be distinguished by its incomplete leaf borders and smaller leaf cells.

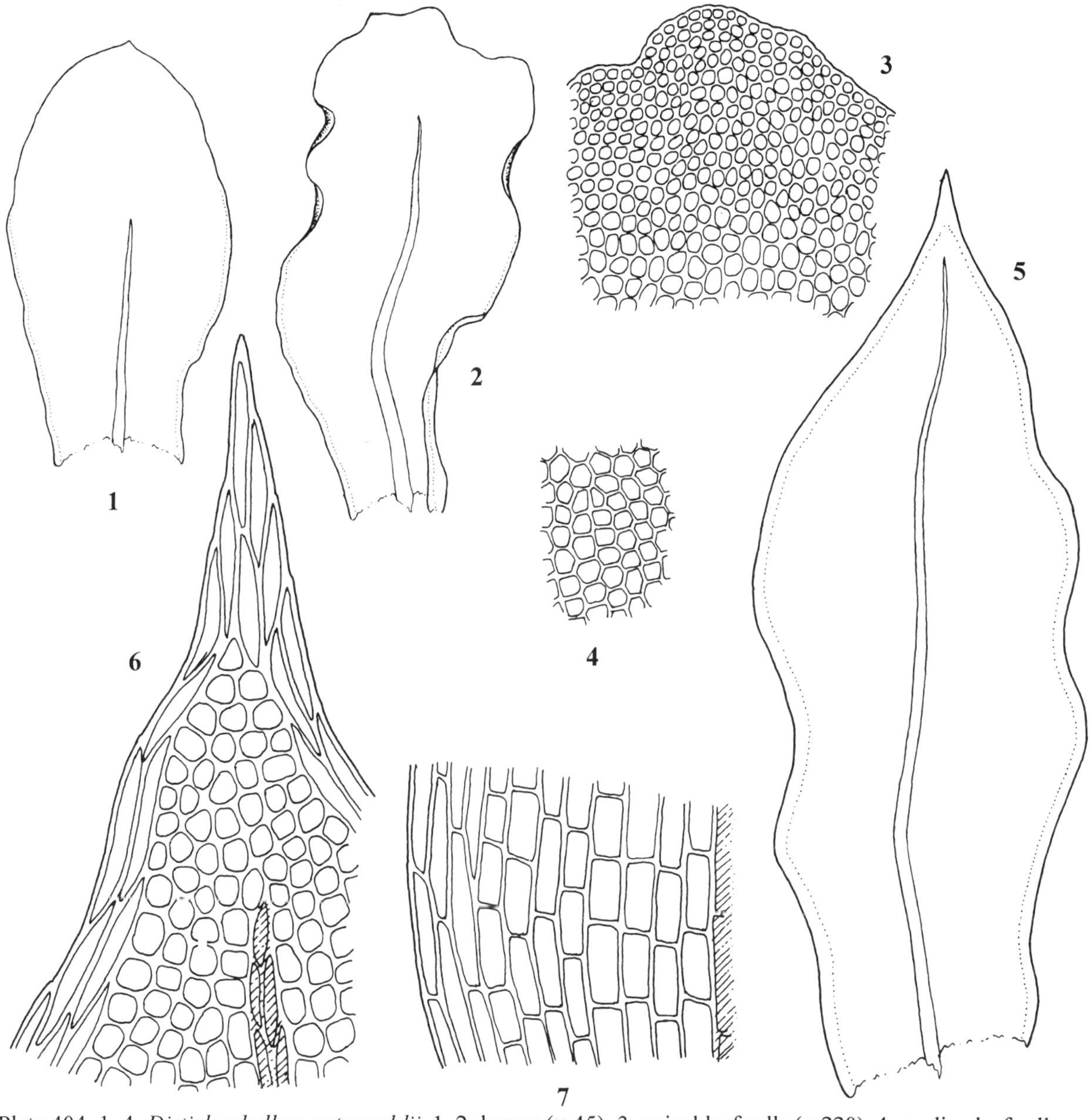

Plate 404. 1–4, *Distichophyllum osterwaldii*. 1–2, leaves (× 45). 3, apical leaf cells (× 220). 4, median leaf cells (× 220). 5–7, *Disticophyllum tortile*. 5, leaf (× 60). 6, apical leaf cells (× 280). 7, basal leaf cells (× 280). (1–4, drawn from *M. Zang 190*, HKAS; 5–7, from *P.-C. Wu et al. 85112*, PE.)

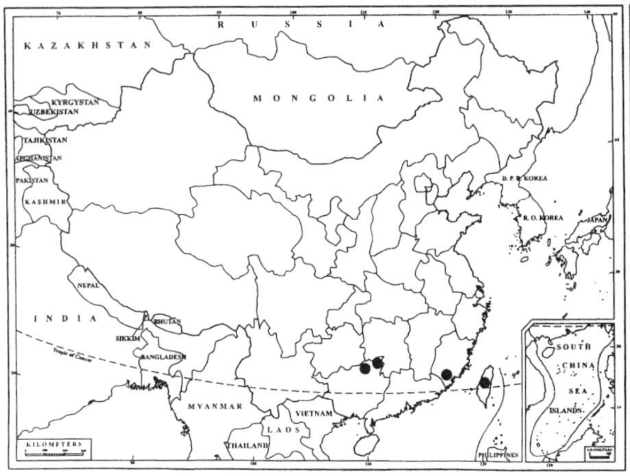

9. *Distichophyllum osterwaldii*

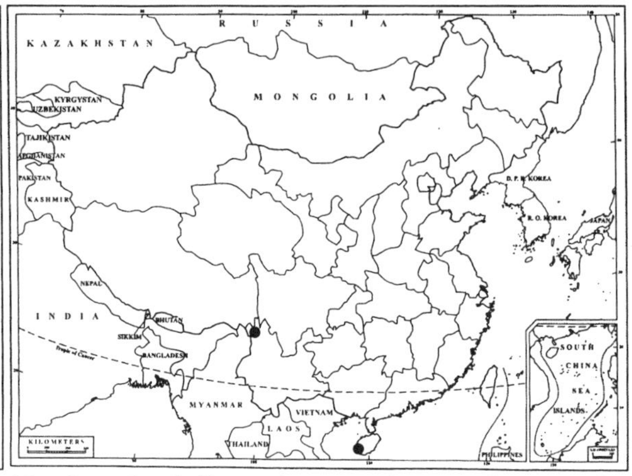

10a. *Distichophyllum subnigricaule* var. *subnigricaule*

10a. ***Distichophyllum subnigricaule*** Broth. var. ***subnigricaule*** 黑茎黄藓原变种 hei-jing huang xain yuan bian-zhong

Philipp. J. Sci. 31: 289. 1926. Type. The Philippines: Mindanao, Zamboanga, *Merrill 8324*.

Plants small to medium-sized, in dense mats. *Stems* dark brown, 1–2 cm long, 2–4 mm wide with leaves. *Leaves* slightly contorted when dry, broadly obovate and spathulate, 1–2 mm × 0.5–0.8 mm; the leaf apices rounded to broadly obtuse, sometimes weakly apiculate; leaf borders very narrow, consisting of 1(–2) rows of thin-walled, linear cells; costa single, short, reaching only about half of the leaf length; leaf cells uniformly round-hexagonal, large, 34–45(–67) µm wide, lax, thin-walled. *Perichaetial leaves* small, about 0.5 mm long, broadly ovate, acute to obtuse; laminal cells mostly short rectangular. *Setae* papillose, 6–8 mm long; capsules not seen. *Calyptrae* papillose.

Illustrations: Plate 405, figs. 1–4.
Habitat: on decaying logs; alt. 900–1400 m.
Distribution: China, Malaysia, Indonesia, and the Philippines.

Chinese specimens examined: HAINAN: Mt. Jianfengling, *P.-C. Wu et al. 85-343*, *85-369a* (H, IBSC, PE, SYS). YUNNAN: Gong Shan (Mt.), *D.-C. Zhang 2503* (FH, IFSBH).

10b. ***Distichophyllum subnigricaule*** var. ***hainanense*** P. J. Lin & B. C. Tan 黑茎黄藓海南变种 hei-jing huang xian hai-nan bian-zhong

Harvard Pap. Bot. 7: 43. 1995. Type. China: Hainan, Mt. Diao-luo, ca. 1050 m, *P.-J. Lin et al. 945A* (holotype IBSC, isotype FH).

Plants small to medium-sized, mat-forming. *Stems* procumbent, 1–2 cm long. *Leaves* mostly broadly obovate to spathulate with short, apiculate leaf apices; laminal cells small, 18–23 µm wide, lax, thin-walled; one or two rows of submarginal leaf cells slightly smaller than paracostal cells; leaf borders very narrow, usually consisting of one row of cells. *Perichaetial leaves* acute. *Setae* short, 5–7 mm long, strongly papillose; capsules small, oblong-ovoid, ca. 1 mm long.

Illustrations: Plate 405, figs. 5–9.
Habitat: on root stumps of trees; alt. ca. 1050 m.
Distribution: endemic to China.

Chinese specimen examined: see the type information cited above.

This variety differs from var. *subnigricaule* in having smaller leaf cells and one or two rows of slightly smaller submarginal leaf cells.

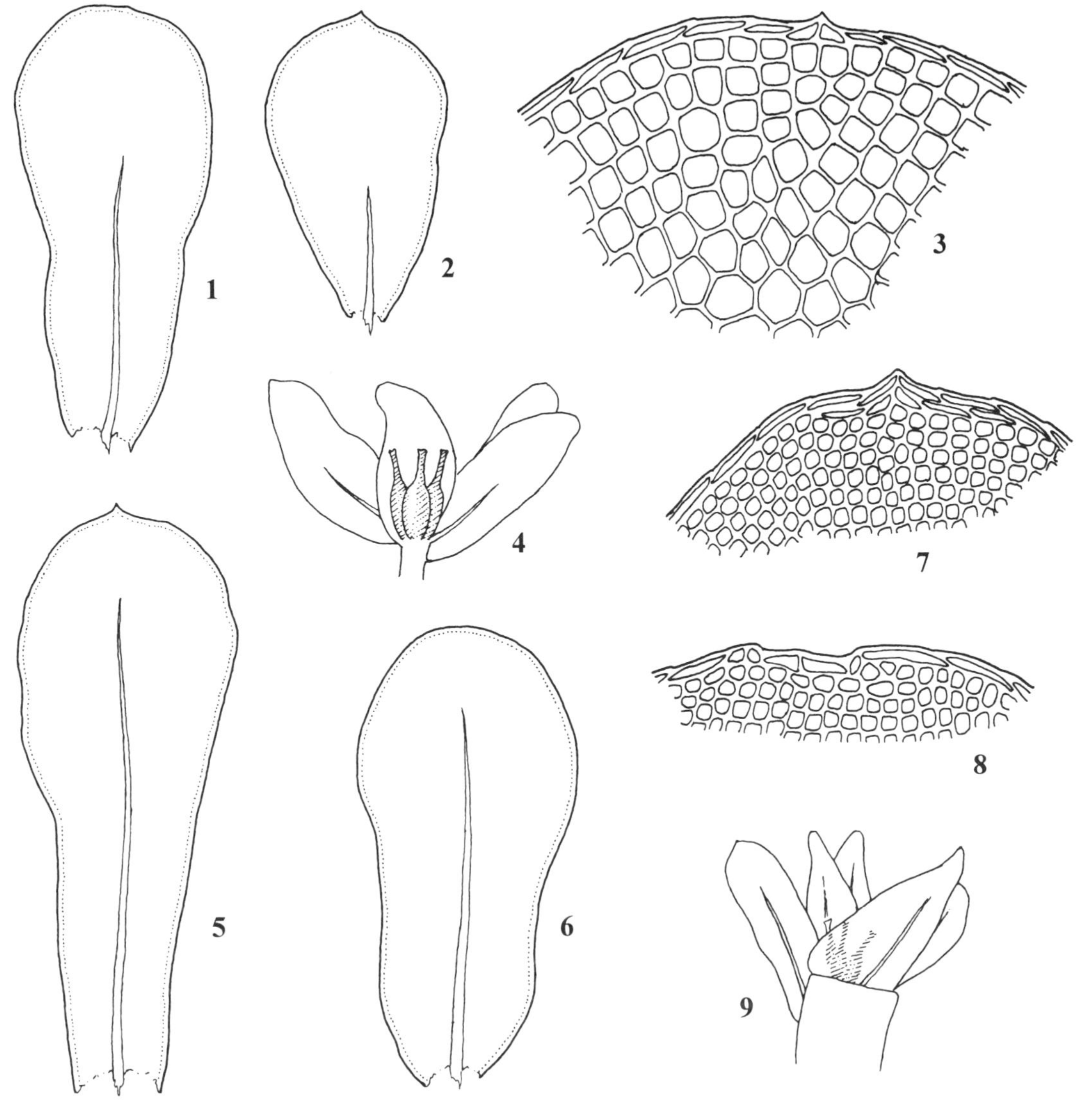

Plate 405. 1–4, *Distichophyllum subnigricaule* var. *subnigricaule*. 1–2, leaves (× 47). 3, apical leaf cells (× 220). 4, perichaetium (× 47). 5–9, *Distichophyllum subnigricaule* var. *hainanense*. 5–6, leaves (× 47). 7–8, apical leaf cells (× 200). 9, perichaetium (× 47). (1–4, drawn from *P.-C. Wu et al. 85-343*, PE; 5–9, *P.-J. Lin et al. 945a*, IBSC.)

11. ***Disticophyllum tortile*** Dozy & Molk. ex Bosch & Sande Lac. 卷叶黄藓 juan-ye huang xian
Bryol. Jav. 2: 27. 1862. Type. Indonesia: Java, *Amann s.n.*

Plants large. *Stems* to 4 cm tall. *Leaves* oblong-lingulate to slightly obovate, 2.5–4.0 mm long; leaf apices apiculate to mucronate, with acumens (27–)50–60(–100) µm long; leaf borders often wavy, with 2–3 rows of linear cells; laminal cells nearly homogeneous, round-hexagonal, large, 40–60(–70) µm wide, thin-walled. *Setae* papillose, less than l cm long. *Calyptrae* smooth, fringed below.

Illustrations: Plate 404, figs. 5–7.
Habitat: on soil and wet seepy and shaded banks.
Distribution: China, Thailand, Vietnam, Malaysia, Indonesia, Kalimantan, and the Philippines.

Chinese specimen examined: HAINAN: Mt. Jianfengling, *Z.-H. Li 85-112* (SYS).

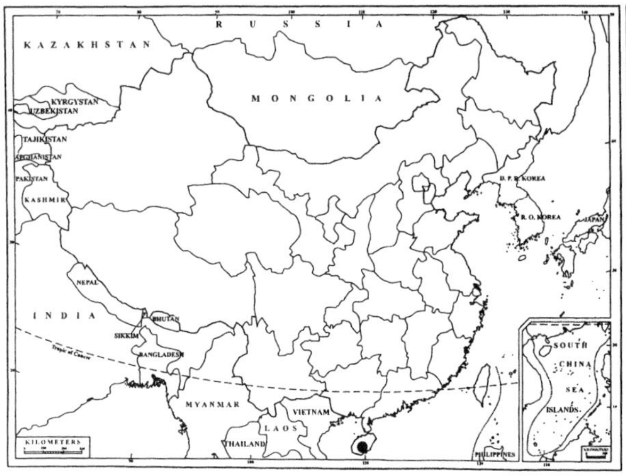

10b. *Distichophyllum subnigricaule* var. *hainanense*

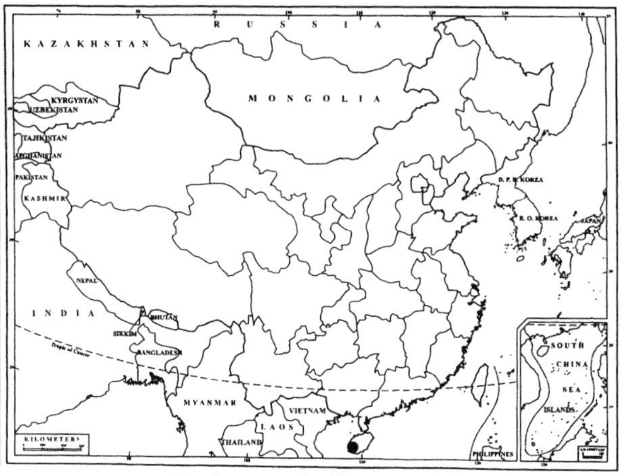

11. *Disticophyllum tortile*

Distichophyllum tortile is easily identified by its usually large plants and the largest leaf cells among all Chinese species of the genus. The lax and thin-walled leaf areolation is comparable to *D. subnigricaule*, but the leaf apices of *D. tortile* have stout acumens measuring 50–100 µm long. The leaf apices of *D. subnigricaule* are broadly rounded, only occasionally becoming weakly apiculate. Sometimes, smaller plants of *D. tortile* can be confused with *D. collenchymatosum*. Nonetheless, the large and lax leaf cells and the papillose setae of *D. tortile* are two reliable distinguishing characters.

12. ***Distichophyllum wanianum*** B. C. Tan & P.-J. Lin 王氏黄藓 wang-shi huang xian
Trop. Bryol. 10: 57. 1995. Types. China: Yunnan, Lv-chun Co., *M. Zang 550* (holotype IBSC; isotypes FH, HKAS); ibid, *M. Zang 653* (paratypes FH, HKAS, IBSC).

Plants small, mat-forming. *Stems* up to 1.6 cm tall, 2 mm wide with leaves. *Leaves* strongly crisped or cirrate when dry, broadly spathulate with a long narrow basal part when wet, 2.0–2.5 mm × 1.0–1.2 mm, the leaf apices broadly rounded, at times with a minute mucro; leaf margins plane, wavy in the upper half even when wet, with strong borders all around; the border about 10–15 µm wide, consisting of 2–3 rows of thick-walled, linear cells, but becoming one row around the leaf apex; costa single, reaching near the apex; laminal cells nearly homogeneously small, quadrate-polygonal, 11–13(–18) µm wide, becoming smaller (9–13 µm wide) apically and rectangular [33–45(–67) × 11–22 µm] basally. *Perichaetial leaves* small, oblong-spathulate, distinctly mucronate, with costa well developed. *Sporophytes* lateral; setae 5–10 mm long, smooth below, gradually becoming weakly papillose distally; capsules oblong-ovoid, 2 mm long.

Illustrations: Plate 406, figs. 1–6.
Habitat: on branches in dense forests; alt. 100–2300 m.
Distribution: endemic to China.

Specimens examined: GUANGDONG: Ru-yuan Co., *P.-J. Lin 865* (FH, IBSC). HAINAN: Mt. Diaoluo, *P.-J. Lin et al. 945b* (FH, IBSC).

Its broadly spathulate leaves are reminiscent of a small form of *Distichophyllum mittenii*. In contrast, *D. wanianum* has strong and thick leaf borders without submarginal rows of smaller cells. Its closest affinity is with the two varieties of *D. oblongum*, the other Chinese endemics in the same region from Guizhou.

Species Not Seen or Studied:

1. *Distichophyllum jungermannioides* (C. Müll.) Bosch & Sande Lac.
Bryol. Jav. 2: 22. 1861. *Mniadelphus jungermannioides* C. Müll., Syn. Musc. Frond. 2: 660. 1851.

The Chinese record of this species is from Botel Tobago or Orchid Island of Taiwan (B.-Y. Yang and W.-C. Lee 1964). The species is distinctive in being very small and having concave, broadly spathulate leaves with short cuspidate leaf apices, ca. 1 mm long; leaf cells are small, round-hexagonal, measuring (8–)10–20(–25) µm wide; the leaf borders are strong and well developed. Most importantly, the leaf costa is short reaching halfway below midleaf.

2. *Distichophyllum obtusifolium* Ther.
Monde Pl., ser. 2, 9: 22. 1907.

This species was reported from Hainan (P.-J. Lin et al. 1992), Hong Kong (M.-L. So 1995) and Taiwan (M.-J. Lai and J.-R. Wang-Yang 1976). The Hainan specimens were shown to be either *D. mittenii* or *D. cirratum* var. *elmeri.* Unfortunately, the collections from Taiwan and Hong Kong were not available for confirmation.

Distichophyllum obtusifolium can be identified by its incompletely differentiated border at leaf apex that becomes slightly denticulate because of the short, rectangular marginal cells. Furthermore, the submarginal rows of leaf cells in *D. obtusifolium* are not noticeably smaller than the paracostal cells, as is the situation in *D. mittenii* or *D. cirraum* var. *elmeri.*

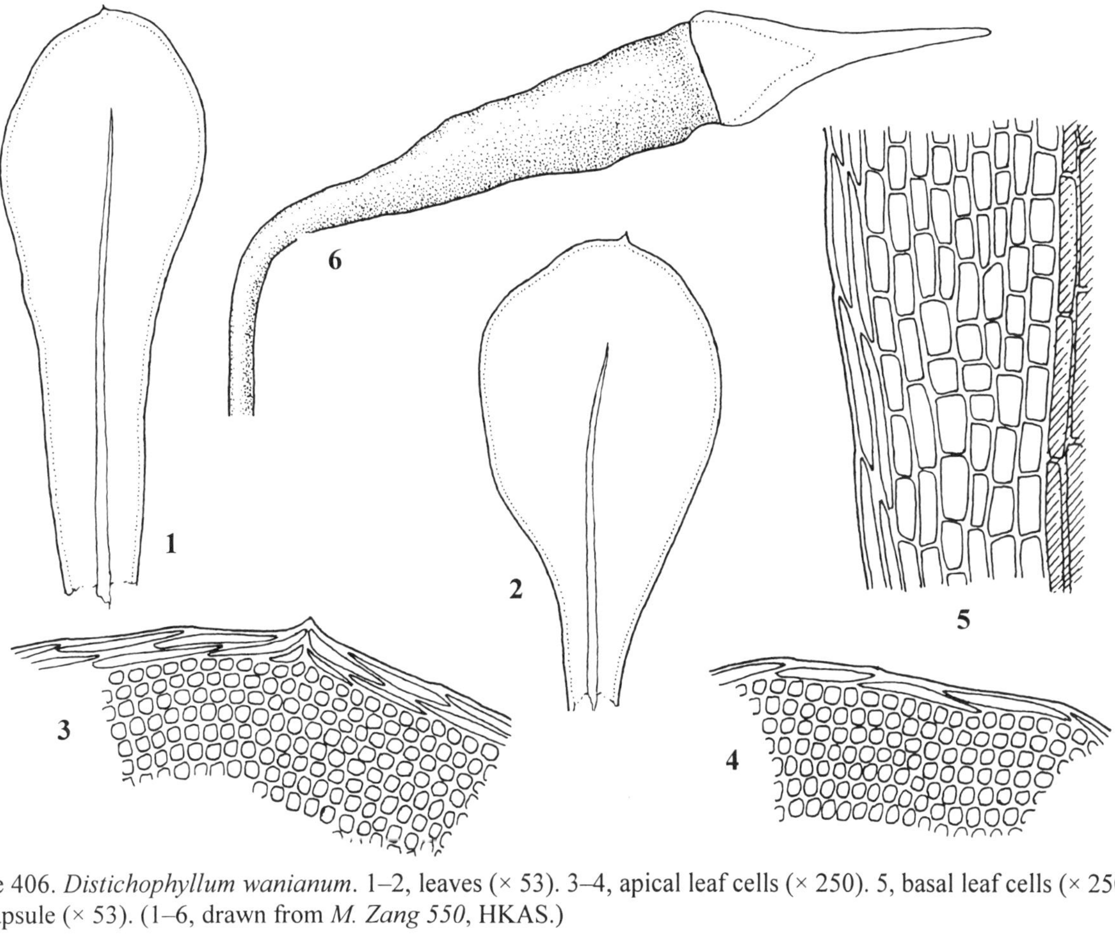

Plate 406. *Distichophyllum wanianum*. 1–2, leaves (× 53). 3–4, apical leaf cells (× 250). 5, basal leaf cells (× 250). 6, capsule (× 53). (1–6, drawn from *M. Zang 550*, HKAS.)

8. ***Hookeria*** Smith 油蘚屬 you xian shu
Trans. Linn. Soc. London 9: 275. 1808.

Plants soft, complanate, in pale green mats. *Stems* simple or branched, prostrate on ground. *Leaves* heteromorphic, forming irregular rows; dorsal leaves symmetrical, lateral leaves asymmetrical, slightly wrinkled when dry, flattened when wet, ovate, broadly lanceolate to lanceolate, broadly acute to acuminate, often radiculose at the leaf tips, borders not differentiated or weakly differentiated with only one row of elongate cells; costa none; leaf cells large, hyaline, oval-hexagonal to shortly rectangular, becoming shorter toward the apex, thin-walled. *Autoicous*. *Perichaetia* lateral. *Perigonia* gemmiform. *Setae* long, smooth; capsules oblong, horizontal to pendulous; opercula long rostrate; peristome teeth papillose. *Calyptrae* mitriform. *Spores* slightly papillose.

The genus has about ten species in the world. Although three species were reported in the literature for China, only *H. acutifolia* is accepted here. *Hookeria lucens* (Hedw.) Sm., which differs from *H. acutifolia* in having a round-obtuse leaf apex, is a western North American and European species and is probably not present in China. The Chinese records of *H. lucens* are based on collections named *H. nipponensis* (Besch.) Broth., which nas a somewhat broadly acute leaf apex. This latter taxon is treated as a synonym of *H. acutifolia* (Iwatsuki 1991).

1. ***Hookeria acutifolia*** Hook. & Grev. 尖叶油藓 jian-ye you xian
Edinburgh J. Sci. 2: 225.1825. Type. Europe.
Hookeria nipponensis (Besch.) Broth., Nat. Pflanzenfam. I(3): 934. 1907. *Pterygophyllum nipponense* Besch., Ann. Sci. Nat., Bot., ser. 7, 17: 362. 1893.

Plants soft, complanate, in pale green mats. *Stems* 2–3(–5) cm long, simple or branched, prostrate on ground, in cross section with nearly homogeneous parenchyma cells with a slightly differentiated central strand. *Leaves* heteromorphic, forming irregular rows; dorsal leaves symmetrical, lateral leaves asymmetrical, slightly wrinkled when dry, flattened when wet, ovate, broadly lanceolate to lanceolate, 4–7 mm × 4 mm, broadly acute to acuminate, often radiculose at the leaf tips, borders not differentiated or weakly differentiated with only one row of longer cells; costa none; leaf cells large, hyaline, oval-hexagonal to short rectangular, 94–188 µm × 40–60 µm, becoming shorter toward apex, thin-walled. *Autoicous. Perichaetial leaves* smaller, ovate-lanceolate. *Perigonia* gemmiform. *Setae* long, 10–20 mm long, reddish brown to yellow, smooth; capsules oblong-ovoid, horizontal to pendulous, 1–2 mm long; opercula long rostrate; exostome teeth lanceolate, striate below, papillose above; endostome segments as long as the exostome, basal membrane high. *Calyptrae* mitriform. *Spores* 12–16 µm, slightly papillose.

Illustrations: P.-J. Lin & B. C. Tan 1995 (Fig. 36, A–F).
Habitat: on wet and shaded ground, rocks, decaying logs with humic covers; alt. 550–2500 m.
Distribution: China, widespread in Asia, Africa, and eastern North America, scattered along coastal northwestern North America and reaching northern South America.

Chinese specimens examined and reported: ANHUI: Huang Shan (Mt.), *P.-C. Chen 7132* (IBSC, PE). FUJIAN: Mt. Wuyi, *P.-J. Lin 2980* (FH, IBSC). GUANGDONG: Ru-yang Co., *P.-J. Lin 874* (FH, IBSC); Yang-chun Co., *B.-H. Liang 89785* (FH, IBSC). GUANGXI: Long-sheng Co., *P.-C. Wu et al. 894* (FH, IBSC, PE); Tian-lin Co., *P.-J. Lin et al. 2412* (FH, IBSC); Xing-an Co., *P.-J. Lin et al. 1328* (FH, IBSC). GUIZHOU: Chi-shui Co., *Crosby 16135* (MO); Jiang-kou Co., *X.-J. Li 96* (FH, IBSC, MO); Lei-shan Co., *Crosby 15520* (MO); Mt. Fanjing, *Crosby 15880* (MO); Sui-yang Co., *Crosby 15956* (MO). HAINAN: Diaoluoshan Forestry Station, *Reese 17860* (FH, IBSC, LAF, MO); Jianfengling Forestry Station, *Reese 17800* (IBSC, LAF, MO); Mt. Limu, *P.-J. Lin et al. 68* (FH, IBSC); Mt. Wuzhi,, *P.-J. Lin 3567* (FH, IBSC, MO). HONG KONG: Mt. Tai- Mo, *P.-J. Lin et al. 284, 369* (IBSC); Yin Ngan, *P.-J. Lin et al. 899* (IBSC). HUNAN: Bin-zhou City, *P.-J. Lin et al. 1082* (IBSC); Da-yong City, *D.-K. Li 18056* (FH, IBSC); Liu-yang Co., *P.-J. Lin et a l. 552* (FH, IBSC). JIANGSU: Yi-xing Co., *M. Zang 589* (FH, IBSC). JIANGXI: Lu Shan (Mt.), *C. Gao 40831* (FH, IFSBH); Nan-chang City, *M.-X. Gong s.n.* (IBSC). SICHUAN: He-jiang Co., *Q. Li 9829* (FH); Lu-ding Co., *Y. Jia 02151* (MO); Mt. Wo-lang, *C. Gao 18432* (FH, IFSBH); Mt. Omei, *P.-J. Lin 3285* (FH, IBSC), *Redfearn 34739* (MO); Mu-chuan Co., *P.-C. Chen 5843* (FH, IBSC, PE); Ya-an Co., *Q. Li 854* (FH, IBSC). TAIWAN: Chia-yi Co., *M.-J. Lai 2938* (MO); Mt. Ali, *M.-J. Lai 10121* (FH, MO); Mt. Daxue, *Z. Iwatsuki 883* (FH, IBSC, NICH); Ilan Co., *M.-J. Lai 11361* (FH, IBSC, NY); Nan-tou Co., *C.-C. Chuang 367* (NY). XIZANG: Lin-zhi Co., *Y.-G. Su 2361* (FH, IBSC); Mo-tuo Co., *Y.-G. Su 3058* (FH, IBSC). YUNNAN: Da-guan Co., *X.-J. Li 4267* (FH, IBSC); Dulongjiang, *D.-C. Zhang 1013* (FH, IBSC); Fu-gong Co., *D.-C. Zhang 1013* (FH, IBSC), Gongshan (Mt.), *M. Zang 5404* (FH, IBSC); Jing-dong Co., *J.-K. Zhang 1329* (FH, IBSC); Wei-xi Co., *D.-C. Zhang 286* (FH, IBSC); Xi-chou Co., *S.-H. Huang 1* (FH, IBSC); Yuan-yang Co., *M. Zang 4734* (FH, IBSC); Zhen-kang Co., *Q.-W. Wang 7977* (FH, IBSC). ZHEJIANG: Lin-an Co., *P.-C. Wu 7996* (MO); Mt. Jiulong, *R.-L. Hong et al. 215* (FH); Mt. Tianmu, *P.-C. Wu 8004* (IBSC, PE); Mt. Tian-tai, *D.-K. Li 3972* (FH, IBSC).

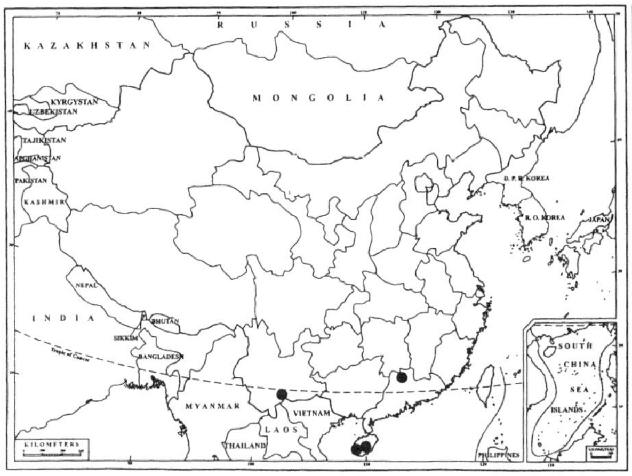

12. *Distichophyllum wanianum*

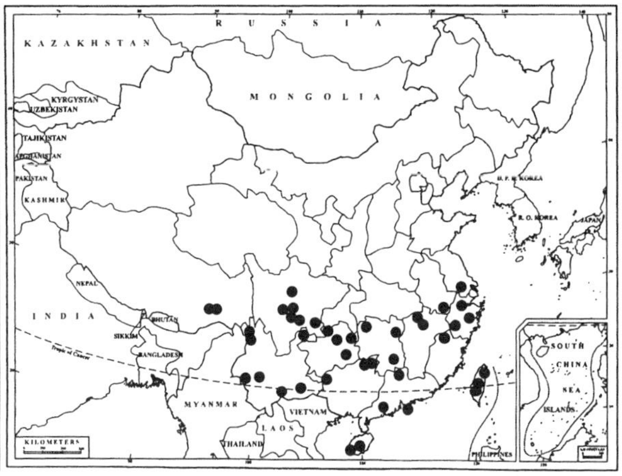

1. *Hookeria acutifolia*

Hookeria acutifolia is easily identified by its large, net-like leaf areolation, mostly without a leaf border. Leaf apical propagules produced by a few, shortly oval laminal cells are unique. In Mainland China, the distribution of this species occurs mostly south of the Yangtze River. Plants from Taiwan populations are larger than their mainland and Hainan counterparts.

9. ***Hookeriopsis*** (Besch.) Jaeg. & Sauerb. 拟油藓属 ni you xian shu

Ber. Thätigk. St. Gallischen Naturwiss. Ges. 1875–76: 358. 1877. *Hookeria* sect. *Hookeriopsis* Besch., Ann. Sci. Nat., Bot., ser. 6, 3: 240. 1876.

Plants medium-sized to larger, mat-forming. *Stems* creeping, elongated. *Leaves* complanate, forming dorsal, ventral and lateral rows, slightly heterophyllous, flexuous when dry, concave when wet; lateral leaves mostly ovate-oblong, asymmetrical, margins not bordered, costae long, double, reaching beyond midleaf; dorsal and ventral leaves smaller than lateral leaves, more acuminate; leaf cells narrowly rhomboidal, becoming shorter at apex and rectangular at base, thin-walled, smooth. *Polygamous*. *Setae* lateral; capsules horizontal; opercula conic-rostrate; peristome double, exostome and endostome typically hookerioid; basal membrane high; cilia absent. *Calyptrae* mitriform. *Spores* spherical, papillose.

The genus has more than a hundred species with a high diversity center in Africa and Central and South America. There is only one species in China.

1. ***Hookeriopsis utacamundiana*** (Mont.) Broth. 并齿拟油藓 bing-chi ni you xian

Nat. Pflanzenfam. I(3): 942. 1907. *Thamniopsis utacamundiana* (Mont.) Buck, Brittonia 39: 219. 1987. *Hookeria utacamundiana* Mont., Ann. Sci. Nat., Bot., ser. 2, 17: 247. 1842. Type. Sri Lanka.

Hookeriopsis geminidens Broth., Philipp. J. Sci. 5: 156. 1910.

Hookeriopsis sumatrana (Bosch & Sande Lac.) Broth., Nat. Pflanzenfam. I(3): 942. 1907. *Lepidopilum sumatranum* Bosch & Sande Lac., Bryol. Jav. 2: 42. 1862.

Thamniopsis secunda (Griff.) Buck, Brittonia 39: 219. 1987. *Hookeriopsis secunda* (Griff.) Broth., Nat. Pflanzenfam. I(3): 942. 1907. *Hookeria secunda* Griff., Calcutta J. Nat. Hist. 3: 280. 1843 [1842].

Plants mat-forming, light green to reddish brown. *Stems* creeping, 3–7 cm long, 3–4 mm wide with leaves when dry, in cross section showing an outermost row of hyalodermal cells surrounding a layer of thick-walled, reddish colored cortical cells, a well differentiated central region of thin-walled parenchyma cells, without a central strand. *Leaves* complanate forming dorsal, ventral and lateral rows, slightly heterophyllous, flexuous when dry, concave when wet; lateral leaves mostly ovate-oblong, asymmetrical, 1.5–2.0 mm × 1.0 mm, broadly acute to acuminate, at times mucronate; margins not bordered, involute on one side, irregularly serrate, with bifid, unicellular teeth toward the apex; costae long, double, reaching beyond midleaf, spinose abaxially; dorsal and ventral leaves smaller than lateral ones, more acuminate; leaf cells narrowly rhomboidal, 56–90 µm × 13–16 µm, becoming shorter

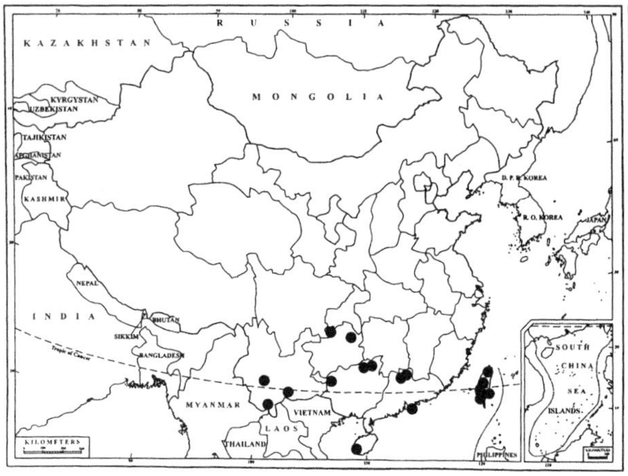

1. *Hookeriopsis utacamundiana*

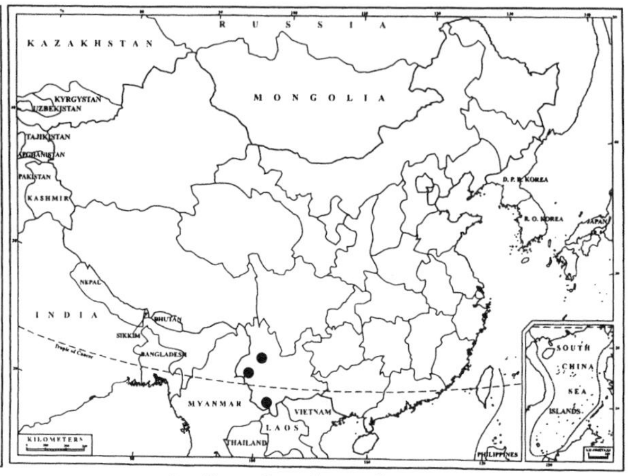

1. *Symphyodon echinatus*

at apex, rectangular at base, thin-walled, smooth, sometimes porose. *Polygamous*. *Perichaetial leaves* small, ovate, acuminate, with a few marginal teeth. *Setae* lateral, 20–30 mm long, reddish brown; capsules horizontal, 1.0–1.5 mm long; opercula conic-rostrate; peristome double, exostome and endostome typically hookerioid, basal membrane high; cilia absent. *Calyptrae* mitriform. *Spores* spherical, 8–10 µm in diameter, papillose.

Illustrations: P.-J. Lin & B. C. Tan 1995 (Fig. 37, A–I).
Habitat: on ground, tree base, decaying logs, boulders under partially dense forest canopy; alt. 400–2400 m.
Distribution: China, widespread in tropical East Asia (including Yakushima Is. of Japan) and the Pacific.

Chinese specimens examined: GUANGDONG: Ru-yuan Co., *P.-J. Lin 1291* (FH, IBSC), *P.-J. Lin 869* (IBSC, MO); Shi-xing Co., *P.-J. Lin et al. 4878* (FH, IBSC). GUANGXI: Long-sheng Co., *P.-C. Wu et al. 11496* (FH, IBSC, PE); Rong-shui Co., *M.-Z. Wang 46035* (PE), *46224* (MO, PE); Tian-lin Co., *P.-J. Lin et al. 2423* (FH, IBSC). GUIZHOU: Chi-shui Co., *B. C. Tan 91-1049* (FH); Shizangdong waterfall, *Crosby 16151A* (MO); Mt. Fanjing, *Crosby 15690* (MO). HAINAN: Mt. Jianfengling, *P.-J. Lin et al. 691* (FH, IBSC, MO). HONG KONG: Mt. Tai-Mo, *P.-J. Lin et al. 1357, 2042* (IBSC). TAIWAN: Chia-yi Co., *S.-H. Lin 62* (H, MO); Hsin-chu Co., *Shevock 14407* (MO); Kao-shiung Co., *C.-C. Chuang 1020* (FH, IBSC, NY); Nan-tou Co., *M.-J. Lai 11538* (H), *M.-J. Lai 8229* (MO); Mt. Daxue, *Iwatsuki & Sharp 3121* (FH, IBSC); Tai-tung Co., *S.-H. Lin s.n.* [Bryophytes of Taiwan Exsiccatae, ser. 3, 115, as *Eriopus parviretus* in part] (MO); Mt. Tianchang, *K. Sawada s.n.* (FH, IBSC). YUNNAN: Jing-dong Co., *J.-K. Zhang 1836-2* (FH, IBSC); Lv-chun Co., *M. Zang 212* (FH, HKAS, IBSC); Xishuangbana, *Redfearn et al. 33907* (IBSC, MO).

The variable nature of leaf morphology in the present species, especially the leaf apex and leaf marginal serration, is extreme. Plants with broadly obtuse leaf apices have been called *Hookeriopsis secunda* and those with very strong serration of leaf margins have been known as *H. geminidens*. There exists, however, an intergrading series of variations between these two extremes, an observation discussed by Tan and Robinson (1990) who recognized one species of the genus in East Asia. Incidentally, the specimen (no. 62 of S. H. Lin's Exsiccatae Series 2, Bryophytes of Taiwan) was misidentified as *Cyclodictyon blumeanum*. It is *Hookeriopsis utacamundiana*.

SYMPHYODONTACEAE

刺果藓科 ci-guo xian ke

by Si He & Lin Bang-juan

Plants small to large, to 20 cm long, dull to glossy, yellowish green, golden yellow to brownish red when older. *Stems* prostrate, irregularly pinnately branched to regularly bipinnately or tripinnately branched; leafy stems appressed or subcomplanate to complanate; pseudoparaphyllia none; central strand absent, cortical cells lax, thin-walled, epidermal cells thick-walled, substereid. *Stem leaves* and branch leaves similar or dimorphic, plane to deeply concave, sometimes undulate when dry, oval, ovate, oblong, oblong-ovate, oblong-lanceolate or lingulate; leaf apices truncate, rounded, obtuse, apiculate, or acute to acuminate; margins plane to slightly inflexed, finely serrulate to coarsely serrate above, serration nearly always formed by a row of slightly enlarged cells, subentire to entire below, not or narrowly decurrent at the base; costae double, unequal, extending to 1/2 total leaf length; apical leaf cells undifferentiated or conspicuously shorter than the adjacent cells; median cells linear, prorulose to strongly prorate; alar cells more or less differentiated, quadrate, subquadrate, or irregularly rectangular, thin- to thick-walled, of 3–5 rows of cells extending 3–6 cells up the margins. *Axillary hairs* brownish or hyaline, uniseriate, two- to (rarely) three-cells long, terminal cells larger. *Brood bodies* sometimes present in leaf axils, filamentous, uniseriate, papillose, of 12–24 quadrate cells, in dense, reddish brown clusters. *Dioicous*. *Perichaetial leaves* typically sheathing at the base, suddenly narrowed from ovate to long acuminate to subulate acumens, laminal cells often thick-walled, porose, 2–3-costate (or costae occasionally absent). *Setae* 1.0–4.5 cm long, roughened to strongly papillose distally, smooth proximally; capsules suberect to erect, oblong-ovoid to cylindrical, symmetric to faintly asymmetrical, sparsely to densely echinate, spines to 150 μm long, formed as extensions of exothecial cell walls; exothecial cells subquadrate to elongate-rectangular, thin- to thick-walled, not collenchymatous; stomata present, immersed, few to numerous on the capsule neck; columella persistent, frequently extending beyond the mouth of capsules; annuli differentiated, of 3–5 rows of small, thick-walled, subquadrate to quadrate cells; opercula conic, apiculate to obliquely long rostrate, smooth to occasionally prickly papillose; peristome hygrocastique; exostome teeth 0.28–0.70 mm long; endostomial segments 0.10–0.36 mm long, sometimes strongly reduced; basal membrane 15–75 μm high. *Calyptrae* cucullate, smooth. *Spores* spherical, 13–20 μm in diameter, yellow to yellowish brown, finely to strongly papillose.

1. ***Symphyodon*** Mont. 刺果藓属 ci-guo xian shu
   Ann. Sci. Nat. Bot., ser. 2, 16: 279. 1841.

For description of the genus, see that of the family. The genus *Symphyodon* is composed of 15 species in the world (He & Snider 2000). Five species are recognized in China.

1. Leaves spreading, complanate; leaf apices rounded to truncate; exostome densely papillose throughout .... 3. *S. pygmaeus*
1. Leaves imbricate or appressed, not complanate; leaf apices broadly apiculate, acute to shortly acuminate or bluntly acuminate; exostome smooth or finely papillose above 2
   2. Leaves imbricate, distinctly concave; leaf apices broadly apiculate; leaf margins subentire; leaf cells somewhat projecting .... 4. *S. weymouthioides*
   2. Leaves appressed, not or slightly concave; leaf apices acute to shortly acuminate or bluntly acuminate; leaf cells moderately to strongly prorate 3
3. Stem and branch leaves clearly differentiated; leaf cells distinctly prorate .... 1. *S. echinatus*
3. Stem and branch leaves somewhat differentiated; leaf cells prorulose 4
   4. Stem leaves gradually acuminate; median leaf cells broader, ca. 12:1 .... 2. *S. perrottetii*
   4. Stem leaves bluntly acuminate, median leaf cells narrower, ca. 20:1 .... 5. *S. yuennanensis*

1. ***Symphyodon echinatus*** (Mitt.) Jaeg. 异叶刺果藓 yi-ye ci-guo xian
   Ber. Thätigk. St. Gallischen Naturwiss. Ges. 1876–77: 296. 1878. *Stereodon echinatus* Mitt., J. Proc. Linn. Soc., Bot., Suppl. 1: 110. 1859. Type. India: Sikkim, *Hooker 734* (holotype NY; isotypes BM, S, W).

Plants medium-sized, to 3 cm long, yellowish green to brownish green, not glossy. *Stems* with somewhat dendroid shoots bearing many bipinnate to tripinnate branchlets. *Leaves* appressed, clearly dimorphic; stem leaves ovate-lanceolate, 1.3–1.4 mm × 0.5–0.6 mm, symmetric, gradually tapered, narrowly acuminate at the apex; margins irregularly serrate above, serrulate or subentire below, often clearly decurrent at the base; branch leaves 0.7–0.9 mm × 0.30–0.40 mm wide, slightly incurved on one side or on both sides, oval, broadly ovate to oblong-ovate, cordate or oblong-lanceolate; all leaf apices broadly to moderately broad acute; margins irregularly serrate in the upper 2/3, serration formed by a row of enlarged cells, subentire below; costae unequal, extending to 2/5–1/2 total leaf length; apical leaf cells not particularly shorter than the adjacent cells; median cells linear, 30–40 µm × 2.5–3.5 µm, strongly prorate at the upper ends, especially in branch leaves; alar cells differentiated, quadrate to rectangular, thick-walled, of 2–4 rows extending 2–4 cells up the margins. *Perichaetial leaves* sheathing at the base, abruptly narrowed to a long acumen, cells thick-walled, porose, costae very short, double. *Sporophytes* not seen.

Illustrations: Plate 407, figs. 1–7.
Habitat: on tree trunks.
Distribution: China, Nepal, India, Sri Lanka, and Thailand.

Chinese specimens examined: YUNNAN: Jing-hong Co., *J.-K. Zhang 780–3, 1236–1* (MO); Lu-xi Co., *Q. Li 1243a* (IBSC); Yang-bi Co., west side of Mt. Diancang, *Redfearn & Y.-G. Su 793* (MO, NY, SMS).

Symphyodon echinatus appears to be closely related to *S. imbricatifolius* (Mitt.) S. P. Churchill from Central and South America, but it differs from the latter in having strongly dimorphic stem and branch leaves, distinctly decurrent stem leaves, shorter and strongly prorate leaf cells, and longer and stronger double costae. The presence of brood bodies is also a distinguishing feature of *S. echinatus*.

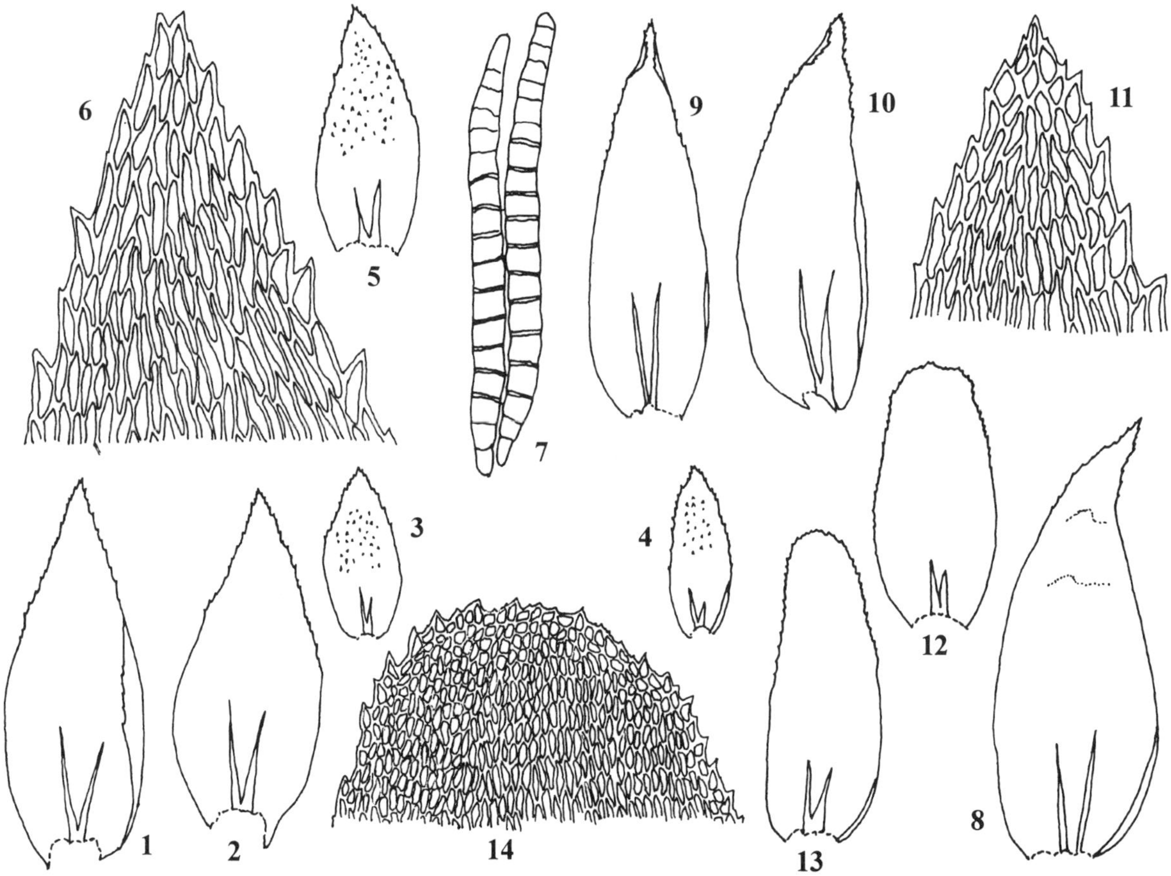

Plate 407. 1–7, *Symphyodon echinatus*. 1–2, stem leaves (× 40). 3–5, branch leaves (× 40). 6, apical leaf cells (× 400). 7, gemmae (× 200). 8–11, *Symphyodon perrottetii*. 8, stem leaf (× 40). 9–10, branch leaves (× 40). 11, apical leaf cells (× 400). 12–14, *Symphyodon pygmaeus*. 12, stem leaf (× 40). 13, branch leaf (× 40). 14, apical leaf cells (× 400). (1–7, drawn from *Redfearn et al. 793*, MO; 8–11, from *J.-M. Jin & S.-S. Hu 386*, PE; 12–14, from *Magill 7722*, MO.)

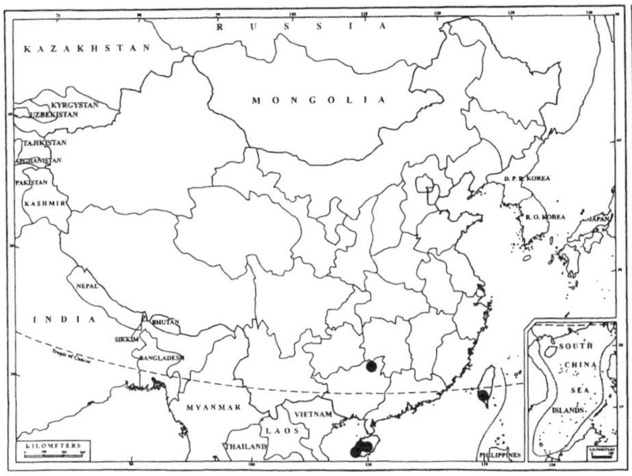

2. *Symphyodon perrottetii*

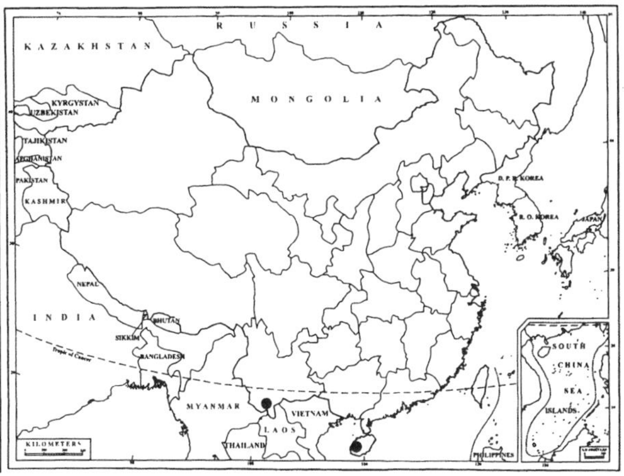

3. *Symphyodon pygmaeus*

2. ***Symphyodon perrottetii*** Mont. 刺果藓 ci-guo xian
Ann. Sci. Nat. Bot., ser. 2, 16: 279, f. 17. 1841. Type. India: Nilghiri, *Perrottet s.n.* (holotype RO; isotypes BM, FH, G, NY).

Plants medium-sized to rather large, slender, to 10 cm long, yellowish green to brownish green or golden reddish, somewhat glossy. *Stems* elongate, pinnately to bipinnately branched. *Leaves* appressed, slightly dimorphic; stem leaves oblong-ovate to oblong-lanceolate, ca. 2.0 mm × 0.8 mm; apices broadly acuminate; margins serrate to serrulate; branch leaves smaller, sometimes more or less undulate when dry, oblong-oval to oblong-lanceolate, 1.1–1.3 mm × 0.35–0.45 mm; apices narrowly acuminate, often incurved on both sides at the apex; margins somewhat incurved, regularly serrate above the middle, serration formed by a row of enlarged cells, subentire below; all leaves with costae unequal, extending to 1/3–1/2 total leaf length; apical leaf cells not clearly shorter than the adjacent cells; median cells linear, 40–60 µm × 3–5 µm, faintly prorate on stem leaves, often clearly prorate on branch-leaves; alar cells differentiated, rectangular, thick-walled, of 3–5 rows extending 3–5 cells up the margins. *Perichaetial leaves* sheathing at the base, gradually or abruptly narrowed from the middle to a long acumen, cells thick-walled, strongly porose, 2–3-costate, costae extending to 1/4 total perichaetial leaf length. *Setae* 1.2–1.5 cm long, papillose distally, smooth proximally; capsules erect, cylindrical, nearly symmetric, 2.8–3.0 mm × 1.0–1.1 mm, densely echinate, spines to 75 µm long; exothecial cells oblong-rectangular, 90–100 µm × 17–18 µm; annuli differentiated, of 3–4 rows of small cells; opercula long conic-rostrate, ca. 1.6 mm long, sometimes echinate; exostome teeth ca. 0.4 mm long, smooth or papillose, the basal cells of outer plates separated by a weakly furrowed line; endostome segments reduced, 0.10–0.15 mm long, adnate to the exostome teeth at the base; basal membrane obscure. *Spores* ca. 17 µm in diameter, distinctly papillose.

Illustrations: Plate 407, figs. 8–11.
Habitat: on branches or saplings.
Distribution: China, Japan, India, Sri Lanka, Laos, Thailand, Vietnam, Indonesia, Malaysia, Singapore, and the Philippines.

Chinese specimens examined: GUANGXI: Long-sheng Co., *J.-M. Jin 386* (MO). HAINAN: Chang-jiang Co., *Reese 17646* (LAF, MO); Mt. Jiefengling, *P.-C. Chen 473a* (MO, PE); Mt. Limuling, *B.-J. Lin 90* (IBSC, MO); Tong-shi Co., *B.-J. Lin 2170* (IBSC, MO). TAIWAN: Kao-hsiung Co., *C.-C. Chuang 1078* (NY); Ping-tung Co., *C.-C. Chuang 1460* (NY).

Symphyodon perrottetii is a primarily tropical Asian species, but its distribution range reaches northward to China (Hainan, Guangxi, Taiwan) and southern Japan. This species is distinguished by its often elongated plants (to 15 cm long), rather large capsules (2.8–3.3 mm × 1.0–1.2 mm), and strongly serrate, acuminate, often undulate branch leaves. It usually grows on tree branches.

3. ***Symphyodon pygmaeus*** (Broth.) He & Snider 钝叶刺果藓 dun-ye ci-guo xian
Bryobrothera 1: 285. 1992. *Homalia pygmaea* Broth., Nat. Pflanzenfam. I(3): 849. 1906. Type: Madagascar: Diego Suarez, *Chenagon s.n.* (lectotype PC, designated by He & Snider 1992; isolectotypes H, REN).

Plants medium-sized, to 5 cm long, yellowish green to golden green, glossy. *Stems* bipinnately to tripinnately branched, subcomplanate. *Leaves* similar in shape, stem leaves smaller than branch leaves, 1.0–1.1 mm × 0.4–0.5 mm, spreading, complanate, oblong-lanceolate to oblong-ovate to lingulate; branch leaves 1.2–1.5 mm × 0.4–0.5 mm; all leaves with apices rounded to truncate, margins coarsely serrate in the upper 2/3, serration formed by a row of enlarged cells (occasionally weakly serrulate at the apex), subentire to serrulate at the base; costae unequal, extending to 1/2 total leaf length; apical leaf cells distinctly shorter than the adjacent cells; median cells linear, 35–50 µm × 3–4 µm, weakly to moderately prorate; alar cells differentiated, subquadrate to shortly rectangular, of 3–5 rows extending 3–4 cells up the margins. *Sporophytes* not seen.

Illustrations: Plate 407, figs. 12–14.
Habitat: on rotten stumps, base of trees or tree trunks and buttresses.
Distribution: China, Nepal, India, Thailand, Madagascar, Réunion, Mozambique, and Hawaii (U.S.A).

Chinese specimens examined: HAINAN: Chang-jiang Co., *Reese 17541* (LAF, MO). YUNNAN: Jing-hong Co., *Magill et al. 7722* (MO), *Redfearn et al. 34281* (MO), *W.-X. Xu 11035*, *11365* (MO); Meng-hai Co., *Magill et al. 7868* (MO), *Redfearn et al. 34236* (MO); Nan-ru Shan (Mt.), *W.-X. Xu 6329* (MO).

Symphyodon pygmaeus is characterized by its truncate leaf apices and abortive branch tips. Some of the sterile plants of this species have previously been misplaced in *Glossadelphus* and *Homalia*.

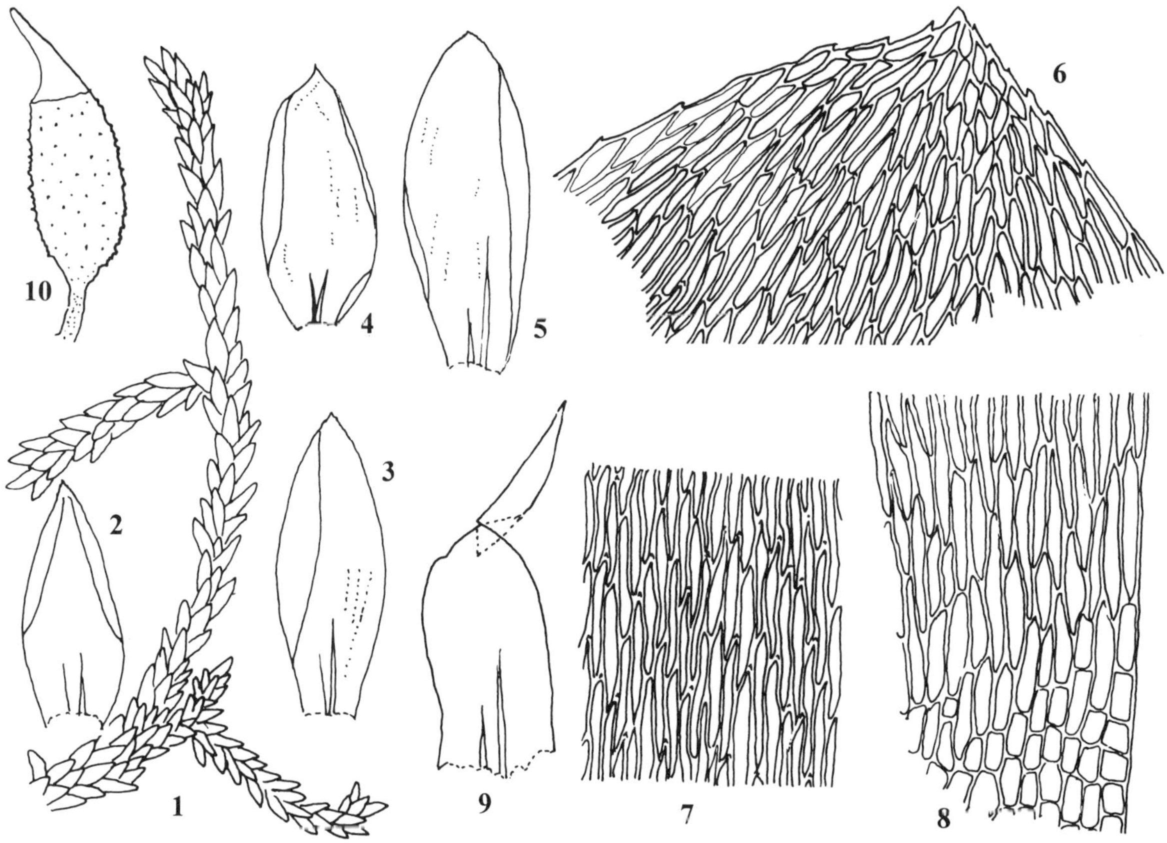

Plate 408. *Symphyodon weymouthioides*. 1, plant (× 38). 2–3, stem leaves (× 45). 4–5, branch leaves (× 45). 6, apical leaf cells (× 500). 7, median leaf cells (× 500). 8, basal leaf cells (× 500). 9, perichaetial leaf (× 45). 10, capsule (× 16). (1–10, drawn from *Cavalerie s.n.*, PC.)

4. *Symphyodon weymouthioides*

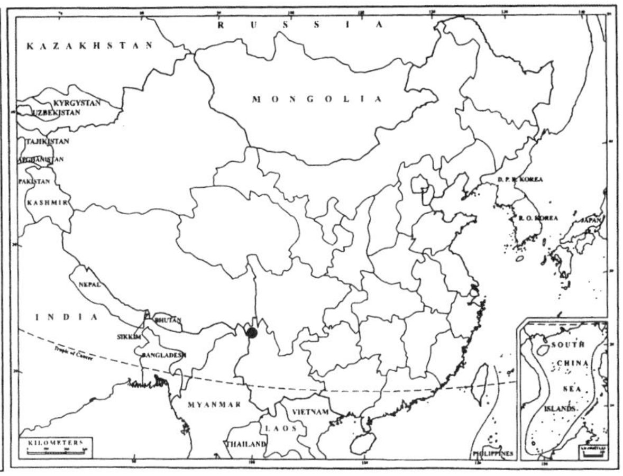

5. *Symphyodon yuennanensis*

4. ***Symphyodon weymouthioides*** Card. & Thér. 贵州刺果藓 gui-zhou ci-guo xian
Bull. Acad. Int. Géogr. Bot. 21: 271. 1911. Type. China: Guizhou, (Kouy-Tcheou), *Cavalerie s.n.* (holotype PC; isotype S).

Plants slender, medium-sized, 3–6 cm long, yellowish to golden brown, glossy. *Stems* irregularly branched. *Stem leaves* and branch leaves similar, moderately imbricate, clearly concave, ovate to broadly ovate; (0.9–)1.0–1.3 mm × (0.4–)0.6–0.8 mm; leaf apices apiculate; margins faintly serrulate to subentire above, entire below; costae unequal, extending to 1/4–1/3 total leaf length; apical leaf cells slightly shorter than adjacent cells; median cells linear, 50–60 µm × 3.5–4.5 um, faintly prorate; alar cells differentiated, quadrate, of 3–5 rows extending 3–4 cells up the margins. *Perichaetial leaves* sheathing at the base, abruptly acuminate at the apex; cells moderately thick-walled, porose, costae double, short or indistinct. *Setae* ca. 1.5 cm long, papillose distally, smooth proximally; capsules erect, oblong-ovoid, symmetric, ca. 2.2 mm × 1.1 mm, sparsely to moderately echinate; walls of exothecial cells moderately thickened, rectangular, 60–85 µm × 20–30 µm; spines to 50 µm long; annuli differentiated, of 3–4 rows of small, quadrate cells; opercula ca. 1.25 mm long, conic-rostrate; exostome teeth 0.4–0.5 mm long, faintly papillose at tips; endostome segments 0.20–0.25 mm long, narrow, smooth, perforate, keeled; basal membrane low, 35–45 µm high. *Spores* ca. 18 µm in diameter, papillose.

Illustrations: Plate 408, figs. 1–10.
Habitat: unknown.
Distribution: China and India.

Chinese specimens examined: see type information cited above.

Symphyodon weymouthioides is distinguished from its closely related species, *S. angustus*, *S. erinaceus*, and *S. gollanioides*, by having leaf apices subentire and apiculate.

5. ***Symphyodon yuennanensis*** Broth. 云南刺果藓 yun-nan ci-guo xian
Symb. Sin. 4: 90. 1929. Type. China: Yunnan, Djiou-djiang, 27°55′N, alt. 2400–2800 m, *Handel-Mazzetti 9473* (holotype H; isotypes S, W).

Plants slender, medium-sized, to 6 cm long, dull, yellowish green to brown. *Stems* pinnately to irregularly pinnately branched. *Leaves* appressed, not clearly dimorphic; stem leaves ovate-lanceolate, 1.6–2.0 mm × 0.65–0.85 mm; branch leaves smaller, oblong to oblong-lanceolate, 1.1–1.4 mm × 0.35–0.50 mm; all leaf apices bluntly acuminate; margins sometimes incurved on one side, serrate above, serration formed by a row of enlarged cells, subentire below; costae unequal, extending to 2/5–1/2 total leaf length; apical leaf cells shorter and wider than the adjacent cells; median cells linear, 40–60 µm × 2–3 µm, prorulose on stem leaves, distinctly prorate on branch leaves; alar cells differentiated, subquadrate to irregularly short rectangular, thick-walled, of 4–5 rows extending 4–5 cells up the margins. *Sporophytes* unknown.

Illustrations: Plate 409, figs. 1–11.
Habitat: unknown.
Distribution: endemic to China.

Specimens examined: see the type information cited above.

Symphyodon yuennanensis is known only from the type locality. Sterile plants of this species are dull, yellowish green to brown. The apices of stem and branch leaves are usually bluntly acuminate, which is a character distinguishing it from its closely related species, *S. perrottetii* and *S. copelandii* Broth. from the Philippines.

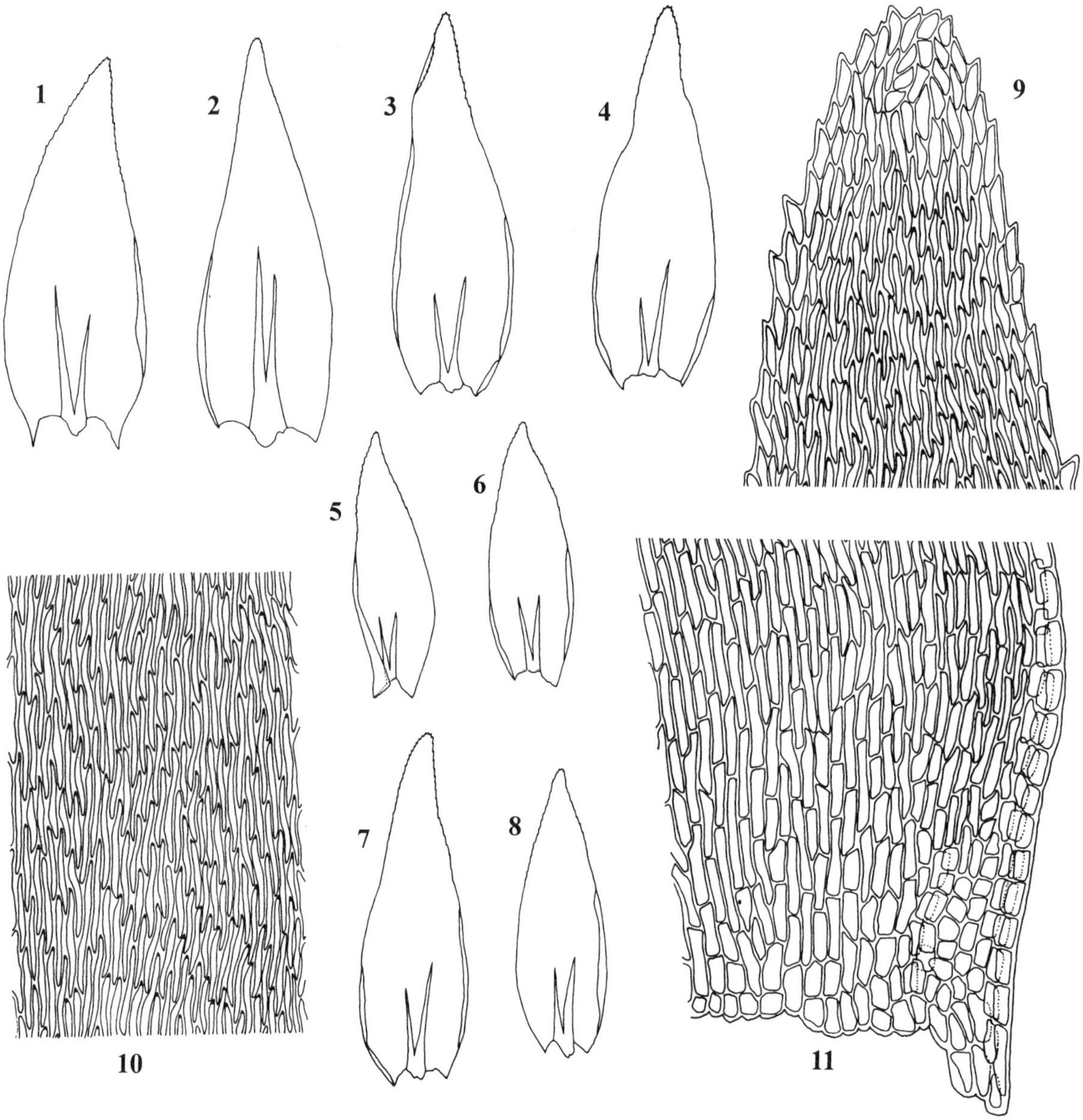

Plate 409. *Symphyodon yuennanensis*. 1–4, stem leaves (× 28). 5–8, branch leaves (× 28). 9, apical leaf cells (× 280). 10, median leaf cells (× 280). 11, basal leaf cells (× 280). (1–11, drawn from *Handel-Mazzetti 9473*, H.)

LEUCOMIACEAE

白藓科 bai xian ke

by Jia Yu and Li Zhi-hua

Plants slender, soft, whitish to yellowish green, somewhat shiny, in loose mats. *Stems* creeping, rarely pendulous, sometimes with clusters of rhizoids, irregularly or rarely pinnately branched, slightly complanate; central strand lacking; paraphyllia absent. *Leaves* contorted when dry, more or less flattened when wet, slightly asymmetric, becoming narrow at base, oblong-ovate, gradually lanceolate above, apices acute to acuminate; leaf margins entire; ecostate, occasionally with indistinct short double costae; laminal cells loose, thin-walled, elongate-hexagonal or lineariform, short at base; alar cells not differentiated. *Monoicous*, dioicous or autoicous. *Perichaetial leaves* broad at base, abruptly narrowed to a short acuminate apex, leaf margins entire. *Setae* slender, slightly coarse above, smooth below, reddish brown; capsules ovoid or oblong-ovoid, horizontal or nearly cernuous, stomata superficial; annuli large, usually persistent; peristome double; exostome teeth narrowly lanceolate, thin, hyaline, longitudinal furrows in the center, with dense ridges and stripes outside, densely striate; endostome segments densely papillose, folded; cilia absent or poorly developed, slightly higher than the mouth; opercula conic, with short or slender beak. *Calyptrae* cucullate, smooth, covered with spare hairs, entire at base or lobed. *Spores* minute, nearly smooth.

1. ***Leucomium*** Mitt. 白藓属 bai xian shu
   J. Linn. Soc. Bot. 10: 181. 1868.

The description of the genus is the same as for *Leucomium strumosum* (Hornsch.) Mitt.

Leucomium is a genus with a pantropical distribution. There are 26 names that had been placed in the genus prior to the revisionary study undertaken by Allen (1987). Two species are now recognized: *L. strumosum* and *L. steerei* Allen & Veling. Only *L. strumosum* is distributed in China.

1. ***Leucomium strumosum*** (Hornsch.) Mitt. 白藓 Bai xian
   J. Linn. Soc. Bot. 12: 502. 1869. *Hypnum strumosum* (Hornsch.) C. Müll., Syn. 2: 238. 1851. *Hookeria strumosa* Hornsch., Fl. Bras. 1(2): 69. 1840. Type. Brazil: Serra dos Orgaô, prope Tijucam, *Olfers s.n.*
   Leucomium aneurodictyon (C. Müll.) Jaeg., Ber. Thätigk. St. Gallischen Naturwiss. Ges. 1877–78: 275. 1880. *Hypnum aneurodictyon* C. Müll., Syn. Musc. Frond. 2: 681. 1851.

Plants slender, soft, whitish to yellowish green, in loose mats. *Stems* creeping, irregularly or sparsely branched. *Leaves* contorted when dry, more or less flattened when wet, leaf shape extremely variable, but usually 1.5–2.0 mm long, slightly asymmetric, becoming narrow at base, oblong-ovate, gradually lanceolate above, apices acute to acuminate; leaf margins entire; costa none; laminal cells loose, thin-walled, smooth, elongate-hexagonal or lineariform, short at base; alar cells not differentiated. *Autoicous* or synoicous. *Setae* reddish brown, 8–18 mm long, incurved, somewhat papillose at top; capsules ca. 0.8 mm long, ovoid or oblong-ovoid, horizontal to pendulous; stomata superficial; opercula long-rostrate, the same length as hypophysis; exostome teeth narrowly lanceolate, thin, hyaline, longitudinal furrows in the center, with dense ridges and stripes outside, densely striate; endostome segments densely papillose, folded; basal membrane high; cilia absent or reduced. *Calyptrae* long-cucullate, smooth, sparely pilose. *Spores* 8–10 μm in diameter.

Illustrations: Plate 410, figs. 1–9.
Habitat: on the forest floors, decaying logs, rocks, and moist soil.
Distribution: China, India, Sri Lanka, Thailand, Malaysia, Indonesia, the Philippines, Papua New Guinea, New Caledonia, Fiji, Central America, Caribbean, South America, and Africa.

Chinese specimens examined: HAINAN: Mt. Jianfengling, *P.-C. Chen et al. 590* (PE). XIZANG: Mo-tuo Co., *Y.-G. Su 2837* (HKAS, MO), *2960* (HKAS). YUNNAN: Nan-jian Co., *J.-K. Zhang 1616a* (YUN); Xishuangbanna, *L. Zhang 320*, *565*, *572*, *810*, *850*, *856* (HKAS); Yun-xian Co., *J.-K. Zhang 1563b*, *1579b*, *1587b* (YUN).

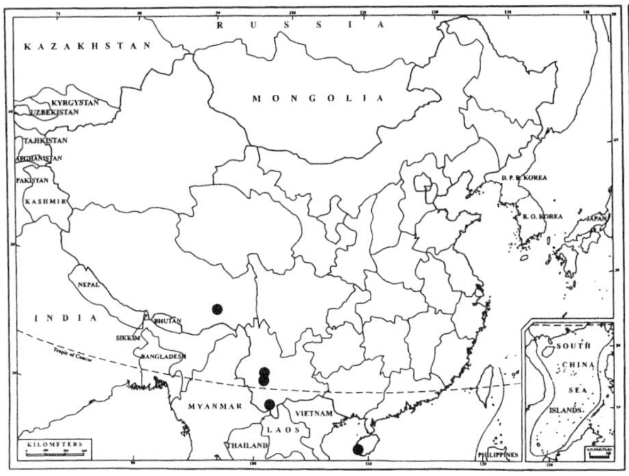

1. *Leucomium strumosum*

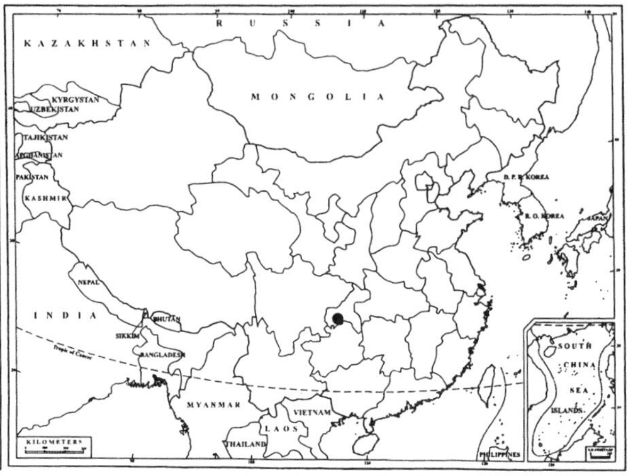

1. *Cyathophorella burkillii*

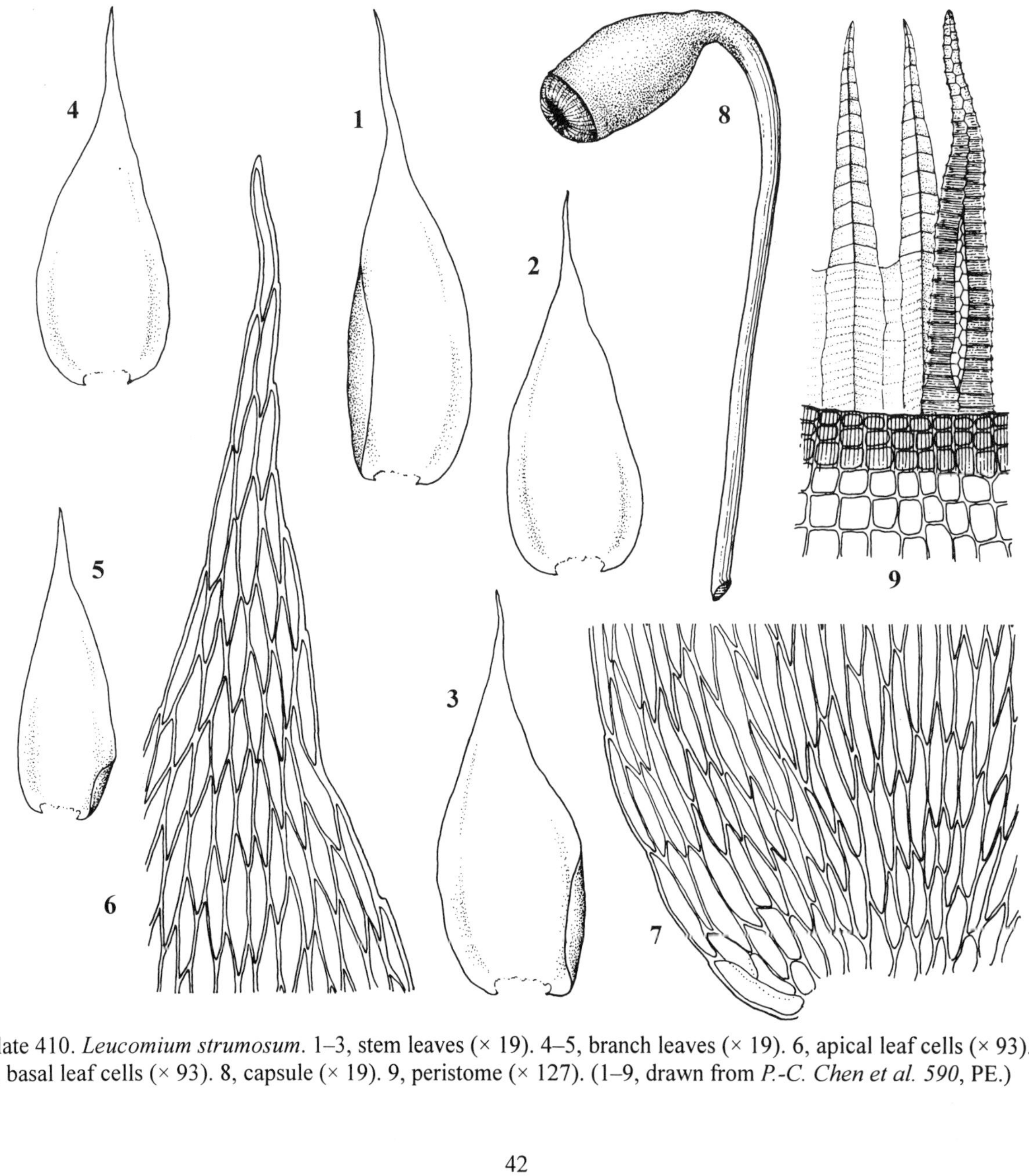

Plate 410. *Leucomium strumosum*. 1–3, stem leaves (× 19). 4–5, branch leaves (× 19). 6, apical leaf cells (× 93). 7, basal leaf cells (× 93). 8, capsule (× 19). 9, peristome (× 127). (1–9, drawn from *P.-C. Chen et al. 590*, PE.)

HYPOPTERYGIACEAE

孔雀藓科 kong-que xian ke

by Jia Yu, Li Zhi-hua, and Wu Peng-cheng

Plants small to medium-sized, soft, complanate, erect-spreading or appressed, lightly yellowish green to dark green, usually not glossy. *Primary stems* slender, complanately explanate, rhizoids brown; secondary stems simple or rarely branched, dendroid, pinnately branched above, tail-like or peacock-like at the apex; cross section of stems oval or rounded, epidermal cells small, incrassate, colored, cortical cells larger, thin-walled; central strand mostly present, thick-walled. *Leaves* in three rows, including two lateral and one ventral rows, dimorphous; lateral leaves plane, ovate to elongate-ovate, rarely ovate-lanceolate, asymmetric; margins usually bordered; amphigastria smaller and rounded; costa uniform cells, single, sometimes forked above; laminal cells isodiametric, smooth; alar cells not differentiated. *Dioicous* or autoicous; male plants similar to female plants, perichaetia only in secondary stems or branches, paraphyses usually absent; perigonial buds smaller. *Setae* elongate; capsules exserted, mostly suberect or inclined, rarely erect; stomata phenotype, rare at base of capsules; peristomes double; exostome teeth occasionally reduced, usually densely striate, keeled, furrowed, with a median zig-zag line, lamellae well developed inside; endostome segments keeled, basal membrane developed; opercula rostrate. *Calyptrae* cucullate or conic, smooth. *Spores* very small, finely papillose.

Hypopterygiaceae, consisting of six genera, can be classified into two subfamilies: Hypopterygioideae and Cyathophoroideae. Species of the family are mainly distributed in the tropical or subtropical regions of the world, and are usually growing on tree trunks or moist soil slopes. Four genera are recognized in China.

The Hypopterygiceae are characterized by their leaf arrangement in three rows, with two lateral rows of large, asymmetrical leaves and one row of rather small, symmetrical amphigastria. There are different opinions concerning their systematic position. Crosby (1974) considered the Hypopterygiaceae to be not a natural group, merging the genera of this family into the Daltoniaceae or Hookeriaceae. Buck and Vitt (1986) recognized Hypopterygiceae and placed them in the Bryales. Buck (1987) redefined this family to be monotypic, with *Hypopterygium* only, and transferred *Cyathophorum*, *Cyathophorella*, and *Dendrocyathophorum* to the Hookeriaceae. Our opinions as well as that of Akiyama (1990a, 1990b, 1992) maintained the traditional classification of the family with two subfamilies: Hypopterygioideae (*Catharomnion*, *Dendrocyathophorum*, *Hypopterygium*, and *Lopidium*) and Cyathophoroideae (*Cyathophorella* and *Cyathophorum*). Kruijer (2002) appeared too late to be taken into consideration for our treatment. All the genera except *Catharomnion* and *Cyathophorum* occur in China.

1. Plants simple or loosely irregularly branched, not flatly dendroid or peacock tail-like; setae short; capsules usually erect, ovoid or spherical .... 1. *Cyathophorella*
1. Plants 1–2 pinnately branched, flatly dendroid, peacock tail-like; setae slender; capsules horizontal or pendulous, oblong-ovoid or obovoid, rarely spherical 2
    2. Lateral leaves long-ligulate; costa percurrent or slightly excurrent; leaf cells mostly papillose, cell walls collenchymatous .... 4 *Lopidium*
    2. Lateral leaves usually ovate; costa not percurrent; leaf cells smooth, cell walls not collenchymatous 3
3. Secondary stems pinnately branched; leaf margins not differentiated .... 2. *Dendrocyathophorum*
3. Secondary stems fan-like; leaf margins differentiated .... 3. *Hypopterygium*

1. ***Cyathophorella*** (Broth.) Fleisch. 雉尾藓属 zhi-wei xian shu

Musci Buitenzorg 3: 1088. 1908. *Cyathophorum* sect. *Cyathophorella* Broth., Nat. Pflanzenfam. I(3): 965. 1907.

Plants slender, green or dark green, not glossy. *Primary stems* short, creeping, usually with rhizoids; secondary stems erect, spreading, simple or rarely branched; branch apex tail-like; gemmae present, consisting of uniseriate cells in leaf axils. *Lateral leaves* ovate or ovate-lanceolate, sharply serrate above; costa short, simple, sometimes forked or absent; laminal cells oval or elongate-hexagonal, smooth, rather thick-walled, often pitted; leaf margins bordered by narrow elongate cells; amphigastria appressed, broadly ovate or ovate-round, entire or serrate at the apex; costa short or absent. *Dioicous*. *Inner perichaetial leaves* sheathing at base. *Setae* elongate; capsules ovoid; exostome teeth lanceolate, with indistinct zigzag furrows, keeled, hyaline, densely finely papillose outside, shortly striate inside; endostome segments narrowly lanceolate; basal membrane high; cilia absent. *Calyptrae* long-rostrate, lobed along the margin, slightly coarse, sometimes with hairs.

There are 16 species of *Cyathophorella* in the world, and 11 of them are well understood (Crosby et al. 2000). Eight species are treated for China.

1. Leaf margins entire or weakly serrate 2
1. Leaf margins grossly spinose-dentate 6
  2. Leaf margins entire 3
  2. Leaf margins serrate 4
3. Costa of amphigastria only 1/2 the leaf length .... 2. *C. hookeriana*
3. Costa of amphigastria almost percurrent .... 7. *C. taiwaniana*
  4. Lateral leaves distinctly bordered .... 1. *C. burkillii*
  4. Lateral leaves indistinctly bordered or only bordered at base 5
5. Plants small, less than 7.5 mm high, yellowish green; costa short .... 3. *C. intermedia*
5. Plants large, 20 mm high, dark color; ecostate or costa indistinct .... 4. *C. kyusyuensis*
  6. Leaf margins bordered .... 8. *C. tonkinensis*
  6. Leaf margins indistinctly bordered 7
7. Amphigastria ecostate .... 6. *C. subspinosa*
7. Amphigastria shortly costate .... 5. *C. spinosa*

1. ***Cyathophorella burkillii*** (Dix.) Broth. 黄雉尾藓 huang zhi-wei xian

Nat. Pflanzenfam. (ed. 2), 11: 278. 1925. *Cyathophorum burkillii* Dix., Rec. Bot. Surv. India 6: 67. pl. 1, f. 1; pl. 2, f. 9. 1914. Type. India: Abor District, *Burkill 37737*.

Primary stems short, tomentose; secondary stems yellowish, erect, usually simple, branches ca. 5 cm high, 8 mm wide, often caudate at apices, sometimes clusters of filamentous gemmae present at branch apices, ca. 0.5 mm

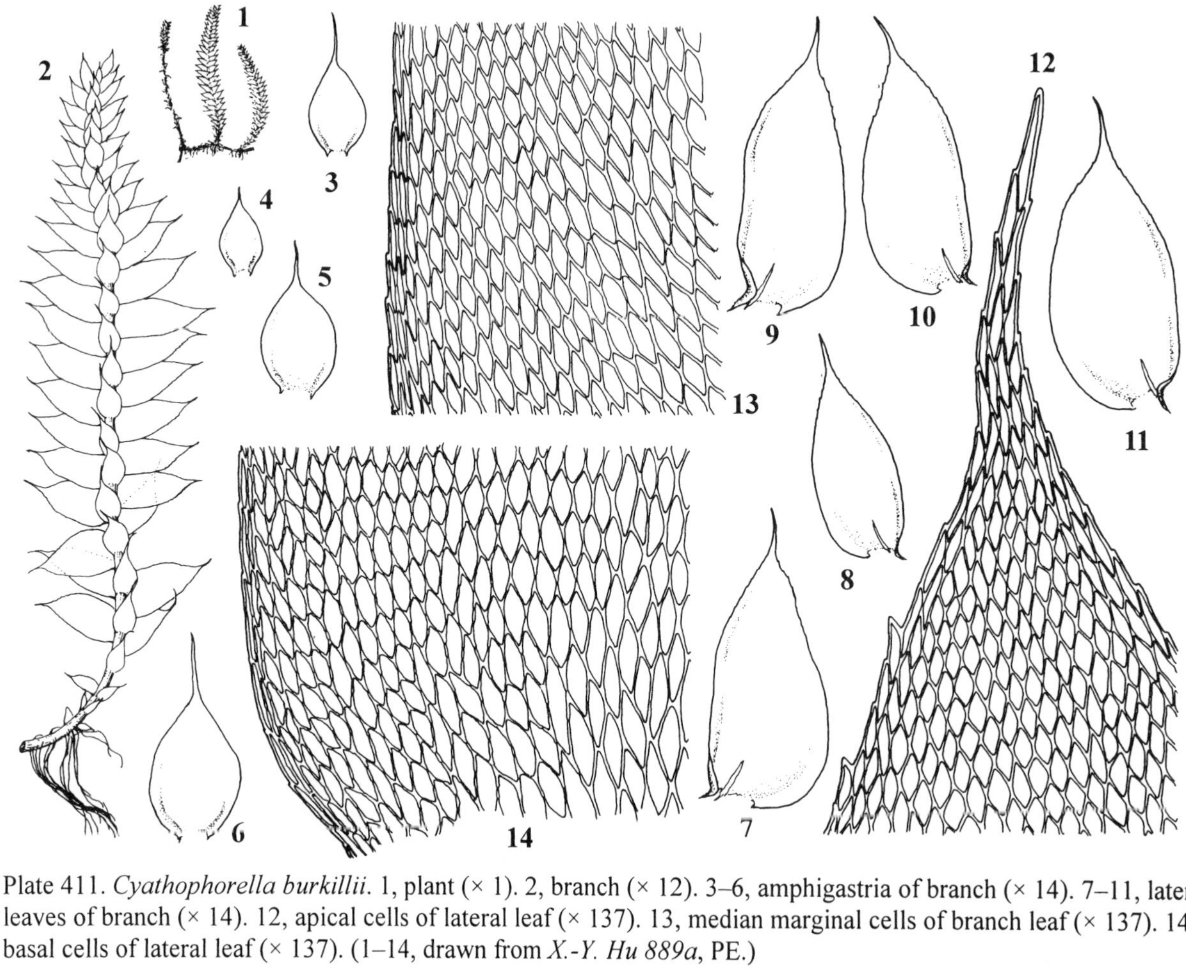

Plate 411. *Cyathophorella burkillii*. 1, plant (× 1). 2, branch (× 12). 3–6, amphigastria of branch (× 14). 7–11, lateral leaves of branch (× 14). 12, apical cells of lateral leaf (× 137). 13, median marginal cells of branch leaf (× 137). 14, basal cells of lateral leaf (× 137). (1–14, drawn from *X.-Y. Hu 889a*, PE.)

long. *Lateral leaves* flattened when wet, ovate, long acuminate, asymmetric, ca. 4.8 mm × 1.9 mm, densely arranged; leaf margins distinctly bordered, denticulate at the apex; costa single, short, bifid above; amphigastria in one row, rounded, with a long acuminate apex, similar to lateral leaves in shape, ca. 2.8 mm × 1.8 mm; costa single, very short; laminal cells elongate-rhomboidal, with porose walls, polygonal toward leaf base; leaf margins bordered by 2 rows of elongate cells. *Perichaetial leaves* gradually narrowed to an acuminate apex. *Setae* 4–5 mm long, curved; capsules ca. 2.8 mm long, 0.58 mm in diameter; opercula long rostrate; peristome well developed. *Spores* 15–18 μm in diameter.

Illustrations: Plate 411, figs. 1–14.
Habitat: on tree trunks.
Distribution: China and India.
Chinese specimen examined: CHONGQING: Nan-chuan Co., *X.-Y. Hu 0873* (PE).

This species is very close to *Cyathophorella hookeriana*, but the leaf apices of *C. burkillii* are denticulate. Gangulee (1977) suggested that *C. burkillii* be treated as a variety of *C. hookeriana* (Griff.) Fleisch. X.-Y. Hu (1991) first reported *C. burkillii* from Mt. Jinfu, Chongqing City.

2. ***Cyathophorella hookeriana*** (Griff.) Fleisch. 短肋雉尾藓 duan-le zhi-wei xian
Musci Buitenzorg 3: 1094. 1908. *Neckera hookeriana* Griff., Notul. Pl. As. 2: 48. f. 2a. 1849. Type. Indonesia: Java.
Cyathophorella densifolia Horik., Bot. Mag. (Tokyo) 48: 460, f. 3. 1935.
Cyathophorella rigidula P.-C. Chen, Gen. Musc. Sin. 2: 135. 1978, *nom. inval.*

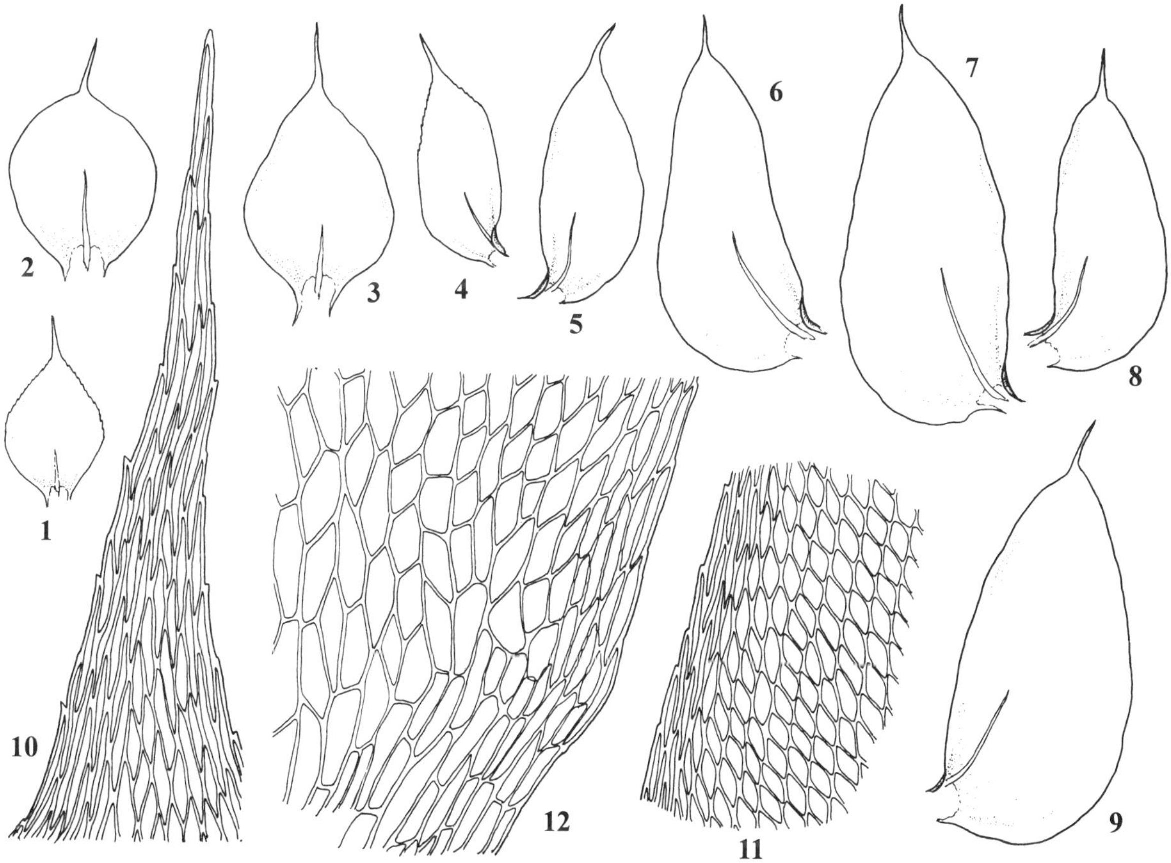

Plate 412. *Cyathophorella hookeriana*. 1, amphigastrium of upper branch (× 11). 2, amphigastrium of middle branch (× 11). 3, amphigastrium of lower branch (× 11). 4–5, latera; leaves of upper branch (× 11). 6–7, lateral leaves of middle branch (× 11). 8–9, lateral leaves of lower branch (× 11). 10, apical cells of lateral leaf (× 137). 11, median cells of lateral leaf (× 137). 12, basal cells of lateral leaf (× 137). (1–12, drawn from *Y.-G. Su 2833*, HKAS.)

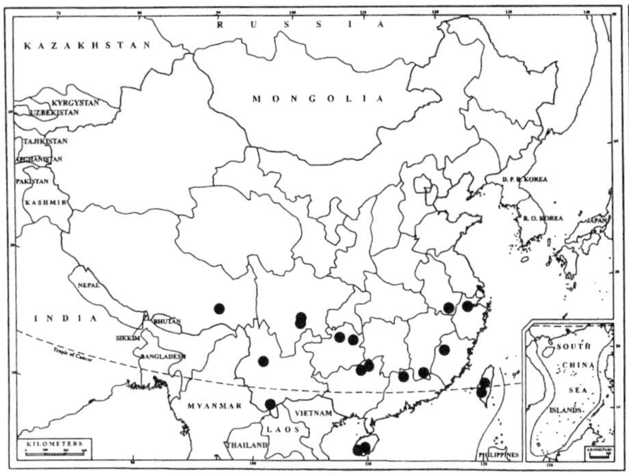

2. *Cyathophorella hookeriana*

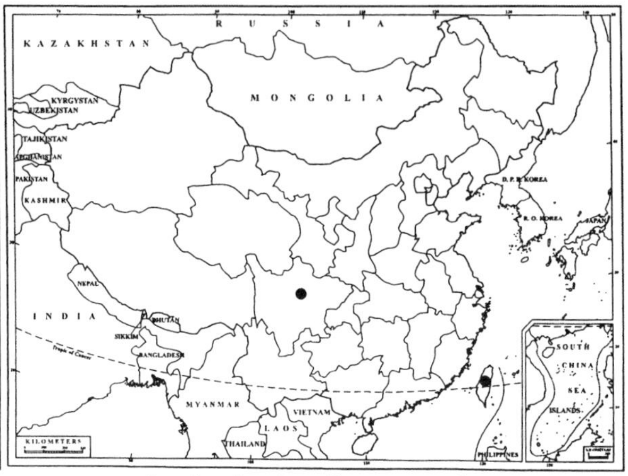

3. *Cyathophorella intermedia*

Plants small, yellowish green, ca. 2.5 cm high, 5 mm wide with leaves, simple, rarely branched. *Leaves* dense, caudate at tips, reddish brown, gemmae forked, often abundant in leaf axils of the caudate tips of some branches; rhizoids tomentose at base; lateral leaves bifarious, ovate-lanceolate, asymmetric, ca. 4.5 mm × 2 mm; leaf margins entire, bordered throughout by 3 rows of narrowly elongate cells; costa single, short and weak; laminal cells rhomboidal, mostly pitted; amphigastria smaller, ovate, symmetric, plicate along margins. *Setae* ca. 1.4 cm long; capsules cylindrical, ca. 2.9 mm long, 1 mm in diameter.

Illustrations: Plate 412, figs. 1–12.
Habitat: on tree trunks, rotten logs, rocks, or slopes.
Distribution: China, Japan, Laos, and the Philippines.

Chinese specimens examined: ANHUI: Huang-shan (Mt.), *P.-C. Chen et al. 7188* (MO, PE, SYS), *B.-J. Lin 38* (IBSC, SYS). FUJIAN: Jiang-le Co., *M.-Z. Wang 47885* (PE, SYS), *X.-L. He 1156* (PE, SYS). GUANGDONG: Ru-yuan Co., *B.-J. Lin 898* (IBSC, SYS), *L. Zhang 147* (IBSC, SYS), *G.-Q. Zeng 153* (IBSC, SYS). GUANGXI: Long-sheng Co., *J.-M. Jin & S.-S. Hu 761* (PE, SYS); Rong-shui Co., *X.-L. He H00366* (PE, SYS). GUIZHOU: Mt. Fanjing, *Crosby 15719, 15897* (MO); Sui-yang Co., *Crosby 15987* (MO). HAINAN: Mt. Wu-zhi, *Z.-H. Li 77-2191* (SYS), *Y.-X. Lin 297* (IBSC, SYS); Mt. Jianfengling, *P.-C. Chen 338b* (IBSC, SYS), *P.-Z. Zheng 183* (IBSC, PE). JIANGXI: Xun-wu Co., *M.-X. Gong 560010* (PE, SYS). SICHUAN: Ma-bian Co., *P.-C. Chen 5800* (PE, SYS); Mt. Omei, *P.-C. Chen et al. 5460* (PE, SYS), *B.-J. Lin 3269* (IBSC, SYS), *Redfearn 34789* (MO). TAIWAN: Chia-yi Co., *S.-H. Lin s.n.* [Bryophytes of Taiwan Exsiccatae ser. 5, no. 210] (MO); Nan-tou Co., *M.-J. Lai 1548* (IBSC, MO, SYS, TAI); Ping-dong Co., *M.-J. Lai 9890* (IBSC, MO, SYS, TAI); Maoanshan (Mt.), *Iwatsuki & Sharp 902B* (IBSC, NICH, SYS). XIZANG: Mo-tuo Co., *M.-Z. Wang 880139* (PE, SYS), *M.-Z. Wang 800300* (PE), *K.-Y. Lang 570A* (PE, SYS). YUNNAN: Xishuangbanna, *W.-X. Xu 6788* (PE, SYS); Meng-lun Co., *B.-J. Lin 3078* (IBSC, SYS); Yang-bi Co., *Redfearn & Y.-G. Su 801a* (MO). ZHEJIANG: Mt. Tian-mu, *P.-C. Wu 7818* (PE, SYS).

This species is characterized by its entire leaf margins and the single costa of amphigastria, which ends below the middle of the leaf.

3. ***Cyathophorella intermedia*** (Mitt.) Broth. 小雉尾藓 (新拟) xiao zhi-wei xian

Nat. Pflanzenfam. (ed. 2), 11: 277. 1925. *Cyathophorum intermedium* Mitt., J. Proc. Linn. Soc., Bot., Suppl. 1: 148. 1859. Type. India (West Himalayas): Kumaon, *Strachey & Winterbottom s.n.*
Dendrocyathophorum intermedium Herz., J. Hattori Bot. Lab. 14: 65. 1955, *nom. inval.*

Plants small, yellowish-green. *Primary stems* prostrate, rhizoids short, tomentose; secondary stems erect, usually simple; branches ca. 7.5 mm long, 0.8 mm wide, terminal of branches tail-like; clusters of filamentous gemmae sometimes present at tops of branches, gemmae about 0.5 mm long. *Lateral leaves* ovate-lanceolate, asymmetric, ca. 2.6 mm × 1 mm, apex long acuminate; leaf margins slightly serrate above; amphigastria ovate-lanceolate, symmetric, ca. 1.4 mm × 0.9 mm; costa single, short, weak, often forked at the apex; laminal cells

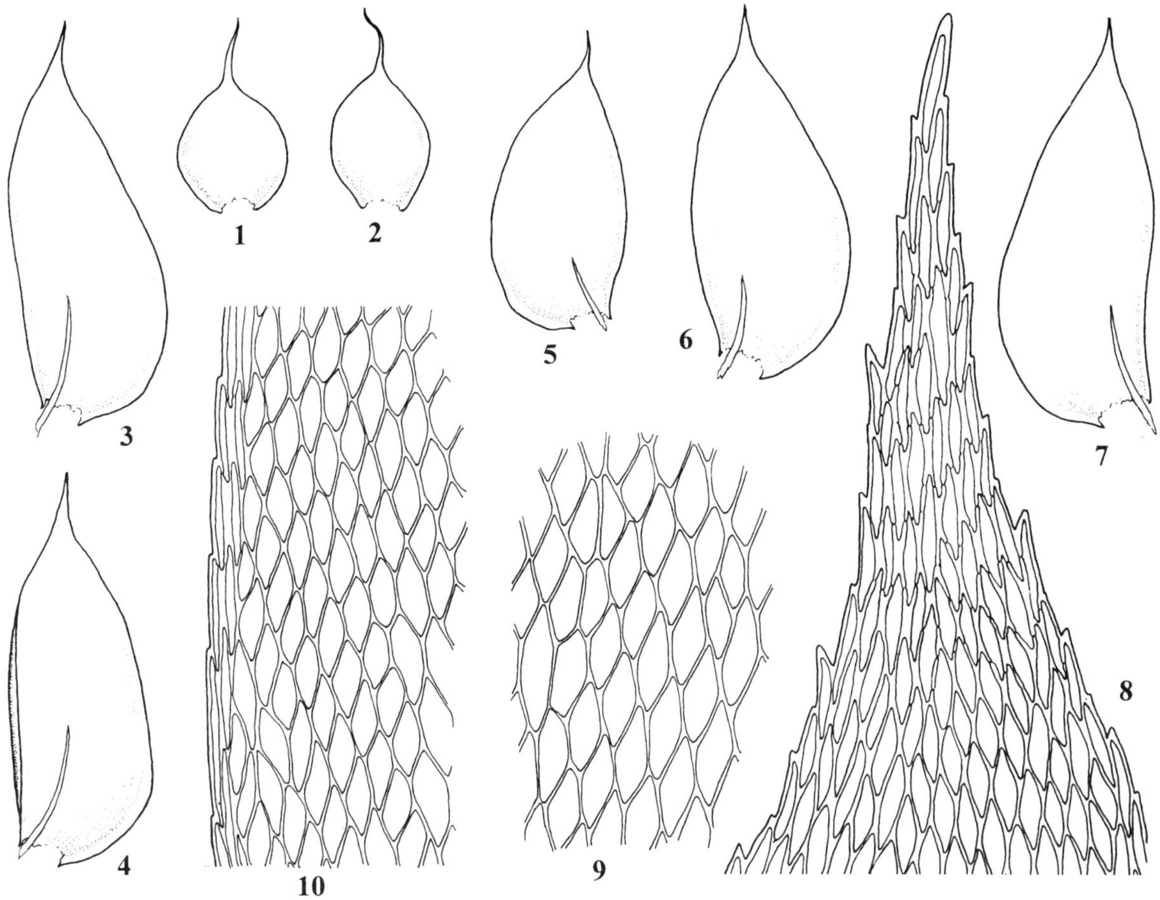

Plate 413. *Cyathophorella intermedia*. 1–2, amphigastria of branch (× 12). 3–7, lateral leaves (× 12). 8, apical cells of lateral leaf (× 127). 9, median marginal cells of lateral leaf (× 127). 10, basal cells of lateral leaf (× 127). (1–10, drawn from *Iwatsuki & Sharp 3211c*, TEN.)

irregularly rhomboidal, becoming shortly rhomboidal toward leaf base; leaf margins bordered by 2 rows of elongate cells near base.

Illustrations: Plate 413, figs. 1–10.
Habitat: on soil.
Distribution: China and India.

Chinese specimens examined: SICHUAN: Wo-long Nature Reserve, *K.-Y. Lang 13167c* (PE), *Z.-S. Qin 073* (PE). TAIWAN: Daxueshanmian, Anmashan, *Iwatsuki & Sharp 3211c* (IBSC, NICH, SYS).

This species is characterized by the very small plants that are only about 0.75 cm high.

4. ***Cyathophorella kyusyuensis*** Horik. & Nog. in Nog. 九州雉尾藓 jiu-zhou zhi-wei xian
J. Sci. Hiroshima Univ., Ser. B, Div. 2, Bot. 2 (3): 25. 2. 1936. Type. Japan: Kyushu, Mt. Aoidake, *Noguchi 236*.

Plants slender, not glossy, in loose mats. *Secondary stems* simple, ca. 2 cm long, 5.5 mm wide; branch apices gradually slender, curved when dry; lateral leaves twisted when dry, plane when wet, ovate, with a long acuminate apex, asymmetric, ca. 3.5 mm × 1.5 mm; amphigastria oblong or ovate, with a long acuminate apex, ca. 2.5 mm × 1.5 mm; leaf margins irregularly serrulate, each tooth consisting of 1–2 cells; costa slender. *Gemmae* numerous, often growing in the upper part of branches, green, branched, ca. 0.9 mm long.

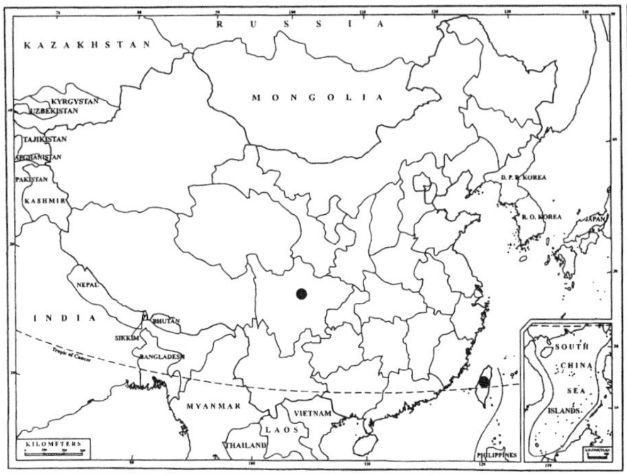

4. *Cyathophorella kyusyuensis*

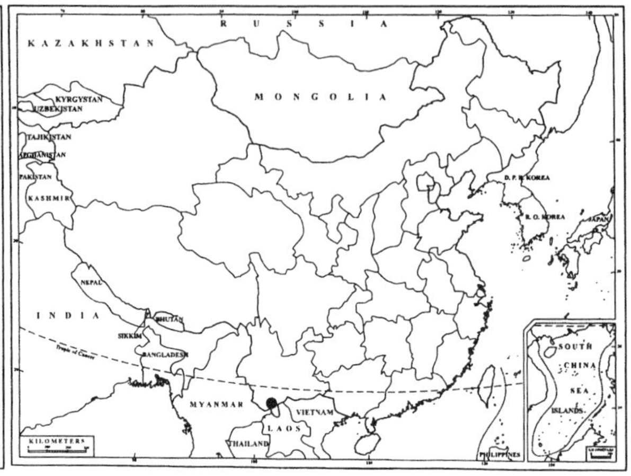

5. *Cyathophorella spinosa*

Illustrations: Plate 414, figs. 1–12.
Habitat: on tree trunks and shaded rocks.
Distribution: China and Japan.

Chinese specimens examined: FUJIAN: Jiang-le Co., *M.-Z. Wang 48355*, *48414* (PE); Mt. Wuyi, *P.-C. Chen et al. 523* (PE, SYS). TAIWAN: Mt. Tayuanshan, *C.-K. Wang 342* (IBSC). ZHEJIANG: Long-quan Co., *P.-C. Wu 359* (PE, SYS).

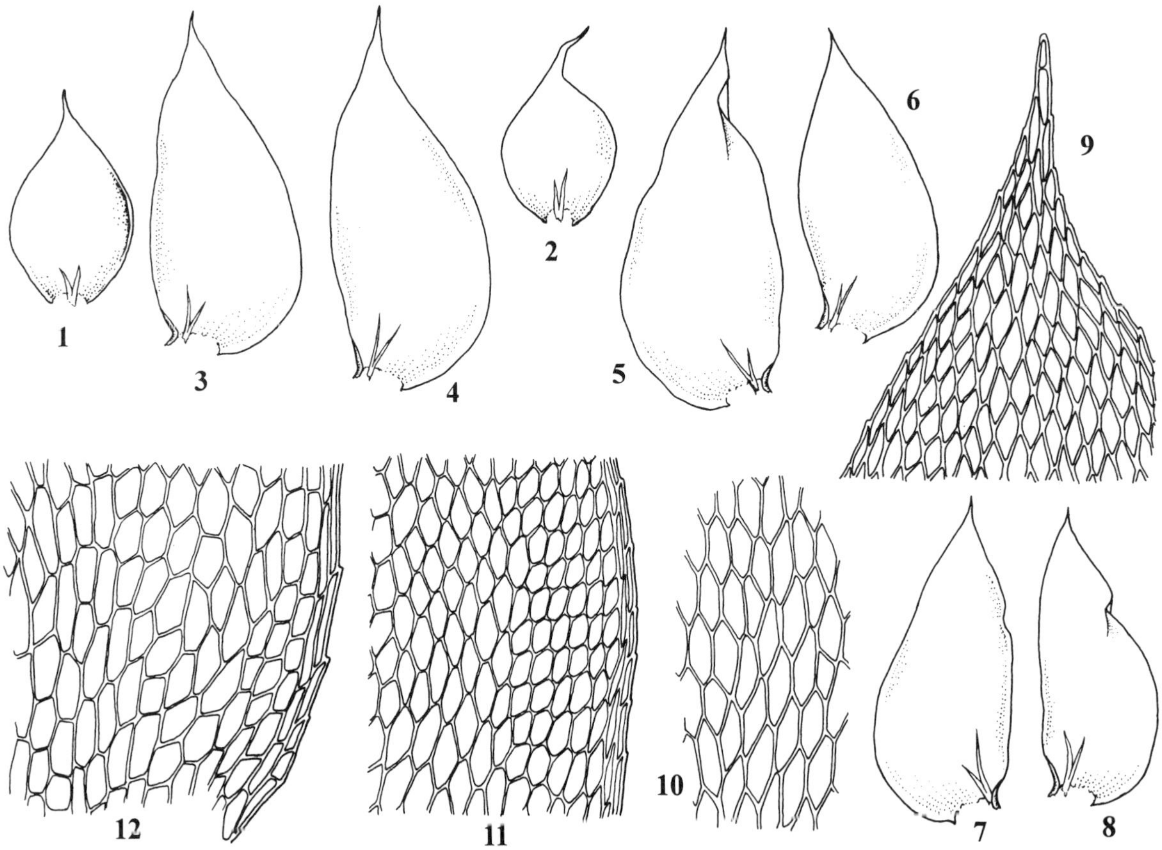

Plate 414. *Cyathophorella kyusyuensis*. 1–2, amphigastria of branch (× 9). 3–8, lateral leaves of branch (× 9). 9, apical cells of lateral leaf (× 93). 10, median cells of lateral leaf (× 93). 11, median marginal cells of lateral leaf (× 93). 12, basal marginal cells of lateral leaf (× 93). (1–12, drawn from *M.-Z. Wang 48414*, PE.)

Cyathophorella kyusyuensis var. *grandistipulacea* (Dix. & Sak.) Nog. reported from Taiwan (M.-J. Lai 1976) cannot be confirmed.

5. ***Cyathophorella spinosa*** (C. Müll.) Fleisch. 雉尾藓 zhi-wei xian
Musci Buitenzorg 3: 1091. 1908. *Cyathophorum spinosum* (C. Müll.) Fleisch., Musci Frond. Archip. Ind. Exsic. 5: 249. 1902. *Hookeria spinosa* C. Müll., Syn. Musc. Frond. 2: 677. 1851. Type. Indonesia: Java.

Plants robust. *Secondary stems* simple, up to 6 cm long, 12 mm wide with leaves, caudate at apices, usually covered with clusters of yellowish red filamentous gemmae. *Lateral leaves* plane when moist, broadly ovate, asymmetric, shortly acuminate at apex, ca. 7 mm × 3 mm; leaf margins spinosely serrate at the upper part of leaf, each tooth usually consisting of several cells; amphigastria costa short, single or forked; laminal cells polygonal, thin-walled, leaf marginal cells narrow, forming inconspicuously bordered margin. *Perichaetial leaves* wider, distinctly spinosely serrate above; costa short, single, unequally forked above.

Illustrations: Bartram 1939 (Pl. 21, fig. 357).
Habitat: on soil.
Distribution: China, Indonesia, and the Philippines.

Chinese specimen examined: YUNNAN: Xishuangbanna ? *C.-W. Wang 1257* (PE).

6. ***Cyathophorella subspinosa*** P.-C. Chen. 刺叶雉尾藓 ce-ye zhi-wei xian
Repert. Spec. Nov. Regni Veg. 58: 31. 1955. Type. China: Guangxi (Kwangsi), Li-kiang (Lung-tschou), Da-ching-schan, ca. 800 m, *P.-C. Chen 25* (holotype PE).

Plants robust, rigid, dark green. *Primary stems* prostrate; secondary stems usually simple, only branched at tips, caudate; filamentous gemmae often growing on leaves. *Lateral leaves* crisped when dry, spreading when wet, lanceolate, becoming narrower near base, asymmetric, ca. 5 mm × 1.5 mm, leaf margins irregularly spinosely teethed, each tooth consisting of 1–3 cells; costa weak, single or forked; laminal cells oval-polygonal, becoming narrower toward leaf margins; amphigastria small, symmetric, ecostate.

Illustrations: Plate 415, figs. 1–9.
Habitat: on rocks.
Distribution: endemic to China.

Chinese specimens examined: FUJIAN: De-hua Co., *P.-Z. Zheng 646* (IBSC, SYS). GUANGDONG: Dian-bai Co., *Z.-F. Wei et al. 02* (IBSC, SYS); Huai-ji Co., *S.-Q. Chen 2* (PE); Lian-ping Co., *B.-J. Lin 1405* (IBSC, SYS); Yang-chuan Co., *H.-G. Ye 001* (IBSC, SYS). HAINAN: Mt. Jianfengling, *B.-J. Lin & L. Zhang 696*, *725* (IBSC). YUNNAN: He-kuo Co., *W.-M. Zhu 57* (PE).

This species is similar to *Cyathophorella spinosa* in having indistinct borders at leaf margins, but differs in having ecostate amphigastria.

7. ***Cyathophorella taiwaniana*** M.-J. Lai 台湾雉尾藓 (新拟) tai-wan zhi-wei xian
Taiwania 21(2): 152. 1976. Type. China: Taiwan, Mt. Ali, *M.-J. Lai 333* (holotype TAI).

Plants slightly yellowish green, loosely tufted. *Primary stems* creeping, brown rhizoids tomentose; secondary stems simple, up to 4 cm high, rarely pinnately branched. *Lateral leaves* ovate, cuspidate at the apex, symmetric, ca. 3.5 mm × 2.0 mm; leaf margins nearly entire, distinctly bordered by 2–4 rows of narrow cells; costa reaching half the length of leaf; laminal cells rhomboidal-hexagonal; amphigastria larger, ovate, symmetric, ca.1.0 mm × 1.5 mm; leaf margins nearly entire, distinctly bordered; costa single, reaching the middle of leaves or percurrent. *Setae* light yellow, 0.8–1.0 cm long; capsules erect, oblong-cylindrical, usually 2–3 mm long.

Illustrations: M.-J. Lai 1976 (Pl. 1, figs. 1–5).
Habitat: on tree trunks or rocks.

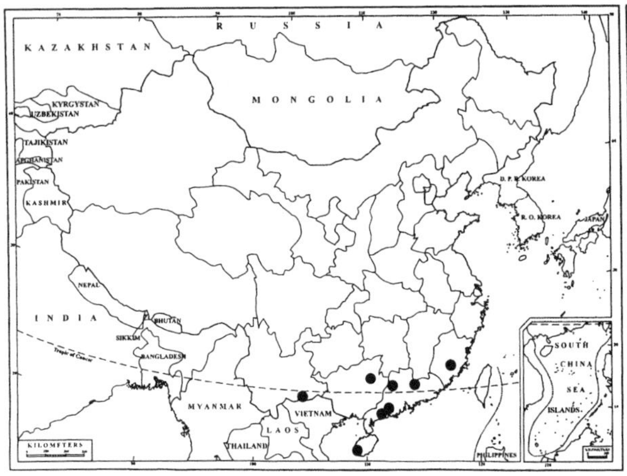

6. *Cyathophorella subspinosa*

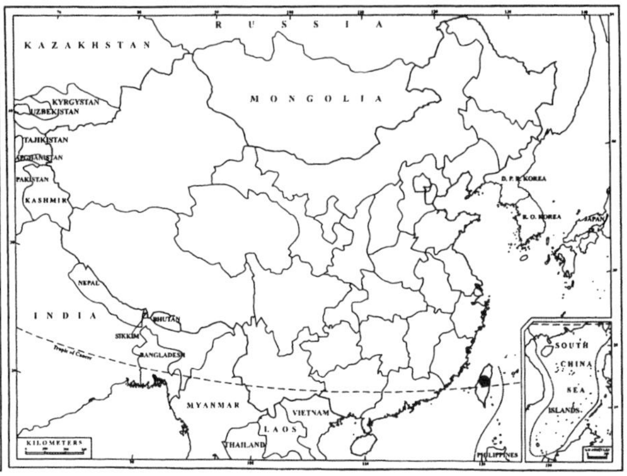

7. *Cyathophorella taiwaniana*

Distribution: endemic to China.

Chinese specimens examined: see the type information cited above.

This species is similar to *Cyathophorella hookeriana* by having distinct borders on leaves and hairy apices, but differs by the costae of the amphigastria almost reaching the apices and by having secondary stems sometimes branched.

8. ***Cyathophorella tonkinensis*** (Broth. & Par.) Broth. 粗齿雉尾藓 cu-chi zhi-wei xian

Nat. Pflanzenfam. (ed. 2), 11: 278. 1925. *Cyathophorum tonkinense* Broth. & Par., Rev. Bryol. 35: 46. 1908. Type. Vietnam: Tonkin, *Eberhardt s.n.*

Cyathophorella japonica (Broth.) Broth., Oefvers. Förh. Finska Vetensk.-Soc. 62A(9): 31. 1921. *Cyathophorum japonicum* Broth. in Card., Bull. Soc. Bot. Genève ser. 2, 3: 279. 1911.

Cyathophorella serrulata P.-C. Chen, Gen. Musc. Sin. 2: 134. 1978, *nom. inval.*

Plants large, dark green. *Primary stems* prostrate, with tomentose rhizoids at base; secondary stems erect or erect-spreading, usually simple, sometimes branched near apex, ca. 6 cm high, 1 cm wide; filamentous, orange-red gemmae sometimes in clusters near branch apices. *Leaves* plane, lateral leaves ovate-rounded, slightly asymmetric, ca. 4.8 mm × 1.9 mm, acuminate; leaf margins bordered, hyaline, serrate, entire at base; amphigastria ovate, acuminate, ca. 2.7 mm × 1.4 mm; costa short, often forked; laminal cells rhomboidal; leaf marginal cells often with sharp teeth near the apex, the length of teeth up to 150 μm, gradually elongate toward margins at base. *Perichaetial leaves* small, narrow. *Setae* short, ca. 2 mm long; capsules cylindrical, ca. 2.6 mm long, 1.1 mm in diameter; peristome teeth two layers, well developed. *Spores* papillose, ca. 30 μm in diameter.

Illustrations: Plate 415, figs. 10–18.
Habitat: on bases of trees or tree trunks and on rocks or soil.
Distribution: China, Japan, Himalayas, Thailand, and Vietnam.

Chinese specimens examined: FUJIAN: Nan-jing Co., *M. Zang 84* (PE, SYS). GUANGDONG: Long-men Co., *Z.-H. Li 83, 125, 133, 174, 211* (SYS); Mt. Luo-fu, *Metcalf 17679* (SYS). GUANGXI: Long-shen Co., *P.-C. Wu & Y.-X. Lin 182, 782* (PE, SYS). HAINAN: Mt. Bawangling, *P.-J. Lin & L. Zhang s.n.* (IBSC, MO); Mt. Diaoloushan, *Z.-H. Li 77-2978, 3265* (SYS); Mt. Jianfengling, *P.-C. Chen et al. 338a* (PE, SYS). TAIWAN: Anmashan (Mt.), *Iwatsuki & Sharp 3149* (IBSC, NICH); Tai-bei Co., *M.-J. Lai 8344* (MO, SYS, TAI); Tao-yuen Co., *M.-J. Lai & J. E. Lewis 0293* (MO). YUNNAN: Xishuangbanna, *C.-W. Wang 9576, 9579, 9583* (PE, SYS); *W.-X. Xu 31, 11105, 11404* (PE, SYS); Gaoligongshan, *M.-Z. Wang 9091c* (PE); Ying-jiang Co., *J.-X. Luo & M.-Z. Wang 813001f* (PE).

This species is characterized by its lateral leaves with spines or gross teeth and by its leaf margins that are distinctly bordered.

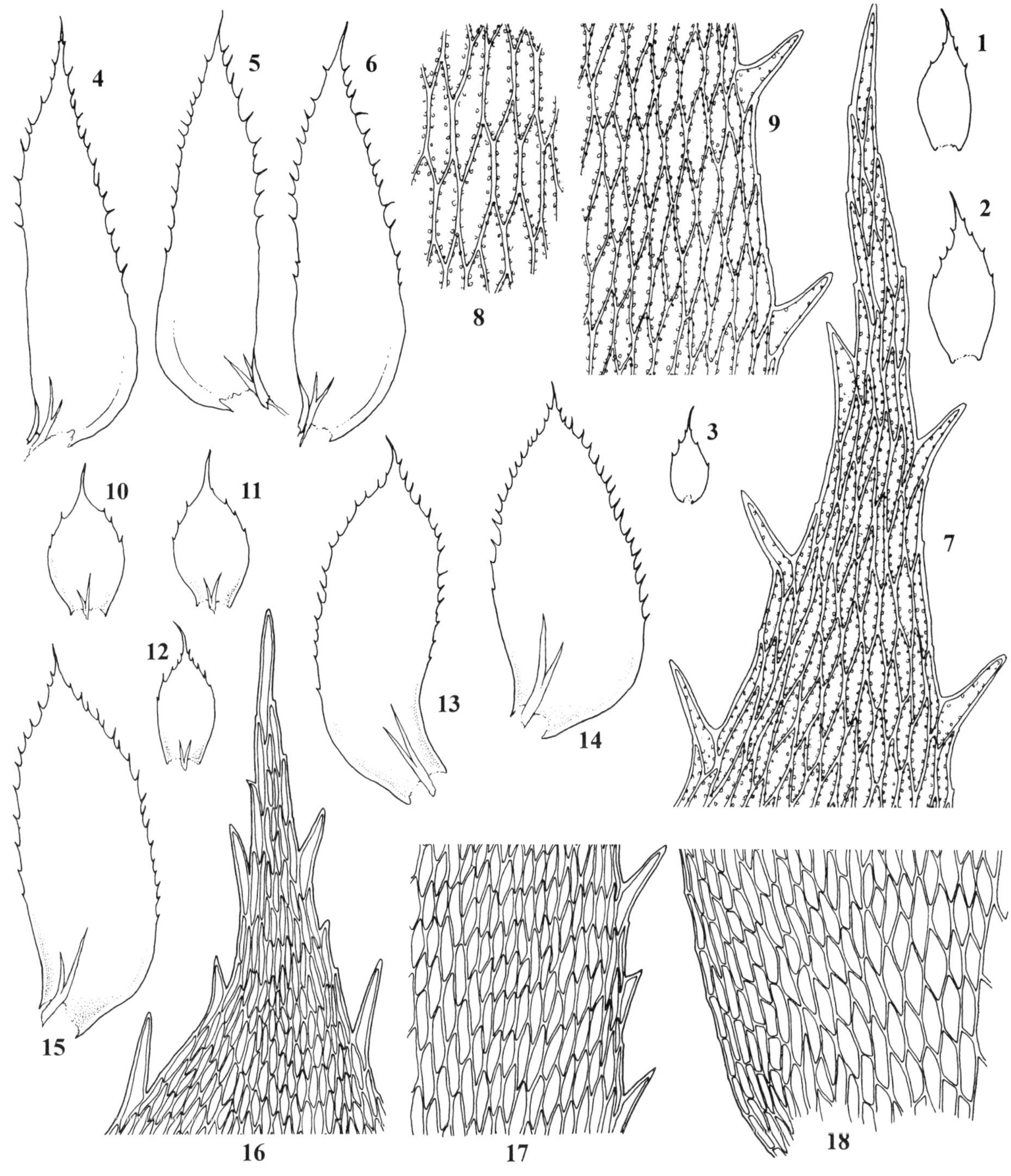

Plate 415. 1–9, *Cyathophorella subspinosa*. 1–3, amphigastria of branch (× 9). 4–6, lateral leaves (× 9). 7, apical cells of lateral leaf (× 127). 8, median cells of lateral leaf (× 127). 9, median marginal cells of lateral leaf (× 127). 10–18, *Cyathophorella tonkinensis*. 10–12, amphigastria of branch (× 9). 13–15, lateral leaves (× 9). 16, apical cells of lateral leaf (× 93). 17, median marginal cells of lateral leaf (× 93). 18, basal marginal cells of lateral leaf (× 93). (1 –9, drawn from *P.-C. Chen 25*, PE; 10–18, from *M. Zang 327*, HKAS.)

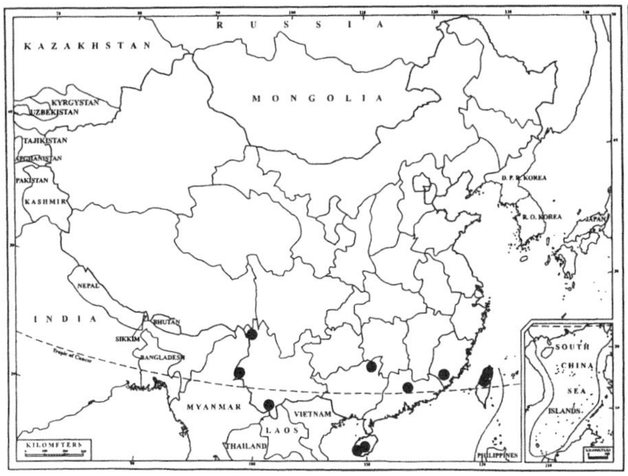

8. *Cyathophorella tonkinensis*

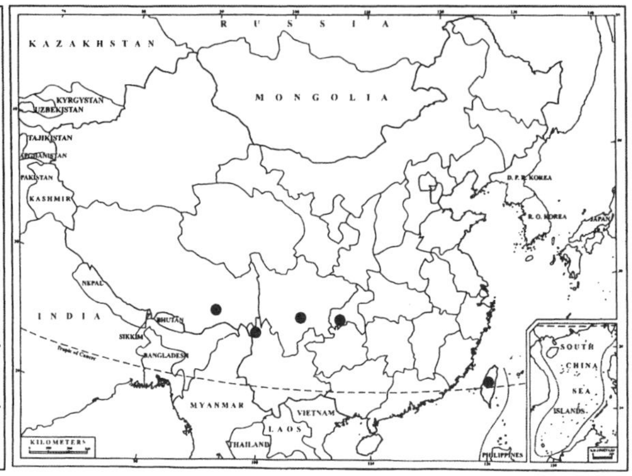

1. *Dendrocyathophorum paradoxum*

2. ***Dendrocyathophorum*** Dix. 树雉尾藓属 shu zhi-wei xian shu
J. Bot. 74: 7. pl. 610, f. 5. 1936.

Plants delicate, yellowish green, not glossy. *Primary stems* prostrate, with loosely scale-like leaves and dense brown rhizoids; secondary stems simple in lower part, loosely with small leaves, irregularly pinnately branched in upper part, complanate, blunt at branch terminal. *Lateral leaves* of stems dense, broadly ovate, asymmetric, acuminate; serrulate in upper part, not bordered; costa single, short, weak or forked; laminal cells elongated hexagonal, thin-walled, becoming larger toward base; amphigastrial leaves broadly ovate, symmetric, lanceolate at apex; serrulate in upper part of margins; costa single, short; laminal cells elongated hexagonal; lateral leaves and amphigastria of branches similar to those of stems, but smaller. *Autoicous*. *Perichaetial leaves* ovate-lanceolate, entire; ecostate. *Setae* elongate, slender, smooth, erect; capsules reddish brown, horizontal or pendent; peristome double, exostome teeth lanceolate, densely striate on outer surfaces in lower part, minutely papillose in upper part; endostome segments as long as the teeth, lanceolate, folded, papillose; basal membranes high; opercula conical, inclined, beaked. *Calyptrae* mitrate. *Spores* spherical, minutely papillose.

Morphological characters of this genus are between *Hypopterygium* and *Cyathophorum*. Herzog & Noguchi (1955) transferred *Cyathophorum intermedium* Mitt. to this genus as *D. intermedium* (Mitt.) Herz. and reported it from Taiwan. However, the concept of *Cyathophorum intermedium* described by Mitten (1859) was not the same as that of *D. intermedium* by Herzog & Noguchi (1955), which has slender, flattened, dendroid branches. Therefore, the latter species was renamed as *Dendrocyathophorum herzogii* by Gangulee (1977), which is very closely related to *Cyathophorum intermedium*. Both species occur in China. *Dendrocyathophorum herzogii* was considered to be a synonym of *D. decolyi* (Kruijer 1995), but both are treated as synonyms of *D. paradoxum* in this study.

1. ***Dendrocyathophorum paradoxum*** (Broth.) Dix. 树雉尾藓 shu zhi-wei xian
J. Bot. 75: 126. 1937. *Eurydictyon paradoxum* (Broth.) Horik. & Nog., J. Sci. Hiroshima Univ., Ser. B, Div. 2, Bot. 2 (3): 22. Pl. 1. 1936. *Hypopterygium paradoxum* Broth. in Card., Bull. Soc. Bot. Genève ser. 2. 4: 378. 1912. Type. Japan.
Dendrocyathophorum decolyi (Broth. ex Fleisch.) Kruijer, Lindbergia 20: 90. 1995–96 [1996]. *Hypopterygium decolyi* Broth. ex Fleisch., Musci Buitenzorg 3: 1079. 1908.
Dendrocyathophorum herzogii Gangulee, Mosses E. India 6: 1540. 1977.

Plants slender, up to 2–5 cm high, 1.0–1.5 cm wide with leaves. *Primary stems* and secondary stems with tomentose rhizoids; secondary stems simple in lower part with loose, small leaves, irregularly pinnately branched above, complanate, planely inclined; leaves in 3 rows, loose, reflexed when dry, lateral leaves ovate-lanceolate, asymmetric, acuminate, ca. 3.0 mm × 1.2 mm, distinctly serrulate above; laminal cells elongate-rhomboidal, thin-walled, porose, gradually wider and longer toward base, bordered by 2 rows of narrow and elongate cells at base; amphigastria in 1 row, smaller, broadly ovate or oblong-ovate, symmetric, caudate, ca. 1.6 mm × 0.6 mm; costa single, or forked; laminal cells elongate-hexagonal, smooth. *Perichaetial leaves* small, narrowly elongate. *Setae* 6–8 mm long; peristome the same as described in the genus.

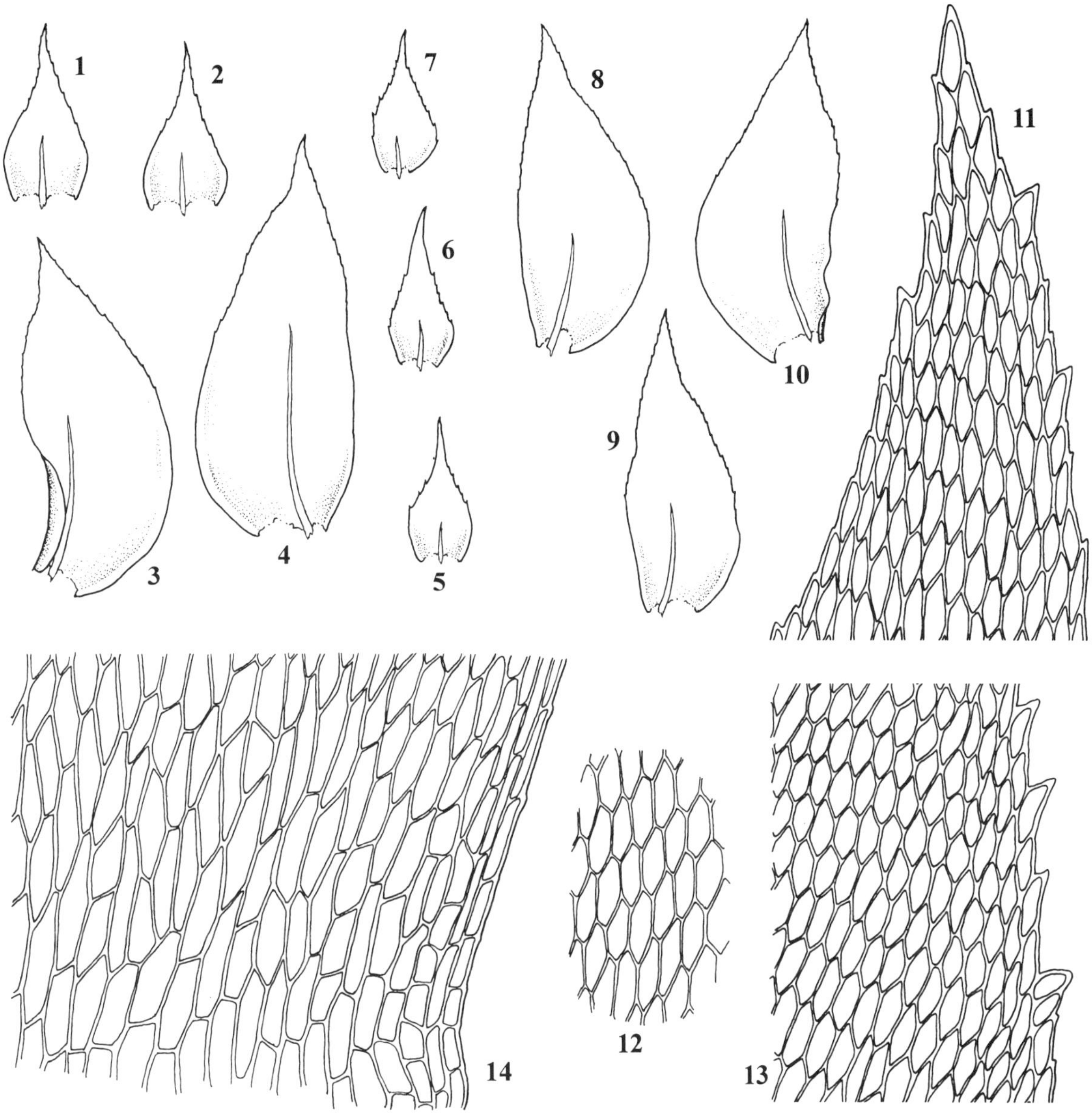

Plate 416. *Dendrocyathophorum paradoxum*. 1–2, amphigastria of stem (× 12). 3–4, lateral leaves of stem (× 12). 5–7, amphigastria of branch (× 12). 8–10, lateral leaves of branch (× 12). 11, apical cells of lateral leaf (× 127). 12, median cells of lateral leaf (× 127). 13, median marginal cells of lateral leaf (× 127). 14, basal cells of lateral leaf (× 127). (1–14, drawn from *M.-Z. Wang 10822a*, PE.)

Illustrations: Plate 416, figs. 1–14.
Habitat: on thin soil over rocks.
Distribution: China, Japan, India, Thailand, Vietnam, the Philippines, and New Guinea.

Chinese specimens examined: CHONGQING: Mt. Jinfu, *X.-Y. Hu H0859* (PE). SICHUAN: Mt. Omei, *J.-X. Luo & B.-D. Xin 3161a* (PE). Taiwan: Nan-tou Co., *M-J. Lai 8199*, *8203* (MO). XIZANG: Mo-tuo Co., *M.-Z. Wang 6591* (PE), *Y.-G. Su 16349* (KUN). YUNNAN: Gong-shan Co., *M.-Z. Wang 10822a* (PE).

3. ***Hypopterygium*** Brid. 孔雀藓属 kong-que xian shu
Bryol. Univ. 2: 709. 1827.

Plants creeping, usually with brown rhizoids. *Primary stems* creeping; secondary stems erect with remote scale-like leaves or nudicaulous, rarely with dense brown rhizoids, erect-spreading above, 1–2 pinnately branched, rarely 3 pinnately branched, often peacock tail-like. *Lateral leaves* broadly ovate, oblong or ovate-ligulate, asymmetric; leaf margins mostly bordered by narrowly elongate cells, serrate above; costa single, vanishing below leaf apex; laminal cells rhomboidal or oval-hexagonal, loose, smooth, thin- or thick-walled, gradually becoming longer and looser toward leaf bases; amphigastria appressed, broadly ovate or ovate, symmetric, with long apex, margins bordered. *Dioicous* or autoicous. *Inner perichaetial leaves* slightly longer or sheathing at the base, with acuminate apex, margins entire. *Setae* elongate, slender, smooth, sometimes twisted; capsules ovoid or elongate-ovoid, horizontal or pendulous; annuli wide, caducous or persistent; peristome double; exostome teeth weakly keeled, densely striate; endostome segments wide, well developed, basal membranes high; cilia 2–3. *Calyptrae* smooth, mitriform or conical. *Spores* minute. *Gemmae* numerous, often growing on branches.

The species of *Hypopterygium* are mainly distributed in the tropical and subtropical regions of the world. The genus consists of about 55 species, with 22 of them better understood. Seven species are presently recognized in this treatment.

1. Leaf costa evidently setulosely long excurrent .... 1. *H. aristatum*
1. Leaf costa not excurrent or only shortly excurrent 2
  2. Leaf cells somewhat hyaline; leaf margins entire or only serrulate 3
  2. Leaf cells distinctly hyaline; leaf margins dentate 4
3. Leaf margins entire .... 3. *H. flavolimbatum*
3. Leaf margins serrulate .... 7. *H. tibetanum*
  4. Underleaves widely cordate; costa reaching the middle of leaves .... 6. *H. tenellum*
  4. Underleaves ovate-rotund; costa vanishing near leaf apex or slightly excurrent 5
5. Leaf costa nearly reaching leaf apex; median leaf cells rather short; capsules and setae reddish .... 5. *H. japonicum*
5. Leaf costa reaching 2/3 the leaf length; median leaf cells rather long; capsules and setae reddish brown 6
  6. Underleaves with setulose apices .... 2. *H. fauriei*
  6. Underleaves with short apices .... 4. *H. formosanum*

1. ***Hypopterygium aristatum*** Bosch & Sande Lac. 毛尖孔雀藓 mao-jian kong-que xian
   Bryol. Jav. 2: 12. 141. 1861. Type. Indonesia: Java, Pangerango, *Motley s.n.*

Primary stems prostrate, rhizoids tomentose; secondary stems erect, nudicaulous near the base, irregularly pinnately branched. *Lateral leaves* ovate-rounded, asymmetric, with abruptly shortly acute apices; leaf margins bordered by 2 rows of narrowly elongate cells, serrate at the upper part; costa reaching 2/3 the leaf length; amphigastria nearly rounded, symmetric; leaf margins bordered by 1 row of cells, nearly entire; costa long excurrent.

Illustrations: Dozy & Molkenboer 1861, Pl. 141.
Habitat: on rocks.
Distribution: China, Japan, Himalayas, Thailand, and Indonesia.

Chinese specimens examined: We have not seen any specimens of this species. It was reported from Mt. Fan-jing, Guizhou province by B.-G. Zhong and Y.-X. Xiong (1990) and from Taiwan by M.-J. Lai (1976).

2. ***Hypopterygium fauriei*** Besch. 拟东亚孔雀藓 ni dong-ya kong-que xian
   Ann. Sci. Nat. Bot., ser. 7, 17: 391. 1893. Type. Japan: *Textor s.n.*

Primary stems creeping; secondary stems erect, erect-spreading above, flatly branched, dendroid, ca. 4–5 cm high, ca. 3 cm wide with leaves, leafy branches loosely twisted at the base of secondary stems, rhizoids brown, tomentose in the axils of leaves. *Lateral leaves* dense, crisped when dry, erect-spreading when wet, elongate-ovate or broadly ovate, asymmetric, ca.1.2 mm × 0.6 mm; leaf margins bordered by 1–2 rows of narrowly elongate cells, mostly serrulate from apex to base along antical ones; costa single, reaching 2/3 the leaf length; laminal cells

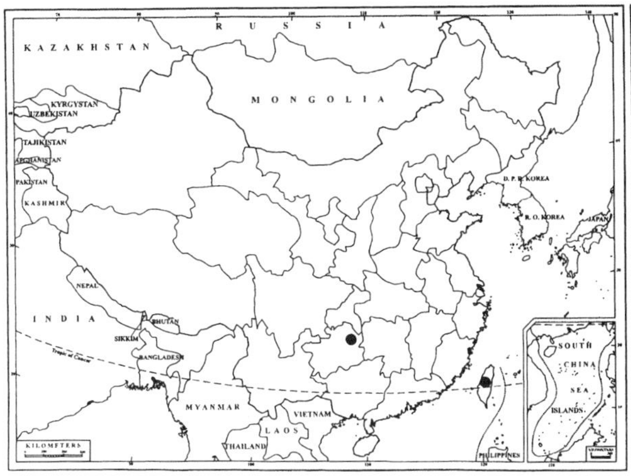

1. *Hypopterygium aristatum*

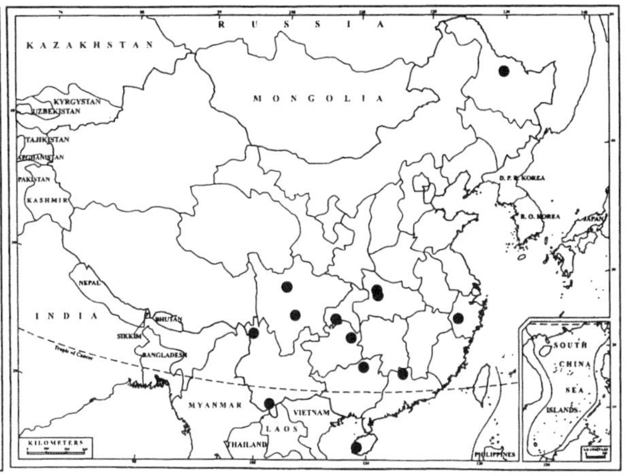

2. *Hypopterygium fauriei*

rhomboidal-hexagonal, 22–40 µm × 13–20 um; amphigastria oblate or ovate-rounded, symmetric, ca. 0.7 mm × 0.5 mm wide, with long acuminate apices; leaf margins bordered, minutely serrulate; costa ending below the apex. *Setae* ca. 2–3 cm long, cygneous above; capsules generally reddish brown, ovoid, asymmetric, horizontal or inclined; opercula rostrate.

Illustrations: Plate 417, figs. 1–11.
Habitat: on tree trunks or rocks.
Distribution: China, Eastern Asia, and Western North America.

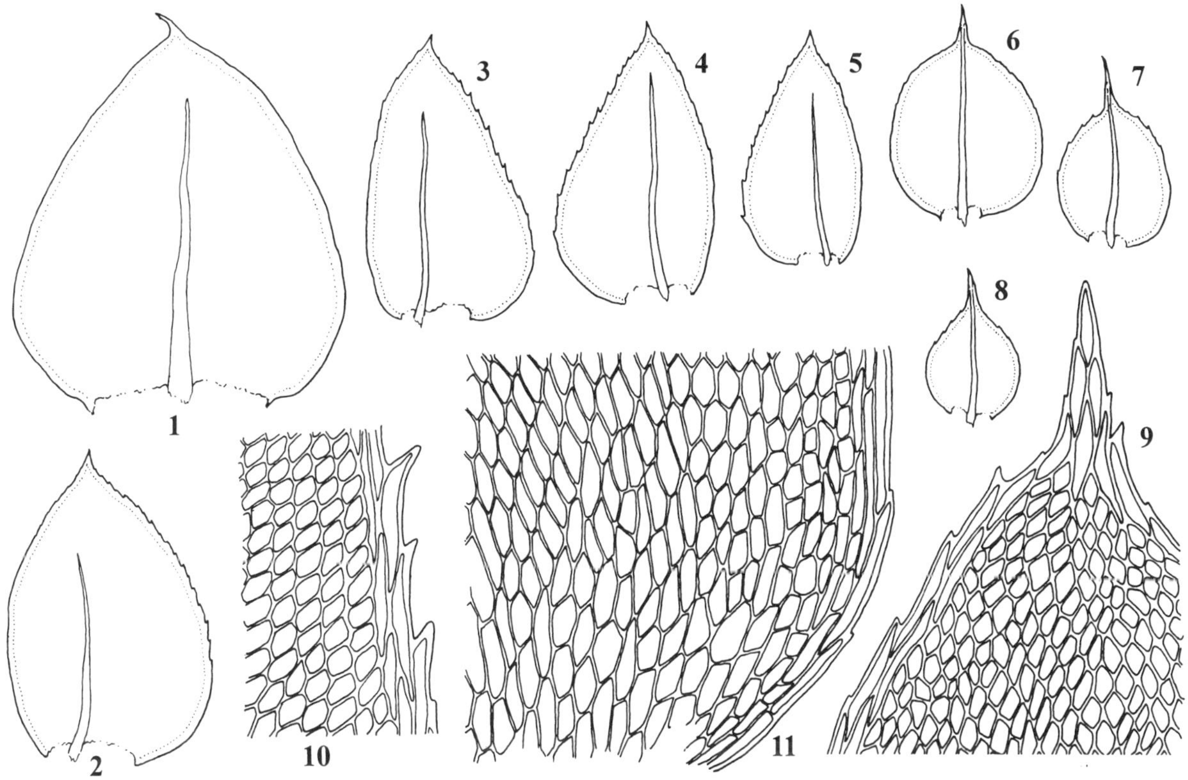

Plate 417. *Hypopterygium fauriei*. 1, lower secondary stem leaf (× 19). 2–5, lateral leaves of branch (× 19). 6–8, amphigastria of branch (× 19). 9, apical cells of lateral leaf (× 127). 10, median marginal cells of lateral leaf (× 127). 11, basal cells of lateral leaf (× 127). (1–11, drawn from *T. 310*, PE.)

Chinese specimens examined: CHONGQING: Mt. Jinfu, *P.-C. Chen 1621* (PE, SYS). GUANGDONG: Le-chang Co., *S.-Q. Chen 190* (MO); Rui-yuan Co., *B.-J. Lin 907* (IBSC, SYS). GUANGXI: Rong-shui Co., *M.-Z. Wang 46310* (PE). GUIZHOU: Mt. Fanjing, *Y. Jiang 7814* (PE, SYS). HAINAN: Chang-jiang Co., *Redfearn 35888* (MO). HEILONGJIANG: Xiaoxinanling (Lesser Khingan Mt.), *P.-C. Chen & C. Gao 632* (PE). HUBEI: Mt. Shennongjia, *P.-C. Wu 293* (MO, PE); Xing-shan Co., *J.-X. Luo 961182* (MO). SICHUAN: Maer-kang Co., *Redfearn 35052* (MO, PE); Mt. Omei, *Allen 6612* (MO). YUNNAN: Xishuangbanna, *Crosby 14820* (MO), *W.-X. Xu 6615, 11103, 11346* (PE, SYS); Gaoligongshan, *M.-Z. Wang 9264* (PE). ZHEJIANG: Sui-chang Co., *Z.-L. Liu 877, 1506* (SHM).

The present species is closely related to *Hypopterygium japonicum*, but the two are different by the following features: 1) the costae of lateral leaves of *H. fauriei* reaching 2/3 the leaf length, while those of *H. japonicum* ending below the apex; 2) the margins of lateral branch leaves of *H. fauriei* serrate throughout, while those of *H. japonicum* have serration not over the middle of the leaf; 3) the median leaf cells of *H. fauriei* longer than 35 μm, while those of *H. japonicum* shorter than 30 μm.

3. ***Hypopterygium flavolimbatum*** C. Müll. 黄边孔雀藓 huang-bian kong-que xian
Syn. Musc. Frond. 2: 10. 1850. Type. Nepal.
Hypopterygium emodi C. Müll. in Kindb., Hedwigia 40: 292. 1901, *nom. nud.*

Plants medium-sized, yellowish green. *Primary stems* creeping, with dense rhizoids; secondary stems 2–3 cm high, erect at the base, loosely leaved, regularly branched above. *Lateral leaves* broadly ovate, patent, asymmetric, ca. 1.4 mm × 0.8 mm, with shortly acute apices; leaf margins bordered by 1–2 rows of narrowly elongate cells, entire or serrulate at apex; costa single, reaching 3/5 the leaf length; laminal cells rhomboidal-hexagonal, ca. 25 μm × 11 μm; amphigastria nearly rounded, excurrent at apex; leaf margins bordered by 1–2 rows of elongate cells. *Setae* erect, curved in the upper part, ca. 1.8 cm long; capsules cylindrical, ca. 1.3 mm long, ca. 1 mm in diameter.

Illustrations: Plate 418, figs. 1–12.
Habitat: on tree trunks, soil or rocks.
Distribution: China and Himalayas.

Chinese specimens examined: CHONGQIN: Chong-qing City, Bei-pei, *J.-Z. Rao 057146* (PE), *P.-C. Chen 5120* (PE, SYS). FUJIAN: Jian-yang Co., *C.-H. Gao 07190* (SHM, SYS); Jiang-le Co., *X.-L. He 866*, 1156 (PE, SYS); Mt. Wuyi, *D.-K. Li 13789* (SHM, SYS), *P.-C. Chen et al. 139* (PE, SYS). GUANGDONG: Feng-kai Co., *Z.-H. Li 84-34* (SYS), Qixing forest station, *Fengkai Expedition Group 26, 27* (IBSC, SYS); He-shan Co., *H.-G. Ye s.n.* (IBSC, SYS); Rui-yuan Co., *L. Zhang 224* (IBSC, SYS), *Z. Huang 44336* (IBSC, SYS), *G.-Q. Zeng 196* (IBSC, SYS); Yang-dong Co., *G.-C. Zhang et al. 17* (IBSC, SYS); Yang-shan Co., *L. Deng 1300* (IBSC, SYS). GUANGXI: Rong-shui Co., *M.-Z. Wang 46311* (PE, SYS); Yang-shuo Co., *P.-C. Wu & Y.-X. Lin 1673* (PE, SYS). GUIZHOU: Mt. Fanjing, *Y. Jiang 7815* (PE, SYS). HAINAN: Mt. Bawangling, *P.-J. Lin et al. 3224* (IBSC, SYS). HUBEI: Jian-shi Co., *H.-C. Zhou 54* (PE, SYS). HUNAN: Yi-zhang Co., *S.-Q. Chen 1815* (IBSC, SYS). SHAANXI: Mt. Qinling, *P.-C. Chen et al. 767* (PE). SICHUAN: Guan-xian Co., *P.-C. Chen 5214* (PE, SYS); Lei-bo Co., *Z.-T. Guan 6468, 7860* (PE, SYS); Mu-chuan Co., *P.-C. Chen 5757* (PE, SYS); Wo-long Co., *J.-X. Luo 3380* (PE). YUNNAN: Dulongjiang, *M.-Z. Wang 9716-b* (PE, SYS); Xishuangbanna, *C.-C. Wang 9843* (PE, SYS), *W.-X. Xu 6631-1* (PE, SYS); Yang-bi Co., *Redfearn & Y.-G. Su 986* (MO). ZHEJIANG: Hang-zhou City, *P.-C. Chen 6014* (PE, SYS); Mt. Tian-mu, *D.-K. Li 3553* (IBSC, SHM, SYS).

4. ***Hypopterygium formosanum*** Nog. 台湾孔雀藓 tai-wan kong-que xian
Trans. Nat. Hist. Soc. Formosa 26: 148. pl. 3, f. 9–13. 1936. Type. China: Taiwan, Tai-nan Co., Mt. Kodama, *Noguchi 7200*.

Primary stems creeping, secondary stems loosely tufted, yellowish green, not glossy, 3–5 cm high, complanately branched, dendroid. *Stem leaves* ovate, symmetric, with a short acute apex. *Lateral leaves* elongate-ovate, asymmetric, ca. 1.2 mm × 0.8 mm, with shortly acute apices; leaf margins bordered by 1–2 rows of narrowly elongate cells, serrate throughout; costa single, reaching 2/3 the leaf length; laminal cells rhomboidal-hexagonal, ca. 22 μm × 11 μm, hyaline; amphigastria small, broadly ovate, symmetric, ca. 0.8 mm × 0.5 mm, with slender and short acumens, leaf margins bordered, serrulate above; costa thin, soft, mostly ending below the apex. *Setae* long, curved at tops; capsules shortly cylindrical, inclined; exostome teeth elongate-lanceolate; endostome segments papillose; basal membrane high; cilia well developed; opercula shortly conical. *Calyptrae* mitriform, smooth. *Spores* not seen.

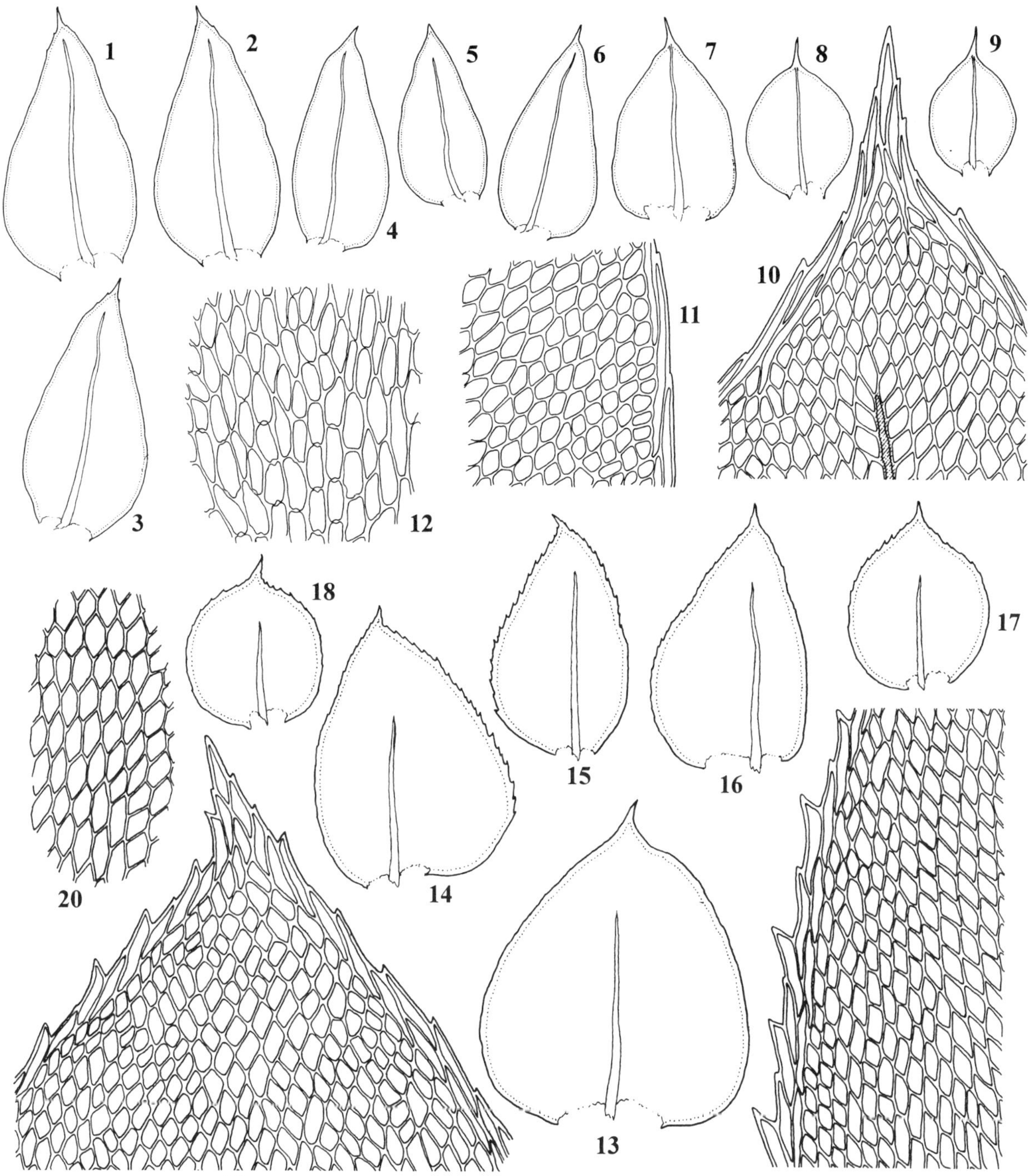

Plate 418. 1–12, *Hypopterygium flavolimbatum*. 1–3, lateral leaves of stem (× 28). 4–6, lateral leaves of branch (× 28). 7–8, amphigastria of stem (× 28). 9, amphigastrium of branch (× 28). 10, apical cells of lateral leaf (× 259). 11, median marginal cells of lateral leaf (× 259). 12, basal cells of lateral leaf (× 259). 13–21, *Hypopterygium formosanum*. 13, lateral leaf of stem (× 19). 14–16, lateral leaves of branch (× 19). 17–18, amphigastria of branch (× 19). 19, apical cells of lateral leaf (× 127). 20, median cells of lateral leaf (× 127). 21, median marginal cells of lateral leaf (× 127). (1–12, drawn from *W.-M. Zhu 31*, PE; 13–21, from *S. He 30756-b*, PE.)

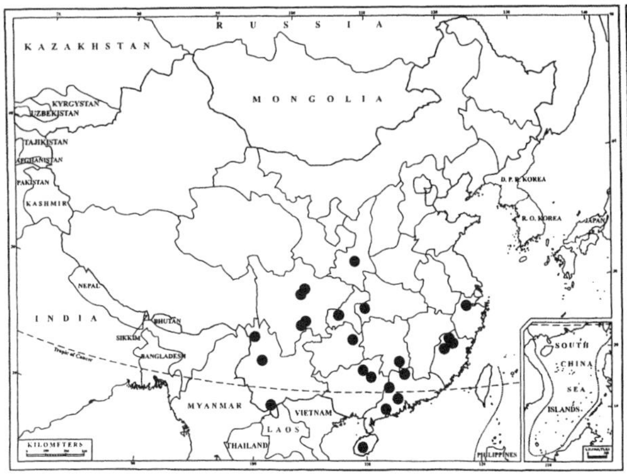

3. *Hypopterygium flavolimbatum*

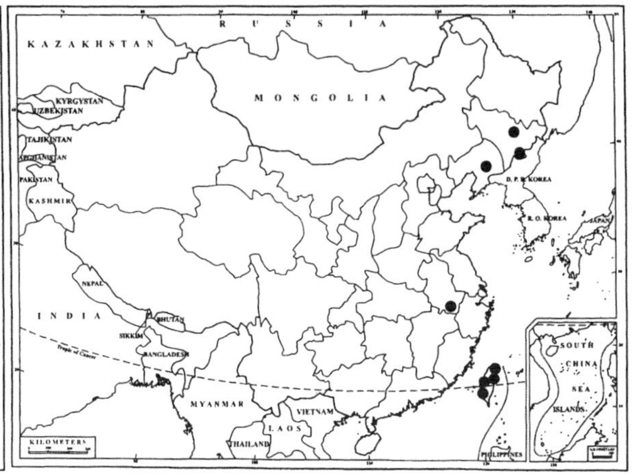

4. *Hypopterygium formosanum*

Illustrations: Plate 418, figs. 13–21.
Habitat: on soil or rocks.
Distribution: China and Japan.

Chinese specimens examined: ANHUI: Huang Shan (Mt.), *P.-C. Chen et al. 6468* (MO, PE). JILIN: Jiao-he Co., *C. Gao 1599*, *1667* (IFSBH, PE, SYS); Mt. Changbai, *C. Gao 632* (IFSBH, SYS). LIAONING: Ben-xi Co., *Y.-L. Zhang 251* (PE). TAIWAN: Hua-lian Co., *J.-Y. Peng 83-56* (IBSC, MO, SYS); Ilan Co., *S.-H. Lin s.n.* [Bryophytes of Taiwan Exsiccatae ser. 4, no. 173] (MO); Tai-chung Co., *S.-H. Lin s.n.* [Bryophytes of Taiwan Exsiccatae ser. 4, no. 174] (MO).

Hypopterygium formosanum is very closely related to *H. fauriei*; however, it differs from the latter by having long acuminate apices of stem and branch leaves and shorter apices and costae (ending below the apex) of the amphigastria.

5. ***Hypopterygium japonicum*** Mitt. 东亚孔雀藓 dong-ya kong-que xian
J. Proc. Linn. Soc., Bot. 8: 155. 1864. Type. Japan.

Plants loosely tufted. *Primary stems* prostrate, with dense rhizoids; secondary stems erect, erect-spreading above, complanately branched, dendroid, ca. 3–4 cm high, ca. 3 cm wide with leaves. *Lateral leaves* broadly ovate, slightly asymmetric, 0.9–1.6 mm long, with acute apex, leaf margins bordered by 1–2 rows of narrowly elongate cells, serrulate above; costa single, reaching 3/5 the leaf length; laminal cells rhomboidal-hexagonal, 18–25 µm × 11–14 µm; amphigastria 0.4–1 mm × 0.2–0.8 mm, with short acuminate apex, margins bordered by 1–2 rows of narrowly elongate cells, nearly entire; costa ending below leaf apex or percurrent. *Setae* erect, ca. 1.5 cm long; capsules oblong-cylindrical, ca. 2.8 mm long; opercula conical, long-rostrate.

Illustrations: Plate 419, figs. 1–9.
Habitat: on tree trunks, rocks, soil or decaying logs.
Distribution: China, Korea, and Japan.

Chinese specimens examined: ANHUI: Qi-men Co., *M.-B. Deng et al. 3* (PE, SYS); between Huizhou and Tangjiang, *P.-C. Chen et al. 4868* (PE, SYS). FUJIAN: Jiang-le Co., *M.-Z. Wang 47637* (PE). CHONGQING: Nan-chuan Co., *S.-X. Tan 2880* (PE), *M.-Z. Wang 860100* (MO, PE). GUANGXI: Long-sheng Co., *Y.-C. Chen 01655* (PE); Rong-shui Co., *M.-Z. Wang 46981*, *46993* (PE). GUIZHOU: Tong-zhou Co., Xiguanyao, *Y. Jiang 4327* (PE, SYS). JIANGXI: Lushan (Mt.), *G.-X. Liang 102* (PE, SYS); Wulingshanmian, *R.-L. Zhao 511* (PE, SYS). HUBEI: Mt. Shennongjia, *P.-C. Chen 213b* (MO, PE); Xing-shan Co., *J.-X. Luo 961206* (MO). SHANNXI: Mt. Qinling, *P.-C. Chen et al. 417*, *612* (PE); Mt. Tai-bai, *Z.-P. Wei 5272* (IWH). SICHUAN: Mt. Omei, *Redfearn 34732* (MO); Ruo-er-gai Co., *S. He 30754a* (MO). TAIWAN: Ilan Co., Shevock 18245 (MO); Nan-tuo Co., *M.-J. Lai 8201* (MO). YUNNAN: Gong-shan Co., *C.-W. Wang 6998*, *7023* (PE). ZHEJIANG: Mt. West Tianmu, *R.-L. Hu 22* (MO).

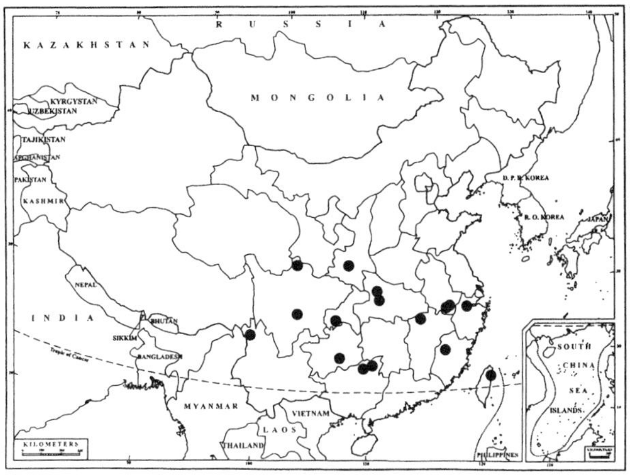

5. *Hypopterygium japonicum*

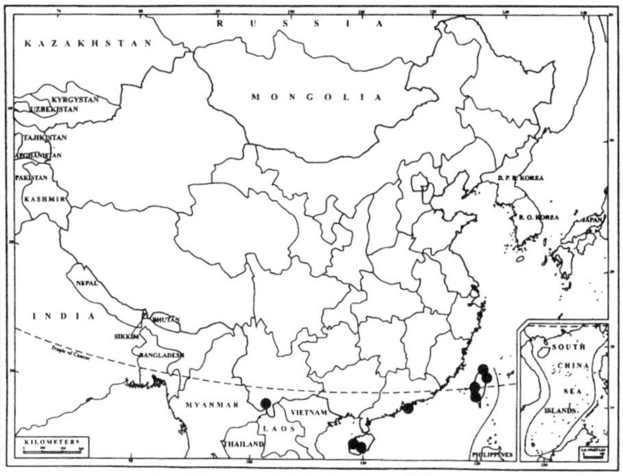

6. *Hypopterygium tenellum*

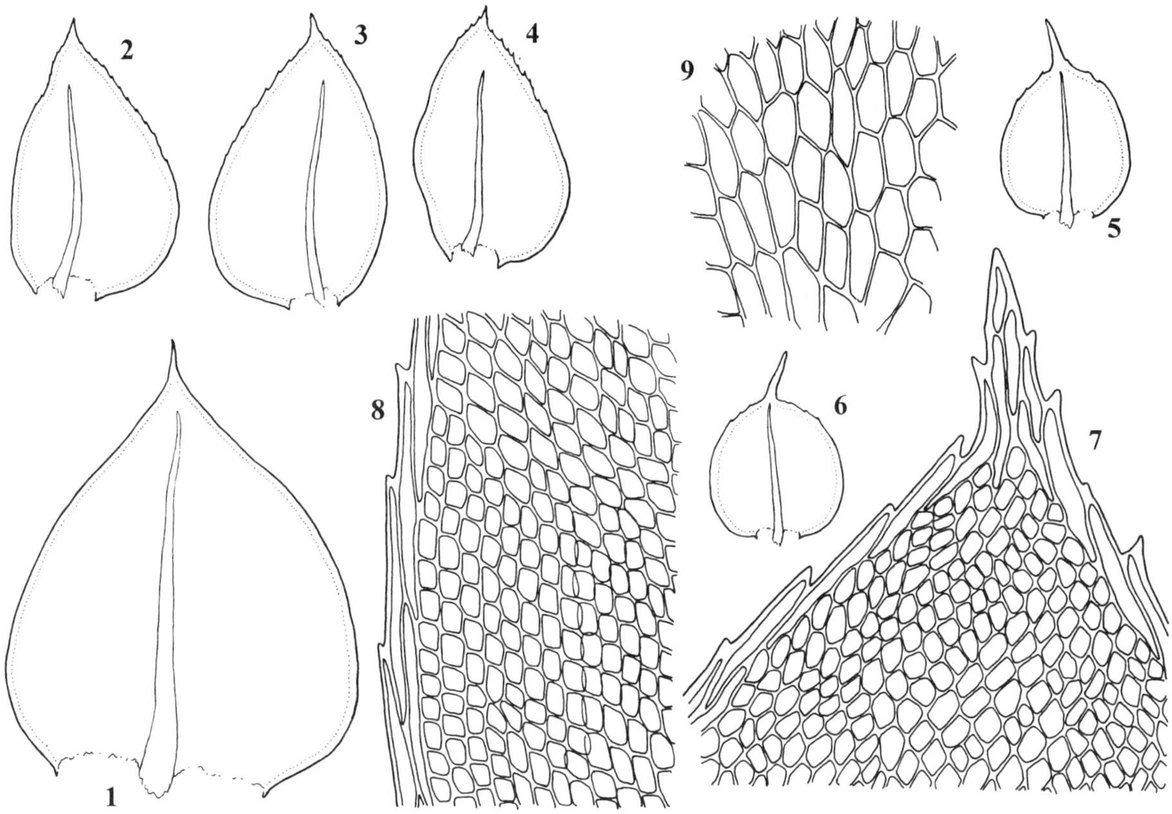

Plate 419. *Hypopterygium japonicum*. 1, stem leaf (× 19). 2–4, lateral leaves of branch (× 19). 5–6, amphigastria of branch (× 19). 7, apical cells of lateral leaf (× 173). 8, median marginal cells of lateral leaf (× 173). 9, basal cells of lateral leaf (× 173). (1–9, drawn from *Z.-S. Qin et al. 14*, PE.)

6. ***Hypopterygium tenellum*** C. Müll. 南亚孔雀藓 nan-ya kong-que xian
Bot. Zeitung (Berlin) 12: 557. 1854. Type. India: Nilghiri.
Hypopterygium ceylanicum Mitt., J. Proc. Linn. Soc., Bot., Suppl. 1: 148. 1859.

Primary stems creeping; secondary stems ca. 3 cm high, dark yellowish green, loose, erect, densely branched above, rhizoids more or less tomentose. *Lateral leaves* broadly ovate, asymmetric, ca. 2.0 mm × 1.5 mm, with abruptly shortly acute apices; leaf margins bordered by 1–2 rows of elongate cells, dentate from the apex to the middle of leaves; costa single, reaching 1/2 the leaf length; laminal cells rhomboidal-polygonal, 20–40 μm × 15–20 µm; amphigastria broadly cordate, symmetric, abruptly acute at the apex; leaf margins bordered by 1 row of elongate cells; costa reaching half of the length of leaves. *Setae* ca. 0.7–1.5 cm long, curved above; capsules inclined, ovoid; opercula long-rostrate, almost as long as the capsules.

Illustrations: Plate 420, figs. 1–10.
Habitat: on tree trunks, fallen logs or rocks.
Distribution: China, Japan, India, Indonesia, Vietnam, Sir Lanka, the Philippines, New Guinea, New Caledonia, and Africa.

Chinese specimens examined: HAINAN: Mt. Bawangling, *P. Zeng 42C, 1962* (SYS, PE), *Z.-H. Li 74-3220, 74-3224, 74-3289, 74-3345* (SYS), *B.-J. Lin & L. Zhang 45, 56* (IBSC, SYS); Mt. Diaoluo, *Z.-H. Li 77-3074* (SYS); Mt. Wuzhi, *Y.-X. Lin 234B, 333* (IBSC, SYS). HONG KONG: Mt. Dadongshan, *M.-Z. Wang 65662* (PE). TAIWAN: Ilan Co., *Shevock 18255* (MO); Miao-li Co., *S.-H. Lin s.n.* [Bryophytes of Taiwan Exsiccatae ser. 3, no. 125] (IBSC, MO, SYS); Ping-tung Co., *Shevock 15058* (MO); Tai-chung Co., *M.-J. Lai 3200* (MO); Tai-pei Co., *Iwatsuki 3620* (IBSC, NICH, SYS). YUNNAN: Xishuangbanna, *C.-W. Wang 9712, 9733* (PE, SYS).

This species is characterized by its stem amphigastria with short costa ending below the middle of leaves.

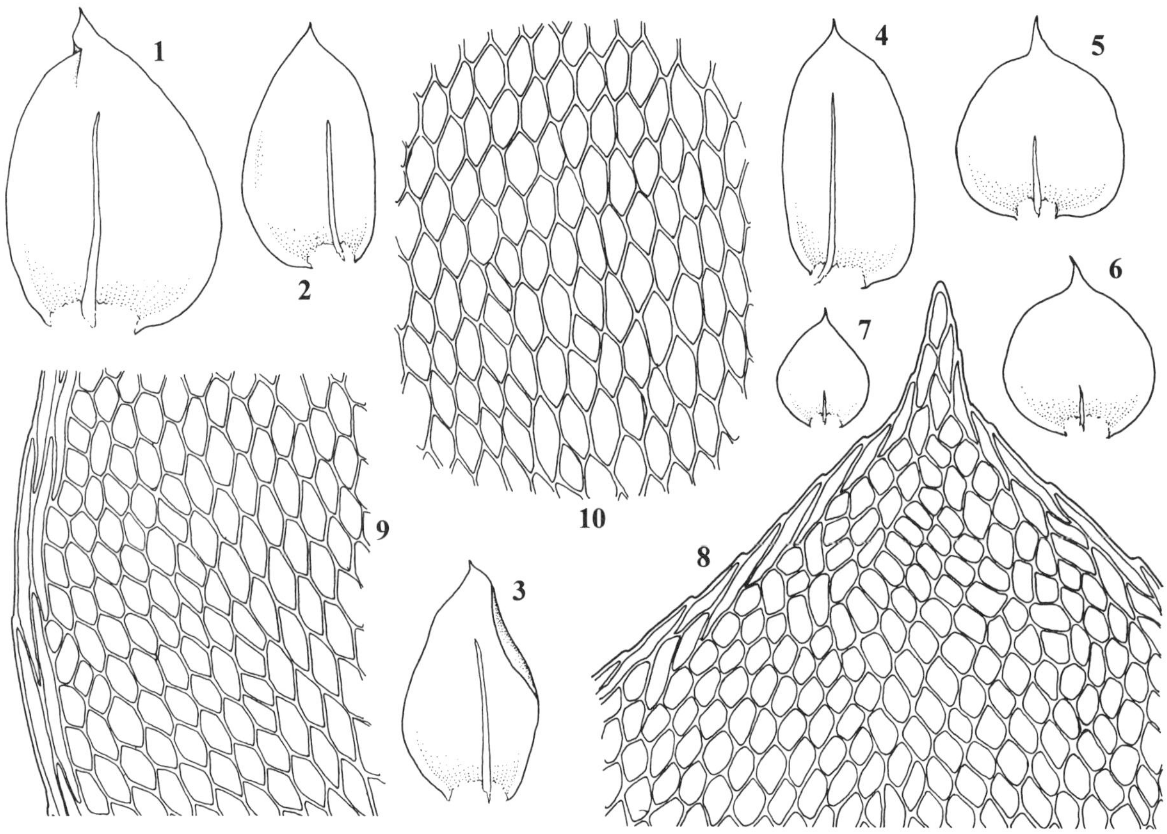

Plate 420. *Hypopterygium tenellum*. 1, lateral leaf of secondary stem (× 28). 2–4, lateral leaves of branch (× 28). 5–7, amphigastria of branch (× 28). 8, apical cells of lateral leaf (× 259). 9, median marginal cells of lateral leaf (× 259). 10, basal cells of lateral leaf (× 259). (1–10, drawn from *Z.-H. Li 74-3220*, SYS.)

7. ***Hypopterygium tibetanum*** Mitt. 西藏孔雀藓 xi-zang kong-que xian
J. Proc. Linn. Soc., Bot., Suppl. 1: 148. 1859. Type. China: Southern Xizang (Tibet), *Thomson 682*.

Primary stems prostrate, rhizoids tomentose; secondary stems erect, 2.5–4.0 cm high. *Lateral leaves* broadly ovate, asymmetric, ca. 1.4 mm × 0.7 mm, acuminate; leaf margins bordered by 1–2 rows of narrowly elongate cells, serrulate at the apex; amphigastria ovate-rounded, symmetric, ca. 0.8 mm × 0.6 mm, cuspidate at the apex; leaf margins bordered by 1–2 rows of narrowly elongate cells, nearly entire; costa ending at the apex. *Setae* 1.4–1.6 cm long, curved in the upper part; capsules inclined, shortly cylindrical, ca. 1.3 mm long, 0.3 mm in diameter, slightly curved; opercula conical at base, rostrate.

Illustrations: No reference.
Habitat: on rocks or moist soil.
Distribution: endemic to China.

This species was published by Mitten (1859) based on a specimen collected from Xizang (Tibet). We have not been able to study the type specimen. P.-C. Chen et al. (1978) and P.-J. Lin and Z.-H. Li (1985) recorded this species in China. There were several specimens that were misidentified as *H. tibetanum*. These specimens belong to other species of *Hypopterygium*.

Uncertain species:

1. *Hypopterygium sinicum* Mitt. 中华孔雀藓 zhong-hua kong-que xian
Trans. Linn. Soc. London, Bot. 3: 169. 1891. Type. China: Hong Kong, *J. C. Bowring s.n.*

Hypopterygium sinicum was reported from Guizhou (Q.-W. Lin 1989; Q.-W. Lin & J. Zhou 1991). Unfortunately, neither the type nor any specimens of this species have been examined. The presence of this species in China remains uncertain.

4. ***Lopidium*** Hook. f. & Wils. 雀尾藓属 que-wei xian shu
Fl. Nov.-Zel. 2: 119. 1854.

Plants slender, creeping. *Secondary stems* erect, simple or regularly pinnately branched, yellowish green or dark green, not glossy; central strand absent. *Lateral leaves* ovate-ligulate, asymmetric, mostly obtuse, mucronate at apex; costa stout, ending below the apex; lamina cells small, oval or rounded, markedly mammillose or unipapillose, walls incrassate, collenchymatous; amphigastria small, triangular or ovate-lanceolate, symmetric, long acuminate toward apex, margins nearly entire. *Monoicous* or autoicous. *Inner perichaetial leaves* long, gradually narrowly acuminate at apex, with oblong leaf base, margins nearly entire; costa strong, ending near leaf apex or almost percurrent; leaf cells hexagonal, walls incrassate, porose. *Setae* short, finely papillose above; capsules small, erect or slightly suberect; annuli absent; peristome double; exostome teeth narrowly lanceolate, papillose above, densely striate below; endostome segments narrowly elongate; basal membrane low; cilia absent. *Calyptrae* small, cucullate. *Spores* small, minutely papillose.

The genus *Lopidium* consists of about sixteen species in the world and generally grows on tree trunks in tropical and subtropical regions. Three species have been known from China.

1. Leaf margins distinctly differentiated throughout.................................2. *L. struthiopteris*
1. Leaf margins only partially differentiated.................................2
 2. Trigones of leaf cells strongly thickened.................................1. *L. nazeense*
 2. Trigones of leaf cells slightly thickened.................................3. *L. trichocladon*

1. ***Lopidium nazeense*** (Thér.) Broth. 东亚雀尾藓 dong-ya que-wei xian
Nat. Pflanzenfam. (ed. 2), 11: 271. 1925. *Hypopterygium nazeense* Thér., Bull. Acad. Int. Géogr. Bot. 19: 17. 1909. Type. Japan: Naze.

Plants yellowish green or dark green, not glossy. *Primary stems* slender, creeping; secondary stems erect, regularly pinnately branched; central strand absent. *Lateral leaves* ligulate, asymmetric, 1.0–2.2 mm × 0.2–0.5 mm, with a short, acute apex; costa reaching the apex; leaf margins nearly entire, often bordered by 1–2 rows of cells in lower parts; laminal cells mostly rounded-oval, usually 7–15 µm wide, distinctly collenchymatous; amphigastria small, ovate-lanceolate, symmetric, with a long acuminate apex; margins nearly entire. *Monoicous* or autoicous. *Inner perichaetial leaves* long, oblong-ovate at leaf base, nearly entire; costa stout, ending at apex or percurrent;

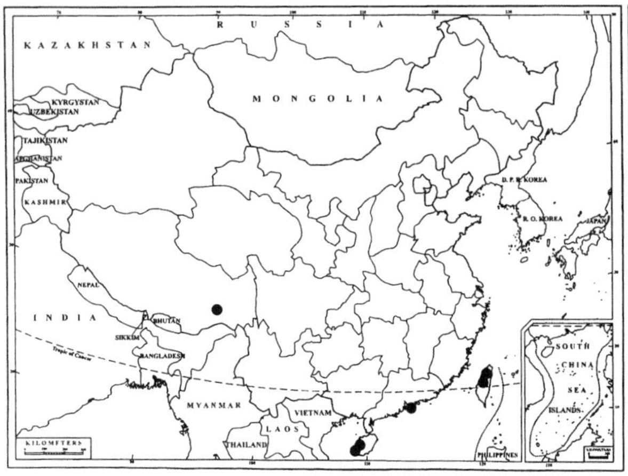

1. *Lopidium nazeense*

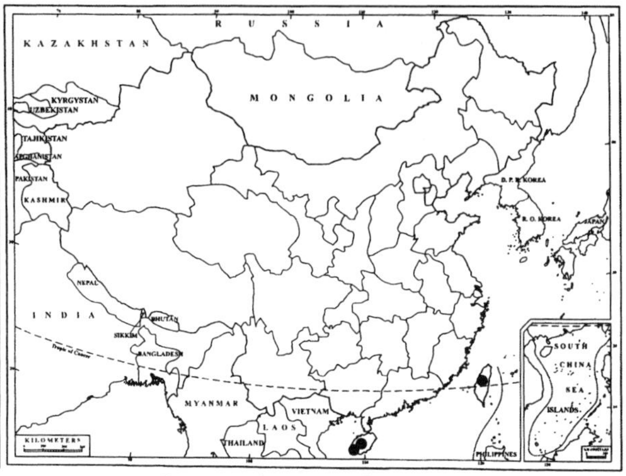

2. *Lopidium struthiopteris*

laminal cells similar to lateral cells. *Setae* short, mammillose or papillose above; capsules small, erect or suberect annuli absent; exostome teeth narrowly lanceolate, papillose above, densely striate in lower parts; endostome segments narrowly elongate; cilia absent. *Calyptrae* small, mitriform. *Spores* small, minutely papillose.

Illustrations: Plate 421, figs. 1–12.
Habitat: on tree trunks, soil or boulders.
Distribution: China and Japan.

Chinese specimens examined: HAINAN: Chang-jiang Co., *Redfearn 35858* (MO); Mt. Jianfengling, *S.-N. Lin 1562* (PE), *P.-C. Chen et al. 330a* (PE). HONG KONG: *M.-L. So 93824f* (BUHK). TAIWAN: Nan-tou Co., *M.-J. Lai 9210* (MO); Tao-yuen Co., *M.-J. Lai & J. E. Lewis 0295* (MO). XIZANG: Mo-tuo Co., *M.-Z. Wang 800301* (PE).

2. ***Lopidium struthiopteris*** (Brid.) Fleisch. 爪哇雀尾藓 zhao-wa que-wei xian

Musci Buitenzorg 3: 1073. 1908. *Hypopterygium struthiopteris* (Brid.) Brid., Bryol. Univ. 2: 716. 1827. *Hypnum struthiopteris* Brid., Muscol. Recent. Suppl. 2: 87. 1812. Type. Réunion: Borbon.

Hypopterygium javanicum (Hampe) Jaeg., Ber. Thatigk. St. Gallischen Naturwiss. Ges. 1874–75: 150. 1876. *Lopidium javanicum* Hampe, Linnaea 38: 672. 1874.

Hypopterygium subpennaeforme Kindb., Hedwigia 40: 282. 1901.

Primary stems creeping; secondary stems erect, slender, yellowish green or dark green, not glossy, mostly regularly branched; central strand absent. *Lateral leaves* ovate-ligulate, asymmetric, 1.0–1.5 mm × 0.3–0.5 mm, acuminate to acute at apex; leaf margins slightly incurved; distinctly bordered, nearly entire, only serrulate at apex; costa stout, reaching the apex; laminal cells irregularly rounded, 11–14 µm wide, cell walls porose, strongly collenchymatous; amphigastria small, ovate-lanceolate, symmetric, 0.6–1.2 mm × 0.2–0.4 mm, long acuminate at apex, entire, slightly bordered. *Dioicous. Inner perichaetial leaves* long slender, oblong-ovate at base, gradually narrowed to an acuminate apex, nearly entire; costa stout, ending at apex or percurrent; lamina cells hexagonal, incrassate, walls porose. *Setae* short, mammillose or papillose above; capsules small, erect or slightly inclined; annuli absent; exostome teeth narrowly lanceolate, papillose above, densely striate in lower parts; endostome segments linear; cilia absent. *Calyptrae* small, mitriform. *Spores* small, finely papillose.

Illustrations: Plate 421, figs. 13–24.
Habitat: on tree trunks.
Distribution: China, Japan, Indonesia, Malaysia, Sir Lanka, Indian Ocean, the Philippines, New Guinea, and New Caledonia.

Chinese specimens examined: HAINAN: Mt. Jianfengling, *P.-C. Chen et al. 142* (PE), *P.-C. Wu et al. 85058* (PE); Mt. Diaoluo, *Z.-H. Li 77-2665, 77-3089* (SYS); Mt. Bawangling, *Redfearn et al. 36229* (IBSC, MO). TAIWAN: Tai-pei Co. (see M.-J. Lai 1976).

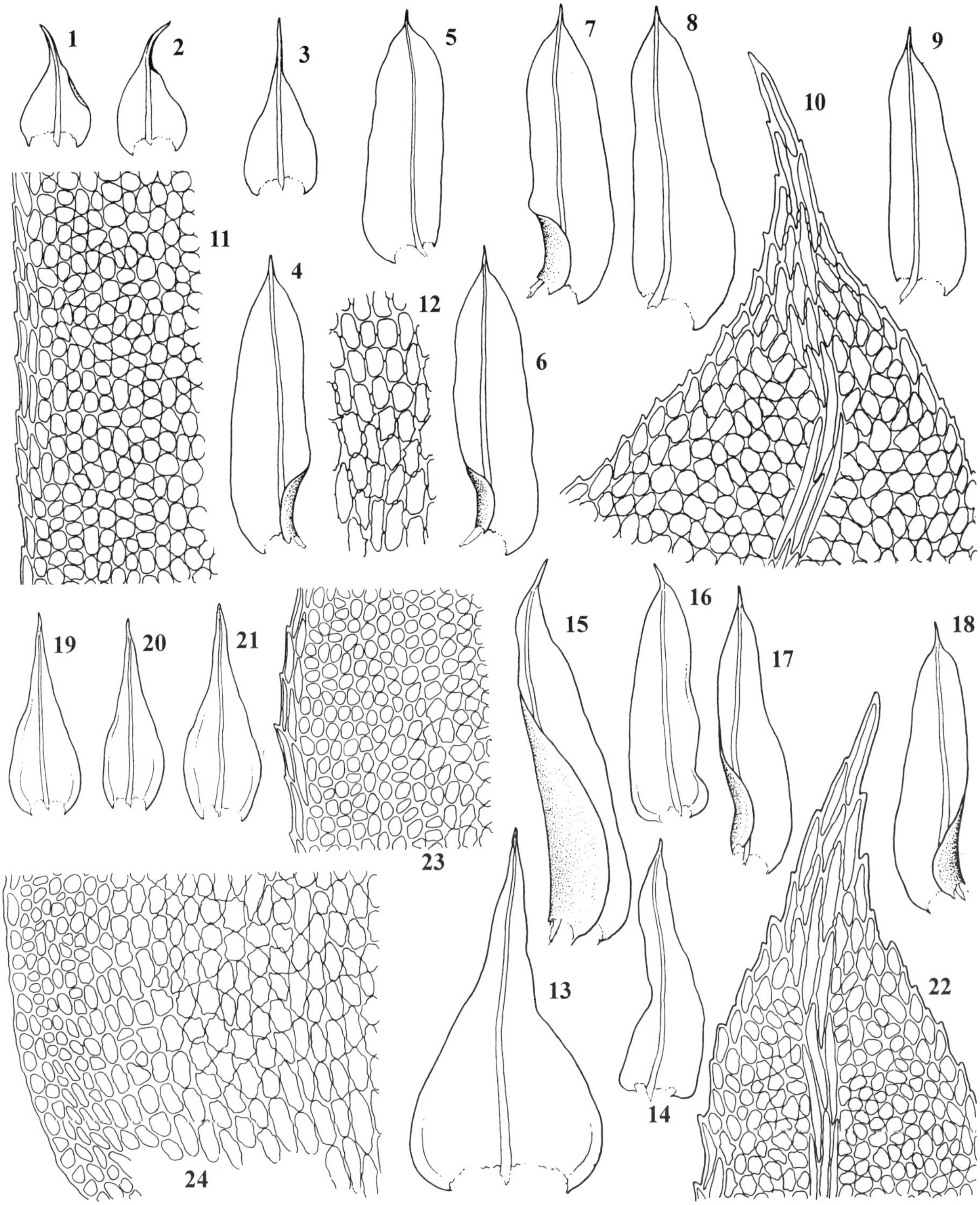

Plate 421. 1–12, *Lopidium nazeense*. 1–3, amphigastria of branch (× 19). 4–9, lateral leaves of branch (× 19). 10, apical cells of lateral leaf (× 173). 11, median marginal cells of lateral leaf (× 173). 12, basal cells of lateral leaf (× 173). 13–24, *Lopidium struthiopteris*. 13–18, lateral leaves (× 37). 19–21, amphigastria (× 37). 22, apical cells of lateral leaf (× 367). 23, median marginal cells of lateral leaf (× 367). 24, basal cells of lateral leaf (× 367). (1–12, drawn from *Iwtasuki & Sharp 1576*, TEN; 13–24, from *Forestry Ministry Exped. Team 3004*, PE.)

3. ***Lopidium trichocladon*** (Bosch & Sande Lac.) Fleisch. 毛枝雀尾藓 mao-zhi que-wei xian
Musci Buitenzorg 3: 1069. 1908. *Hypopterygium trichocladon* Bosch & Sande Lac., Bryol. Jav. 2: 9. Pl. 138. 1861. Type. Indonesia: Java.

Plants yellowish green or dark green, not glossy. *Primary stems* elongate, often appressed; secondary stems pendulous, regularly branched, central strand absent. *Lateral leaves* ligulate, asymmetric, 1.0–2.2 mm × 0.2–0.5 mm, with a short apex; costa reaching the apex; leaf margins nearly entire, often bordered by 1–2 rows of linear cells in lower parts; laminal cells mostly rounded-oval, usually 7–15 µm wide, less porose, slightly collenchymatous; amphigastria small, ovate-lanceolate, symmetric, with elongate acute apex; margins nearly entire. *Monoicous* or autoicous. *Inner perichaetial leaves* elongate, with oblong-ovate bases, gradually narrowed into an acuminate apex, nearly entire; costa stout, ending at apex or percurrent; laminal cells similar to those of lateral leaves. *Setae* short, mammillose or papillose above; capsules small, erect or slightly inclined; annuli absent; exostome teeth lanceolate, papillose above, densely striate below; endostome segments narrow; cilia absent. *Calyptrae* small, mitriform. *Spores* minute, papillose.

Illustrations: Plate 422, figs. 1–14.
Habitat: on tree trunks or boulders.
Distribution: China, Japan, Vietnam, Thailand, Kampuchea, Laos, Indonesia, Malaysia, and the Philippines.

Chinese specimens examined: HAINAN: Ling-shui Co., *Reese 17574*, *17944* (MO); Mt. Diaoluo, *B.-J. Lin & L. Zhang 1078* (IBSC); Mt. Jianfengling, *B.-J. Lin 3841* (IBSC). TAIWAN: Ilan Co., Tailuge, *Z. Iwatsuki & A. J. Sharp 1576* (TEN); YUNNAN: Jing-hong Co., *Redfearn et al. 33684* (MO).

This species is very closely related to *Lopidium nazeense*, but differs in having less porose and slightly collenchymatous leaf cells.

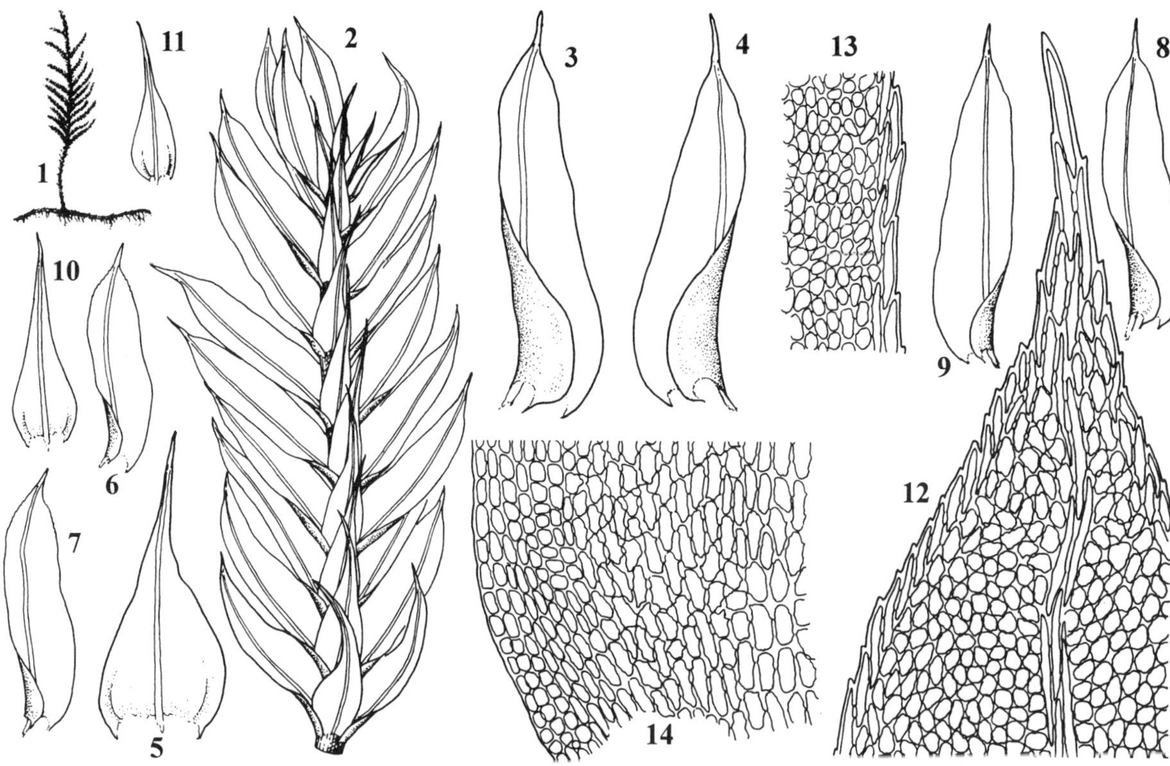

Plate 422. *Lopidium trichocladon*. 1, plant (× 1). 2, branch (× 28). 3–4, lateral leaves of primary stem (× 35). 5, amphigastrium of primary stem (× 35). 6–9, lateral leaves of branch (× 35). 10–11, amphigastria of branch (× 35). 12, apical cells of lateral leaf (× 218). 13, median marginal cells of lateral leaf (× 218). 14, basal cells of lateral leaf (× 218). (1–14, drawn from *B.-J. Lin et al. 850*, IBSC.)

THELIACEAE

鳞藓科 lin xian ke

by Gao Chien and Fu Xing

Plants slender, densely tufted, yellowish green, not glossy or weakly glossy. *Stems* creeping, apices ascending, irregularly pinnately branched or irregularly densely branched, branches often julaceous. *Leaves* homogeneous, cymbiform or rounded-cucullate, apices obtuse or apiculate; leaf margins unistratose, incurved, entire or ciliated or serrate; costae short and weak, single or branched, inconspicuous or absent; leaf cells oval or oblong-elliptic, often grossly unipapillose. *Dioicous. Male inflorescences* terminally at the apices of branches; female inflorescences growing on stems. *Perichaetial leaves* differentiated, mostly broadly lanceolate. *Setae* short or elongate, slender, smooth, red or reddish brown; capsules erect, oblong-ovoid, rarely curved, erecto-patent or horizontal; stomata rare, phenotype; peristome double; exostome teeth lanceolate, combined at base, lightly yellow above, keeled, with zig-zag median line, lamellae conspicuous, sometimes bordered, lamellae high on inner side; endostome segments nearly as long as exostome teeth, yellowish or hyaline, linear, perforate, or sometimes endostome rudimentary; basal membrane high; cilia none or short to elongate; opercula conic, with short or long beaks. *Calyptrae* cucullate, smooth. *Spores* finely roughened or nearly smooth.

1. Plants irregularly pinnately branched; leaves erecto-patent, oblong-ovate or ending with long hairs; costae forked or absent; capsules inclined .... 1. *Fauriella*
1. Plants rarely branched; leaves imbricate, ovate-cucullate, apiculate or obtuse; costa single, inconspicuous; capsules erect .... 2. *Myurella*

1. ***Fauriella*** Besch. 粗疣藓属 cu-you xian shu

Ann. Sci. Nat., Bot., ser. 7, 17: 363. 1893.

Plants slender, delicate, lightly yellowish green or light yellow, weakly glossy. *Stems* creeping or ascendant, irregularly pinnately branched, with loosely tufted rhizoids; paraphyllia rare, linear or nearly foliose. *Leaves* erect, ovate or cochleariform, concave, sometimes ciliate at apex, not decurrent at base; margins involute, grossly dentate or serrate; costa absent or shortly forked; leaf cells oblong-hexagonal, unipapillose, basal cells narrower; alar cells often quadrate. *Dioicous. Perichaetial leaves* sheathing at base. *Setae* slender and elongate; capsules small, ovoid, erect to suberect, or usually horizontal when old; opercula conic; annuli differentiated; peristome double; exostome teeth lanceolate, yellowish, keeled, with zig-zag median line, with conspicuous lamellae on both sides; endostome segments yellowish, nearly as long as the teeth, perforate; basal membrane high; cilia 3, shorter than segments. *Spores* not seen.

There are four species of *Fauriella* in the world. Brotherus (1924, 1929) reported 2 species, *F. robustiuscula* Broth. and *F. tenerrima* Broth., from China. Three species are presently recognized in this treatment.

1. Plants rather large; leaves lanceolate, acuminate at apex; leaf cells linear, often finely papillose at the front corners .... 1. *F. robustiuscula*
1. Plants rather small; leaves rounded ovate, shortly acute or apiculate at apex; leaf cells rhomboidal, often grossly papillose in the centers 2
    2. Plants slender, delicate; leaves remote, scale-like, abruptly apiculate; leaf cells shortly oblong ..... 2. *F. tenerrima*
    2. Plants slightly rigid; leaves densely imbricate, acute; leaf cells widely elliptic .... 3. *F. tenuis*

1. ***Fauriella robustiuscula*** Broth. 大粗疣藓 da cu-you xian

Akad. Wiss. Wien Sitzungsber., Math.-Naturwiss. Kl., Abt. 1, 133: 576. 1924. Type. China: Sichuan (Setschwan), Yan-yuan Co., alt. 2750 m, *Handel-Mazzetti 2764* (holotype H).

Plants somewhat stout, in tufted cushions, lightly yellowish green, not glossy. *Stems* creeping, densely pinnately branched, branches short, erect, julaceous, obtuse at the apex; rhizoids dense, brown. *Leaves* loosely imbricate when dry, erect-spreading when moist, concave, cochleariform or oblong-ovate, tapering to a slender apex; margins involute, serrate; costa absent; leaf cells linear, unipapillose; alar cells quadrate, smooth. *Dioicous. Perichaetial leaves* sheathing at base, acuminate above, serrate near apex. *Setae* slender, ca. 1.5 cm long, reddish brown; capsules erect when young, inclined or horizontal when mature; opercula conic.

Illustrations: Plate 423, figs. 1–9.
Habitat: usually on tree trunks in forests.
Distribution: endemic to China.

Chinese specimen examined: see the type citation above.

2. ***Fauriella tenerrima*** Broth. 小粗疣藓 xiao cu-you xian
Akad. Wiss. Wien Sitzungsber., Math.-Naturwiss. Kl., Abt. 1, 131: 217. 1923. Type. China: Hunan, Yunshan, near Wu-gang Co., alt. 1150 m, *Handel-Mazzetti 12199* (holotype H).

Plants slender, light green or yellowish, somewhat glossy, in small patches. *Stems* prostrate, loosely irregularly branched, branches not in equal length, remotely foliate at apex. *Leaves* ovate, concave, imbricate when dry, erect-spreading when moist, shortly acute or apiculate at apex; margins plane, serrate; costa absent; leaf cells rhomboidal, slightly thick-walled, grossly papillose on the back, basal cells short, rectangular; alar cells quadrate. *Sporophytes* not seen.

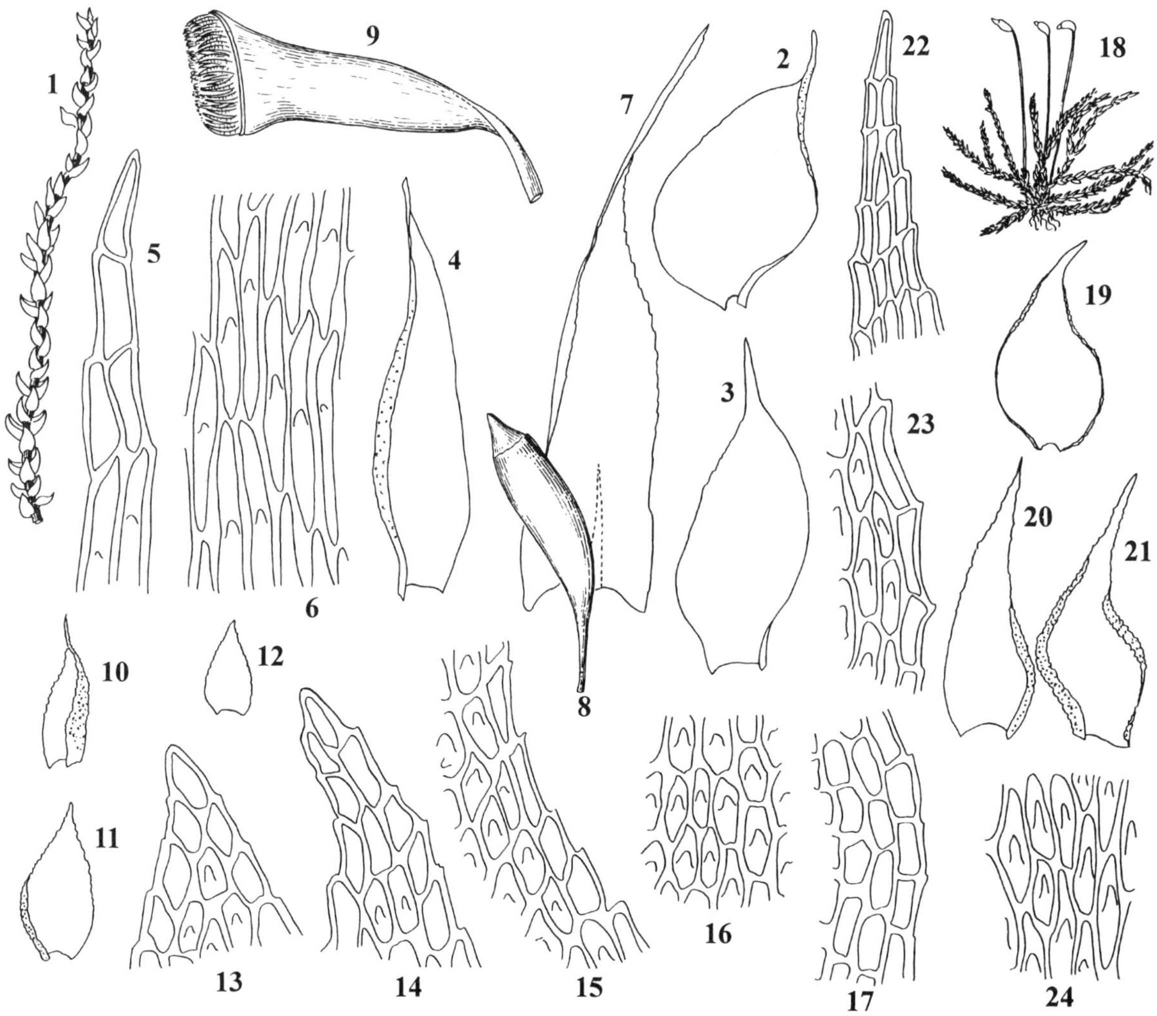

Plate 423. 1–9, *Fauriella robustiuscula*. 1, branch (× 8). 2–4, leaves (× 78). 5, apical leaf cells (× 320). 6, median leaf cells (× 320). 7, perichaetial leaf (× 78). 8, capsule (× 28). 9, opened capsule (× 35). 10–17, *Fauriella tenerrima*. 10–12, leaves (× 78). 13–14, apical leaf cells (× 320). 15, upper leaf cells (× 320). 16, median leaf cells (× 320). 17, median marginal leaf cells (× 320). 18–24, *Fauriella tenuis*. 18, plant habit (× 5). 19–21, leaves (× 78). 22, apical leaf cells (× 320). 23, upper leaf cells (× 320). 24, median leaf cells (× 320). (1–9, drawn from *Handel-Mazzetti 2764*, H; 10–17, from *Handel-Mazzetti 12199*, H; 18–24, from *C. Gao & G.-C. Zhang 8950*, IFSBH.)

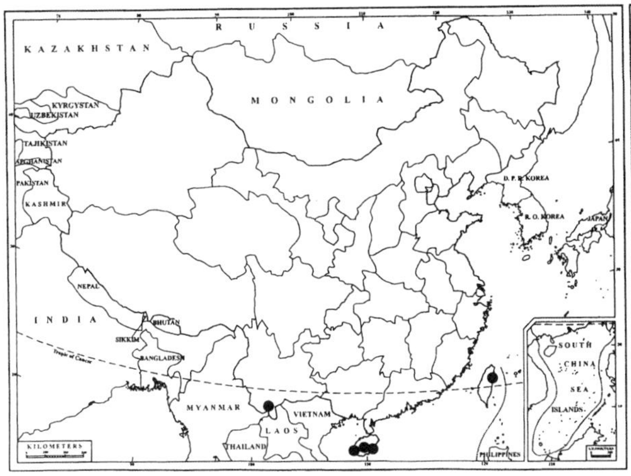

3. *Lopidium trichocladon*

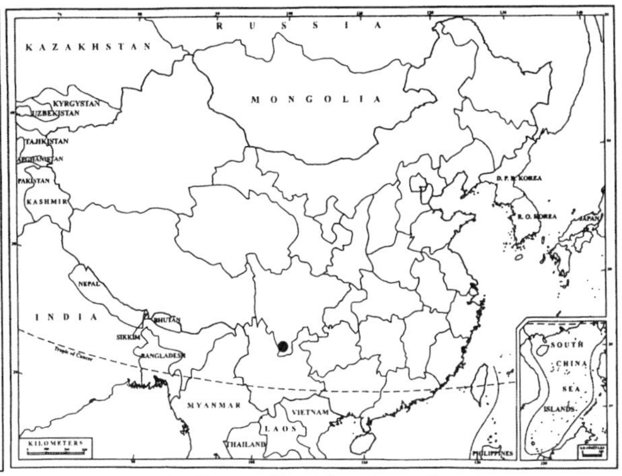

1. *Fauriella robustiuscula*

Illustrations: Plate 423, figs. 10–17.
Habitat: often on rock surfaces, tree bark or decaying logs in forests; alt. 1100–1200 m.
Distribution: China and Japan.

Chinese specimens examined: ANHUI: Huangshan (Mt.), *P.-C. Chen 7449* (PE). GUANGXI: Long-sheng Co., *P.-C. Wu & Y.-X. Lin 811* (PE). GUIZHOU: Gui-yang City, *Guiyang Normal Univ. 77, 166* (PE). HUNAN: Hengshan (Mt.), *G.-Z. He 19* (PE). TAIWAN: Hsin-chu Co., *Shevock 14396* (MO); Nan-tou Co., *Shevock 14454* (MO). ZHEJIANG: Lin-an Co., *P.-C. Wu 24877* (MO).

This species is very close to *F. tenuis* (Mitt.) Card., but differs from the latter by its rather slender plants, ovate and loosely arranged leaves with a short apex, and by having rhomboidal leaf cells with gross papillae.

3. ***Fauriella tenuis*** (Mitt.) Card. in Broth. 粗疣藓 cu-you xian
Nat. Pflanzenfam. (ed. 2), 11: 282.1925. *Heterocladium tenue* Mitt., Trans. Linn. Soc. London, Bot. 3: 176. 1891. Type. Japan, Nikko, *Bisset s.n.*

Plants light green or yellowish green, not glossy, in small dense cushions. *Stems* prostrate, irregularly branched, branches slender and unequal in length, obtuse at apex. *Leaves* imbricate, rounded ovate, concave, acuminate; margins involute, serrate; costa absent or shortly double; leaf cells rounded rhomboidal, 18–25 µm × ca. 5 µm, thick-walled, grossly unipapillose; alar cells quadrate. *Dioicous*. *Setae* 1.0–1.5 cm long, reddish brown; capsules small, suberect or horizontal, pendulous after broken; opercula conic; annuli not differentiated; peristome double; exostome teeth lanceolate, keeled, with zig-zag median line, papillose above, densely striate at lower part; endostome segments as long as exostome teeth; basal membrane high; cilia 1–2. *Calyptrae* cucullate. *Spores* 10–13 µm in diameter.

Illustrations: Plate 423, figs. 18–24.
Habitat: on tree trunks or rotten logs in forests; alt. 1150–1700 m.
Distribution: China and Japan.

Chinese specimens examined: ANHUI: Huangshan (Mt.), *P.-C. Chen 6944, 6947* (MO, PE). GUANGXI: Rong-shui Co., *Long & Zhang GSY026d* (MO). JILIN: Wang-qing Co., *C. Gao & G.-C. Zhang 895* (IFP). TAIWAN: Hsin-chu Co., *S.-H. Lin s.n.* [Bryophytes of Taiwan Exsiccatae ser. 4, no. 165] (MO); Ilan Co., *C.-C. Chuang* 2092 (MO); Miao-li Co., *Shevock 14336* (MO); Ping-tung Co., *S.-H. Lin s.n.* [Bryophytes of Taiwan Exsiccatae ser. 2, no. 65] (MO). ZHEJIANG: Long-quan Co., *P.-C. Wu 201* (PE).

This species is recognized by its hyaline, long leaf apices and narrowly rhomboidal leaf cells.

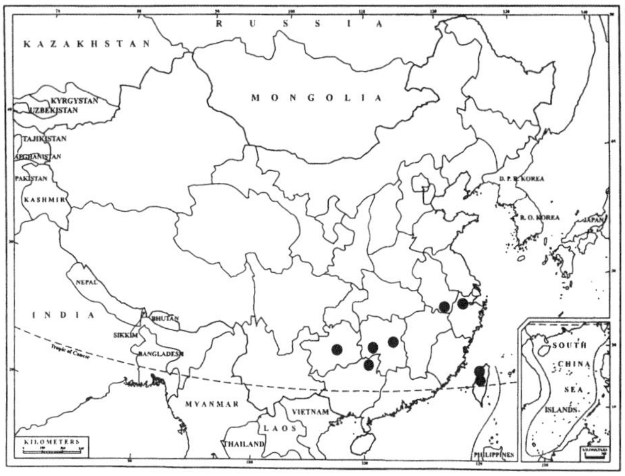

2. *Fauriella tenerrima*

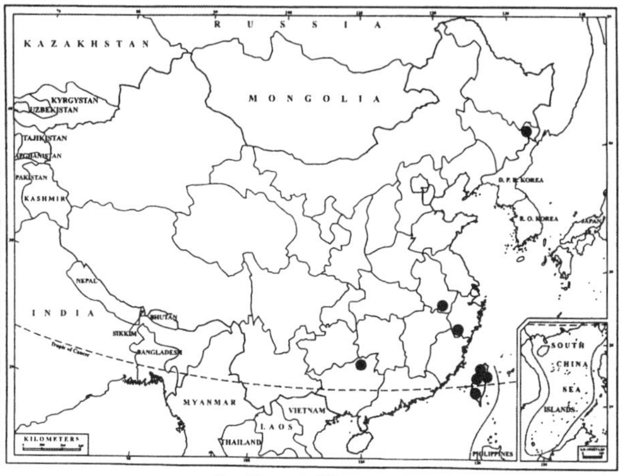

3. *Fauriella tenuis*

2. ***Myurella*** Bruch & W. P. Schimper in B.S.G. 小鼠尾藓属 xiao shu-wei xian shu
Bryol. Eur. 6: 39 (fasc. 52–54, Mon. 1). 1853.

Plants slender, in cushions or loosely caespitose, green or dark green. *Stems* creeping, densely inclined or erect, irregularly branched, forkedly branched or caespitosely branched; often with bundles of rhizoids; branches obtuse at apex, often flagellate; paraphyllia absent. *Leaves* rounded ovate, imbricately arranged, abruptly acute, acuminate or rounded at apex; leaf margins incurved, entire or irregularly serrate; costa single, forked or absent; leaf cells minute, slightly thick-walled, inconspicuously pitted, rounded or rhomboidal, basal cells rectangular or quadrate. *Dioicous*. *Inner perichaetial leaves* reddish brown, lanceolate, acuminate above, margin plane, serrate; costae absent; cells rhomboidal. *Setae* 1–2 cm long; capsules ovoid or shortly cylindrical, yellowish brown, erect, curved after spores disposal; annuli differentiated; exostome teeth lanceolate, yellow or lightly yellowish, hyaline at the apex, keeled, with zig-zag median line, lamellae conspicuous on inner surfaces; endostome segments linear-lanceolate, finely papillose, nearly as long as the teeth; basal membrane high; cilia short or elongate, but shorter than segments; opercula conic, with a short beak. *Calyptrae* minute, cucullate.

Brotherus (1925) recognized four species of *Myurella* in the world. P.-C. Chen et al. (1978) reported 3 species: *M. julacea* (Schwaegr.) B. S. G., *M. tenerrima* (Brid.) Lindb., and *M. sibirica* (C. Müll.) Reim. with 2 synonyms: *M. sinensi-julacea* C. Müll. = *M. julacea* (Schwaegr.) B. S. G. and *M. gracilis* Lindb. = *M. sibirica* (C. Müll.) Reim. from China. Recently, J.-X. Lou and P.-C. Wu (1980) published *M. brevicosta* J.-X. Lou & P.-C. Wu [= *Glossadelphus prostratus* (Dozy & Molk.) Fleisch.] from Xizang. Three species are recognized here for China.

1. Plants densely caespitose, in small cushions; leaves densely imbricate, rounded or oblong-rounded, obtuse, rarely acute at apex .... 1. *M. julacea*
1. Plants loosely caespitose, not in cushions; leaves loosely imbricate, cucullate, shortly ciliate at apex 2
  2. Leaf cells finely papillose on the back; leaf margins finely serrulate; leaf apex short........ 3. *M. tenerrima*
  2. Leaf cells grossly papillose on the back; leaf margins grossly serrate; leaf apex 1–4 cells long........ 2. *M. sibirica*

1. ***Myurella julacea*** (Schwaegr.) Schimp. in B.S.G. 小鼠尾藓 xiao shu-wei xian
Bryol. Eur. 6: 41. (fasc. 52–54. Mon. 3). 1853. *Leskea julacea* Schwaegr. in Schultes, Reise Glockner 2: 363. 1804. Type. France. *Villars s.n.*
Myurella gracillima Kindb. ex Par., Ind. Bryol. Suppl. 253. 1900.
M. sinensi-julacea C. Müll., Nuovo Giorn. Bot. Ital., n. ser. 5: 206. 1898. Type. China: Shaanxi (Schen-si), in monte Kuan-tou-san, Mt. Qinling, *Giraldi 2283d* (isotype H).

Plants densely caespitose, in cushions, light green, weakly glossy, yellowish green when dry. *Stems* erecto-patent, branches erect or erecto-patent, stoloniform; rhizoids in tufts. *Leaves* rounded ovate or oblong, deeply concave, 0.4–0.5 mm × 0.30–0.35 mm, obtusely rounded or apiculate at apex; margins incurved, smooth or serrulate; costa inconspicuous, or short and double; upper leaf cells rhomboidal or rounded-oblong, lower leaf cells short,

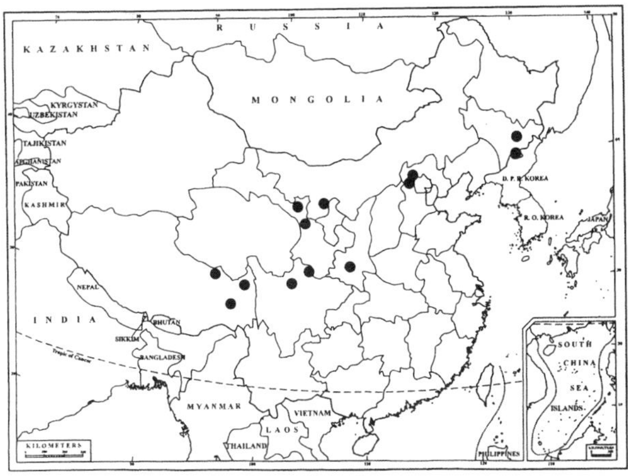

1. *Myurella julacea*

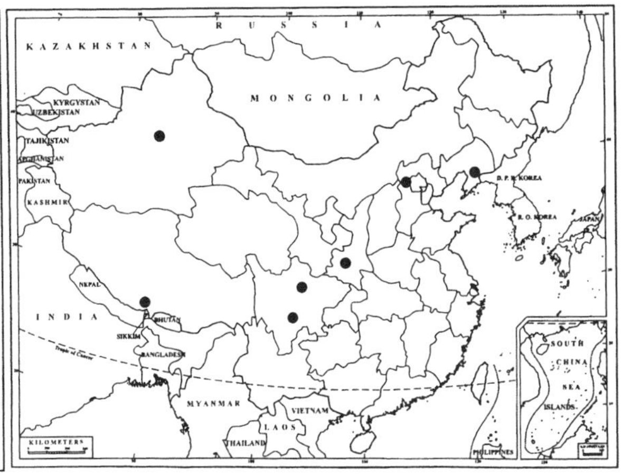

2. *Myurella sibirica*

rectangular or rounded-quadrate, smooth or finely papillose; alar cells quadrate. *Gemmae* axillary, clavate, tufted. *Dioicous*. *Perichaetial leaves* narrowly oblong, acute above, light gray. *Setae* up to 1.5 cm long, reddish brown; capsules shortly cylindrical or cylindrical; opercula shortly conic; exostome teeth lanceolate, conspicuously striate; endostome segments well developed, nearly as long as the teeth; basal membrane high; cilia 2–3. *Spores* 10–14 µm in diameter, minutely papillose.

Illustrations: P.-C. Wu (ed.) 2001 (Pl. 37, figs. 14–20).
Habitat: on sandy stones or moist rocks in cold high land; alt. 1240–4900 m.
Distribution: China and widely in northern temperate regions.

Chinese specimens examined: GANSU: He-xi Co., *S.-Z. Hu 1837* (MO, PE); Tian-zhu Co., *Z.-D. Zhu 75* (PE). HEBEI: Long-guan Co., *Hebei Exped. Team 4533* (PE); Mt. Xiaowutai, *C.-W. Wang 1227* (PE). JINLIN: An-tu Co., *C. Gao 1195b* (IFSBH); Chang-bai Co., *C. Gao 7551* (IFSBH). NEI MONGOL: Mt. Helan, *Z.-G. Tong 20106* (PE). SICHUAN: Ma-er-kang Co., *Allen 6809* (MO); Nan-ping Co., *Redfearn 35377c* (MO). XIZANG: Ba-qing Co., *K.-Y. Zhao 49d* (PE); Bo-mi (Zha-mu) Co., *S.-G. Wu 1124* (HKAS); Chang-du Co., *M. Zang 5304a* (HKAS).

Morphologically, this species is close to *Myuroclada maximoviczii*; however, it differs from the latter by its small plants and indistinct, short costae.

2. ***Myurella sibirica*** (C. Müll.) Reim. 刺叶小鼠尾藓 ci-ye xiao shu-wei xian
Hedwigia 76: 292. 1937. *Hypnum sibiricum* C. Müll., Syn. Musc. Frond. 2: 418. 1851. Type. Russia: Siberia.
Myurella gracilis Lindb., Meddeland. Soc. Fauna Fl. Fenn. 13: 254. 1886, *nom. illeg*.

Plants yellowish green, creeping, in loose tufts. *Stems* ascendant above, flagelliform, irregularly loosely branched; rhizoids tufted. *Leaves* cochleariform, concave, suddenly narrowed to a long apiculus or a short, linear acumen, up to 0.5 mm × 0.30–0.45 mm, constricted at base; margins irregularly dentate, incurved; costa short and double, forked or often indistinct; leaf cells hexagonal or rounded rhombic above, shortly elliptic at base; alar cells rectangular or rounded quadrate, each cell with single papilla. *Dioicous*. *Setae* slender, reddish brown; capsules oblong-ovoid, erect; opercula conic.

Illustrations: P.-C. Wu (ed.) 2001 (Pl. 37, figs. 7–13).
Habitat: usually in bogs or on moist sandy soil in cold high land.
Distribution: China, Japan, Russia, Europe, and North America.

Chinese specimens examined: HEBEI: Mt. Xiaowutai, *Q.-W. Wang 623*, *821* (MO, PE); Mt. Xiling, *Q.-W. Wang 1437*, *1519* (PE). LIAONING: Feng-cheng Co., *C. Gao & G.-C. Zhang 8459*, *8462* (IFSBH). SHAANXI: Mt. Taibai, *Z.-P. Wei 10432* ((PE). SICHUAN: Mao-wen Co., *Redfearn 35542* (MO); Mt. Omei, *C. Gao 19543* (IFSBH). XINJIANG: He-jing Co., *S. He 951185* (MO). XIZANG: Ding-jie Co., *K.-Y. Lang 1157* (PE).

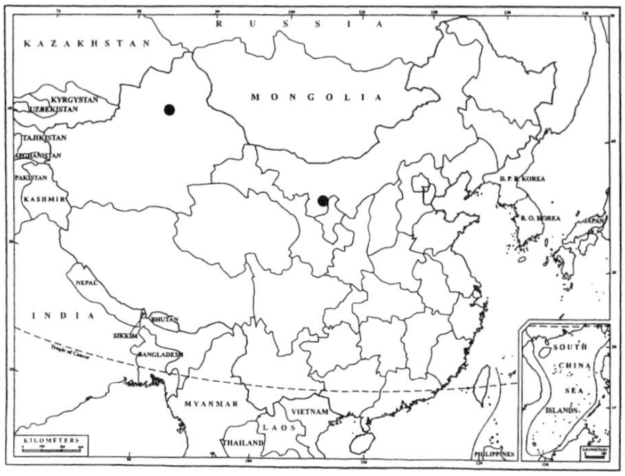

3. *Myurella tenerrima*

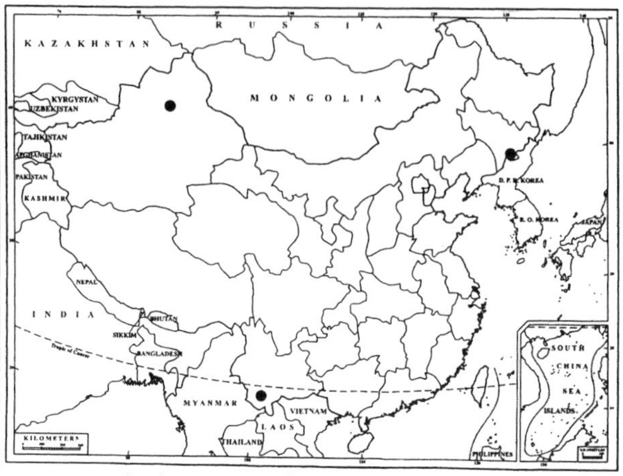

1. *Anacamptodon amblystegioides*

This species is very similar to *Myurella tenerrima* (Brid.) Lindb.; however, differs from the latter by its leaf cells with gross papillae.

3. ***Myurella tenerrima*** (Brid.) Lindb. 柔叶小鼠尾藓 rou-ye xiao shu-wei xian

Musci Scand. 37. 1879. *Pterigynandrum tenerrimum* Brid., Muscol. Recent. Suppl. 4: 132. 1819 [1818]. Type. Southern Switzerland: "in Helvetiae montibus."

Hypnum moniliforme Wahlenb. var. *apiculatum* Somm., Fl. Lapp. Suppl. 62. 1826. *Isothecium apiculatum* (Somm.) Hüb., Muscol. Germ. 598. 1833.

Plants green or light green, weakly glossy, loosely caespitose . *Stems* creeping, slender, julaceous; irregularly branched, often curved, with rhizoids at base. *Leaves* imbricate, cochleariform, concave, ca. 0.6 mm × 0.3 mm, broadly ovate, long or shortly lanceolate, often curved at apex; margins incurved, serrate; costa absent or with inconspicuous single costa or short forked costae; leaf cells short and broad, hexagonal, ca. 9–10 µm in diameter, irregularly shaped at lower part; alar cells quadrate or rectangular, slightly thick-walled, each cell with a single gross papilla on the back. *Perichaetial leaves* widely lanceolate, sheathing at base, yellowish brown, serrate above. *Setae* 1.0–1.5 cm long, reddish brown; capsules erect, obovoid or shortly cylindrical; annuli consisting of 2 rings of cells, deciduous. *Spores* 12–16 µm in diameter, finely papillose.

Illustrations: P.-C. Wu (ed.) 2001 (Pl. 37, figs. 1–6).
Habitat: on sandy soil or moist rocks in cold highland.
Distribution: China, Russia, Europe, and North America.

Chinese specimens examined: NEI MONGOL: Mt. Helan, *Z.-G. Tong 1293* (PE). QINGHAI: *Q.-C. Wu 276c* (PE). XINJIANG: Tianshan (Mt.), *R.-C. Qing 3597b* (PE).

This species is recognized by its slender plants with twisted and julaceously arranged leaves.

FABRONIACEAE

碎米藓科 sui-mi xian ke

by Gao Chien and Fu Xing

Plants soft or slender, yellowish green or bright green, mostly glossy. *Stems* creeping, 1– 2 irregularly pinnately branched, branches erect patent; in cross section central strand absent, cortex consisting of thin-walled cells, epidermal cells slightly thick-walled; rhizoids red, growing on stems; paraphyllia rare. *Leaves* dense, appressed when dry, erect-spreading when moist, rarely secund, ovate or ovate-lanceolate, slightly concave, not plicate, mostly long-acuminate at apex, decurrent at base; margins plane, entire or serrate or ciliate; costa single, short and weak, rarely absent; leaf cells mostly rectangular, smooth, thin-walled; alar cells quadrate or flat quadrate. *Dioicous* or monoicous. *Female inflorescences* growing on lateral branches, paraphyses linear or not developed, regularly arranged. *Inner perichaetial leaves* similar to stem leaves, rarely sheathing at base, ciliate at apex. *Setae* slender and elongate; capsules exserted, erect, ovoid or cylindrical, apophysis short and thick, wrinkled when dry, constricted near mouth; exothecial cells lax, thin-walled, stomata on apophysis; annuli persistent, rarely deciduous; peristome deeply inserted in capsule mouth, single or double; exostome teeth in pair, lanceolate, loosely striate, without lamellae inside, rarely exostome absent; endostome segments narrowly lanceolate or absent; basal membrane low or high; opercula conic, obtuse or rostrate. *Calyptrae* small, cucullate, smooth, rarely remotely ciliate. *Spores* minute.

Brotherus (1925) treated four subfamilies and 16 genera in this family. P.-C. Chen et al. (1978) recognized three subfamilies and 8 genera in China. In this treatment, we include 10 genera in the Fabroniaceae.

1. Plants rather stout, not glossy or somewhat glossy, regularly pinnately branched; leaf cells short, hexagonal or elliptic, alar cells inconspicuous, subquadrate 2
1. Plants rather delicate, glossy, irregularly pinnately branched; leaf cells rather large, rhomboidal, alar cells conspicuous, in several rows of rectangular cells 4
    2. Leaves without costa .... 10. *Schwetschkeopsis*
    2. Leaves with single costa, ending above midleaf 3
3. Exostome teeth densely striate .... 4. *Helicodontium*
3. Exostome teeth papillose .... 9. *Schwetschkea*
    4. Leaves ecostate or costa inconspicuous; peristome double, endostome very rudimentary 5
    4. Leaves single costate; peristome single layer or absent or double, endostome relatively well developed 6
5. Leaves shortly piliferous; endostome segments virtually lacking .... 3. *Habrodon*
5. Leaves long piliferous; endostome rudimentary .... 5. *Iwatsukiella*
    6. Peristome single or absent 7
    6. Peristome double 8
7. Leaf margins serrate or ciliate .... 2. *Fabronia*
7. Leaf margins mostly entire, only serrulate at apex .... 7. *Levierella*
    8. Leaf margins involute, serrate near leaf apex .... 8. *Rozea*
    8. Leaf margins plane, entire 9
9. Exostome teeth widely lanceolate, papillose; endostome segments linear .... 1. *Anacamptodon*
9. Exostome teeth narrowly lanceolate, striate; endostome segments keeled .... 6. *Juratzkaeella*

1. ***Anacamptodon*** Brid. 反齿藓属 fan-chi xian shu

Muscol. Recent. Suppl. 4: 136. Pl. 2: f. 11. 1819 [1818].

Plants slender, yellowish green or green, not glossy or weakly glossy, in lax tufts. *Stems* prostrate, apex ascending, with many rhizoids; simple or rarely branched, branches short; branches erect or ascending, densely foliate. *Leaves* erect, often secund, ovate or ovate-lanceolate, concave, acuminate or suddenly narrowed to a long apex; margins plane, entire or toothed; costa rigid, ending above midleaf, rarely percurrent; upper leaf cells green, oblong or oblong-hexagonal; alar cells quadrate. *Autoicous*. *Perichaetial leaves* lanceolate, acuminate, not sheathing at base; costa slender. *Setae* slender, erect, purple red or dark brown, twisted when dry; capsules erect, ovoid, symmetrical, apophysis short and distinct, constricted below capsule mouth; annuli large and broad, thin-walled, persistent; peristome double; exostome teeth often in pairs, erect when moist, reflexed when dry, widely lanceolate, brown, smooth or finely papillose, with lamellae inside; endostome segments shorter than teeth, linear, brown, nearly smooth; basal membrane absent; opercula conic, with a straight or oblique beak. *Calyptrae* cucullate.

There are about 10 species of this genus in the world. P.-C. Chen et al. (1978) reported 4 species of *Anacamptodon* in China, including *A. amblystegioides* Card., *A. fortunei* Mitt., *A. sublatidens* Card. = *A. latidens* (Besch.) Broth. and *A. subulatus* Broth. = *A. latidens* (Besch.) Broth. Three species are recognized in this treatment.

1. Leaves erect; costa stout, percurrent to shortly excurrent; leaf marginal cells bistratose .... 2. *A. fortunei*
1. Leaves mostly secund; costa slender, extending above the midleaf, but ending below the apex; leaf marginal cells unistratose 2
    2. Plants rather rigid; leaf costa ending in the leaf apex .... 1. *A. amblystegioides*
    2. Plants slender; leaf costa ending near the midleaf .... 3. *A. latidens*

1. ***Anacamptodon amblystegioides*** Card. 柳叶反齿藓 liu-ye fan-chi xian
Bull. Soc. Bot. Genève ser. 2, 3: 279. 1911. Type. Japan: Pref. Nagano, Agematsu, *Faurie 3482*.

Plants soft and slender, yellowish green or green, in flat patches. *Stems* prostrate, apex ascending, irregularly branched, branches short. *Leaves* not appressed when dry, erect-spreading when moist, widely lanceolate, tapering to a long acuminate apex; leaf margins plane, entire or finely serrulate; costa thick, vanishing below the leaf apex; leaf cells oblong-elliptic, shorter near the leaf base; alar cells quadrate, in several rows. *Capsules* small, shortly cylindrical; opercula conic with a short beak; peristome features the same as for the genus.

Illustrations: Plate 424, figs. 1–10.
Habitat: on tree trunks in broad-leaved forests; alt. 1360–3600 m.
Distribution: China and Japan.

Chinese specimens examined: JILIN: Mt. Changbai, *Z.-W. Ao 4696* (IFSBH). XINJIANG: Fu-kang Co., *J.-C. Zhao 325* (IFSBH). YUNNAN: Lancanjing (Doker-la), *Handel-Mazzetti 8173* (H).

2. ***Anacamptodon fortunei*** Mitt. 华东反齿藓 hua-dong fan-chi xian
J. Proc. Linn. Soc., Bot. 8: 152. 1864. Type. China: Zhejiang (Chekiang), *Fortune s.n.*
Anacamptodon japonicus Broth., Öfvers. Förh. Finska Vetensk.-Soc. 62A(9): 29. 1921.

Plants slender, yellowish green or green, in flat patches. *Stems* ascending, irregularly pinnately branched, branches dense, erect, up to 3 mm long. *Stem leaves* and branch leaves similar, but the latter slightly larger than the former, appressed when dry, slightly secund to one side, 0.8–1.0 mm long, widely lanceolate, gradually narrowed to an acute apex; margins entire, bistratose; costa thick, percurrent or excurrent; upper and median leaf cells oblong-elliptic or hexagonal, 20–30 µm wide, minutely papillose; alar cells quadrate. *Autoicous*. *Setae* 5–7 mm long, reddish brown; capsules ovoid, erect; peristome double; exostome teeth lanceolate, finely papillose; endostome linear. *Spores* 10–13 µm in diameter.

Illustrations: Plate 424, figs. 11–17.
Habitat: on tree trunks in broad-leaved forests.
Distribution: China and Japan.

Chinese specimen examined: HEBEI: Mt. Wu-ling, *G.-C. Zhang C9* (IFSBH).

Anacamptodon fortunei is characterized by having bistratose leaf margins and stout, percurrent to shortly excurrent costa.

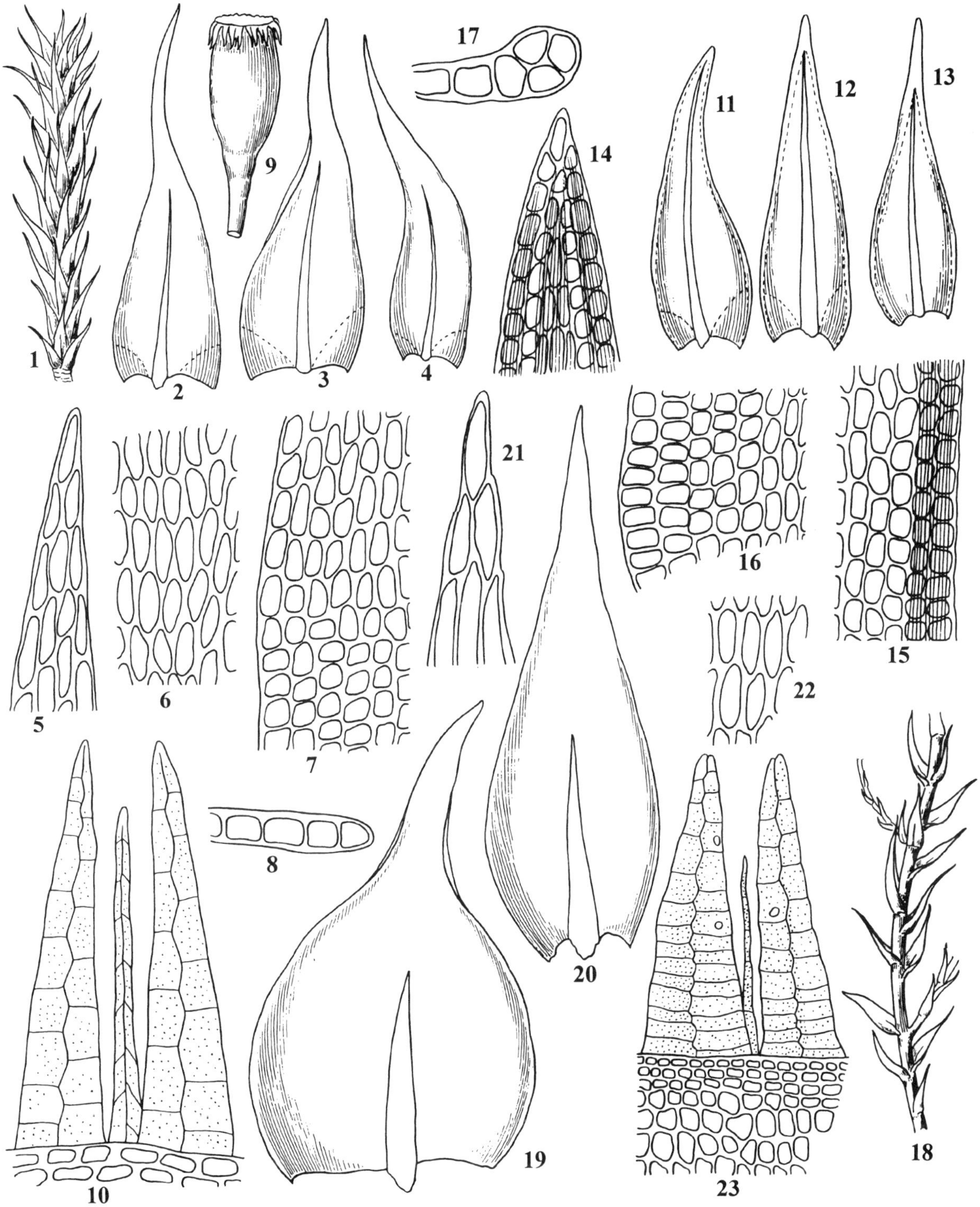

Plate 424. 1–10, *Anacamptodon amblystegioides*. 1, stem (× 15). 2–4, leaves (× 25). 5, apical leaf cells (× 320). 6, median leaf cells (× 320). 7, basal leaf cells (× 320). 8, cross-section of leaf margin (× 420). 9, capsule (× 15). 10, peristome (× 320). 11–17, *Anacamptodon fortunei*. 11–13, leaves (× 25). 14, apical leaf cells (× 320). 15, marginal leaf cells (× 320). 16, basal leaf cells (× 320). 17, cross-section of leaf margins (× 420). 18–23, *Anacamptodon latidens*. 18, stem (× 50). 19, stem leaf (× 70). 20, branch leaf (× 70). 21, apical cells of branch leaf (× 400). 22, median leaf cells (× 400). 23, peristome (× 400). (1–10, drawn from *Z.-W. Ao 4696*, IFSBH; 11–17, from *G.-C. Zhang C9*, IFSBH; 18–23, from *C. Gao 22218*, IFSBH.)

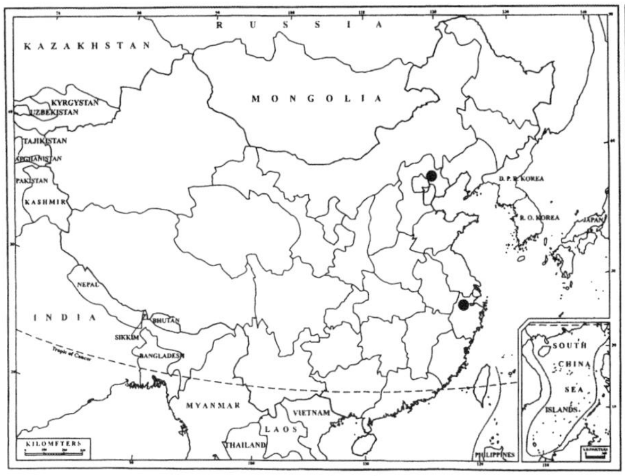

2. *Anacamptodon fortunei*

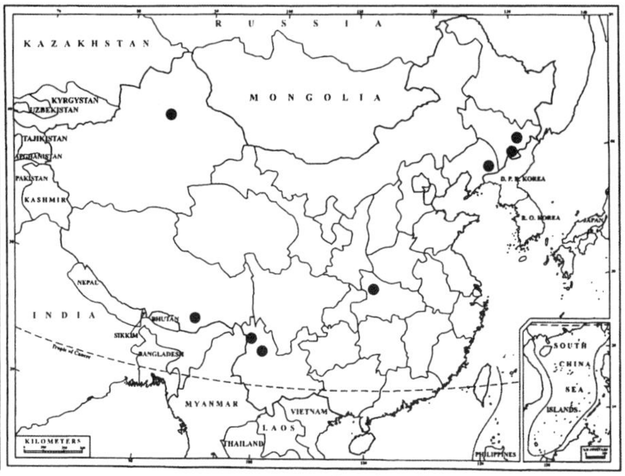

3. *Anacamptodon latidens*

3. ***Anacamptodon latidens*** (Besch.) Broth. 阔叶反齿藓 kuo-ye fan-chi xian

Nat. Pflanzenfam. I(3): 906. 1907. *Schwetschkea latidens* Besch., J. Bot. (Morot) 13: 41.1899. Type. Japan: Pref. Aomori, Mimmaya, *Faurie 14068*.

Anacamptodon sublatidens Card., Bull. Soc. Bot. Genève, ser. 2, 3: 279. 1911.

Anacamptodon subulatus Broth., Symb. Sin. 4: 93. 1929. Type. China: Yunnan, Lancan River, *Handel-Mazzetti 5892* (holotype H).

Plants slender, dark green, yellowish green or brown when old, glossy, in small patches. *Stems* covered with dense rhizoids, irregularly branched; branches short, erect or erect-spreading. *Leaves* caespitose, erect, sometimes secund, ovate at base, ovate-lanceolate toward the apex, 0.35–0.50 mm long; margins plane, entire; costa thick, extending above the midleaf; leaf cells green, elliptic or narrowly hexagonal, 18–28 µm wide; alar cells quadrate. *Dioicous*. *Inner perichaetial leaves* narrowly lanceolate, not sheathing at base, acuminate; costa weak. *Setae* elongate; capsules erect, ellipsoidal, symmetrical, apophysis short, constricted under mouth when dry; annuli 3–4 cells wide; peristome double, inserted into deep position below the mouth; exostome teeth broadly lanceolate, grayish brown, reflexed when dry, finely papillose; endostome segments shorter than exostome teeth, not keeled, reddish brown, smooth; basal membrane lacking; opercula conic, shortly beaked.

Illustrations: Plate 424, figs. 18–23.
Habitat: on tree trunks or rocks in broad-leaved forests; alt. 1200–3800 m.
Distribution: China, Japan, and Russia.

Chinese specimens examined: HUBEI: Mt. Shennongjia, *P.-C. Wu 193*, *264* (MO, PE). JILIN: An-tuo Co., *C. Gao 3394* (IFSBH); Mt. Changbai, *C. Gao 2218* (IFSBH, MO). LIAONING: Heng-ren Co., *C. Gao & G.-C. Zhang 128* (IFSBH). XINJIANG: Mt. Bogeda, *Y. Zhao 1502* (IFSBH). XIZANG: Long-zi Co., Zhun-ba, *M. Zang 1316* (HKAS). YUNNAN: Bin-chuan Co., *M.-K. Cui 6114* (HKAS); Ilowadi River, *Handel-Mazzetti 9164* (H).

2. ***Fabronia*** Raddi 碎米藓属 sui-mi xian shu

Atti Accad. Sci. Siena 9: 231. 1808.

Plants small, slender, soft, bright green, yellowish green or dark green, mostly glossy, in flat patches. *Stems* prostrate, with clustered rhizoids; irregularly branched, branches mostly densely foliate, julaceous, sometimes flat; central strand not well differentiated. *Leaves* imbricate or remotely foliate, sometimes secund toward one side, ovate or ovate-lanceolate, mostly piliferous, straight, smooth or serrate, sometimes ciliate; costa slender and weak, ending at midleaf, sometimes dentate on the dorsal side of the costa; leaf cells rhombic or rhomboidal; alar cells differentiated, quadrate or flat quadrate, in numerous rows, occasionally not differentiated. *Autoicous*. *Inner perichaetial leaves* sheathing at base, slenderly long-acuminate, ciliate or serrate, costa absent. *Setae* slender, yellowish or brownish red, twisted when dry, smooth; capsules erect, obovoid, apophysis short, sometimes not distinct; opercula conic, shortly beaked; peristome single, rarely absent, broadly lanceolate, not clearly keeled, densely cross-striate or papillose. *Calyptrae* cucullate, smooth. *Spores* spherical, nearly smooth or finely papillose.

Index Muscorum (Wijk et al. 1962) cited 110 species of *Fabronia* in the world. Crosby et al. (2000) listed 61 species in the genus with 17 well understood species. P.-C. Chen et al. (1978) reported 7 species from China. Three species are recently published in China: *Fabronia anacamptodens* C. Gao, *F. augustifolia* C. Gao & X. Fu, and *F. papillidens* C. Gao. In this treatment 10 species of *Fabronia* are recognized.

1. Leaves serrulate, teeth formed by projecting cells 2
1. Leaves serrate or ciliate, teeth formed by 1 large cell to multi-cells 6
   2. Leaves short, broadly ovate 3
   2. Leaves elongate, ovate-lanceolate 4
3. Peristome teeth absent .... 5. *F. matsumurae*
3. Peristome teeth present .... 3. *F. ciliaris*
   4. Opercula short-conic, with a curved long beak .... 8. *F. rostrata*
   4. Opercula curved, shortly obtuse 5
5. Stems and branches secund; leaves symmetrical .... 10. *F. secunda*
5. Stems and branches erect; leaves asymmetrical .... 4. *F. curvirostris*
   6. Leaf costa smooth on the back .... 7. *F. pusilla*
   6. Leaf costa spinulose on the back 7
7. Leaves ovate; leaf cells papillose by projections at the upper ends; leaf margins unicellularly serrate .... 9. *F. schensiana*
7. Leaves lanceolate; leaf cells smooth; leaf margins serrate with 1–2 cells 8
   8. Plants rather small; leaves narrowly lanceolate; leaf margins long ciliate .... 2. *F. angustifolia*
   8. Plants rather large; leaves widely lanceolate; leaf margins shortly ciliate 9
9. Peristome teeth densely papillose, not reflexed .... 6. *F. papillidens*
9. Peristome teeth striate, reflexed .... 1. *F. anacamptodens*

1. ***Fabronia anacamptodens*** C. Gao in C. Gao & G.-C. Zhang 反齿碎米藓 fan-chi sui-mi xian
   Acta Bot. Yunnan. 3(4): 397. 1981. Type. China: Xizang (Tibet), Lasa City, *M. Zang 15* (holotype IFSBH; isotype HKAS).

Plants delicate, yellowish green or brown green, glossy, densely tufted. *Stems* creeping, irregularly branched, branches erect or ascending; rhizoids in tufts. *Leaves* widely lanceolate or ovate-lanceolate, hair-pointed at apex; margins plane, teeth formed by ciliate cells; costa slender, ending just above the midleaf, usually with spinulose papillae on the back; leaf cells rhomboidal or narrowly rhomboidal, 28– 40 μm × 7–9 μm, smooth; alar cells quadrate, 9–12 μm × 9–14 μm. *Monoicous*. *Perichaetial leaves* widely lanceolate, serrate above. *Setae* yellowish green, brownish when mature, 2.5–3.0 mm long; capsules ovoid to shortly cylindrical; peristome single, often in pairs, lanceolate, densely striate, reflexed when dry.

Illustrations: Plate 425, figs. 1–8.
Habitat: on tree trunks or rock surfaces in forests; alt. 2900–3900 m.
Distribution: endemic to China.

Chinese specimens examined: XIZANG: Long-zhi Co., Sananqulin, *M. Zang 1442a* (HKAS).

This species is very similar to *F. pusilla* Raddi; however, it differs from the latter by having spinulose papillae on the back of costa and densely striate peristome teeth that are reflexed when dry.

2. ***Fabronia angustifolia*** C. Gao & X. Fu in P.-C. Wu 狭叶碎米藓 xia-ye sui-mi xian
   Fl. Bryophyt. Sin. 6: 88. 2001. Type. China: Xizang, Long-zhi, Jia-yu, *M. Zang 1126* (holotype HKAS; isotype IFSBH).

Plants slender, yellowish green, weakly shinning. *Stems* creeping, 3–5 mm long, ascending at apex, irregularly branched, branches 1–2 mm long, ascending; rhizoids sparse, loosely on stems. *Leaves* imbricate, remote at lower part of stems, broadly ovate-lanceolate, 0.3–0.6 mm × 0.10–0.15 mm, ciliate at apex; margins plane, ciliate, with cilia consisting of 2–5 cells; costa slender, up to 3/4 the leaf length or sometimes percurrent, long spinulose on the back; upper leaf cells linear or narrowly rhomboidal, 30 μm × 95 μm; alar cells quadrate or rectangular, in 3–4 rows of cells. *Autoicous*. *Male inflorescences* mostly on lateral branches; perigonial leaves few, shortly ciliate at margin. *Female inflorescences* on lateral branches; perichaetial leaves broad at base, rarely ciliate at margin. *Setae* slender, yellowish green, 1.0–1.5 mm long; capsules ovoid to shortly cylindrical; peristome and annuli absent; opercula conic, shortly rostrate. *Spores* 20–25 μm in diameter, smooth.

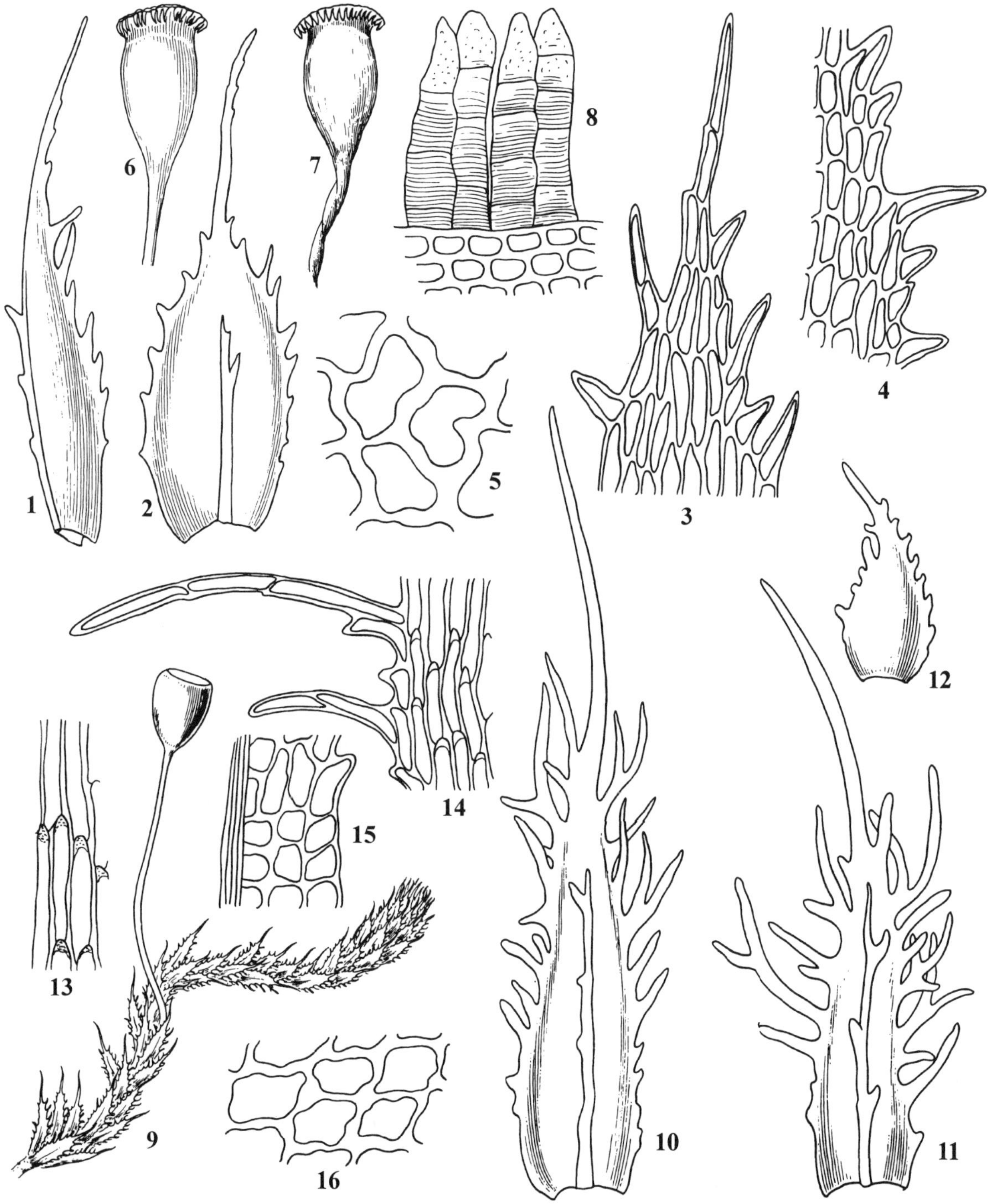

Plate 425. 1–8, *Fabronia anacamptodens*. 1–2, leaves (× 90). 3, apical leaf cells (× 400). 4, marginal leaf cells (× 400). 5, exothecial cells (× 520). 6–7, opened capsules (× 65). 8, peristome (× 400). 9–16, *Fabronia angustifolia*. 9, plant with sporophyte (× 39). 10–11, leaves (× 90). 12, apical leaf of stem (× 90). 13, apical marginal leaf cells (× 400). 14, marginal ciliae (× 400). 15, basal leaf cells (× 400). 16, exothecial cells (× 400). (1–8, drawn from *M. Zang 1442*, HKAS; 9–16, from *M. Zang 1126*, HKAS.)

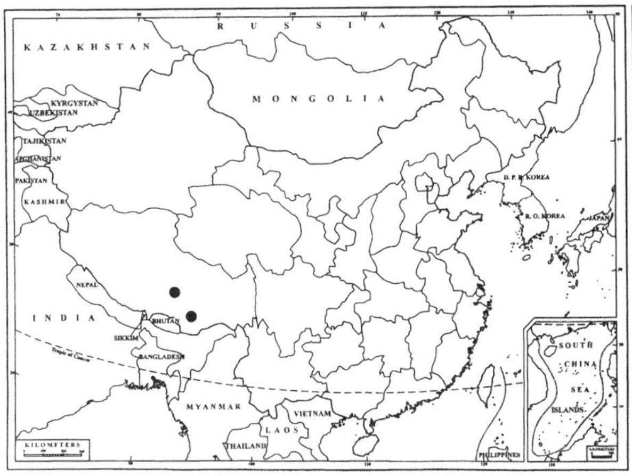

1. *Fabronia anacamptodens*

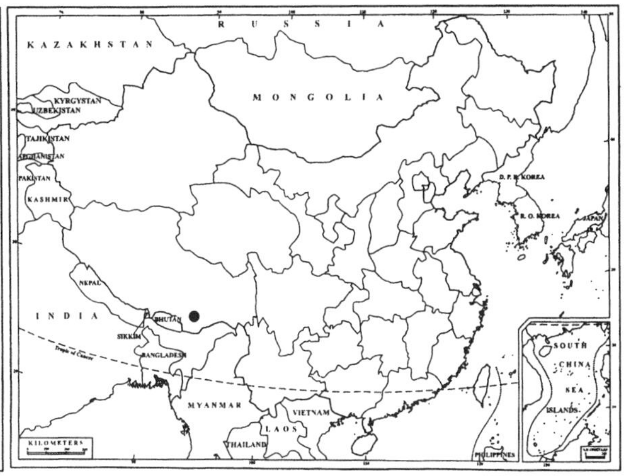

2. *Fabronia angustifolia*

Illustrations: Plate 425, figs. 9–16.
Habitat: on rock surfaces or tree trunks in cold alpine mountain regions.
Distribution: endemic to China.

Chinese specimen examined: see the type information cited above.

This species is characterized by having numerous male or female inflorescences on lateral branches, only 3–4 rows of quadrate or rectangular cells at base, narrowly rhomboidal or linear and projecting upper leaf cells, and having spinulose papillae on the back of costa.

3. ***Fabronia ciliaris*** (Brid.) Brid. 八齿碎米藓 ba-chi sui-mi xian
Bryol. Univ. 2: 171. 1827. *Hypnum ciliare* Brid., Muscol. Recent. Suppl. 2: 155. 1812. Type. Switzerland: "in Helvetiae montibus."
Fabronia octoblepharis Schwaegr., Sp. Musc. Frond., Suppl. 1(2): 338. 1816, *nom. illeg.*

Plants slender, green or yellowish green, glossy, loosely or densely tufted. *Stems* creeping, ascending, irregularly branched; rhizoids in tufts. *Leaves* dense, erect or loosely appressed when dry, erect ascending when moist, 0.3–0.8 mm long, ovate-lanceolate, constricted at base, wider in the middle, gradually narrowed to a yellowish hair-pointed at apex; margins plane, irregularly dentate; costa slender, ending at 2/3 the leaf length, smooth on the back; upper leaf cells shortly rhomboidal, 20–40 µm × 8–10 µm; alar cells quadrate, in several rows. *Dioicous*. *Perichaetial leaves* lanceolate, irregularly toothed above. *Setae* 2–4 mm long, straight or slightly curved; capsules 0.5–1.0 mm long, obovoid to shortly cylindrical, yellowish brown; peristome teeth in 8 pairs, brown or reddish brown, lingulate, irregularly striate. *Spores* 10–15 µm in diameter, smooth or finely papillose.

Illustrations: P.-C. Wu (ed.) 2001 (Pl. 41, figs. 8–15).
Habitat: on tree trunks or rotten logs in forests, sometimes on wet rocks; alt. 1700–2000 m.
Distribution: China and widely distribution in the world.

Chinese specimens examined: GUANGXI: Tian-lin Co., *C. Gao 2309* (IFSBH). HEBEI: Xing-long Co., *J.-X. Ji 1088* (CNU). HUNAN: Heng-shan Co., *T. Cao 22091* (IFSBH). JILIN: Hui-nan Co., *X.-G. Kang 3566* (IFSBH). SICHUAN: Nan-ping Co., *S. He 30077* (MO). XIZANG: Bo-mi Co., *G.-C. Zhang & J.-Y. Fang 14913* (IFSBH); Cha-yu Co., *M.-Z. Wang 7891* (PE). YUNNAN: An-ning Co., *M. Zang 81-108* (HKAS); Du-long-jiang (river), *Q.-T. Zhang 3898* (HKAS); Qiao-jia Co., *Q.-T. Zhang 6080* (HKAS). ZHEJIANG: Mt. West Tianmu, *M. Zang 265* (HKAS).

This species is very similar to *F. matsumura,* but it differs in having narrow leaves and peristome teeth.

4. ***Fabronia curvirostris*** Dozy & Molk. 弯喙碎米藓 wan-hui sui-mi xian
Ann. Sci. Nat., Bot., ser. 3, 2: 304. 1844. Type. Indonesia: Java.

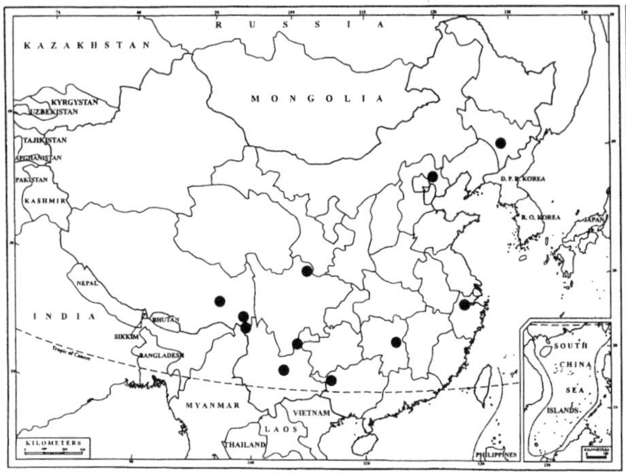

3. *Fabronia ciliaris*

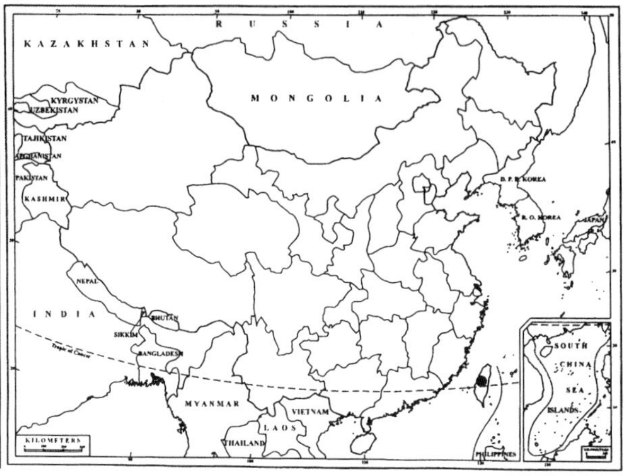

4. *Fabronia curvirostris*

Plants slender, yellowish green, glossy, densely tufted. *Stems* prostrate, irregularly branched, branches erect or ascending, up to 5 cm long. *Stem leaves* widely lanceolate, narrowed at base, gradually narrowed from middle to a hair-pointed apex; branch leaves lanceolate, ciliate at apex, flexuous when dry, up to 0.7 mm long; margins plane, sharply serrate; costa slender, vanishing in the middle of leaf, smooth on the back; upper leaf cells elliptic or narrowly rhomboidal, 24–48 µm × ca. 8 µm, marginal cells smaller, basal cells short; alar cells quadrate, in several rows. *Monoicous*. *Female inflorescences* growing on short branches. *Setae* up to 3 mm long, brownish; capsules erect, shortly cylindrical or ovoid, ca. 0.6 mm long; opercula conic, rostrate; annuli undifferentiated; peristome teeth lingulate, in pairs, reddish brown, striate. *Spores* 16–18 µm in diameter, finely papillose.

Illustrations: Bartram 1939 (Pl. 21, fig. 361).
Habitat: on tree trunks or moist rocks.
Distribution: China, Indonesia, and the Philippines.

Chinese specimens examined: TAIWAN: Nan-tou Co., *S.-H. Lin s.n.* [Bryophytes of Taiwan Exsiccatae, ser. 1, no. 16] (MO).

5. ***Fabronia matsumurae*** Besch. 东亚碎米藓 dong-ya sui-mi xian
J. Bot. (Morot) 13: 40. 1899. Type. Japan: Tokyo Pref.

Plants minute, green, glossy, loosely tufted. *Stems* creeping, irregularly branched; rhizoids in tufts on the ventral surface of stems; branches erect, obtuse at apex. *Leaves* spirally arranged, often in two rows, ovate, apex short, suddenly acute, apical cell yellowish, 0.5–0.8 mm × 0.4 mm; leaf margins plane, upper margin irregularly serrate with single cell; costa single, weak, ending in the middle of leaf, smooth on the back; leaf cells rhomboidal or elongate-hexagonal, 20–40 µm × 10–13 µm; alar cells quadrate, in several rows. *Autoicous*. *Setae* 2–3 mm long, yellowish brown; capsules ovoid, erect, cup-like after broken; opercula with a short beak; peristome absent. *Calyptrae* cucullate. *Spores* 14–17 µm in diameter, minutely papillose.

Illustrations: Plate 426, figs. 1–7.
Habitat: on tree trunks or rocks in broad-leaved forests or coniferous mixed forests; alt. 460–1020 m.
Distribution: China, Korea, Japan, and Russia.

Chinese specimens examined: BEIJING: Xiang-shan, *F.-D. Chen 041* (CNU). HEBEI: Lai-yuan Co., *M.-J. Li & J.-C. Zhao 172-A* (MO). HELONGJIANG: Da-xing-an-ling (Greater Khingan Mt.), *C. Gao 13117* (IFSBH). HUBEI: Li-chuan Co., *1980 Sino-Amer. Exped. 1979* (MO); Wuhan Botanical Garden, *T. Cao 3* (IFSBH). HUNAN: Bao-jing Co., *W.-Z. Zhang 001* (MO, PE). JILIN: Lin-jiang Co., *C. Gao 7720* (IFSBH). NEI MONGOL: Mt. Helan, *Z.-G. Tong 69, 1148, 2159* (HIMC). SHAANXI: Mt. Qingling, *D.-Q. Lu 93* (WUK). SHANDONG: Tai-an City, *C. Gao & E.-Z. Bai 34020* (IFSBH). SICHUAN: Mt. Kouji, *K.-K. Chen 12* (IFSBH). XIZANG: Long-zi Co., Zhunbada, *M. Zang 1018* (HKAS). YUNNAN: Kun-ming City, *T. Cao & Z.-W. Ao 21284* (IFSBH).

6. ***Fabronia papillidens*** C. Gao 疣齿碎米藓 you-chi sui-mi xian
Acta Bot. Yunnan. 3(4): 397. 1981. Type. China: Xizang, Cha-yu Co., *M. Zang 1195* (holotype IFSBH; isotype HKAS).

Plants slender, brownish green, weakly glossy, in delicate cushions. *Stems* creeping, irregularly branched; branches erect; rhizoids in tufts. *Leaves* appressed when dry, erect-spreading when moist, lanceolate, single-cell ciliate at apex; leaf margins plane, serrate with single-celled cilia above; costa slender, ending in the middle of leaf, with single-celled spines on the back; leaf cells narrowly rhomboidal, 55–65 µm × 7–10 µm, thin-walled; alar cells quadrate or rectangular, 10–20 µm × 9–10 µm. *Perichaetial leaves* widely lanceolate, with sharp teeth at apex. *Setae* up to 4 mm long, yellowish brown; capsules shortly cylindrical; exothecial cells oblong-hexagonal; annuli undifferentiated; peristome teeth lingulate, short, papillose. *Spores* not seen.

Illustrations: Plate 426, figs. 8–16.
Habitat: on tree trunks or rock surfaces in open area or in forests; alt. 3600 m.

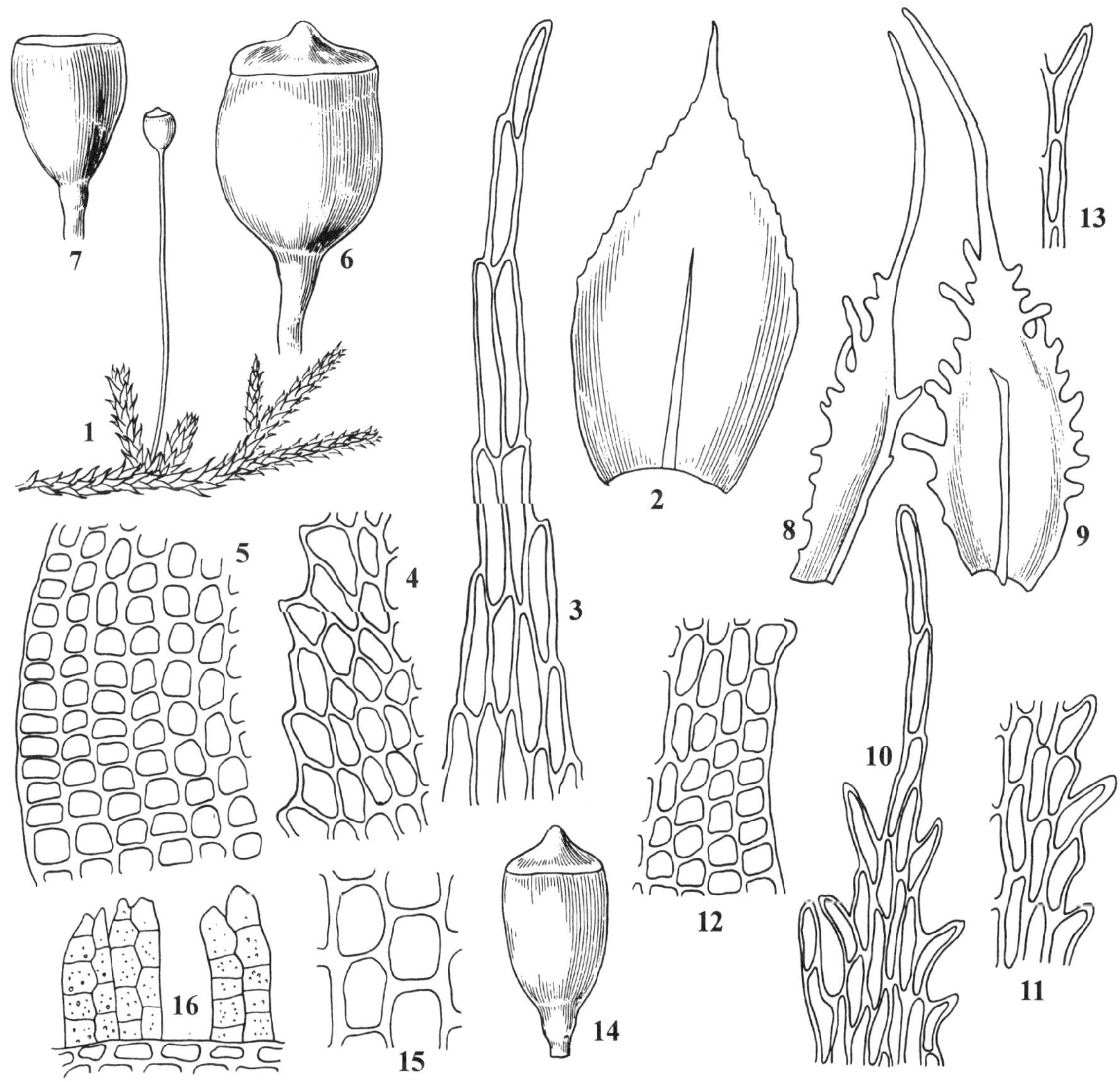

Plate 426. 1–7, *Fabronia matsumurae*. 1, plant with sporophyte (× 20). 2, leaf (× 90). 3, apical leaf cells (× 400). 4, upper marginal leaf cells (× 400). 5, basal leaf cells (× 400). 6–7, capsules (× 65). 8–16, *Fabronia papillidens*. 8–9, stem leaves (× 90). 10, apical leaf cells (× 400). 11, upper marginal leaf cells (× 400). 12, basal marginal leaf cells (× 400). 13, marginal tooth of leaf (× 400). 14, capsule (× 65). 15, exothecial cells (× 400). 16, peristome (× 400). (1–7, drawn from *C. Gao 7720*, IFSBH; 8–16, from *M. Zang 1145*, HKAS.)

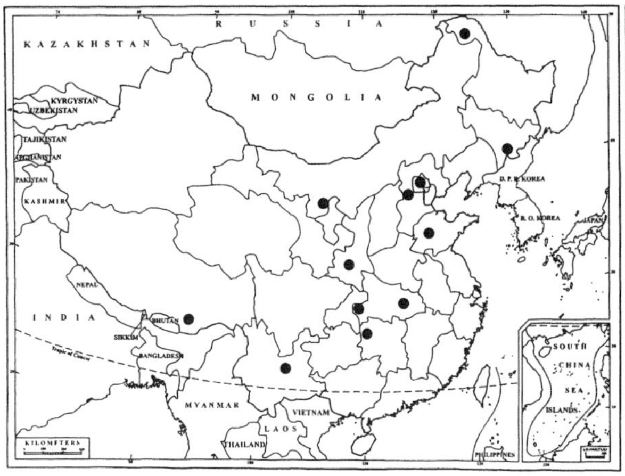

5. *Fabronia matsumurae*

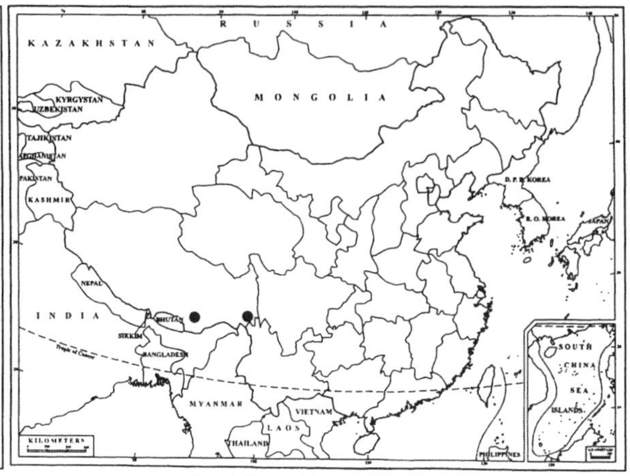

6. *Fabronia papillidens*

Distribution: endemic to China.
Chinese specimen examined: XIZANG: Long-zi Co., Zhunba, *M. Zang 1145* (HKAS).

This species is similar to *Fabronia pusilla*, but differs in having oblong-hexagonal exothecial cells and long-spinulose ending leaf costa.

7. ***Fabronia pusilla*** Raddi 碎米藓 sui-mi xian

Atti Accad. Sci. Siena 9: 231. 1808. *Fabronia octoblepharis* ssp. *pusilla* (Raddi) Kindb., Eur. N. Amer. Bryin. 1: 13. 1897. Type. Europe: Toscana, *Raddi s.n.*

Plants minute, yellowish green, glossy, densely in tufts. *Stems* creeping, ascending at apex, irregularly branched; branches erect-spreading or erect, obtuse at apex; rhizoids sparse. *Leaves* densely imbricate or spreading, not appressed when dry, ovate-lanceolate, acuminate to hairy at apex, ca. 0.50 mm × 0.15 mm, broader in the middle; leaf margins plane, ciliate, each tooth consisting of 1–2 cells; costa single, weak, vanishing in the middle of leaf, smooth on the back; upper leaf cells narrowly rhomboidal or narrowly elliptic, 25–40 μm × ca. 7 μm; basal cells quadrate or rectangular; alar cells quadrate. *Monoicous*. *Female inflorescences* growing on lateral branches. *Perichaetial leaves* widely lanceolate. *Setae* ca. 4 cm long, yellowish brown; capsules curved; exostome teeth lanceolate, ca. 145 μm long, ca. 54 μm wide at base; endostome not developed. *Spores* spherical, 14–15μm in diameter, finely papillose.

Illustrations: P.-C. Wu (ed.) 2001 (Pl. 43, figs. 10–15).
Habitat: on tree trunks or moist rocks in forests, sometimes on rotten logs; alt. 3900 m.
Distribution: China, Bhutan, Central Asia, Europe, Africa, and North America.

Chinese specimens examined: XIZANG: Long-zi Co., Zhunba, *M. Zang 1126*, *1073a* (HKAS); Ya-dong Co., *Z.-X. Tan 16* (IFSBH). YUNNAN: *M. Zang 36* (HKAS).

This species is similar to *Fabronia schensiana* C. Müll., but it differs from the latter by its smooth costa on the dorsal side of leaves.

8. ***Fabronia rostrata*** Broth. 毛尖碎米藓 mao-jian sui-mi xian

Symb. Sin. 4: 92. 1929. Type. China: Yunnan, Mekong, 28°11′N, *Handel-Mazzetti 8018* (holotype H).

Plants weak, yellowish green, somewhat glossy, in dense patches. *Stems* creeping, with brownish rhizoids, irregularly branched; branches erect, up to 3 mm long, obtuse at apex. *Leaves* imbricate, erect-spreading when moist, oblong-ovate, slightly concave, gradually narrowed to a hairy apex; leaf margins plane, serrate; costa weak, ending in the middle of leaf, smooth on the back; upper leaf cells oblong-elliptic or rhomboidal; alar cells quadrate or rectangular, in several rows. *Dioicous*. *Setae* up to 5 mm long, slender; capsules small, ovoid; opercula conic, with stout beak; peristome teeth absent.

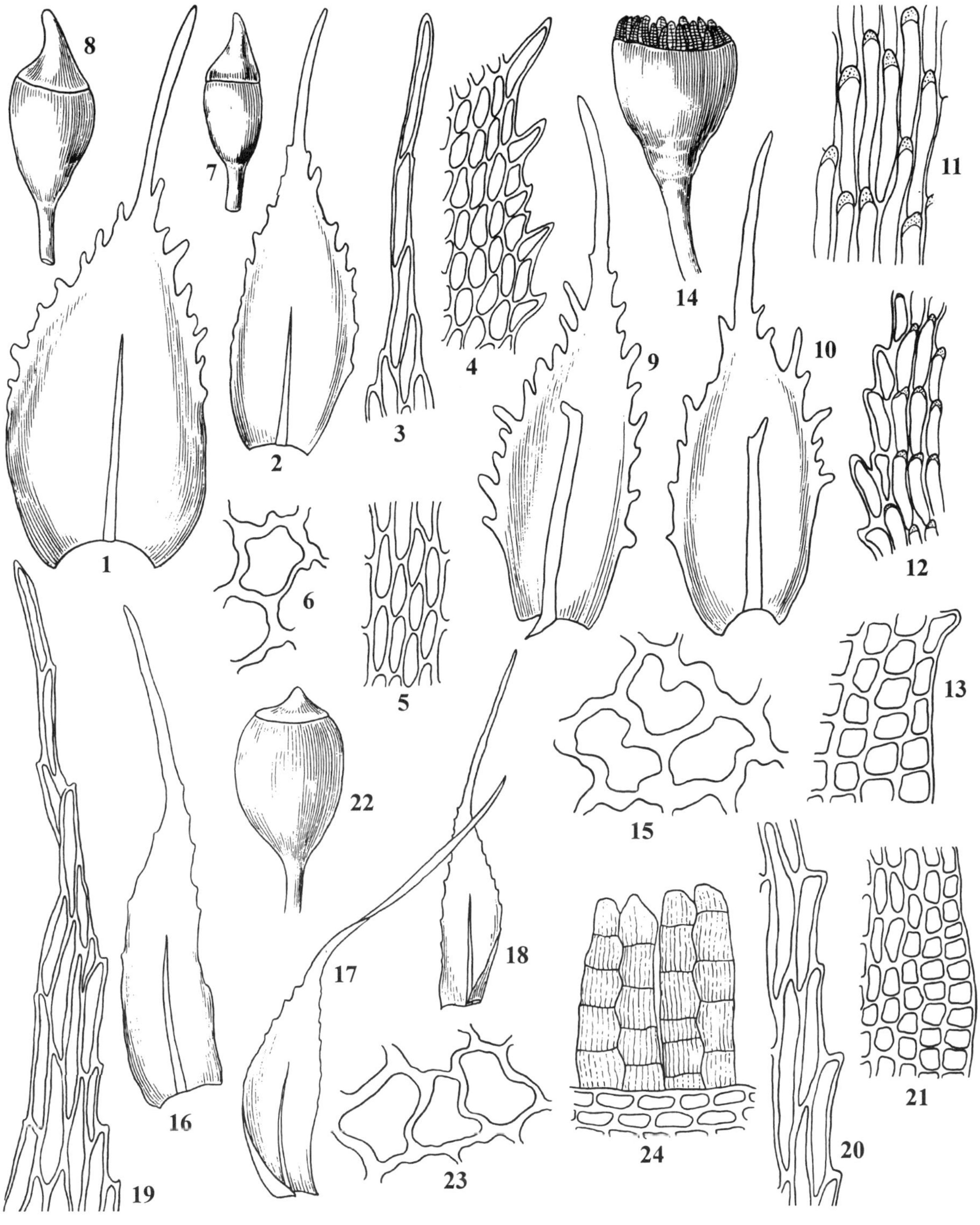

Plate 427. 1–8, *Fabronia rostrata*. 1–2, leaves (× 90). 3, apical leaf cells (× 400). 4, marginal leaf cells (× 400). 5, basal leaf cells (× 400). 6, exothecial cells (× 400). 7–8, capsules (× 65). 9–15, *Fabronia schensiana*. 9–10, leaves (× 90). 11, median leaf cells (× 400). 12, marginal leaf cells (× 400). 13, basal leaf cells (× 400). 14, capsule (× 65). 15, exothecial cells (× 520). 16–24, *Fabronia secunda*. 16–18, leaves (× 90). 19, apical leaf cells (× 400). 20, marginal leaf cells (× 400). 21, basal leaf cells (× 400). 22, capsule (× 65). 23, exothecial cells (× 400). 24, peristome (× 400). (1–8, drawn from *Handel-Mazzetti 8018*, H; 9–15, from *Giraldi 1986*, H; 16, 19–21, from *Sharp et al. 246*, NICH; 17–18, 22–24, from *Handel-Mazzetti 1400*, H.)

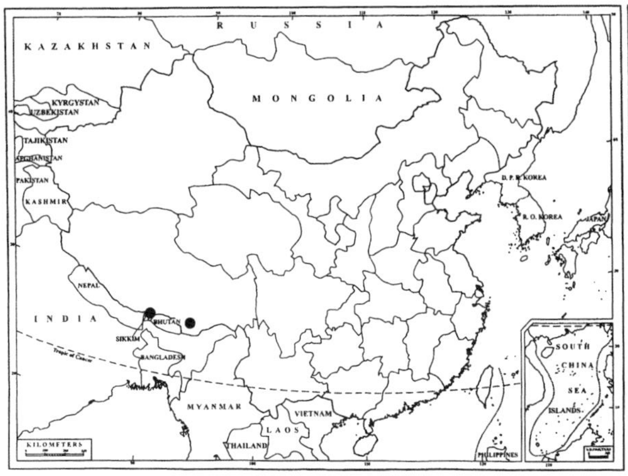

7. *Fabronia pusilla*

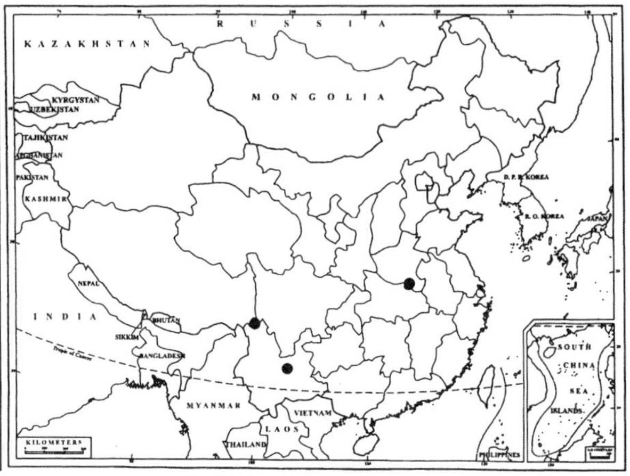

8. *Fabronia rostrata*

Illustrations: Plate 427, figs. 1–8.
Habitat: on rocks or tree trunks in forests.
Distribution: endemic to China.

Chinese specimen examined: HENAN: Mt. Jigong, *J.-X. Luo 66* (MO, PE). YUNNAN: Kun-ming City, *C.-W. Wang 3306* (PE).

9. ***Fabronia schensiana*** C. Müll. 陕西碎米藓 shaan-xi sui-mi xian

Nuovo Giorn. Bot. Ital., n. ser. 4: 262. 1897. Syntypes. China: Shaanxi (Schen-si), Mt. Lao-y-san, May 1896, *Giraldi s.n.*; Sche-kin-tsuen, Dec. 1895, *Giraldi s.n.*

Fabronia microspora C. Gao, Acta Bot. Yunnan. 3(4): 397. 1981. Type. China: Xizang, Long-zi Co., Sananqulin, *M. Zang 1407a* (holotype IFSBH; isotype HKAS).

Plants delicate, yellowish green, glossy, tufted in small patches. *Stems* creeping; rhizoids sparse. *Leaves* remote, erect, not appressed when dry, widely lanceolate, ca. 0.77 mm × 0.23 mm; leaf margins plane, regularly ciliate above; costa single, vanishing in the middle of leaf, sometimes spinulose on the back; upper leaf cells narrowly rhomboidal, 43–52 µm × 7–9(–11) µm; basal cells oblong or rectangular; alar cells quadrate, in several rows. *Monoicous*. *Perichaetial leaves* widely lanceolate. *Setae* erect, 3–5 mm long; capsules erect, ovoid or globose, ca. 0.4 mm in diameter; peristome not seen.

Illustrations: Plate 427, figs. 9–15.
Habitat: on broad-leaved forest ground or rotten logs, sometimes on moist rocks.
Distribution: endemic to China.

Chinese specimens examined: YUNNAN, Gong-shan Co., *M. Zang 3933* (HKAS); Kun-ming City, *X.-J. Li 53* (HKAS).

When C. Müller (1897) published *Fabronia schensiana*, he did not mention spinulose costa on the leaf back. C. Gao et al. (1981) described *F. microspora* as having spinulose costa and differentiated spore morphology. In this treatment, *Fabronia microspora* is treated as a synonym of *F. shensiana*.

10. ***Fabronia secunda*** Mont. 偏叶碎米藓 pian-ye sui-mi xian

Ann. Sci. Nat. Bot., ser. 2. 17: 521. 1842. Type. India: Nilgiris, *Perrottet s.n.*

Plants delicate, brownish green, in small patches. *Stems* prostrate, irregularly branched; branches erect, numerous; rhizoids in bundle. *Leaves* ovate-lanceolate, erecto-patent or more or less secund, ciliate at apex; leaf margins remotely sharply toothed; costa slender, ending above the middle of leaf; leaf cells linear; alar cells quadrate, in 3–4 rows. *Sporophytes* not seen.

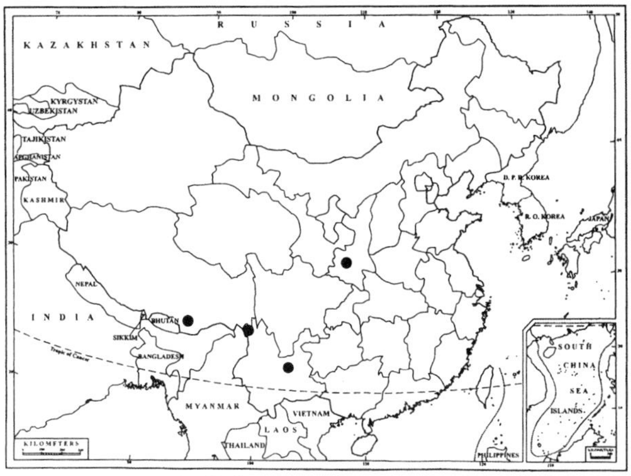

9. *Fabronia schensiana*

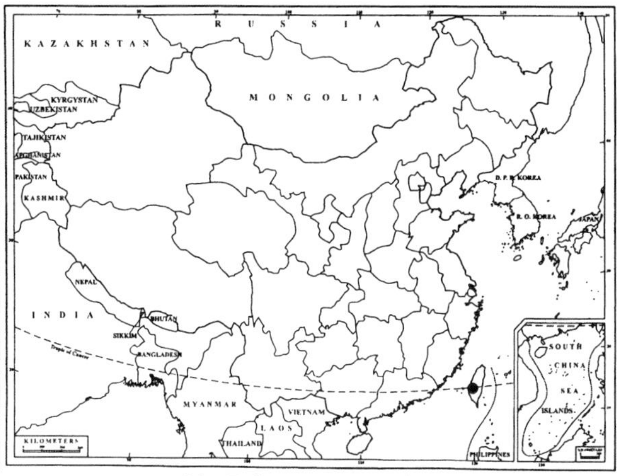

10. *Fabronia secunda*

Illustrations: Plate 427, figs. 16–24.
Habitat: on tree trunks.
Distribution: China, Nepal, India, and Sri Lanka.

Chinese specimen examined: TAIWAN: Jia-yi Co., *Iwatsuki & Sharp 246* (NICH).

3. ***Habrodon*** Schimp. 柔齿藓属 rou-chi xian shu

Syn. Musc. Eur. 505. 1860.

Plants slender, not glossy, in small interwoven patches. *Stems* creeping, elongate, ascending above, irregularly branched; branches short and erect; rhizoids sparse; central strand not differentiated. *Leaves* imbricate when dry, erecto-patent when moist, slightly reflexed, ovate, concave, suddenly acute or acuminate to a short apex; leaf margins straight, entire or serrulate; costa absent; leaf cells oblong-oval, rhomboidal or narrowly oblong-elliptic, basal cells broader; alar cells flat quadrate in several rows. *Gemmae* unicellular, oblong-ovoid. *Dioicous*. *Inner perichaetial leaves* erect, not sheathing at base, narrowly acuminate toward the apex, margins dentate, cells linear. *Setae* slender, yellowish brown or purple red, twisted when dry; capsules erect, symmetrical, oblong-ovoid, not constricted at mouth, hypophysis short; exothecial cells thin-walled, longitudinally plicate; annuli large, in 3–4 rows of cells, rarely one row of cells, deciduous when mature; peristome appearing single; endostome teeth absent; exostome teeth lanceolate, deeply inserted in the mouth, not connected to each other, lamellae conspicuous, keeled, highly lamellate, not papillose, striate at base; opercula conic, with a short straight beak. *Calyptrae* erect, cucullate. *Spores* not seen.

There are only two species remaining in the genus *Habrodon*, *H. kashmiriensis* Vohra and *H. perpusillus* (De Not.) Lindb. Only the latter species is reported from China.

1. ***Habrodon perpusillus*** (De Not.) Lindb. 柔齿藓 rou-chi xian

Öfvers. Forh. Kongl. Svenska Vetensk.-Akad. 20: 401. 1863. *Pterogonium perpusillum* De Not., Musc. Ital. Spec. 84. 1838. Type. Italy: Sardinia, Villacidro.

The description of this species is the same as that of genus.

Illustrations: P.-C. Wu (ed.) 2001 (Pl. 44, figs. 6–10).
Habitat: on bark of trees in needle and broad-leaved mixed forests, sometimes on rocks; alt. 3700 m.
Distribution: China, Korea, Russia, Europe, and North America.

Chinese specimens examined: LIAONING: An-shan City, Qian Shan (Mt.), *C. Gao 1948* (IFP). SICHUAN: Mu-li Co., *C. Gao et al. 20723*, *20901*, *20997* (IFSBH). XIZANG: Long-zhi Co., Sananqulin, *M. Zang 1345* (HKAS); Zhunba, *M. Zang 1234* (HKAS).

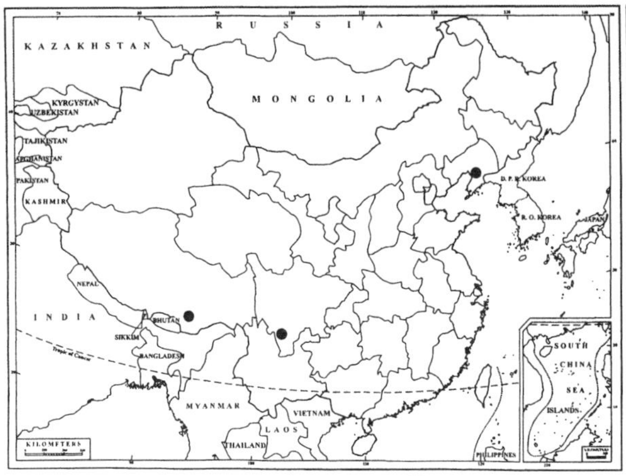

1. *Habrodon perpusillus*

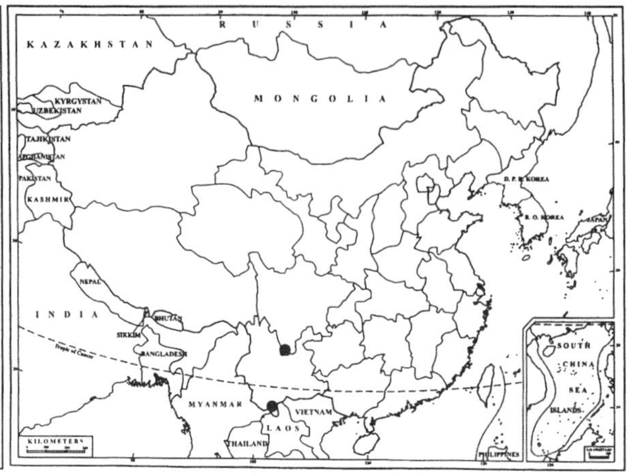

1. *Helicodontium doii*

Morphologically, this species is similar to *Iwatsukiella leucotricha* (Mitt.) Buck & Crum; however, it differs from the latter in having a short leaf apex and single peristome teeth.

4. ***Helicodontium*** (Mitt.) Jaeg. 旋齿藓属 xuan-chi xian shu

Ber. Thatigk. St. Gallischen Naturwiss. Ges. 1876–77: 224. 1878. *Hypnum* sect. *Helicodontium* Mitt., J. Linn. Soc., Bot. 12: 545. 1869.

Plants slender, delicate, dark green, not glossy, in creeping mats. *Stems* prostrate, irregularly branched; branches short, ascending or arcuate, densely foliate; rhizoids sparse; central strand not differentiated. *Leaves* ovate-lanceolate, acuminate above; leaf margins plane, entire below, toothed above; costa slender, ending in the middle or above leaf middle; leaf cells oblong-elliptic or oblong-hexagonal; alar cells quadrate, in several rows. *Autoicous*. *Inner perichaetial leaves* sheathing at base. *Setae* short, 5–7 mm long, slender, reddish brown, twisted when dry, smooth or papillose; capsules ovoid or shortly cylindrical, constricted below the mouth; annuli persistent; opercula shortly rostrate; peristome double; exostome teeth lanceolate, yellowish brown, keeled, with zig-zag median line; endostome segments linear, as long as exostome teeth, smooth or finely papillose; basal membrane low. *Calyptrae* cucullate, smooth. *Spores* spherical, smooth.

Currently, there are about 21 species of *Helicodontium* in the world with eight better-known species (Crosby et al. 2000). Two species are recognized in China.

1. Plants slender; seta papillose; spores 10–15 µm in diameter....................1. *H. doii*
1. Plants delicate; seta smooth; spores 20–25 µm in diameter....................2. *H. formosicum*

1. ***Helicodontium doii*** (Sak.) Taoda 大旋齿藓 da xuan-chi xian

Hikobia 8: 48. 1977. *Schwetschkea doii* Sak., Bot. Mag. (Tokyo) 46: 498. 1932. Type. Japan: Pref. Kagoshima, Mt. Jisso, *Sakurai 2701*.

Helicodontium robustum (Toy.) Shin, Sci. Rep. Kagoshima Univ. 32: 85, 9. 1. 1954. *Schwetschkea robusta* Toy., Acta Phytotax. Geobot. 6: 176. 5. 1937.

Plants slender, yellowish green or dark green, not glossy, in plane interwoven patches. *Stems* prostrate, irregularly branched; branches short, erect or erecto-patent, rarely secondarily branched; rhizoids sparse. *Branch leaves* erecto-patent, appressed when dry, ovate-lanceolate, acuminate above, 1.0–1.7 mm long; leaf margins plane, entire below, slightly serrulate near leaf apex; costa single, weak, ending at 2/3 the leaf length; leaf cells s-shaped linear, 45–75 µm × 6–8 µm, smooth; apical leaf cells short-rhomboidal; alar cells quadrate, in several rows, clearly differentiated. *Autoicous*. *Setae* 5–7 mm long, papillose; capsules ovoid, suberect; peristome double, equal in length; exostome teeth lanceolate, papillose above, densely striate in the middle and below; endostome segments linear; basal membrane low; cilia absent. *Spores* 10–15 µm in diameter.

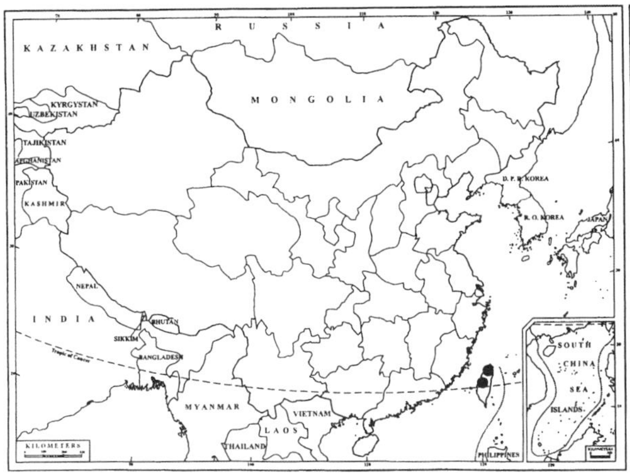

2. *Helicodontium formosicum*

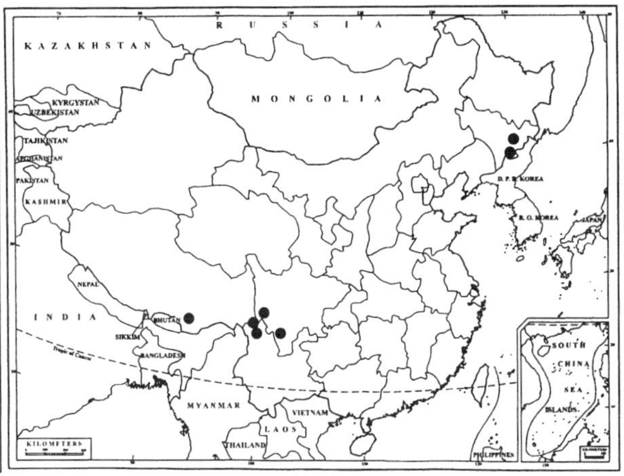

1. *Iwatsukiella leucotricha*

Illustrations: Plate 428, figs. 1–11.
Habitat: on bark of trees and rotten logs in broad-leaved forests; alt. 2500–2800 m.
Distribution: China and Japan.

Chinese specimens examined: SICHUAN: Du-kou City, *L.-S. Wang 83-304* (HKAS). YUNNAN: Meng-hai Co., *X.-J. Li 850008* (HKAS).

2. ***Helicodontium formosicum*** (Card.) Buck 台湾旋齿藓 tai-wan xuan-chi xian
J. Hattori Bot. Lab. 47: 52. 1980. *Schwetschkea formosica* Card., Beih. Bot. Centralbl. 19(2): 125. 1905. Type. China: Taiwan, Mt. Maruyama, *Faurie 6* (isotype H).
Fabronia formosana Sak., Bot. Mag. (Tokyo) 46: 378. 1932. Type. China: Taiwan, Tai-chiu, Noko-gun, Horisha, *Sasaoka 2101*.

Plants delicate, green or yellowish green, in dense patches. *Stems* slenderly weak, prostrate, 3–4 cm long, irregularly branched, erecto-patent or creeping, ascending above. *Leaves* small, erect or erect spreading, not appressed, widely lanceolate, acute, 0.5–0.8 mm × 0.16–0.25 mm; leaf margins plane, mostly entire below, serrulate above; costa slender, weak, disappearing in the middle of leaf; median and upper leaf cells elliptic or narrowly rhomboidal, thin-walled; alar cells quadrate or flat quadrate, in several rows, thin-walled. *Monoicous* or autoicous. *Perichaetial leaves* small, ovate-lanceolate, gradually tapering to a narrow apex. *Setae* slender, 5–6 mm long, yellowish green, smooth, twisted above when dry; capsules ovoid to cylindrical, erect or patent, 0.8–0.9 mm long, 0.4–0.5 mm in diameter; opercula shortly rostrate; peristome double; exostome teeth triangular at base; endostome segments linear, smooth, as long as exostome teeth; basal membrane low. *Spores* 20–25 µm in diameter, smooth.

Illustrations: Taoda 1977 (Fig. 3, A–T, as *Schwetschkea formosica*).
Habitat: on tree trunks, rarely on rock surfaces.
Distribution: China and Japan.

Chinese specimens examined: TAIWAN: Wanshan (Mt.), *Faurie 5* (H).

5. ***Iwatsukiella*** Buck & Crum 小柔齿藓属 xiao rou-chi xian shu
J. Hattori Bot. Lab. 44: 351. 1978.

This is a monotypic genus, consisting of *Iwatsukiella leucotricha* (Mitt.) Buck & Crum. Originally, the genus was a member of *Habrodon* Schimp. However, Buck and Crum (1978) considered it to be an independent genus based on its numerous leaf alar cells, foliose pseudoparaphyllia, and double peristome teeth that are smooth below and papillose above. The generic description is the same as for the species.

1. ***Iwatsukiella leucotricha*** (Mitt.) Buck & Crum 小柔齿藓 xiao rou-chi xian

J. Hattori Bot. Lab. 44: 352. 1978. *Habrodon leucotrichus* (Mitt.) Perss., Svensk, Bot. Tidskr. 40: 319. 1946. *Heterocladium leucotrichum* Mitt., Trans. Linn. Soc. London, Bot. 3: 176. 1891. Type. Japan: Tochigi Pref., Mt. Nantaizan, *Bisset s.n.*

Habrodon piliferus Card., Bull. Soc. Bot. Genève, ser. 2, 3: 280. 1911.

Plants slender, green, slightly glossy, irregularly procumbent. *Main stems* closely attached on substrata, irregularly pinnately branched, branches short and densely covered with leaves; rhizoids sparse; pseudoparaphyllia numerous, foliose; central strand not differentiated. *Leaves* reflexed when moist, imbricate when dry, rounded or rounded ovate at base, not decurrent, abruptly narrowed to a multi-cellular, piliferous apex, straight or curved; leaf margins entire; costa absent or inconspicuous forked; leaf cells elliptic, somewhat thick-walled; alar cells not clearly differentiated. *Dioicous*. *Inner perichaetial leaves* erect, not sheathing at base, gradually narrowed to a slender apex; margins serrulate. *Setae* 5–7 mm long, slender, purple red or reddish brown, twisted when dry; capsules erect, symmetrical, oblong-ellipsoid, apophysis short, mouth broad, not constricted when dry; exothecial cells thin-walled; opercula shortly cylindrical, obtuse; annuli 1–2(3–4) rows of cells; peristome double; exostome teeth lanceolate, smooth or finely striate below, papillose above; endostome segments linear; basal membrane low. *Calyptrae* cucullate. *Spores* tiny.

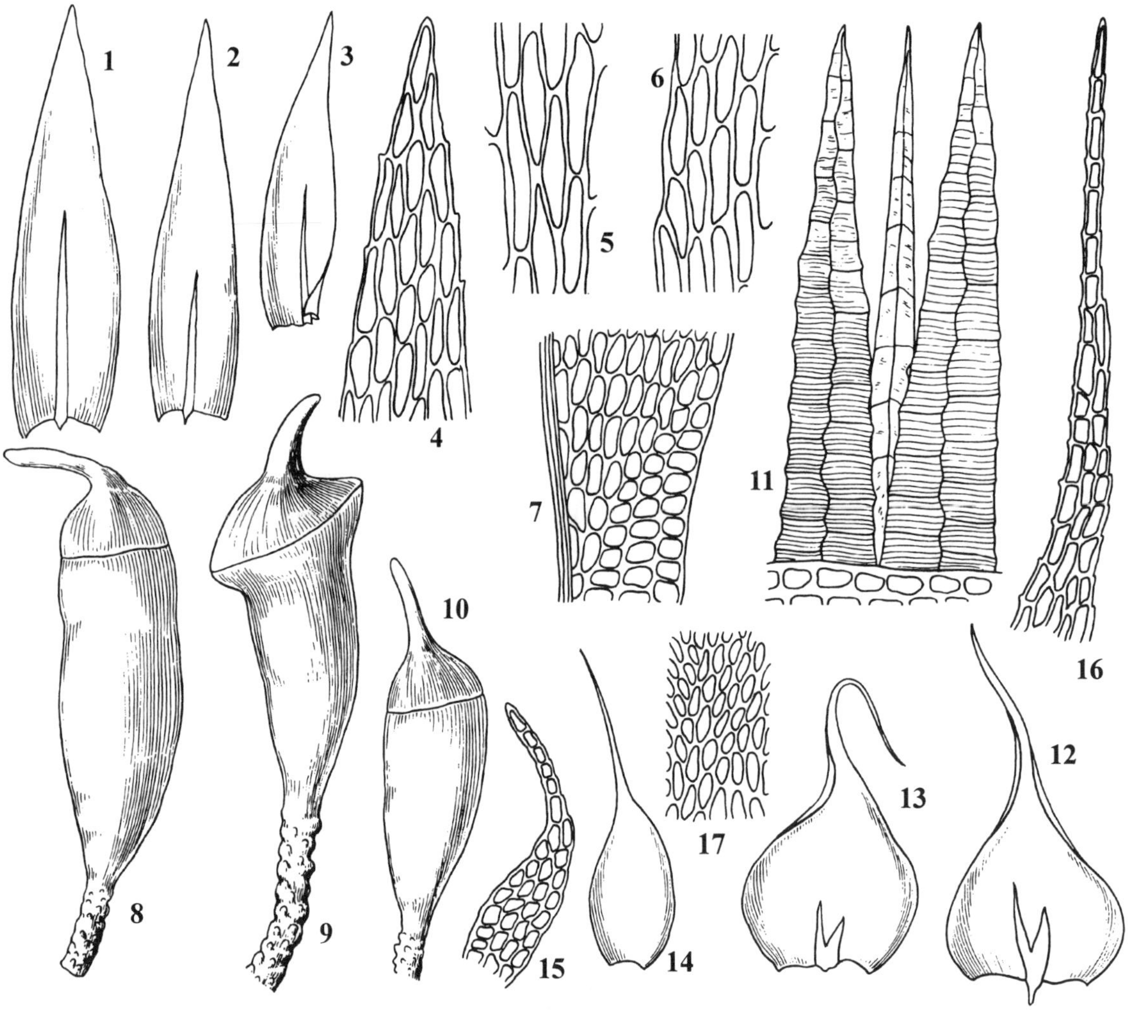

Plate 428. 1–11, *Helicodontium doii*. 1–3, leaves (× 28). 4, apical leaf cells (× 320). 5, median leaf cells (× 320). 6, median marginal leaf cells (× 320). 7, basal leaf cells (× 320). 8–10, capsules (× 16). 11, peristome (× 320). 12–17, *Iwatsukiella leucotricha*. 12–13, leaves (× 50). 14–15, pseudoparaphyllia (× 90). 16, apical leaf cells (× 320). (1–11, drawn from *X.-J. Lin 850008*, HKAS; 12–17, from *C. Gao 7623*, IFSBH.)

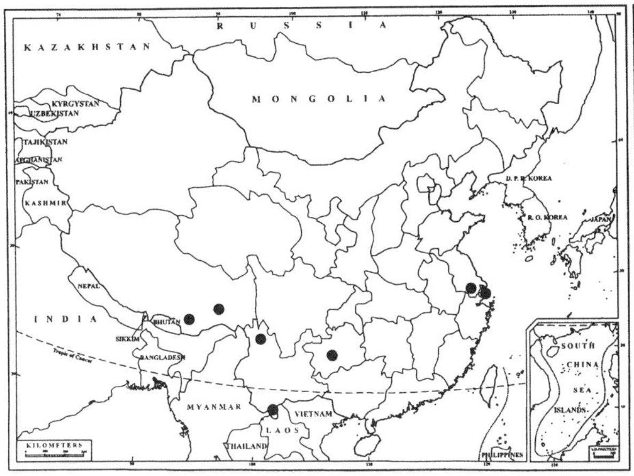

1. *Juratzkaeella sinensis*

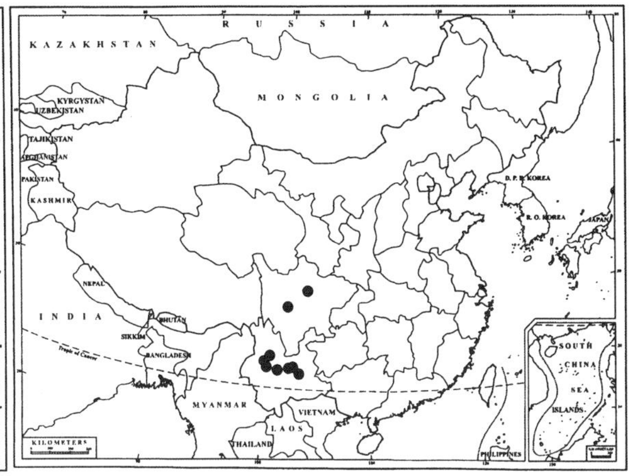

1. *Levierella fabroniacea*

Illustrations: Plate 428, figs. 12–17.
Habitat: on tree trunks of coniferous and broad leaved mixed forests; alt. 1700 m.
Distribution: China and Japan.

Chinese specimens examined: JILIN: An-tu Co., *J.-Y. Feng 26878* (IFSBH); Changbai Co., *C. Gao 7623* (IFSBH). SICHUAN: Mu-li Co., *P.-Q. Li 42166* (PE); Xiang-cheng Co., *Z.-G. Liu 4454*, *4475* (MO, PE). YUNNAN: De-qin Co., *J.-L. Yang 2377* (HKAS); Zhong-dian Co., *X.-J. Li 1391* (HKAS). XIZANG: Long-zi Co., Sananquilin, *M. Zang 1395* (HKAS).

This species is very close to *Habrodon perpusillus* (De Not.) Lindb.; however, it differs from the latter in having piliferous leaf apices and inconspicuous alar cells.

6. ***Juratzkaeella*** Buck 拟无毛藓属 ni wu-mao xian shu
Rev. Bryol. Lichénol. 43: 312. 1977.

Juratzkaeella Buck is a monotypic genus segregated from *Juratzkaea* Lor. (Buck 1977). The former can be separated from the latter on the basis of the following characters: leaves serrulate, annuli present, peristome attached at the mouth; exostome teeth distinctly cross-striolate below, and opercula conic-rostrate in *Juratzkaeella*; whereas leaves entire, annuli absent, peristome inserted somewhat below the mouth, exostome teeth papillose, indistinctly cross-striolate below, and opercula short-conic in *Juratzkaea*. The generic description is the same as that of *J. sinensis* (Fleisch. ex Broth.) Buck.

1. ***Juratzkaeella sinensis*** (Fleisch. ex Broth.) Buck 中华拟无毛藓 zhong-hua ni wu-mao xian
Rev. Bryol. Lichénol. 43: 313. 1977. *Juratzkaea sinensis* Fleisch. ex Broth., Nat. Pflanzenfam. (ed. 2), 11: 290, fig. 640. 1925. Type. China: Guizhou (Kouy-Tcheou, Kweichow), Pin-fa (Ping-ba), *Cavalerie 1935 p.p.* (lectotype H, designated by Buck 1977).

Plants slender, delicate, yellowish green or light green, glossy, in mats. *Stems* creeping, 1.0–3.5 cm long, irregularly sparsely branched; branches short, usually simple, more or less curved; pseudoparaphyllia foliose; in cross section 2–3 layers of thick-walled epidermal layer, central cortical cells large, thin-walled, central strand absent. *Stem leaves* and branch leaves homogeneous, imbricate, narrowly ovate-lanceolate, acuminate at apex, plane or somewhat concave, not decurrent at base; leaf margins plane, serrulate above the middle; costa single, thin, weak, vanishing about in the middle of leaf, often with rhizoids; median and upper leaf cells elongate-rhomboidal or shortly linear, smooth; alar cells quadrate, in several rows. *Autoicous. Perichaetial leaves* elliptic, acuminate, subentire below, serrulate near apex; costa weak or absent. *Setae* slender, erect to flexuous, brownish red, smooth; capsules erect, symmetrical, cylindrical, yellowish brown or reddish brown; exothecial cells thin-walled; opercula conic-rostrate; annuli differentiated; peristome double, attached at the mouth; exostome teeth lanceolate, with zig-zag median line, densely cross-striolate below the middle, papillae above; endostome segments as long as exostome teeth, keeled, perforate, sparsely papillose; basal membrane low; cilia absent. *Calyptrae* cucullate, smooth. *Spores* minute, spherical, finely papillose.

Illustrations: Plate 429, figs. 1–7.
Habitat: on rocks or tree trunks in forests; alt. 1256 m.
Distribution: endemic to China.

Chinese specimens examined: JIANGSU: Su-zhou City, *Z.-L. Wan 36* (PE). SHANGHAI: Jia-ding Distr., *D.-K. Li 01131* (SHM); Gui-lin Park, *S.-M. Xu 032* (PE). XIZANG: Long-zi Co., Jia-yu, *G. He 959* (HKAS), Sananqulin, *M. Zang 1419* (HKAS); Mo-tuo Co., *Y.-G. Su 3547* (HKAS). YUNNAN: Meng-la Co., *Magill et al. 7662* (MO); Zhong-dian Co., *L.-S. Wang 1718* (HKAS).

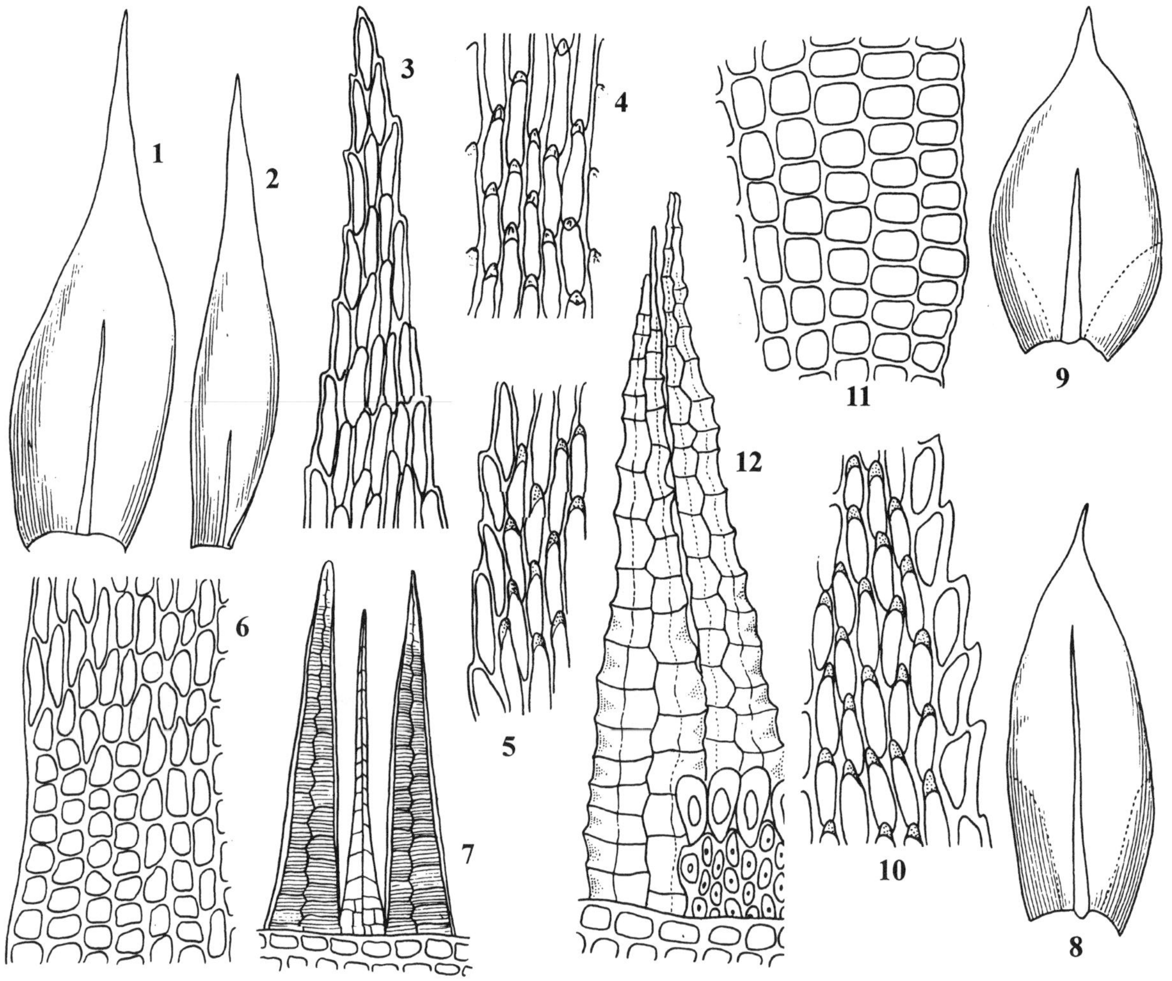

Plate 429. 1–7, *Juratzkaeella sinensis*. 1–2, leaves (× 25). 3, apical leaf cells (× 320). 4, median leaf cells (× 320). 5, median marginal leaf cells (× 320). 6, basal leaf cells (× 320). 7, peristome (× 320). 8–12, *Levierella fabroniacea*. 8–9, leaves (× 28). 10, median marginal leaf cells (× 320). 11, basal marginal leaf cells (× 320). 12, peristome (× 320). (1–7, drawn from *D.-K. Li 1131*, IFSBH; 8–12, from *M.-K. Cui 15*, HKAS.)

7. ***Levierella*** C. Müll. 白翼藓属 bai-yi xian shu
Boll. Soc. Bot. Ital. 1897: 73. 1897.

Plants slender, delicate, green, yellowish green when old, not glossy or weakly glossy, in loose patches. *Main stems* prostrate, loosely foliate; rhizoids brownish, in bundle; irregularly branched, branches caespitose, apex obtuse, curved when dry. *Leaves* erect-spreading, broadly lanceolate to ovate-lanceolate, shortly acuminate at apex; leaf margins plane, serrulate above; costa single, ending in the middle of leaf; leaf cells elliptic or oblong-rhomboidal; alar cells quadrate, in several rows. *Autoicous*. *Female inflorescences* caespitose. *Perichaetial leaves* erect, light green, sheathing at base, tapering to a lanceolate apex. *Setae* slender, erect, yellowish; capsules erect, shortly cylindrical, yellowish brown, apophysis short; opercula conic-rostrate, with a short beak; annuli wide, deciduous when mature; peristome appearing single; exostome teeth lanceolate, yellow, smooth, inserted deeply in the mouth, with zig-zag median line, lamellae remote; endostome absent; cilia none. *Calyptrae* not seen. *Spores* tiny.

There are four species of this genus in the world (Crosby et al. 2000), with two better-understood species. One species, *Levierella fabroniacea* C. Müll., is treated in China. Brotherus (1925) placed this genus in the Entodontaceae; however, P.-C. Chen et al. (1978) considered it as a member of the Fabroniaceae based on its leaves with single costa, leaf cells rhomboidal, and its habitat on tree bark.

1. ***Levierella fabroniacea*** C. Müll. 白翼藓 bai-yi xian
Boll. Soc. Bot. Ital. 1897: 73. 1897. Type. India: 13 XII 1895, *Gollan s.n.*

Plants slender, green or yellowish green, not glossy or weakly glossy, in loosely tufted patches. *Stems* prostrate, irregularly branched; main branches slender, erect or ascending, secondary branches erect, julaceous, secund at the apex when dry. *Leaves* imbricate, concave, ovate or ovate-lanceolate, shortly acuminate at the apex, 1.3 mm × 0.6 mm; margins plane, serrulate near the apex; costa single, ending about midleaf; leaf cells oblong-elliptic or narrowly rhomboidal, 24–38 µm × 4–8 µm, basal cells shorter and broader; alar cells quadrate in numerous rows. *Autoicous*. *Sporophytes* on main branches. *Perichaetial leaves* lanceolate, long acuminate toward the apex. *Setae* erect, 0.8 mm long, yellowish brown; capsules erect, shortly cylindrical; opercula conic; annuli differentiated; peristome appearing single, endostome lacking; exostome teeth lanceolate, not cross-striolate. *Spores* spherical, small.

Illustrations: Plate 429, figs. 8–12.
Habitat: on rocks or tree trunks.
Distribution: China, India, Nepal, and South Africa.

Chinese specimens examined: SICHUAN: Wen-chuan Co., *Allen 6724* (MO), *Redfearn 34854* (MO); Ya-jiang Co., *Handel-Mazzetti 2080* (H). YUNNAN: An-ning Co., *Koponen 37976* (H, MO); Bin-chuan Co., *M.-K. Cui 15* (HKAS); Chu-xiong Co., *Redfearn et al. 1767* (MO); Da-li Co., *Redfearn et al. 1461* (MO); Kun-ming City, *Handel-Mazzetti 8607* (H), *Redfearn et al. 1991* (MO); Lu-nan Co., *Redfearn et al. 2269* (MO); Song-ming Co. *Redfearn et al. 2020* (MO); Wei-shan Co., *Redfearn et al. 1202* (MO); Yang-bi Co., *Redfearn et al. 319* (MO).

8. ***Rozea*** Besch. 小绢藓属 xiao-juan xian shu
Mém. Soc. Sci. Nat. Cherbourg 16: 241. 1872.

Plants slender or somewhat robust, delicate, green, light yellowish green or reddish brown, glossy, in small patches. *Stems* creeping, forkedly or caespitosely branched; branches dense, erect or erecto-patent, obtuse at apex; central strand absent; pseudoparaphyllia foliose. *Leaves* imbricate when dry, sometimes slightly secund, erect or erecto-patent, elliptic-ovate or widely ovate-lanceolate, concave, longitudinally plicate, shortly acute or long acuminate at apex; margins revolute, toothed near the apex; costa single, vanishing in the middle of leaf; leaf cells oblong-elliptic or narrowly rhomboidal, mostly smooth, rarely papillose or projecting; basal cells short, loosely arranged; alar cells quadrate or flat quadrate. *Autoicous* or dioicous. *Inner perichaetial leaves* light green, widely lanceolate, piliferous at apex; costa short. *Setae* slender, erect, reddish; capsules erect, cylindrical, reddish brown, apophysis short; opercula conic, with a short beak; annuli absent; peristome double; exostome teeth lanceolate, yellowish, with zig-zag median line, densely striolate, distinctly trabeculate on inner surfaces; endostome segments light yellow, narrowly lanceolate, as long as exostome teeth, folded, smooth; basal membrane low, rarely high; cilia short, weak. *Spores* spherical, small.

Brotherus (1925) placed *Rozea* in the Entodontaceae. However, P.-C. Chen et al. (1978) considered *Rozea* to be closely related to *Juratzkaea* based on its leaves with slender costa, leaf cells elongate-rhomboidal and double peristome teeth; thus they placed it in the Fabroniaceae. Four species of *Rozea* from China were reported: *R. diversifolia* Broth., *R. microcarpa* Broth. (= *R. fulva* C. Müll. ex Fleisch.), *R. myura* Herz. [= *R. pterogonioides* (Harv.) Jaeg.], and *R. pterogonioides*. Crosby et al. (2000) recognized six species of *Rozea* in the world with five well understood. In this treatment, we recognized three species.

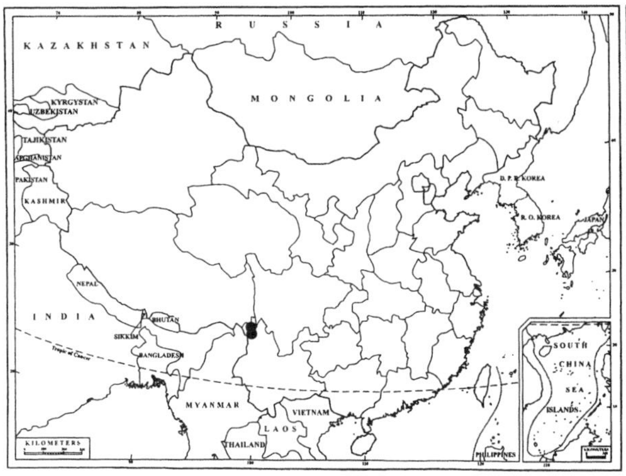

1. *Rozea diversifolia*

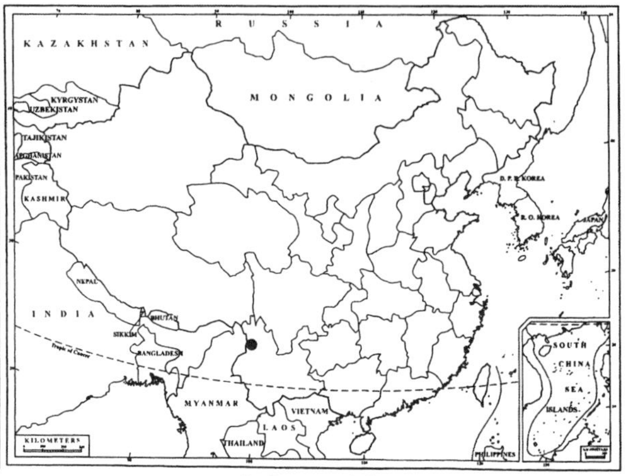

2. *Rozea fulva*

1. Plants medium-sized; leaf apices cucullate, abruptly cuspidate; opercula short and obtuse; endostome weakly developed; basal membrane high .... 3. *R. pterogonioides*
1. Plants relatively delicate; leaf apices acuminate; opercula long rostrate; endostome segments well developed; endostome membrane low 2
   2. Leaves deeply plicate; leaf cells linear .... 1. *R. diversifolia*
   2. Leaves smooth; leaf cells narrowly rhomboidal .... 2. *R. fulva*

1. ***Rozea diversifolia*** Broth. 异叶小绢藓 yi-ye xiao-juan xian
Symb. Sin. 4: 110. 1929. Syntypes. China: Yunnan, Mekong-Salwin-Scheidekette, 28°4′N, *Handel-Mazzetti 8248* (H); Rücken Pongatong, *Handel-Mazzetti 9669* (H).

Plants delicate, yellowish green, glossy, in lax patches. *Stems* prostrate, densely covered with rhizoids, pinnately branched; branches nearly equal length, ca. 1.0 cm long, erecto-patent, densely foliate. *Leaves* imbricate when dry, erect-spreading when moist, concave, semi-tubulose, deeply plicate, oblong-ovate, gradually shortly acute, ca. 1.3 mm long; leaf margins involute, entire; costa ending in the middle of leaf; leaf cells linear; alar cells quadrate or flat quadrate. *Sporophytes* not seen.

Illustrations: Plate 430, figs. 1–5.
Habitat: on branches under *Rhododendron* bushes.
Distribution: endemic to China.

Chinese specimens examined: see the type information cited above.

2. ***Rozea fulva*** C. Müll. ex Fleisch. 小蒴小绢藓 xiao-shuo xiao-juan xian
Hedwigia 61: 405. 1920. Type. Nepal.
Rozea microcarpa Broth. in Handel-Mazzetti, Symb. Sin. 4: 110. 1929. Type. China: Yunnan, Dsuningkou, Dienso and Hodjing, 26°24′N, *Handel-Mazzetti 6589* (holotype H).

Plants slender, slightly yellowish brown, glossy, densely tufted. *Stems* prostrate, densely branched; branches erect or erecto-patent, up to 1 cm long, densely covered with leaves; rhizoids brownish, throughout stems. *Leaves* imbricate when dry, spreading when moist, concave, narrowly ovate, gradually narrowed to a slender apex, up to 1.3 mm long; leaf margins revolute, serrate above; costa slender, ending in the middle of leaf; leaf cells narrowly rhomboidal; basal cells short; alar cells quadrate. *Autoicous*. *Inner perichaetial leaves* erect, lanceolate, with an acuminate apex, serrate above. *Setae* slender, up to 1 cm long, reddish brown; capsules erect, cylindrical, yellowish brown; opercula conic; peristome double; exostome teeth lanceolate, yellowish brown, papillose above, striolate below; endostome segment linear, light color; basal membrane high; cilia none. *Spores* 12–15 μm in diameter, smooth.

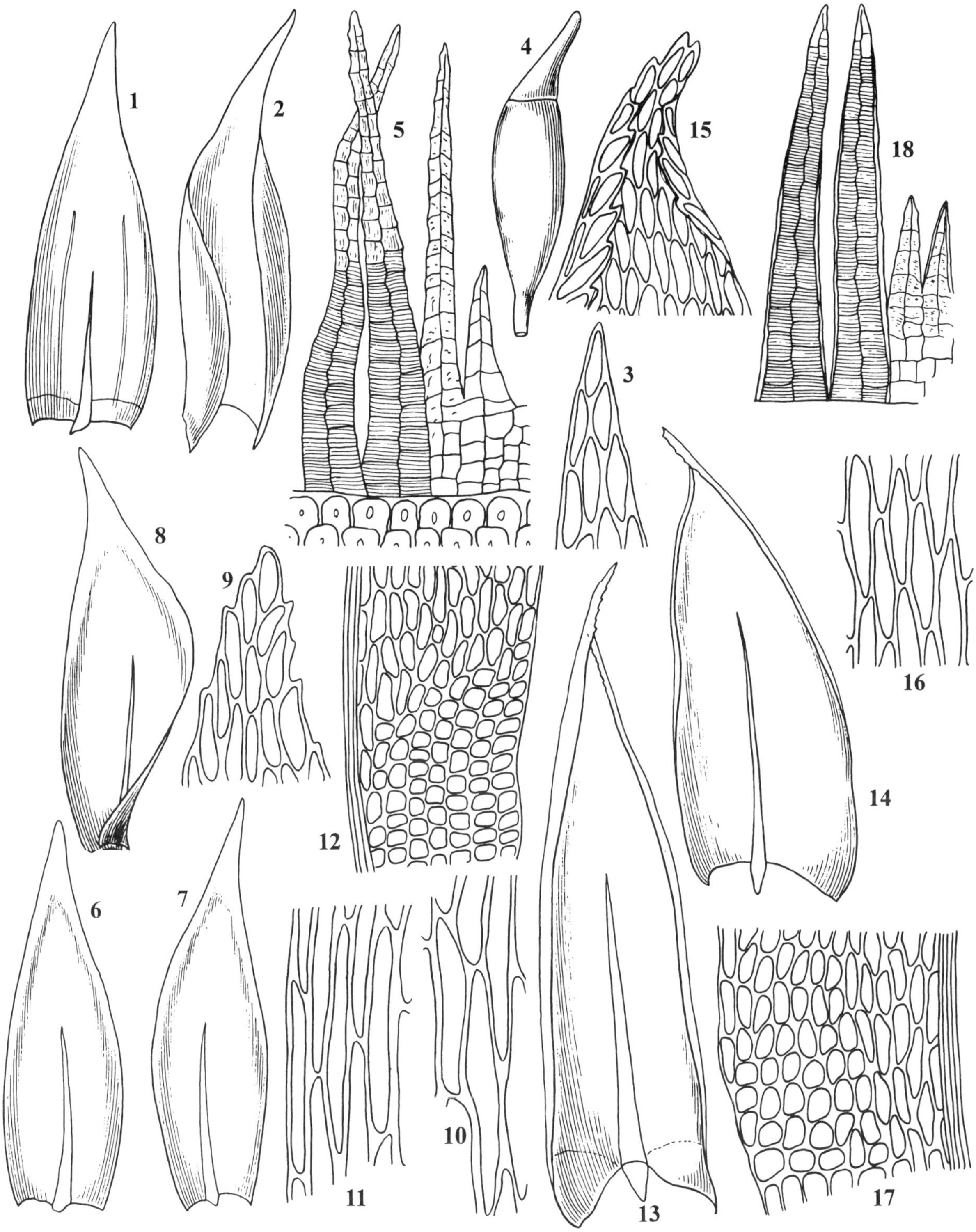

Plate 430. 1–5, *Rozea diversifolia*. 1–2, leaves (× 11). 3, apical leaf cells (× 320). 4, capsule (× 11). 5, peristome (× 320). 6–12, *Rozea fulva*. 6–8, leaves (× 25). 9, apical leaf cells (× 320). 10, median leaf cells (× 320). 11, median marginal leaf cells (× 320). 12, basal leaf cells (× 320). 13–18, *Rozea pterogonioides*. 13–14, leaves (× 11). 15, apical leaf cells (× 320). 16, median leaf cells (× 320). 17, basal leaf cells (× 320). 18, peristome (× 320). (1–5, drawn from *C. Gao 20358*, IFSBH; 6–12, from *M. Zang 5351*, HKAS; 13–18, from *W.-H. Li 25-192*, IFSBH.)

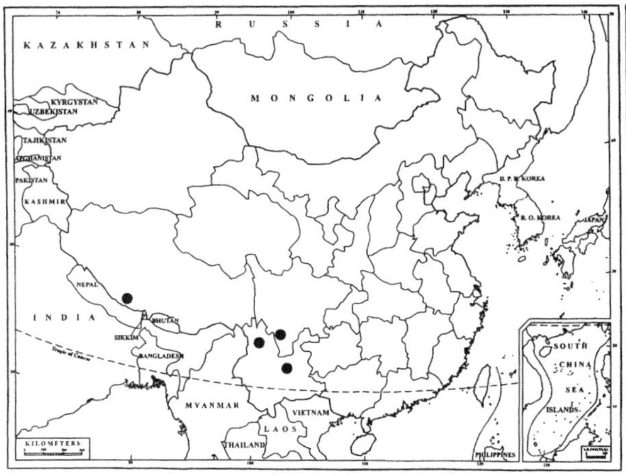

3. *Rozea pterogonioides*

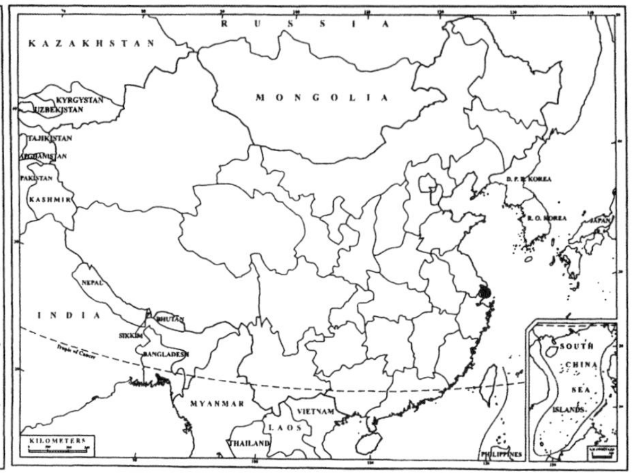

1. *Schwetschkea brevipes*

Illustrations: Plate 430, figs. 6–12.
Habitat: on tree trunks in forests.
Distribution: China, India, and Nepal.

Chinese specimen examined: see the type citation of *Rozea microcarpa* above.

3. ***Rozea pterogonioides*** (Harv. in Hook.) Jaeg. 翼叶小绢藓 yi-ye xiao-juan xian
Ber. Thätigk. St. Gallischen Naturwiss. Ges. 1876–77: 274. 1878. *Leskea pterogonioides* Harv. in Hook., Icon. Pl. 1: 24, f. 8. 1836. Type. Nepal: *Wallich s.n.*
Rozea myura Herz., Hedwigia 65: 161. 1925. Type. China: Yunnan, Kunming?, Pe Tsao, *Ten 83*.

Plants medium-sized, slender, yellowish green, slightly glossy, in dense tufts. *Stems* prostrate, up to 1.2 cm long, irregularly branched; branches short, julaceous; branches ca. 1.0 cm long, densely covered with leaves; rhizoids in bundle. *Leaves* closely imbricate when dry, erect-spreading when moist, oblong-ovate or narrowly ovate, ca. 1 mm long, often concave, bi-plicate; leaf margins incurved or plane, sometimes revolute, serrulate near apex; costa single, ending in the middle of leaf; leaf cells narrowly rhomboidal or narrowly elliptic; alar cells quadrate or laxly arranged, thin-walled. *Dioicous*. *Setae* erect, reddish brown, up to 1.0–1.5 cm long; capsules erect, ovoid, apophysis short; opercula shortly obliquely beaked; peristome double; exostome teeth narrowly lanceolate; endostome segments linear, as long as teeth, keeled; cilia absent. *Spores* 12–16 µm in diameter.

Illustrations: Plate 430, figs. 13–18.
Habitat: on rocks covered in thin layer of soil or tree trunks in forests; alt. 3000–3700 m.
Distribution: China, Bhutan, India, and Nepal.

Chinese specimens examined: SICHUAN: Mu-li Co., *C. Gao 20927* (IFSBH). XIZANG: Ji-long Co., *S.-K. Chen 289* (HKAS). YUNNAN: Li-jiang Co., Mt. Yulong, *C.-W. Wang 3936* (PE).

9. ***Schwetschkea*** C. Müll. 附干藓属 fu-gan xian shu
Linnaea 39: 429. 1875.

Plants slender to very slender, green, slightly glossy, in interwoven patches. *Stems* creeping, ascending at apex; usually pinnately branched; branches short, erect or erecto-patent, curved when dry; rhizoids in bundle; central strand not differentiated. *Leaves* dense, erect, sometimes secund when dry, erecto-patent when moist, ovate-lanceolate, acuminate above; leaf margins plane, serrulate; costa single, weak, ending in the middle of leaf; leaf cells oblong-elliptic or oblong-hexagonal; basal cells broader; alar cells quadrate, in several rows. *Autoicous*. *Inner perichaetial leaves* ovate-lanceolate, suddenly narrowed to a piliferous apex. *Setae* delicate, twisted when dry, reddish brown, rough or smooth above; capsules erect, oblong ovoid, constricted below the mouth when dry;

opercula conic, with a long, oblique beak; annuli differentiated; peristome double; exostome teeth lanceolate, yellowish brown, somewhat lamellate below the middle, smooth, papillose above; endostome segments narrowly lanceolate, as long as exostome teeth, keeled, papillose; basal membrane low or moderately high; cilia absent. *Spores* medium-sized.

Wijk et al. (1967) listed 25 species in *Schwetschkea* in the world. Recently, Crosby et al. (2000) recognized 18 species with only one better understood and 17 less-known species. Most of these species are often briefly mentioned in a checklist or treated by a local flora. P.-C. Chen et al. (1978) reported 8 species of *Schwetschkea* and Koponen (1978) recognized 10 species in China. In this treatment, five species are treated.

1. Leaves and branches erect 2
1. Leaves and branches secund 3
  2. Peristome teeth absent .... 3. *S. gymnostoma*
  2. Peristome teeth present .... 5. *S. sinica*
3. Plants rather thick; branches 4–5 mm long; setae yellowish brown .... 4. *S. laxa*
3. Plants slender; branches 3–4 mm long; setae reddish brown 4
  4. Setae short, up to 1.1 mm long .... 1. *S. brevipes*
  4. Setae rather long, 2–3 mm long .... 2. *S. courtoisii*

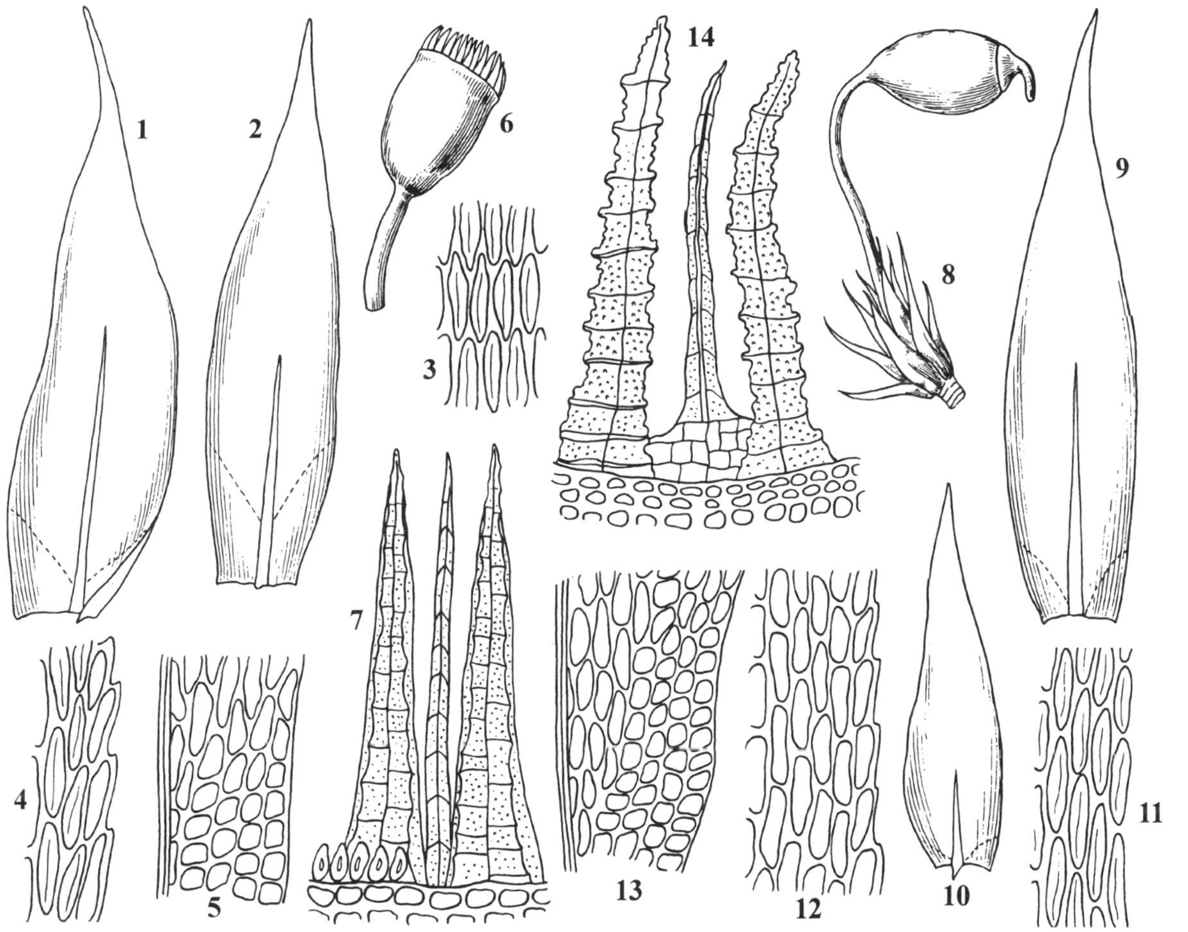

Plate 431. 1–7, *Schwetschkea brevipes*. 1–2, leaves (× 35). 3, median leaf cells (× 230). 4, median marginal leaf cells (× 230). 5, basal leaf cells (× 230). 6, capsule (× 15). 7, peristome (× 230). 8–14, *Schwetschkea courtoisii*. 8, branch with sporophyte (× 15). 9–10, leaves (× 35). 11, median leaf cells (× 230). 12, median marginal leaf cells (× 230). 13, basal leaf cells (× 230). 14, peristome (× 230). (1–7, drawn from *Henry s.n.*, H; 8–14, from *Courtois 119*, H.)

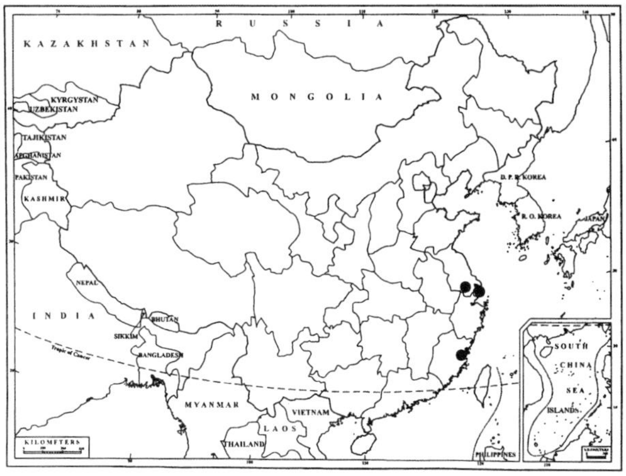

2. *Schwetschkea courtoisii*

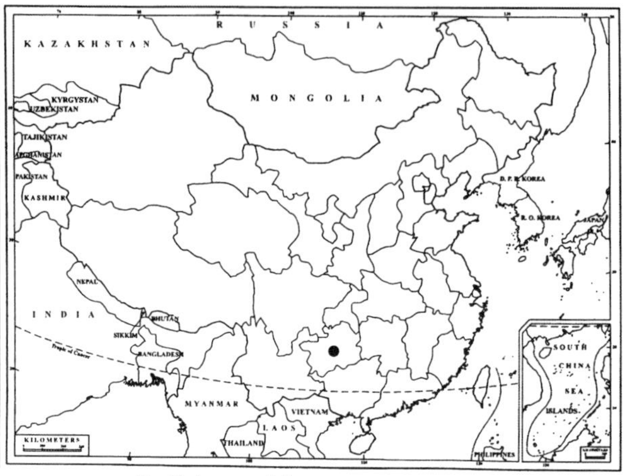

3. *Schwetschkea gymnostoma*

1. ***Schwetschkea brevipes*** (Broth. & Par.) Broth. 短枝附干藓 duan-zhi fu-gan xian
Nat. Pflanzenfam. I(3): 1232. 1909. *Rhynchostegium brevipes* Broth & Par., Rev. Bryol. 33(1): 26. 1906. Type. China: Shanghai, Zi Ka Wei, *Henry s.n.* (holotype H).

Plants slender, yellowish green, not glossy, prostrate, in small patches. *Stems* irregularly branched; branches 3–4 mm long, simple or branched again, erect. *Stem leaves* and branch leaves appressed when dry, secund when moist, narrow at base, ovate-lanceolate, acuminate above, 1.2 mm × 0.5 mm; branch leaves shorter and narrower than stem leaves; leaf margin plane, serrate; costa single, ending at 2/3 length of leaf; median and upper leaf cells elliptic, thin-walled, not clearly pitted; alar cells quadrate, in several rows. *Dioicous*. *Perichaetial leaves* rather small, ovate-lanceolate, acuminate, ecostate. *Setae* 1.1 mm long, laevigate; capsules oval-cylindrical, 1 mm long, curved when dry; opercula with short beaks; peristome double; exostome teeth lanceolate; endostome segments as long as the teeth. *Calyptrae* not seen. *Spores* not seen.

Illustrations: Plate 431, figs. 1–7.
Habitat: on tree trunks.
Distribution: endemic to China.

Chinese specimen examined: see the type citation above.

2. ***Schwetschkea courtoisii*** Broth. & Par. 华东附干藓 hua-dong fu-gan xian
Rev. Bryol. 35: 127. 1908. Type. China: Shanghai, Zi Ka Wei, *Courtois 119* (holotype H).
Schwetschkea incerta Thér., Ann. Crypt. Exot. 5: 178. 1932. Type. China: Fujian, Gu-shan (Mt.), Buong Kang, *H. H. Chung 156* (holotype PC).

Plants slender, yellowish green or brownish green, in creeping patches. *Stems* prostrate, irregularly branched; branches short, 3–4 mm long, mostly ascending, attenuate. *Leaves* spreading, secund when dry, ovate-lanceolate or lanceolate, acuminate; branch leaves small, suddenly acute, 1.2–1.3 mm × 0.5–0.8 mm; leaf margins plane, entire or toothed above; costa single, slender, up to the middle of leaf; median leaf cells oblong-hexagonal, ca. 20 µm × 10 µm; alar cells flat quadrate, in several rows, thin-walled, smooth. *Autoicous*. *Perichaetial leaves* erect, ovate-lanceolate, acute, costa percurrent. *Setae* delicate, erect, 2–3 mm long; capsules ovoid, slightly curved; opercula with a curved beak; peristome double; exostome teeth and endostome segments equal in length. *Spores* ca. 18 µm in diameter.

Illustrations: Plate 431, figs. 8–14.
Habitat: on bark of trees.
Distribution: endemic to China.

Chinese specimens examined: JIANGSU: Su-zhou City, *Z.-L. Wan 35* (MO, PE), *Z.-L. Wan 38* (MO). SHANGHAI: Qing-pu, Zhongshan Park, *S.-P. Xu 317* (PE).

3. ***Schwetschkea gymnostoma*** Thér. 缺齿附干藓 que-chi fu-gan xian
Monde Pl. 9(45): 22. 1907. Type. China: Guizhou (Kouy Tcheou), Tchen-fong, *Esquirol s.n.* (holotype PC).

Plants rather thick, yellowish green or brownish green, slightly glossy, in flat patches. *Stems* creeping, up to 1.5 cm long, irregularly branched; branches 4–5 mm long, branch apices ascending, obtuse. *Leaves* dense, imbricate when dry, erecto-patent when moist, ovate-lanceolate, 0.5–0.7 mm × 0.2–0.3 mm; narrowed at base, acute at apex; leaf margins plane, serrulate above; costa single, ending in the middle of leaf, smooth on the back; upper and median leaf cells narrowly rhomboidal, shorter at leaf base; alar cells flat quadrate, in several rows. *Autoicous*. *Female inflorescences* growing on lateral branches. *Perichaetial leaves* similar to stem leaves. *Setae* 5–6 mm long, brownish, erect or curved; capsules shortly cylindrical, ca. 1.2 mm long, 0.3–0.4 mm in diameter, yellowish brown; opercula flatly conic, with a short, curved beak; annuli undifferentiated, only 2–3 rows of flat quadrate cells along the mouth of capsule; peristome teeth absent. *Spores* not seen.

Illustrations: Plate 432, figs. 1–7.
Habitat: on moist rocks under forest floor.

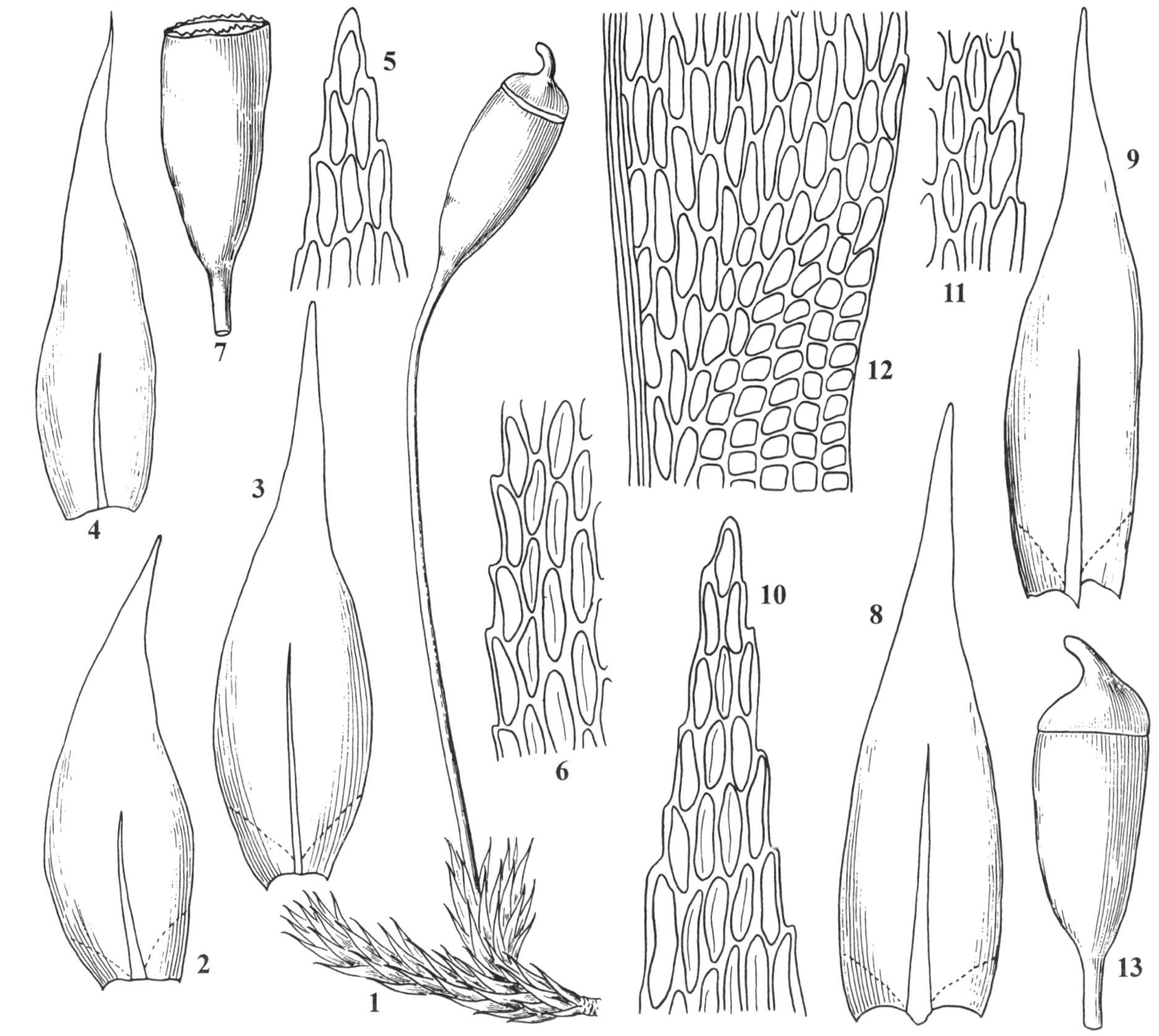

Plate 432. 1–7, *Schwetschkea gymnostoma*. 1, plant (× 8). 2–4, leaves (× 25). 5, apical leaf cells (× 320). 6, median marginal leaf cells (× 320). 7, capsule (× 25). 8–13, *Schwetschkea laxa*. 8–9, leaves (× 25). 10, apical leaf cells (× 320). 11, median marginal leaf cells (× 320). 12, basal leaf cells (× 320). 13, capsule (× 25). (1–7, drawn from *Esquirol s.n.*, PC; 8–13, *X.-Y. Zeng 363*, HKAS.)

Distribution: endemic to China.
Chinese specimen examined: see the type citation above.

This species is characterized by its rather stout plants and the absence of peristome teeth.

4. ***Schwetschkea laxa*** (Wils.) Jaeg. 东亚附干藓 dong-ya fu-gan xian
Ber. Thätigk. St. Gallischen Naturwiss. Ges. 1876–77: 222. 1878. *Neckera laxa* (Wils.) C. Müll., Syn. Musc. Frond. 2: 83. 1850. *Pterogonium laxum* Wils., London Bot. 7: 276. Pl. 10 E. 1848. Type. China: Guangdong, Chuanshan Qundao (Chusan), Islands, *T. Anderson s.n.*
Schwetschkea matsumurae Besch., J. Bot. (Morot) 13: 40. 1899. Type. Japan: Tokyo, Middle Honshu.
Schwetschkea sublaxa Broth. & Par., Rev. Bryol. 35: 41. 1908. Type. China: Shanghai, Zô cè, *P. Henry s.n.*

Plants slender, yellowish green or brownish green, not glossy to weakly glossy. *Stems* prostrate, irregularly branched, ascending; branches 3–5 mm long, julaceous. *Stem leaves* and branch leaves similar, not appressed when dry, not transparent, ovate-lanceolate, 1.0–1.2 mm × 0.3–0.5 mm, narrowed at base, tapering to a shortly acute apex; leaf margins straight, nearly entire; costa thick, vanishing in the middle of leaf; upper and median leaf cells rhomboidal to elliptic, 40–65 μm × 8–13 μm, thin-walled; apical leaf cells shorter; alar cells clearly differentiated, flatly quadrate, in several rows, thin-walled. *Autoicous*. *Setae* short, 2–3 mm long, smooth, yellowish brown; capsules ovoid, erect or slightly inclined; opercula conic-rostrate; peristome double; exostome teeth lanceolate, finely papillose; endostome segments nearly as long as the teeth, narrowly lanceolate, keeled. *Spores* 18–30 μm in diameter.

Illustrations: Plate 432, figs. 8–13.
Habitat: on bark of trees; alt. 750–3600 m.
Distribution: China and Japan.

Chinese specimens examined: CHONGQING: Chong-qing City, *T. Cao & Q. Li 41384* (IFSBH); FUJIAN: Collector unknown *387* (SYS). HUNAN: Chang-sha City, *G.-Z. He 1321* (PE). JIANGSU: Wu-xi City, *Z.-L. Li 38, 72* (PE); Nan-jing City, *Nanjiang Normal College 1955* (PE). SICHUAN: Dao-cheng Co., *X.-J. Li 3025* (HKAS). TAIWAN: Tai-pei Co., *M.-J. Lai 1351* (MO). YUNNAN: An-ning Co., *S.-T. Zeng 363* (HKAS); Da-guan Co., *S. Smith 6095* (HKAS); Kun-ming City: *X.-J. Li 2052* (HKAS).

This species is characterized by having opaque leaves and shorter and broader leaf cells (40–62 × 8–13 μm).

5. ***Schwetschkea sinica*** Broth. & Par. 中华附干藓 zhong-hua fu-gan xian
Rev. Bryol. 38: 56. 1911. Type. China: Jiangsu (Kang Sou), Tsing Kia K'iao, *Courtoisi s.n.*

Plants delicate, brownish green, not glossy, in flat patches. *Stems* prostrate, up to 8 mm long, ascending above; irregularly branched; branches 3–4 mm long, erecto-patent or creeping, apex ascending; rhizoids numerous. *Leaves* imbricate, dense, appressed when dry, erecto-patent when moist, widely ovate or widely ovate-lanceolate, 0.5–0.7 mm × 0.2–0.3 mm, gradually tapering to a slender apex; leaf margins plane, serrulate above; costa single, up to above the middle of leaf, smooth on the back; upper and median leaf cells narrowly rhomboidal or narrowly oblong-elliptic; basal leaf cells shorter; alar cells clearly differentiated, flat-quadrate, in several rows. *Autoicous*. *Perichaetial leaves* erect, similar to stem leaves, somewhat broader, lighter in color. *Setae* 1.2–1.3 cm long, erect, reddish brown; capsules shortly cylindrical, 1.0–1.1 mm × 0.3–0.4 mm; annuli undifferentiated, with only 2–3 rows of smaller cells around the mouth; opercula conic, with a short, oblique beak. *Spores* not seen.

Illustrations: No reference.
Habitat: on tree trunks or moist rock surfaces.
Distribution: endemic to China.

Chinese specimen examined: ANHUI: Dun-si Co., *J.-R. Xu 76* (PE).

The distinct characters of this species are its small and delicate plants with erecto-patent leaves on stems and branches.

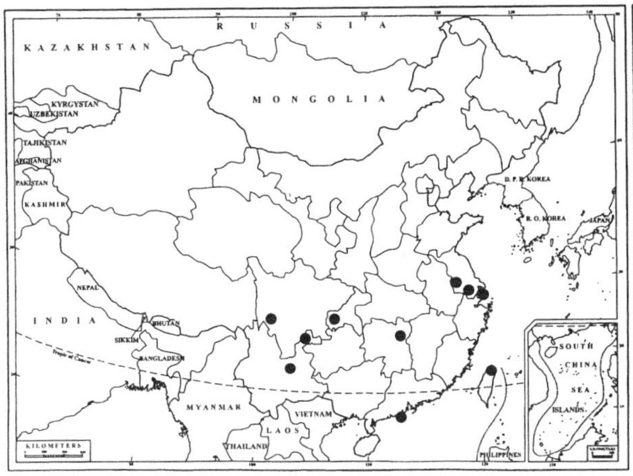

4. *Schwetschkea laxa*

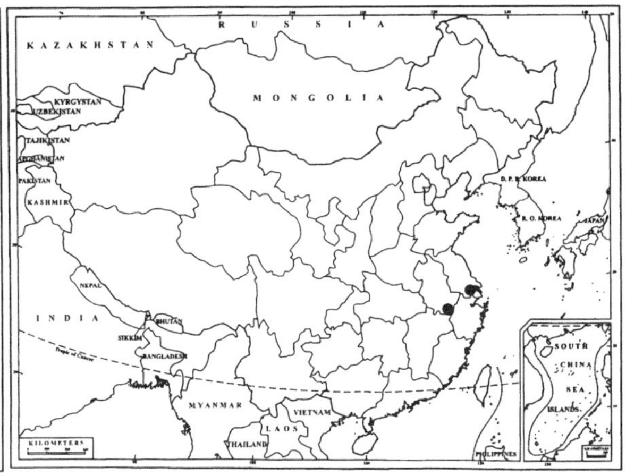

5. *Schwetschkea sinica*

10. ***Schwetschkeopsis*** Broth. 拟附干藓属 ni fu-gan xian shu
Nat. Pflanzenfam. I(3): 877. 1907.

Plants slender, rigid, green or yellowish green, slightly glossy. *Stems* creeping, ascending at apex, pinnately branched or irregularly pinnately branched; branches short, simple or branched again, erect or erecto-patent, densely foliate, julaceous; paraphyllia sparse, lanceolate or linear. *Branch leaves* imbricate, erect or erecto-patent when moist, ovate-lanceolate, concave, acuminate at apex; leaf margins plane, serrulate; costa absent; leaf cells narrowly hexagonal or oblong-elliptic, weakly papillosely projecting in the front ends of cells; basal cells shorter and lax; alar cells flat quadrate, in several rows. *Dioicous* or monoicous. *Inner perichaetial leaves* sheathing at base, tapering to an acuminate or ciliate apex. *Setae* slender, incurved, twisted when dry, orange in color; capsules erect, rounded ovoid, symmetrical or slightly asymmetrical, apophysis short; opercula conic, with a short or long beak; peristome double; exostome teeth lanceolate, yellowish, with zig-zag median line, densely striate below, trabeculate or lamellate inside; endostome segments nearly as long as the teeth, hyaline, narrowly lanceolate, keeled, perforate, densely finely papillose; basal membrane ca. 1/3 length of exostome teeth, smooth; cilia present, but rudimentary. *Calyptrae* cucullate. *Spores* spherical.

Wijk et al. (1967) listed 5 species under *Schwetschkeopsis* in the world. Four species are currently well understood (Crosby et al. 2000). Two species are known from China.

1. Plants densely branched; leaves ecostate .... 1. *S. fabronia*
1. Plants loosely branched; leaves short, double-costate .... 2. *S. formosana*

1. ***Schwetschkeopsis fabronia*** (Schwaegr.) Broth. 拟附干藓 ni fu-gan xian
Nat. Pflanzenfam. I(3): 878. 1907. *Helicodontium fabronia* Schwaegr., Sp. Musc. Frond., Suppl. 3(2): 294. 1830. Type. Nepal: *Wallich s.n.*
Schwetschkeopsis denticulata (Sull.) Broth., Nat. Pflanzenfam. I(3): 878. 1907. *Leskea denticulata* Sull., Musci Allegh. 19. 1846.
Schwetschkeopsis japonica (Besch.) Broth., Nat. Pflanzenfam. I(3): 878. 1907. *Schwetschkea japonica* Besch., Ann. Sci. Nat., Bot., ser. 7, 17: 362. 1893.

Plants slender, green or yellowish green, glossy, in flat patches. *Stems* creeping, pinnately branched, branches in equal length, not circinate when dry, julaceous; rhizoids numerous, scattered. *Leaves* ovate-lanceolate, 0.5–0.7 mm long, concave, shortly acute at apex; leaf margins plane, serrulate; costa absent; leaf marginal cells thin-walled; median leaf cells elliptic, 20–28 µm × 4–7 µm, often mammillose on the back; alar cells clearly differentiated, quadrate or flat quadrate; branch leaves similar to stem leaves, slightly smaller. *Dioicous*. *Setae* 4–8 mm long, erect; capsules oblong-ovoid, symmetrical; peristome double; exostome teeth papillose above, striate below; endostome segments as long as the teeth; cilia present. *Calyptrae* cucullate. *Spores* 13–20 µm in diameter.

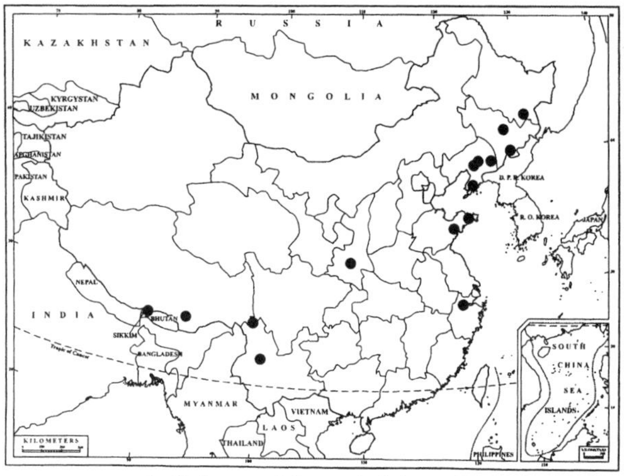

1. *Schwetschkeopsis fabronia*

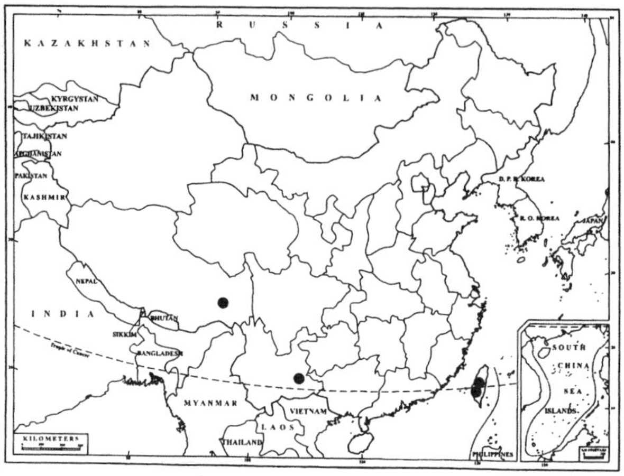

2. *Schwetschkeopsis formosana*

Illustrations: P.-C. Wu (ed.) 2001 (Pl. 48, figs. 1–6).
Habitat: on tree trunks in broad-leaved forests, rarely on moist rocks.
Distribution: China, Korea, Japan, and North America.

Chinese specimens examined: HEILONGJIANG: Ning-an Co., *C. Gao 6159* (IFP). JILIN: Chang-bai Co., *C. Gao 7193* (IFSBH); Jiao-he Co., *X.-G. Kang 1986* (IFSBH). LIAONING: An-shan City, Qianshan, *C. Gao 1904* (IFSBH); Ben-xi, *C. Gao 7055, 7067* (IFSBH); Heng-ren Co., *T. Cao 153* (IFSBH); Zhuang-he Co., *C. Gao 5787* (IFSBH). SHAANXI: Hu-xian Co., *X. Wang 320* (IFSBH), *Z.-P. Wei 4554* (IFSBH). SHANDONG: Qing-dao City, Mt. Laoshan, *C. Gao & N.-Z. Bai 34515, 34255* (IFSBH); Mt. Kun-lun, *E.-Z. Bai 34268, 24501* (IFSBH) . XIZANG: Long-zi Co., *M. Zang 1313* (HKAS); Ya-dong Co., *M. Zang 87* (HKAS). YUNNAN: De-qin Co., *L.-S. Wang 2091* (HKAS); Yang-bi Co., *Redfearn et al. 550a* (MO). ZHEJIANG: Hang-zhou City, *B. C. Tan 93-581* (FH, MO).

This species is characterized by having dense branches and ecostate leaves.

2. ***Schwetschkeopsis formosana*** Nog. 台湾拟附干藓 tai-wan ni fu-gan xian
J. Hattori Bot. Lab. 5: 41. 1951. Type. China: Taiwan, Tai-nan Co., Mt. Niitaka, *Noguchi 6829* (holotype NICH).

Plants delicate, yellowish green, glossy. *Stems* creeping, up to 6 cm long, rarely branched or nearly pinnately branched; branches 5–10 mm long, julaceous. *Leaves* closely imbricate when dry, erect-spreading when moist, narrowly ovate or oblong-lanceolate, 1.0–1.3 mm × 0.4–0.5 mm, acuminate above, concave below; upper leaves smaller, 0.6–0.7 mm × 0.20–0.25 mm; leaf margins plane, serrulate above; costa slender, forked; median leaf cells oblong-elliptic, 35–49 µm × 4.5–6.0 µm, projecting at upper ends; alar cells quadrate or flat quadrate, in several rows; branch leaves smaller, similar to stem leaves, ecostate or costa indistinct. *Sporophytes* not seen.

Illustrations: Noguchi 1951 (Fig. 2, A–H).
Habitat: on bark of trees in alpine regions.
Distribution: endemic to China.

Chinese specimens examined: TAIWAN: Between Nan-tou Co. and Hua-lien Co., *S.-H. Lin s.n.* [Bryophytes of Taiwan, Exsiccatae ser. 3, no. 144] (MO). XIZANG: Bo-mi Co., Tong-mai, *J.-W. Zhang 15031* (PE). YUNNAN: Qiu-bei Co., *X.-J. Li s.n.* (HKAS).

LESKEACEAE

薄罗藓科 bo-luo xian ke

by Cao Tong, Sun Jun, and Gao Chien

Plants mostly slender, dull or slightly shiny, in dense patches. *Stems* creeping, often irregularly branched; branches slender, erect or ascending; paraphyllia few, unbranched, or absent; central strand poorly developed. *Stem leaves* and branch leaves nearly uniform, ovate or ovate-lanceolate; costa strong, mostly single, reaching the middle or the apex of leaves, rarely rather short or lacking; leaf cells mostly parenchymatous, rarely elongate, smooth or unipapillose. *Monoicous* or dioicous. *Perichaetia* on stems, perigonia usually on lateral branches. *Setae* elongate, erect, smooth or coarsely papillose; capsules mostly erect, sometimes pendent, asymmetric; stomata phaneropore; peristome double; exostome teeth lanceolate or shortly lanceolate, lamellate or sometimes keeled; endostome segments varying greatly; basal membrane present, high; cilia usually rudimentary; opercula obtusely conic, with a short rostrum. *Calyptrae* cucullate, usually smooth, rarely with hairs. *Spores* small, spherical.

Species of the Leskeaceae are widely distributed in temperate zones. They are usually epiphytic on tree trunks or growing on surfaces of rocks. There are different viewpoints about the family's delimitation and its systematic positions. We basically adopted the system of Brotherus (1924) and P.-C. Chen et al. (1978). The genus *Okamuraea* is transferred from the Rhytidiaceae to this family. The Leskeaceae have a close affinity with the Thuidiaceae, but the species of the present family can be distinguished from the Thuidiaceae by the following features: 1) paraphyllia absent or few, unbranched, smooth; 2) leaf cells smooth or unipapillose; 3) capsules mostly erect, symmetric; 4) endostomes varying greatly among genera and cilia usually rudimentary, rarely well developed. Nine genera are recognized for the Leskeaceae in this treatment.

1. Plants rigid; endostome evidently longer than the exostome .... 9. *Regmatodon*
1. Plants mostly slender, delicate; endostome as long as the exostome or slightly shorter 2
   2. Leaf costa single, ending below the midleaf or forked, rarely absent; leaf margins entire; leaf cells rounded, rounded ovate or shortly rhomboidal, smooth or mamillose .... 7. *Pseudoleskeella*
   2. Leaf costa single, reaching above midleaf; leaf margins serrulate; leaf cells rhomboidal, hexagonal or elongate-rhomboidal to nearly linear, mostly papillose or projecting at the upper ends, rarely smooth 3
3. Endostome segments rudimentary or absent 4
3. Endostome segments relatively well developed 5
   4. Plants small; leaves not plicate; leaf cells mostly rounded hexagonal or rhomboidal .... 4. *Lindbergia*
   4. Plants rather large; leaves often plicate; leaf cells long rhomboidal .... 5. *Okamuraea*
5. Capsules inclined, not symmetrical; apophysis conspicuous; cilia well developed .... 8. *Pseudoleskeopsis*
5. Capsules erect, symmetrical; apophysis inconspicuous; cilia not rudimentary or absent 6
   6. Endostome segments irregularly shaped, not well developed 7
   6. Endostome segments regularly shaped, well developed 8
7. Plants rigid; leaves ovate-lanceolate, conspicuously plicate; leaf cells rounded hexagonal .... 3. *Leskeella*
7. Plants delicate; leaves triangular-lanceolate, inconspicuously plicate; leaf cells long rhomboidal to nearly linear .... 6. *Orthoamblystegium*
   8. Dioicous; leaves acuminate at apex; leaf cells long rhomboidal or nearly linear, mostly projecting at the upper ends .... 1. *Lescuraea*
   8. Autoicous; leaves abruptly tapering to acute or acuminate at apex; leaf cells rounded to rounded hexagonal, smooth or papillose .... 2. *Leskea*

1. ***Lescuraea*** Bruch & Schimp. in B.S.G. 多毛藓属 duo-mao xian shu

Bryol. Eur. 5: 101. 1851.
Pseudoleskea sect. *Lescuraea* (Bruch & Schimp. in B.S.G.) Best, Bull. Torrey Bot. Club 27: 233. 1900.
Pseudoleskea Bruch & Schimp. in B.S.G., Bryol. Eur. 5: 147. 1852.
Ishibaea Broth. & Okam., Bot. Mag. (Tokyo) 29: 186. 1915.

Plants slender, delicate, bright green to yellowish green, slightly shiny, in mats. *Primary stems* prostrate, regularly to irregularly sparsely branched; branches with dense leaves, erect or curved, obtuse at tips; paraphyllia few to abundant on stems, sparse on branches, short, filiform or triangularly lanceolate. *Stem leaves* oblong-ovate, narrowly decurrent at base, acuminate upward; margins plane or recurved, serrate above; costa single, thick, strong, up to or ending below leaf apex, projecting at tips; leaf cells thin-walled, elongate to linear; upper cells often projecting at upper ends on back surfaces; alar cells quadrate to flatly quadrate; branch leaves similar to stem leaves, slightly narrower and smaller, costa more distinctly projecting on the back. *Dioicous*. *Inner perichaetial leaves* larger than stem leaves, long-sheathing at base, narrowly acuminate above, entire, costa short. *Perigonia* on stems or branches. *Setae* elongate, erect, smooth; capsules erect, symmetric or pendent, asymmetric; annuli differentiated; peristome double; exostome teeth narrowly lanceolate, fused at base, transversely striolate below, smooth or finely papillose; endostome segments membranous, smooth or finely papillose; basal membrane high or low; cilia usually absent; opercula conic, with short and erect rostrum. *Spores* medium-sized, papillose.

Numerous bryologists (e.g. Lawton 1957) considered *Pseudoleskea* to be synonym of *Lescuraea*. This concept is accepted in this treatment. There are about 39 species of *Lescuraea* in the world.

The first record of Chinese *Lescuraea* was published by C. Müller (1896) as *Pseudoleskea papillarioides* C. Müll. (= *Leskeella nervosa*) from Shaanxi Province. Later, Brotherus (1929) described *Pseudoleskea setschwanica* Broth. and *P. yunnanensis* Broth. from Sichuan and Yunnan. Takaki (1955) published *Pseudopleuropus morrisonensis* Tak. [= *Lescuraea morrisonensis* (Tak.) Nog.] from Taiwan. For Taiwan, M.-J. Lai (1976) reported three additional species: *Lescuraea incurvata* (Hedw.) Lawton., *L. julacea* Besch et Card. [= *L. saxicola* (B. S. G.) Milde] and *L. saxicola* (B.S.G.) Milde. P.-C. Chen et al. (1978) listed 6 species of *Lescuraea* from China, of which *L. mutabilis* (Brid.) Lindb. ex Hag. was reported new to the country. *Lescuraea morrisonensis* fo. *sichuanensis* Y.-F. Wang, R.-L. Hu & Redfearn from Sichuan was synonymized under *Actinothuidium*. Six species are recognized in the present treatment. *Lescuraea morrisonensis* (Tak.) Nog. was not treated here because no specimens were available for study.

The genus can be divided into two subgenera based on endostome structures: subg. *Lescuraea* and subg. *Pseudoleskea*. Lawton (1957) revised the taxa of the genus from Europe and North America. Since subdivisional levels under families and genera are not adopted in this flora, no separate key to subgenera is provided.

1. Plants medium-sized to rather large; leaf cells isodiametric or rhomboidal; capsules inclined; endostome segments as long as the teeth, basal membrane high (subg. *Pseudoleskea*) 2
1. Plants small; leaf cells long rhomboidal or nearly linear; capsules erect; endostome segments shorter than the teeth, basal membrane low (subg. *Lescuraea*) 5
   2. Leaves broadly ovate and decurrent at base, acuminate toward apex; branch apices with multi-flagelliform branches; leaf margins plane ..... 5. *L. setschwanica*
   2. Leaves ovate or oblong-ovate and not decurrent at base, lanceolate toward apex; branch apices without flagelliform branches; leaf margins revolute 3
3. Leaf margins clearly revolute from base to near leaf apex on both sides; median leaf cells rectangular to oblong-rectangular ..... 6. *L. yuennanensis*
3. Leaf margins only revolute from base to the midleaf on both sides; median leaf cells oblong-rounded or oblong-rhomboidal 4
   4. Median leaf cells oblong-rounded, 10–12 µm long; leaves secund ..... 1. *L. incurvata*
   4. Median leaf cells oblong-rhomboidal, 15–20 µm long; leaves erect or slightly secund ..... 3. *L. radicosa*
5. Stem leaves ovate-lanceolate, gradually long acuminate toward apex; costa slender; median leaf cells nearly linear, 9–12:1 (length/width) ..... 2. *L. mutabilis*
5. Stem leaves triangular-lanceolate, shortly acute toward apex; costa thick; median leaf cells oblong-rhomboidal, 5–9:1 (length/width) ..... 4. *L. saxicola*

1. ***Lescuraea incurvata*** (Hedw.) Lawton 弯叶多毛藓 wan-ye duo-mao xian

Bull. Torrey Bot. Cl. 84: 290. 1957. *Pseudoleskea incurvata* (Hedw.) Loeske, Hedwigia 50: 313. 1911. *Leskea incurvata* Hedw., Sp. Musc. Frond. 216. 1801. Type. Europe.

Plants medium-sized, dull, in dense mats. *Stems* flexuous, up to 3 cm long, loosely to densely subpinnately branched; branches simple, ca. 5 mm long, with dense leaves; paraphyllia lanceolate, abundant on stems, few on branches. *Stem leaves* appressed on stems when dry, obviously secund, obliquely arranged when moist, obovate-lanceolate, concave, acute or gradually to mucronate, 0.9–1.2 mm × 0.5–0.6 mm; leaf margins recurved, somewhat toothed at leaf apex; costa thick, strong, nearly percurrent, with small teeth at back surfaces; median leaf cells varied greatly in shape and size, from rounded hexagonal to oblong, 10–12 µm × 4.0–4.5 µm, projecting at upper ends,

1. *Lescuraea incurvata*

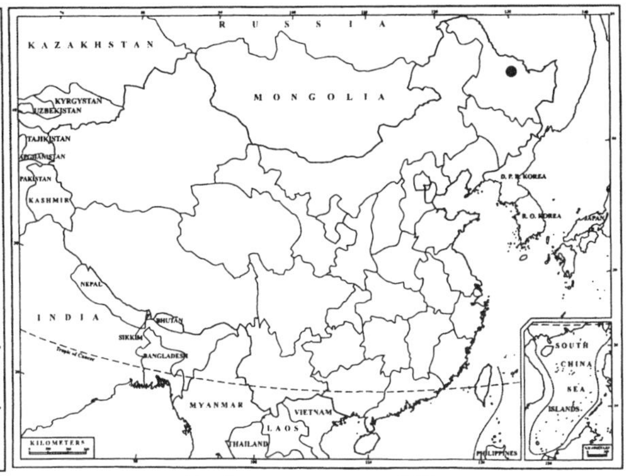

2. *Lescuraea mutabilis*

somewhat thin-walled; upper cells long; basal cells hexagonal, thin-walled; alar cells subquadrate, thin-walled; branch leaves similar to stem leaves. *Dioicous*. *Inner perichaetial leaves* oblong-lanceolate, acuminate, up to 0.8 mm long, costa weak. *Setae* dark brown, ca. 8 mm long; capsules pendent, obovoid to ellipsoid, ca. 1.0 mm long, 0.6 mm in diameter; peristome double; exostome teeth lanceolate, 0.35 mm long, densely striate below, verrucose above, lamella high; endostome segments as long as the exostome teeth, linear-lanceolate, folded, verrucose; cilia absent. *Spores* 9–15 µm in diameter.

Illustrations: P.-C. Wu (ed.) 2001 (Pl. 63, figs. 1–8).
Habitat: on moist soil, rocks or tree bases at high elevation.
Distribution: China, Japan, Europe, and North America.

Chinese specimens examined: NEI MONGOL: Mt. He-lan, *Z.-G. Gong 1338* (HIMC, IFSBH). SICHUAN: Ba-tang, *S. He 31646* (MO); Dao-cheng Co., *M.-Z. Wang 814221* (IFSBH, PE); Xiang cheng Co., *S. He 31799* (MO); Ya-jiang, *S. He 31890*a (MO). XIZANG: Cha-yu Co., *S.-G. Wu 1096, 1097 1118* (HKAS, IFSBH, *M.-Z. Wang 11887* (IFSBH, PE); Chang-Du Co., *J.-W. Zhang 1158* (IFSBH, PE); Mi-lin Co., *S.-K. Chen 148, 150, 153* (HKAS, IFSBH). YUNNAN: Bao-shan Co., *J.-X. Lou 811763* (IFSBH, PE); Bi-luo-xue Shan (Mt.), *M.-Z. Wang 6742* (IFSBH, PE); De-qing Co., *L.-S. Wang 1817* (HKAS, IFSBH); Fu-gong Co., *L.-S. Wang 594* (HKAS, IFSBH); Li-jiang Co., *L.-S. Wang 382* (HKAS, IFSBH).

The leaf shape of this species varies greatly. The leaf apices can be acute or obtuse. The strongly secund leaves toward one side and short leaf cells (10–12 µm long) projecting at upper ends make this species distinct.

2. ***Lescuraea mutabilis*** (Brid.) Lindb. ex Hag. 多毛藓 duo-mao xian
Kongel. Norske Vidensk. Selsk. Skr. (Trondheim) 1908 (9): 53. 1909. *Hypnum mutabile* Brid., Muscol. Recent. 2 (2): 170. 1801. Type. Switzerland: Alps, without specific locality.

Plants small, soft, green, slightly shiny, in dense tufts. *Stems* creeping, slender, irregularly branched; branches short, simple; paraphyllia few, small, linear to lanceolate. *Stem leaves* lanceolate, acuminate, 1.8 mm × 0.45 mm, deeply plicate; margins recurved, dentate at apex; costa thick, strong, percurrent, smooth; median leaf cells sublinear, projecting at upper ends, 40–45 µm × 4–5 µm, thin-walled; alar cells distinctly differentiated, mostly quadrate to flatly quadrate, thin-walled, 8–12 µm wide; branch leaves similar to stem leaves, but rather smaller, upper costa dentate on dorsal side. *Inner perichaetial leaves* lanceolate, up to 1.3 mm long, from oblong-ovate base narrowed to a lanceolate apex; costa ending at middle; margins dentate above. *Setae* slender, 1.8 mm long, dark brown; capsules erect, oblong-ellipsoidal to cylindrical, 1.6–2.0 mm long, 0.55–0.60 mm thick; peristome double; exostome teeth lanceolate, 0.25 mm long, yellow or dark brown, transversely plicate below, hyaline, smooth above, lamellae low; endostome segments fragile, linear, ca. 1/2 the length of teeth; cilia absent or rudimentary; opercula with long rostrum, 0.5 mm long. *Spores* spherical, 16–22 µm in diameter.

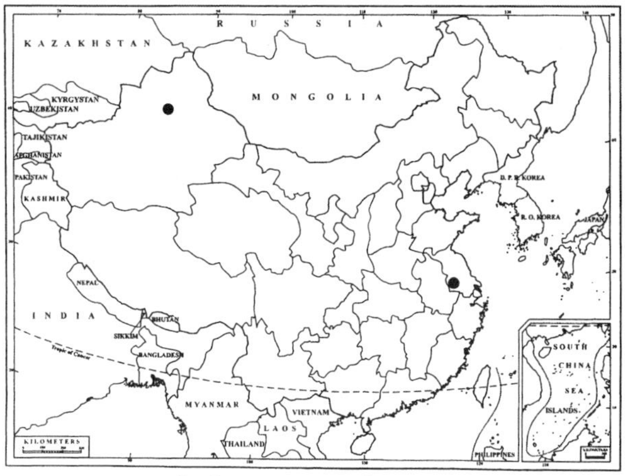

3. *Lescuraea radicosa*

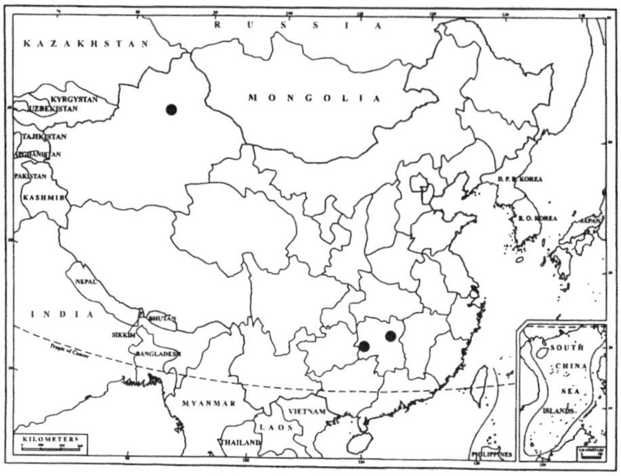

4. *Lescuraea saxicola*

Illustrations: Noguchi 1991 (Fig. 353, A).
Habitat: on rocks
Distribution: China, Japan, India, North Africa, Russia, southern Europe, and North America.

Chinese specimens examined: HEILONGJIANG: Xiao Xing An Ling (Lesser Khingan Mt.), Wu-ying Dist., *C. Gao 0011724* (IFSBH).

This species is similar to *L. saxicola*, but differs by having elongate and thick-walled leaf cells.

3. ***Lescuraea radicosa*** (Mitt.) Moenk. 密根多毛藓 mi-gen duo-mao xian
Laubm. Eur. 693. 1927. *Pseudoleskea radicosa* (Mitt.) Macoun & Lindb., Cat. Canad. Pl., Musci 6: 181. 1892. *Hypnum radicosum* Mitt., J. Linn. Soc. Bot. 8: 31. 1864. Type. Canada: British Columbia, *Drummond 225*.

Plants soft, yellowish green to brownish green, slightly shiny, in dense tufts. *Primary stems* creeping, with abundant fasciculate rhizoids, irregularly branched; branches mostly curved in arch-shape, blunt at tips; paraphyllia short, small. *Leaves* slightly secund, up to 1.5–1.9 mm × 0.5–0.7 mm, gradually tapered from subdecurrent, ovate to elliptic base to an apiculate apex, biplicate at base; margins narrowly recurved, entire or dentate above; costa ending at apex, serrate on back surface; leaf cells usually smooth or faintly projecting at upper ends on back surfaces; median cells hexagonal to long-rhomboidal, 15–20 µm × ca. 5 µm; basal cells rather wide, oblong; alar cells quadrate; branch leaves slightly secund, margins serrate. *Dioicous*. *Setae* 5–7 mm long, slightly curved; capsules pendent, long-ellipsoidal; peristome double; exostome teeth lanceolate, densely striate below, verrucose above; endostome segments as long as exostome teeth, somewhat perforate. *Spores* 14–18 µm in diameter, reddish brown, papillose.

Illustrations: P.-C. Wu (ed.) 2001 (Pl. 63, figs. 9–16).
Habitat: on soil or rocks in forests.
Distribution: China, Japan, Europe, and North America.

Chinese specimens examined: JIANGSU: Nanjing City, *M. Zang 61914* (HKAS, IFSBH). XINJIANG: Tian Shan (Mt.), CAS glacier station, *B.-C. Tan 93-970*(FH).

This species is similar to *L. incurvata*. The difference between the two species is that the leaf cells of *L. radicosa* are longer (more than 15 µm long), while those of *L. incurvata* are shorter (less than 12 µm long).

4. ***Lescuraea saxicola*** (Schimp. in B. S. G.) Mol. in Lor. 石生多毛藓 shi-sheng duo-mao xian
Moosstudien 144. 1864. *Lescuraea striata* var. *saxicola* Schimp. in B. S. G., Bryol. Eur. 5: 103, t. 459. 1851. Type. Europe: Alps, without specific locality.
Lescuraea julacea Besch. & Card., Bull. Soc. Bot. Genève, ser. 2, 3: 284. 1911.

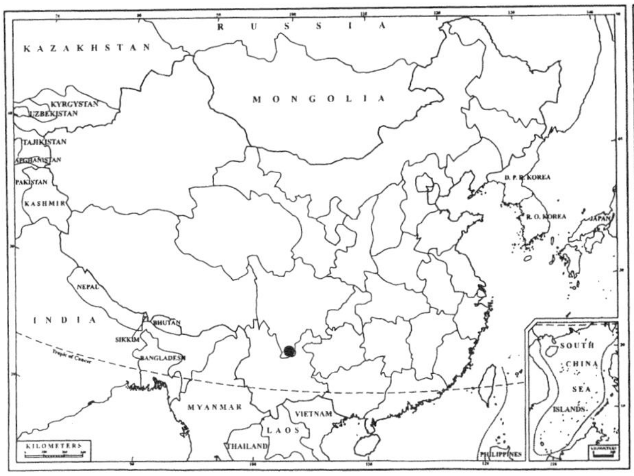

5. *Lescuraea setschwanica*

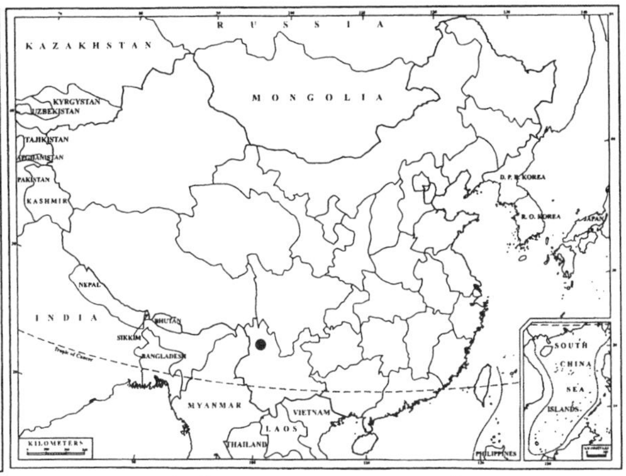

6. *Lescuraea yuennanensis*

Plants in dense tufts. *Stems* creeping, subpinnately or pinnately branched; branches up to 5 cm long, curved or erect; paraphyllia few, triangular to lanceolate. *Stem leaves* appressed when dry, ovate-lanceolate, acuminate to acute or mucronate at apex, 1.0 mm × 0.4 mm, concave, almost without plication; leaf margins recurved almost throughout; costa thick, strong, percurrent, convex at back, usually flexuous; median leaf cells long-rhomboidal to sublinear, thin-walled, projecting at upper ends, 25–35 µm × 4.0–4.5 µm, slightly shorter at lower part, basal cells near margins rectangular, 15–20 µm × 5–9 µm; alar cells flatly quadrate, 8–13 µm wide; branch leaves somewhat smaller than stem leaves with upper costa dentate at back side. *Dioicous*. *Inner perichaetial leaves* ovate to oblong, gradually narrowed up to 1.7 mm long; costa short, entire. *Setae* 7–10 mm long, erect, brown; capsules erect, long-ellipsoidal to cylindrical, 1.3–1.5 mm long, 0.6–0.7 mm thick; peristome double; exostome teeth lanceolate, narrowed, yellowish, striate at lower part, almost smooth at upper part, lamella blurred; endostome segments shorter than exostome teeth, smooth; linear, fragile; opercula 0.4 mm long. *Spores* spherical, 10–15 µm in diameter.

Illustrations: P.-C. Wu (ed.) 2001 (Pl. 65, figs. 1–8).
Habitat: on tree bases.
Distribution: China, Japan, Korea, Europe, and North America.

Chinese specimens examined: HUNAN: Chang-sha City, *C. Gao 34* (IFSBH); Hui-tong Co., *C. Gao 40127, 40128* (IFSBH). XINJIANG: Fu-kang Co., *B.-C. Tan 93-840* (FH).

Lescuraea saxicola is similar to *L. mutabilis*, but the former is distinguished from the latter by having smaller paraphyllia, wider leaves, and short leaf cells (5–9:1).

5. ***Lescuraea setschwanica*** (Broth.) T. Cao & W.-H. Wang 四川多毛藓 si-chuan duo-mao xian
Fl. Bryophyt. Sin. 6: 152. 2001. *Pseudoleskea setschwanica* Broth., Akad. Wiss. Wien Sitzungsber., Math.-Naturwiss. Kl., Abt. 1, 133: 579. 1924. Type. China: Sichuan, Hui-li, *Handel-Mazzetti 931* (holotype H).

Plants slender, yellowish green to reddish brown, in dense mats. *Primary stems* creeping, 4–5 cm long, irregularly branched; branches with abundant flagelliform branchlets on tips; paraphyllia absent. *Leaves* imbricate, appressed on stems when dry, spreading when moist; stem leaves large, loosely arranged, broadly ovate, distinctly decurrent at base, acuminate upward, 1.6–1.9 mm × 1.0–1.1 mm, sometimes with secund apices; branch leaves small, closely inserted, ovate or broadly ovate at base, acuminate upward, 0.8–1.6 mm × 0.5–0.8 mm; flagelliform leaflets dense, ovate-lanceolate, 0.4–0.5 mm × 0.2–0.3 mm; margins plane, entire, sometimes recurved, sparsely dentate above; costa single, thick, strong, broad at base, gradually narrowed and slightly twisted above, ending below leaf apex; upper and median leaf cells irregularly quadrate, 4–7 µm wide, slightly thick-walled, smooth; basal cells irregularly quadrate, 5–15 µm × 7–9 µm, thick-walled. *Dioicous*. *Sporophytes* not seen.

Illustrations: Plate 433, figs. 1–8.
Habitat: on tree trunks in high elevation.

Distribution: endemic to China.
Chinese specimen examined: see the type information cited above.

The distinguishing characters of this species are: 1) plants slender, irregularly branched, branch tips with many flagella; 2) leaf base widely ovate; alar decurrent, often auriculate, consisting of large, quadrate to rectangular cells; and 3) upper leaf cells rather small, irregular in shape, 4–7 μm wide.

Brotherus (1929) listed two specimens under this species. One of them (*Handel-Mazzetti 931*) is the type, and another specimen, *Handel-Mazzetti 9086*, collected from Yunnan, between Bahan (Pehalo) and Ludjing (Salwin), 27°58′N, is not this species. The collection contains poorly developed *Herpetineuron toccoae.*

6. ***Lescuraea yuennanensis*** (Broth.) T. Cao & W.-H. Wang 云南多毛藓 yun-nan duo-mao xian
Fl. Bryophyt. Sin. 6: 154. 2001. *Pseudoleskea yuennanensis* Broth., Akad. Wiss. Wien Sitzungsber., Math.-Naturwiss. Kl., Abt. 1, 133: 579. 1924. Type. China: Yunnan, Li-jiang Co., Yu-long Shan (Mt.), *Handel-Mazzetti 4282* (holotype H).

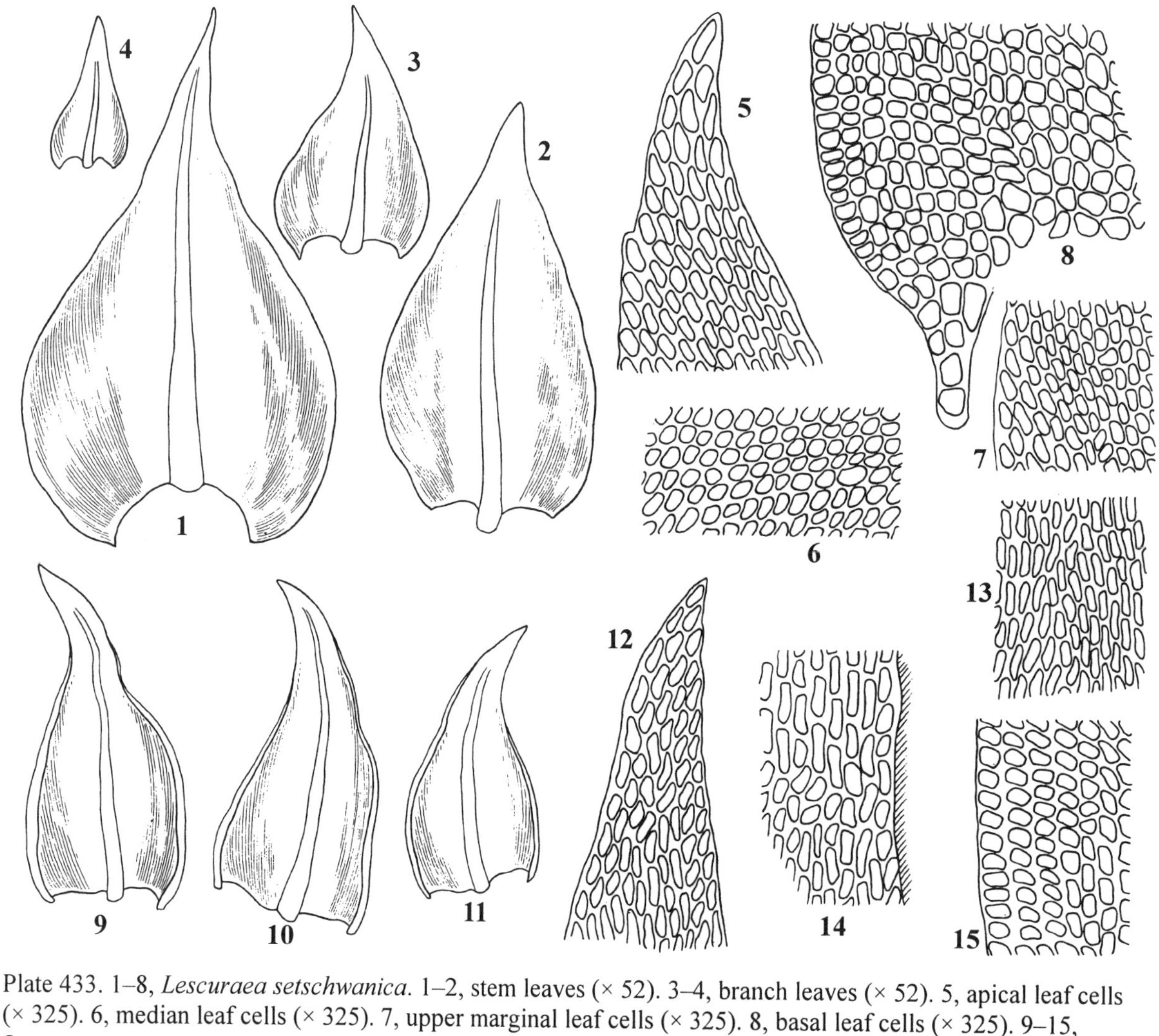

Plate 433. 1–8, *Lescuraea setschwanica.* 1–2, stem leaves (× 52). 3–4, branch leaves (× 52). 5, apical leaf cells (× 325). 6, median leaf cells (× 325). 7, upper marginal leaf cells (× 325). 8, basal leaf cells (× 325). 9–15, *Lescuraea yuennanensis.* 9–11, leaves (× 52). 12, apical leaf cells (× 325). 13, median leaf cells (× 325). 14, juxtacostal leaf cells (× 325). 15, basal leaf cells (× 325). (1–8, drawn from *Handel-Mazzetti 931*, H; 9–15, from *Handel-Mazzetti 4282*, H.)

Plants slender, reddish brown to yellowish green, in dense mats. *Primary stems* prostrate, 4–5 cm long, with irregular, thin, elongate branches; paraphyllia absent. *Leaves* imbricate, secund above when dry, spreading when moist, ovate-lanceolate, acuminate upward, 0.9–1.1 mm × 0.4–0.5 mm; margins obviously recurved from base to apex at both sides; costa thick, strong, broadening below, slightly thin upward, sometimes flexuous, ending near apex; upper leaf cells irregularly rectangular, 10–15 µm × 4–5 µm, thick-walled; median cells rectangular to linear, 10–18 µm × 3–5 µm, thick-walled; basal cells near margins flatly quadrate to subquadrate, 5–10 µm × 7–10 µm; basal cells near costa rectangular, 10–20 µm × 5–7 µm. *Dioicous*? *Sporophytes* not seen.

Illustrations: Plate 433, figs. 9–15.
Habitat: on moist rocks.
Distribution: endemic to China.
Chinese specimen examined: YUNNAN: Tannenund Weidenbstanden, *Handel-Mazzetti 8220* (H).

This species is characterized by having slender plants with abundant irregular branches, ovate-lanceolate and concave leaves with margins obviously recurved from base to upper part, small, irregularly linear to rectangular, thick-walled leaf cells, and quadrate basal cells near margins.

2. ***Leskea*** Hedw. 薄罗藓属 bo-luo xian shu
Sp. Musc. Frond. 211. 1801.

Plants rather slender, dark green, dull. *Stems* creeping, sparsely radiculose, pinnately or subpinnately branched; branches short, ascending or erect; paraphyllia few, linear to shortly lanceolate, rarely lacking. *Leaves* appressed on stems when dry, erect-spreading when moist, sometimes slightly secund; cordate or ovate, slightly decurrent at base, gradually narrowed to a mucronate or acuminate apex, or sometimes blunt to acute at apex, usually slightly plicate; margins often recurved, entire or serrulate near apex; costa single, thick, strong, ending below apex; leaf cells thin-walled; upper cells rounded to hexagonal, unipapillose, rarely bipapillose or multipapillose; median cells rhomboidal; basal cells subquadrate; branch leaves smaller. *Monoicous*. *Inner perichaetial leaves* pale, sheathing at base, with either long or short apices, cells smooth. *Setae* slender, elongate; capsules erect, oblong-cylindrical, sometimes curved or pendent, exothecial cells thin-walled; annuli differentiated, deciduous; peristome double; exostome teeth pale yellow, linear, with a long acuminate apex, margins not differentiated, striate below, papillose above; endostome segments pale yellow, finely papillose, linear, folded, equal to or longer than teeth; basal membrane low; cilia rudimentary; opercula conic. *Spores* small, verrucose.

There are about 15 species of *Leskea* in the world. Species are mainly distributed in the temperate zones on moist soil or rock surfaces, especially common on bases of tree trunks. Müller (1897) published the first Chinese record of *Leskea* (*Leskea magniretis* = *Lindbergia sinensis*). Bartram (1935) recorded *Leskea scabrinervis* Broth. & Par. from Fujian province. M.-J. Lai (1976) recorded *Leskea subacuminata* Nog. from Taiwan. Chen et al. (1978) listed 4 species of *Leskea* from China, of which *L. consanguinea* (Mont.) Mitt. and *L. subfiliramea* Broth & Par. were new to the country. *Leskea polycarpa* Ehrh. ex Hedw. and *L. arenicola* Best were recently reported from China by R.-L. Hu & Y.-F. Wang (1981) and Redfearn and P.-C. Wu (1986), respectively. In this study three species are recognized, with *Leskea gracilescens* Hedw. newly reported from China. The taxonomic status of *Leskea subacuminata* Nog., *L. consanguinea* (Mont.) Mitt., and *L. subfiliramea* Broth. & Par. are uncertain because no specimens were available for study.

1. Leaves broadly ovate, 1.0–1.5:1 (length/width), rounded at apex .... 1. *L. gracilescens*
1. Leaves ovate or ovate-lanceolate; 2:1 (length/width), rather acute or acuminate at apex 2
  2. Plants rather large; leaves ovate, distinctly secund, acute at apex; paraphyllia numerous; leaf cells unipapillose; costa smooth on the back .... 2. *L. polycarpa*
  2. Plants small; leaves ovate-lanceolate, acuminate at apex, not secund or slightly secund; paraphyllia few; leaf cells smooth; costa roughened on the back .... 3. *L. scabrinervis*

1. ***Leskea gracilescens*** Hedw. 细枝薄罗藓 xi-zhi bo-luo xian
Sp. Musc. Frond. 222. 1801. Type. U.S.A.: Pennsylvania, Lancaster, *Muhlenberg s.n.*

Plants tiny, green to yellowish green, in prostrate mats. *Primary stems* creeping, 1–2 cm long, irregularly branched, branches short, dense. *Stem leaves* and branch leaves isomorphous; branch leaves involute, appressed on

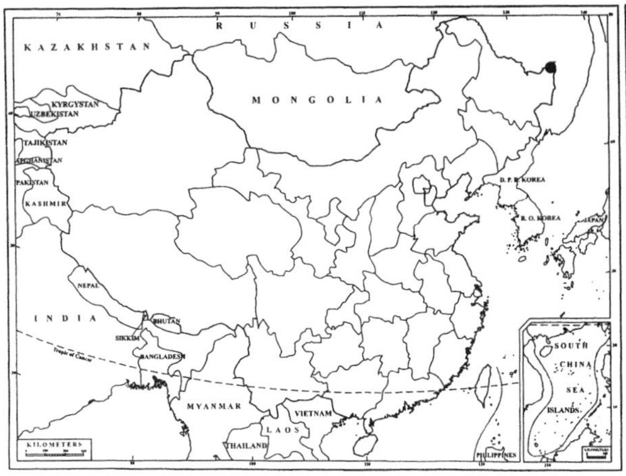

1. *Leskea gracilescens*

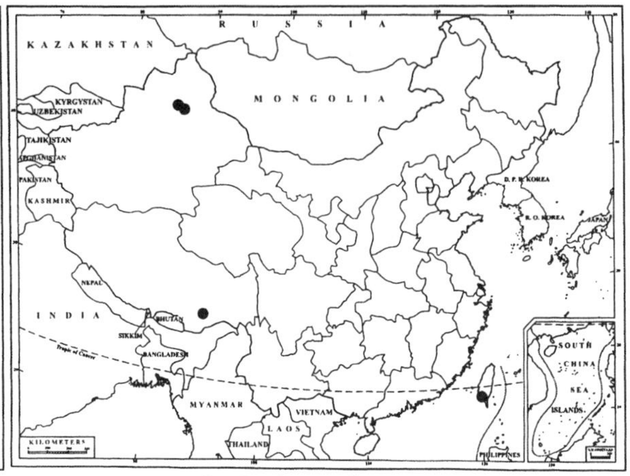

2. *Leskea polycarpa*

branches when dry, erect-spreading upward when moist; broadly ovate to ovate, 0.5–0.8 mm × 0.4–0.6 mm, usually plicate at base, with blunt apex; margins entire, narrowly recurved at the base; costa single, ending below the apex; upper and median leaf cells round-polygonal, 7–10 µm, unipapillose. *Sporophytes* not seen.

Illustrations: P.-C. Wu (ed.) 2001 (Pl. 58, figs. 10–16).
Habitat: on bases of tree trunks.
Distribution: China and North America.

Chinese specimen examined: HEILONGJIANG: Fu-yuan Co., *C. Gao 6364* (IFSBH).

This species is characterized by having broad leaves (1.0–1.5:1) with obtuse apices.

2. ***Leskea polycarpa*** Hedw. 薄罗藓 bo-luo xian
Sp. Musc. Frond. 225. 1801. Syntypes. Europe: Germany and Austria.

Plants green to dark green, in dense or loose tufts. *Stems* creeping, subpinnately branched; paraphyllia lanceolate to widely lanceolate. *Leaves* tightly appressed on stems when dry, ovate, 1.0–1.6 mm long, asymmetric, obviously secund, more or less sickle-shaped, curved, biplicate below; margins plane, entire, sometimes serrulate above, recurved at base; costa single, thick, strong, ending below apex; leaf cells round-polygonal to widely elliptic, 7–10 µm wide, thin-walled, with low papilla on both sides of leaf surfaces; basal cells near margins flatly rectangular, 10–12 µm wide, thin-walled. *Monoicous*. *Setae* erect, ca. 1 cm long; capsules erect, cylindrical; peristome double; exostome teeth transversely striolate at base; endostome segments with low basal membrane; cilia absent; opercula conic. *Spores* 9–13 µm in diameter, faintly papillose.

Illustrations: P.-C. Wu (ed.) 2001 (Pl. 58, figs. 1–9).
Habitat: on tree bases or rock surfaces.
Distribution: China, Japan, Caucasus, Siberia, Europe, and North America.

Chinese specimens examined: TAIWAN: Kao-hsing Co., *C.-C. Chuang 1154* (FH). XINJIANG: Mi-quan Co., *B. Cao s.n.* (IFSBH); Mt. Bo-ge-da, Tianchi (Lake), *J.-C. Zhao 03198* (IFSBH). XIZANG: Mi-lin Co., *Tibet Team 7511* (IFSBH).

This species is distinguished from other species by having unipapillose leaf cells and ovate leaves that are secund and gradually tapered to an acuminate apex.

3. ***Leskea scabrinervis*** Broth. & Par. 粗肋薄罗藓 cu-lei bo-luo xian
Rev. Bryol. 33: 26. 1906. Type. China: Shanghai, Zi Ka Wei, *P. Henry s.n.*

Plants tiny, green to yellowish green, in prostrate tufts. *Primary stems* creeping, up to 1–2 cm long, one or two times irregularly branched; branches short, dense, patent upward, branchlets usually curved. *Stem leaves* ovate-lanceolate, 1.0 mm × 0.4 mm, plane, slightly subsecund above; branch leaves similar to the stem leaves in shape, but with slightly thinner costa; margins entire, strong, wide, ending below the apex, coarse at back surface; median leaf cells round-hexagonal, 7–10 μm, thin-walled; marginal cells somewhat transversely elongate. *Inner perichaetial leaves* lanceolate, 1.5 mm × 0.4 mm, semi-sheathing at base, cells oblong to rectangular, costa strong, percurrent or excurrent, forming a long tip. *Setae* up to 5 mm long; capsules erect. Others features not seen.

Illustrations: Plate 434, figs. 1–14.
Habitat: on tree bark or soil.
Distribution: endemic to China.

Specimens examined: FUJIAN: *Thériot s.n.* (IFSBH, PE); Xia-men City, *H.-H. Chung 6024A*, *B7*, *B380*, *6024* (FH); *B10*, *B3454* (as *Leskea acuminata*, FH). HENAN: Ji-gong Shan (Mt.), *H.-J. YU 46* (HKAS, IFSBH). SHANGHAI: *M. Zong 12* (HKAS, IFSBH); Yu Shan (Mt.), *S.-M. Xu 177* (IFSBH).

This species is easily recognized by its tiny and slender plants, ovate-lanceolate leaves, and strong and wide costa that are coarse at back.

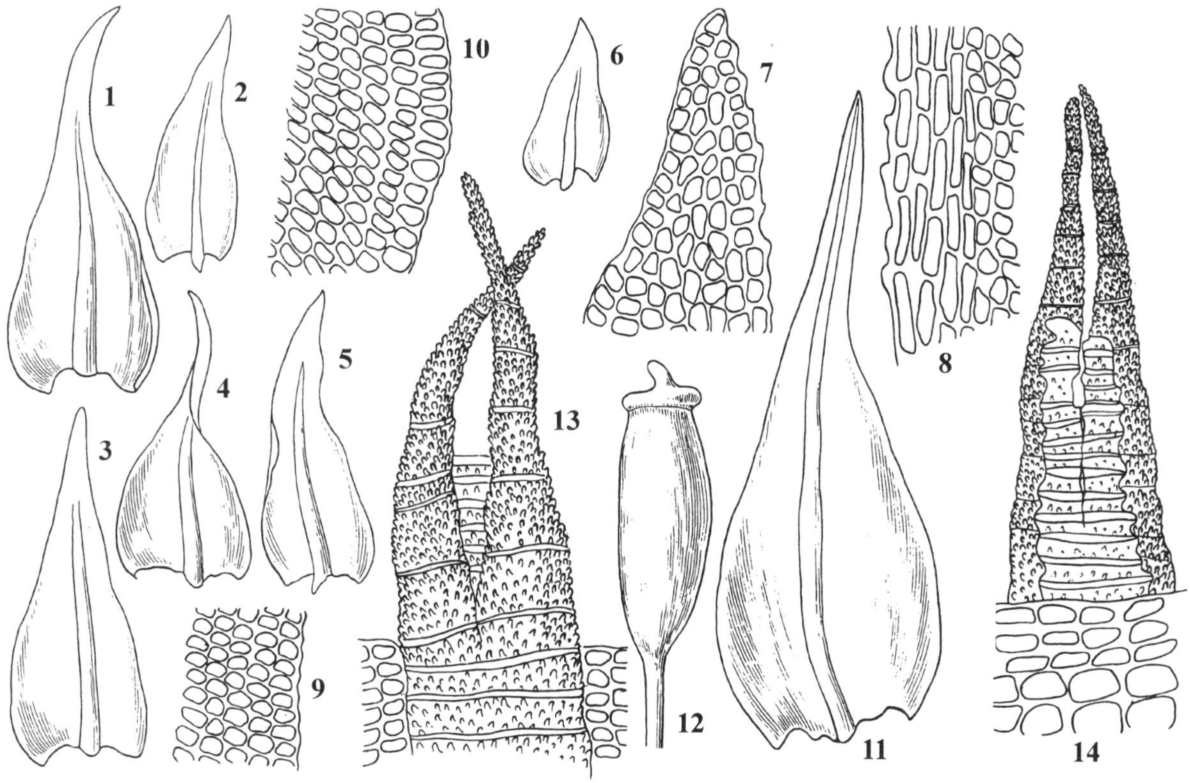

Plate 434. *Leskea scabrinervis*. 1–6, leaves (× 52). 7, apical leaf cells (× 325). 8, lateral view of papillose costa (× 325). 9, median marginal leaf cells (× 325). 10, basal leaf cells (× 325). 11, perichaetial leaf (× 52). 12, capsule (× 21). 13–14, peristome (× 325). (1–14, drawn from *H.-H. Chung M6*, PE.)

3. ***Leskeella*** (Limpr.) Loeske 细罗藓属 xi-luo xian shu
Moosfl. Harz. 255. 1903.

Plants slender, dark green to brownish, dull, in dense small mats. *Primary stems* creeping, with brown reddish fascicled rhizoids; branches short; paraphyllia lacking. *Leaves* imbricate when dry, erect or secund when moist, cordate or oblong-ovate at base, slightly decurrent, usually biplicate; margins slightly recurved below, plane

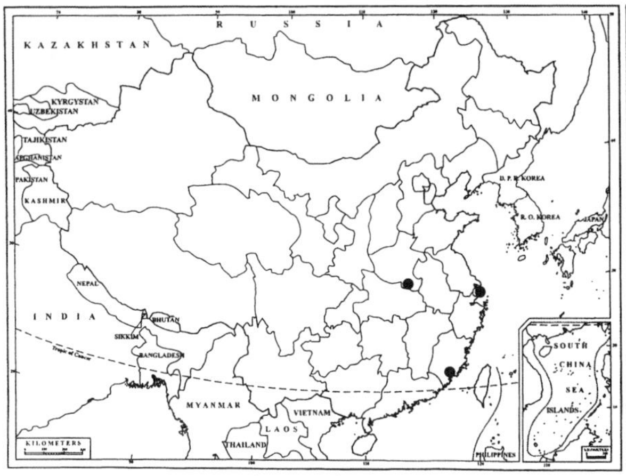

3. *Leskea scabrinervis*

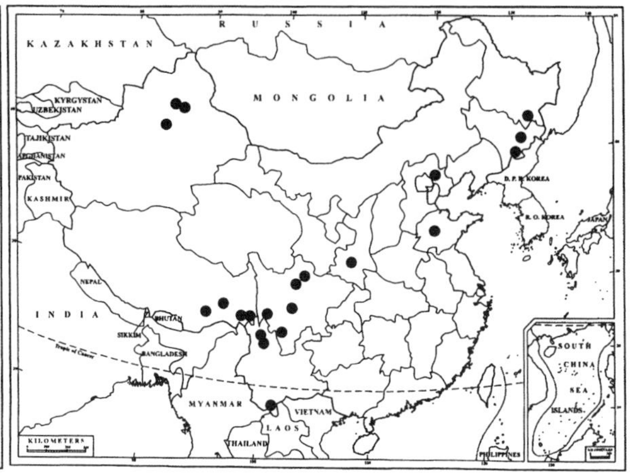

1. *Leskeella nervosa*

above, entire; costa single, strong, yellowish, reaching leaf apices; upper leaf cells round to hexagonal, smooth, slightly thick-walled; median cells oblong; alar cells quadrate; branch leaves rather small, plane at margins, costa shorter and weaker than that of stem leaves. *Dioicous*. *Inner perichaetial leaves* more or less sheathing at base, rapidly narrowed to an acuminate apex; costa thin, weak, percurrent, cells linear. *Setae* erect; capsules erect, rarely somewhat curved, cylindrical to oblong-cylindrical, red or brown; annuli persistent or deciduous; peristome double; exostome teeth shortly lanceolate, yellow, borders differentiated, transversely or obliquely striolate on outer surfaces, papillose or smooth at apex; endostome segments yellow, finely papillose, irregularly thickened, lanceolate or narrowly lanceolate, rarely fragmentary; basal membrane high; cilia absent, rarely rudimentary; opercula stout, conic, obliquely rostrate. *Spores* small, reddish brown, verrucose.

There are about 4 species of *Leskeella* in the world. Only one species is recognized in China. C. Gao (1977) first reported *L. nervosa* (Brid.) Loeske from northeastern China. *Leskeella tectorum* (Funck ex Brid.) Hag. was reported by P.-C. Chen et al. (1978), but it is now a synonym of *Pseudoleskeella tectorum* (Brid.) Kindb.

1. ***Leskeella nervosa*** (Brid.) Loeske 细罗藓 xi-luo xian

Moosfl. Harz. 255. 1903. *Leskea nervosa* (Brid.) Myr., Coroll. Fl. Upsal. 52. 1834. *Pseudoleskeella nervosa* (Brid.) Nyholm, Moss Fl. Fennoscandia. II. Musci 776. 1969. *Pterigynandrum nervosum* Brid., Spec. Musc. Recent. Suppl. 1: 132. 1806. Type. Poland: Glatz (Klodzko), *Seliger s.n.*

Pseudoleskea papillarioides C. Müll., Nuovo Giorn. Bot. Ital., n. ser. 3(1): 118. 1896. Type. China: Shaanxi, in medio monte kuan-tou-san, July 1894, *Giraldi s.n.* (isotype H).

Plants slender, green to brownish green, dull, in prostrate tufts. *Stems* creeping, up to 3–5 cm long, radiculose; irregularly or subpinnately branched; branches dense, ascending, ca. 0.5–1.0 cm long; paraphyllia none. *Leaves* erect or secund, imbricate when dry, spreading when moist; stem leaves cordate at base, suddenly tapered to a long apiculate apex, 1.1–1.5 mm × 0.4–0.5 mm, with few plicae below; margins slightly recurved below; costa thick, strong, brown, ending below leaf apex or only up to 2/3 the leaf length; upper leaf cells round-quadrate to hexagonal, 15–20 µm × 7–10 µm; median cells elliptical, 15–20 µm × 7–9 µm; basal cells subquadrate; branch leaves narrower and smaller than stem leaves, costa thin, weak. *Dioicous*. *Setae* erect, ca. 1.3 cm long, reddish; capsules cylindrical, erect or pendent; peristome double; exostome teeth yellowish brown, transversely striolate at base; endostome segments reduced; basal membrane high; cilia rudimentary; opercula stoutly conic, obliquely rostrate. *Spores* 12–17 µm, brown, finely papillose.

Illustrations: P.-C. Wu (ed.) 2001 (Pl. 60, figs. 1–11).
Habitat: on tree trunks, wet rocks or humus under forests.
Distribution: China, Japan, Europe, North America, and Greenland.

Chinese specimens examined: HEBEI: Xing-long Co., *C. Gao 31* (IFSBH). HEILONGJIANG: Ning-an Co., *C. Gao 6053* (IFSBH). JILIN: An-tu Co, *Koponen 36735*, *36885*, *36722*, *36717*, *36720* (H, IFSBH); Mt. Changbai, *C.*

Gao 22135 (IFSBH, MO). SHAANXI: Mt. Tai-bai, *M.-X. Zhang 123* (IFSBH, IWH). SHANDONG: Tai Shan (Mt.), *C. Gao 34093* (IFSBH). SICHUAN: Hei-shui Co., *Redfearn 35503* (MO); Mu-li Co., *M.-Z. Wang 23776* (IFSBH, PE); Song-pan Co., *B. Allen 7148* (MO); Xiang-cheng Co., *S. He 31741* (MO); Ya-an Co., *C. Gao 17105* (IFSBH, MO). XINJIANG: He-jing Co., *S. He 951183* (MO); Mt. Begeda, *J.-C. Zhao 002431, 2807* (IFSBH), *S.-E. Liu 3361* (MO, PE); Tian Shan (Mt.), *K.-J. Guan 1897* (MO, PE); Wulumuqi City, *B.-C. Tan 93-1011* (NY). XIZANG: Bo-mi Co., *G.-C. Zhang 14598* (IFSBH); Cha-yu Co., *Z.-C. Ni 64-6* (MO); Mi-lin Co., *Ecological Dept. Tibet team 7444* (IFSBH, PE); Men Gong Distr., *M. Zang 6821* (HKAS, IFSBH). YUNNAN: Li-jiang Co., *X.-J. Li 85280* (HKAS, IFSBH); Xishuangbanna, *Magill et al. 7645* (MO); Zhong-dian Co., *J.-K. Yang 1271* (HKAS, IFSBH).

The main distinctive features of this species are: 1) plants slender, irregularly and densely branched (sometimes loosely); 2) stem leaves gradually narrowed from an ovate base to a long apiculate apex, with margins recurved and obviously plicate; 3) costa rather slender, ending below apex or at 2/3 the leaf length; 4) leaf cells short and small, upper and median cells round-polygonal; basal cells near margins subquadrate, smooth; and 5) capsules erect; peristome double, but endstome segment reduced and cilia rudimentary.

4. *Lindbergia* Kindb. 细枝藓属 xi-zhi xian shu

Gen. Eur. N. Amer. Bryin. 15. 1897.

Plants slender, bright green to brownish green, dull. *Stems* slender, numerously irregularly branched; paraphyllia few or lacking. *Leaves* densely imbricate when dry, erect-spreading when moist, ovate to ovate-lanceolate, slightly concave, somewhat decurrent at base; margins plane, usually entire, rarely faintly serrulate above; costa single, strong, ending well below leaf apex; leaf cells thin-walled, loosely arranged, oval to irregularly rhomboidal, smooth or unipapillose; marginal leaf cells smaller, quadrate to subquadrate, with several rows of quadrate to flatly quadrate cells near basal margins. *Monoicous. Inner perichaetial leaves* larger, pale green, erect, sheathing at base, lanceolate above, gradually tapered to subulate at apex. *Setae* erect; capsules oblong-ovoid, erect, rarely curved, with small mouth; annuli sometimes differentiated; peristome double; exostome teeth lanceolate, yellowish, papillose, fused at base, blunt at apex, without striolate, with folded ridges on outer surfaces, trabeculate on inner surfaces; endostome segments finely papillose; basal membrane moderately high; cilia lacking; opercula conic, blunt. *Spores* spherical or ovoid, coarsely and densely papillose.

There are about 17 species of *Lindbergia* known in the world, and 10 of them are well understood (Crosby et al. 2000). Most species of the genus are growing on trees. Six species and one variety have been recorded in China (Redfearn et al. 1996). Levier (1906) first reported *L. brachyptera* (Mitt.) Kindb. from Shandong province. Brotherus (1907) made new combinations of *Lindbergia sinensis* and *L. magniretis* based on *Schwetschkea sinensis* C. Müll. and *Leskea magniretis* C. Müll. Thériot (1926) described a new variety *L. magniretis* var. *yunnanensis*. Dixon (1934) reported 2 species from China, with *L. japonica* Card. new to the country. C. Gao (1977) reported *Lindbergia austinii* (Sull.) Broth. from northeastern China, and he (C. Gao 1985) also published a new species, *L. brevifolia*, from Xizang. Most recently, a new species, *L. serrulatus* C. Gao, T. Cao & W.-H. Wang was described (C. Gao et al. 2001). Four species are here recognized in China. *Lindbergia japonica* Card. is not treated, because no specimens were available for us to study. *Lindbergia magniretis* (C. Müll.) Broth. is synonymized with *L. sinensis* (C. Müll.) Broth. The status of *Lindbergia magniretis* var. *yunnanensis* Thér. & Copp. is uncertain.

1. Leaf cells papillose 2
1. Leaf cells smooth or nearly so 3
  2. Leaves ovate or ovate-lanceolate, gradually or abruptly tapering to a short acumen; leaf cells with blunt and rounded papillae .... 1. *L. brachyptera*
  2. Leaves broadly triangularly ovate, abruptly narrowed to a short acumen; leaf cells with C-shaped papillae .... 2. *L. brevifolia*
3. Stems nearly pinnately branched; costa slender; leaf cells thin-walled; peristome teeth in pairs .... 3. *L. serrulata*
3. Stems irregularly branched; costa stout; leaf cells thick-walled; peristome teeth not in pairs .... 4. *L. sinensis*

1. *Lindbergia brachyptera* (Mitt.) Kindb. 细枝藓 xi-zhi xian

Eur. N. Amer. Bryin. 1: 13. 1897. *Pterogonium brachyptera* Mitt., J. Proc. Linn. Soc., Bot. 8: 37. 1864. Type. British North America.

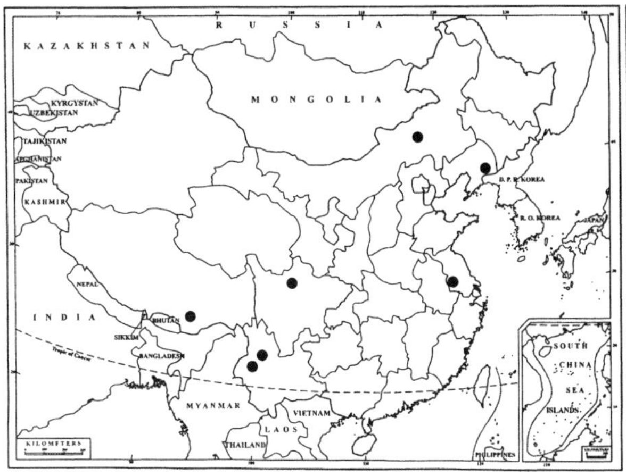

1. *Lindbergia brachyptera*

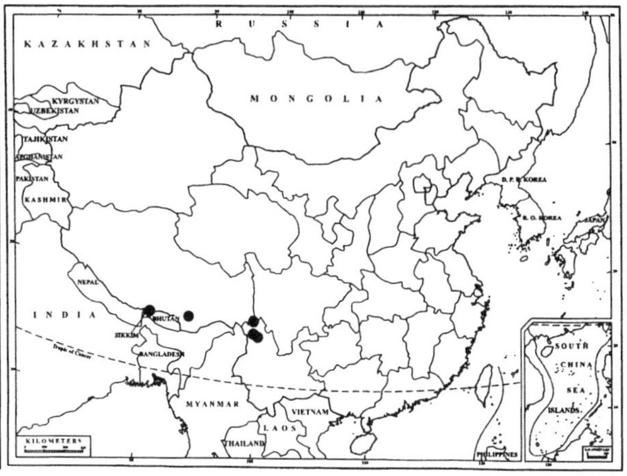

2. *Lindbergia brevifolia*

Lindbergia austinii (Sull.) Broth., Nat. Pflanzenfam. I(3): 993. 1907. *Lindbergia brachyptera* var. *austinii* (Sull.) Grout, Moss. Fl. N. Amer. 3: 196. 1934. *Leskea austinii* Sull., Icon. Musci Suppl. 81. 1874.

Plants medium-sized, dark green to brown, dull, in tufts. *Stems* creeping, up to 2–3(–5) cm long, with abundant rhizoids, irregularly or somewhat pinnately branched; branches slender, straight or sometimes curved, ca. 1(–2) cm long, attenuated at apex; paraphyllia few to abundant, linear to narrowly lanceolate. *Leaves* appressed, slightly secund when dry, spreading when moist, broadly ovate to oblong-ovate at base, gradually or rapidly tapered to a lanceolate apex, (0.6–)0.8–1.1 mm × (0.35–)0.40–0.50 mm; margins plane or slightly incurved, smooth, entire, sometimes slightly revolute at base; costa single, thick, strong, ending at 2/3 the leaf length; upper and median leaf cells rounded quadrate to rounded polygonal, 12–20 µm × 8–12 µm, unipapillose, thick-walled; basal cells near margins flatly rectangular, 12–17 µm × 5–8 µm, in several rows. *Monoicous*. *Setae* erect, 5–7(–10) mm long; capsules erect, symmetric, shortly cylindrical; peristome single, pale yellowish, narrowly to somewhat broadly lanceolate, 200–230 µm long, 40–78 µm wide, papillose; opercula conic, with short blunt beaks. *Spores* not seen.

Illustrations: P.-C. Wu (ed.) 2001 (Pl. 56, figs. 1–9).
Habitat: on bases of tree trunks.
Distribution: China, Japan, Far East of Russia, Europe, and North America.

Chinese specimens examined: JIANGSU: Nan-jing City, *M. Zang 6543* (IFSBH). LIAONING: Kuan-dian Co., *Buck 23769* (NY), *B.-C. Tan 93-642* (FH, NY). NEI MONGOL: Xi-lin-hao-ti City, *Z.-G. Gong 1085*, *1086*, *1096* (IFSBH). SHAANXI: Hu Co., *Z.-P. Wei 4530* (HKAS, IFSBH); Ln-Kia-Po, *Giraldi 2121* (as *Pylaisia entodontea*, H). SICHUAN: Ma-er-kang (Markam) Co., *Allen 6884*, *6932* (MO), *Redfearn 35103* (MO). XIZANG: Long-zi Co., *M. Zang 1358*, *1478*, *1480* (HKAS, IFSBH). YUNNAN: Bao Shan (Mt.), *J.-X. Luo 811821* (IFSBH, PE); Yang-bi Co., *Redfearn 502a*, *872*b (MO).

Lindbergia austinii sometimes is treated as variety of *Lindbergia brachyptera*. They are similar in overall leaf shape, size, and leaf cells; the differences between them are in the leaf margins (plane vs. slightly incurved) and peristome teeth (40 µm wide vs. to 80 µm wide and densely papillose vs. finely papillose).

2. ***Lindbergia brevifolia*** C. Gao 阔叶细枝藓 kuo-ye xi-zhi xian
Bryofl. Xizang. 307. 1985. *Lindbergia ovate* C. Gao, Acta Bot. Yunnan. 3(4): 399. 1981, *hom. illeg*. Type. China: Xizang, Ya-dong Co., Xia-ma-si, *S.-K. Chen 4978* (holotype HKAS; isotype IFSBH).

Plants medium-sized, yellowish green to brownish green, slightly shiny, in tufts. *Primary stems* creeping on substratum, 1–2 cm long, irregularly branched; paraphyllia present, consisting of uni- to multi-seriate cells. *Leaves* imbricate, appressed on stems when dry, reflexed, spreading when moist, abruptly tapered from an ovate base to an acuminate apex, 1.0–1.6 mm × 0.6–0.8 mm; margins plane, entire; costa thick, strong, ending below apex; leaf cells round-polygonal, 10–15 µm wide, each basal cell with a high C-shaped papilla on both surfaces. *Monoicous*.

Perichaetial leaves mostly sheathing at base, suddenly tapered to an apiculate apex, hyaline; cells smooth. *Setae* erect, reddish brown, 0.9–1.0 mm long; capsules oblong-ovoid, 0.15–0.20 mm long; peristome single, only exostome present; teeth long-lanceolate, 0.3 mm long, with lamella below, verrucose on surface; opercula with stout and short beaks. *Spores* large, 25–40 µm in diameter, brown, densely papillose.

Illustrations: Plate 435, figs. 1–14.
Habitat: on tree trunks.
Distribution: endemic to China.

Chinese specimens examined: XIZANG: Long-zi Co. (San An Qu Lin), *M. Zong 1929, 1947* (HKAS, IFSBH); Ya-dong Co., Chun Pi He (River), *M. Zang 57* (IFSBH). YUNNAN: De-qin Co., *J.-K. Yang 7786* (HKAS, IFSBH); Li-jiang Co., X.-J. Li 169 (HKAS); Wei-xi Co., *M. Zang 1929, 1947* (HKAS).

Lindbergia brevifolia is easily distinguished from other species by basal leaf cells with a highly lecotropal papilla, leaves tapered to an apiculate apex from a broadly ovate base, and capsules erect, oblong-ovoid.

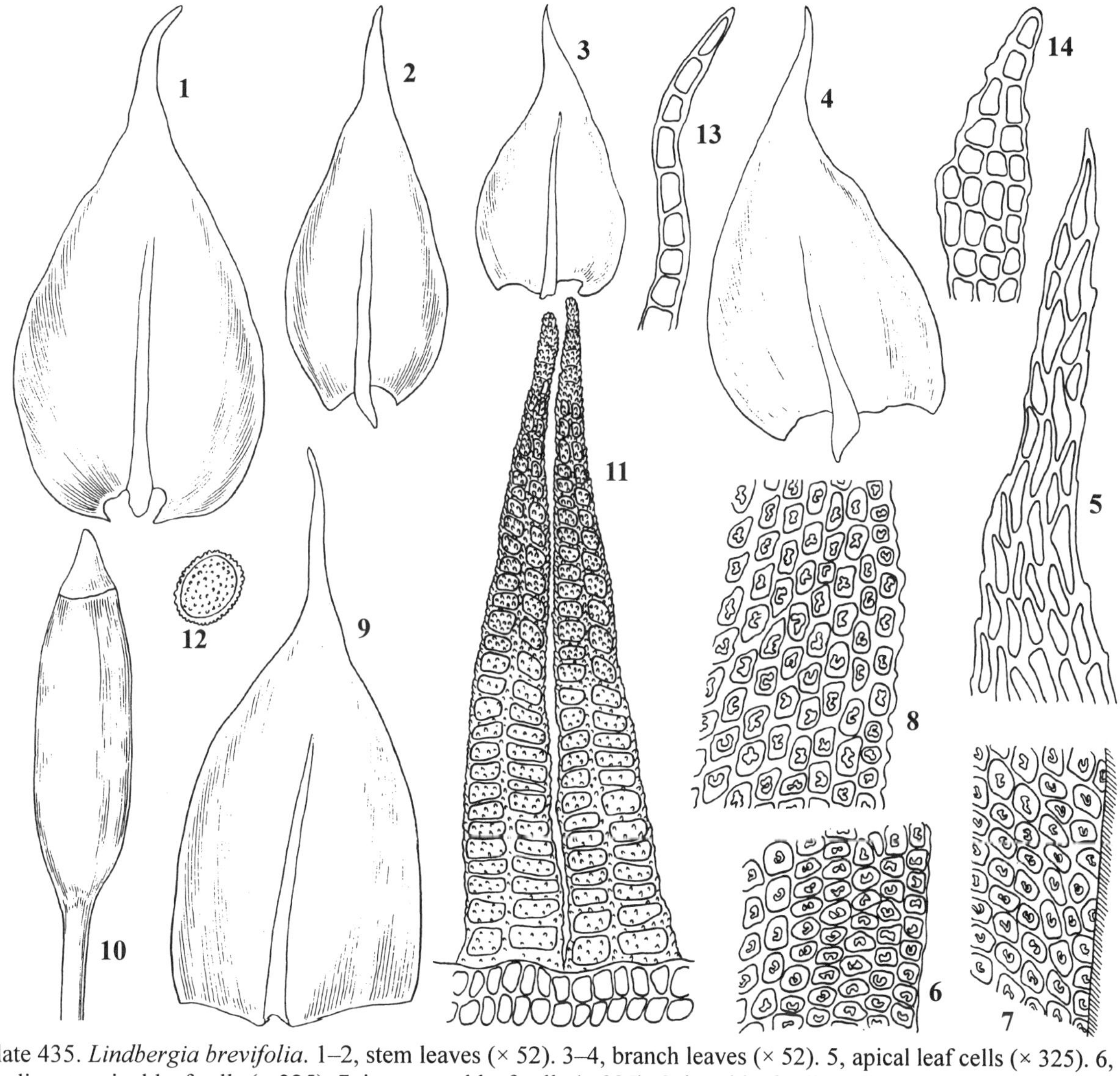

Plate 435. *Lindbergia brevifolia.* 1–2, stem leaves (× 52). 3–4, branch leaves (× 52). 5, apical leaf cells (× 325). 6, median marginal leaf cells (× 325). 7, juxtacostal leaf cells (× 325). 8, basal leaf cells (× 325). 9, perichaetial leaf (× 52). 10, capsule (× 21). 11, peristome (× 325). 12, spore (× 325). 13–14, paraphyllia (× 325). (1–14, drawn from *S.-K. Chen 4978*, HKAS.)

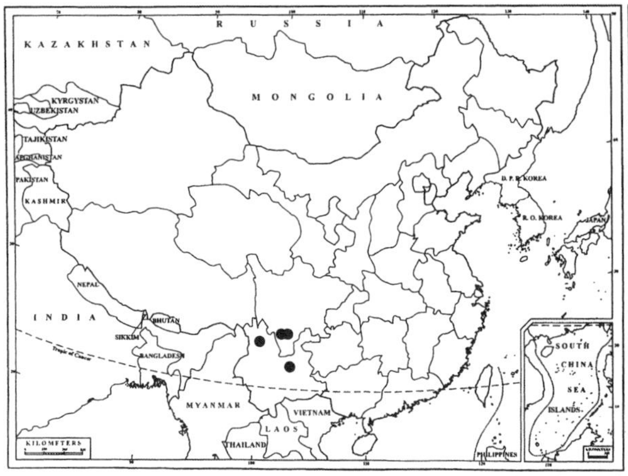

3. *Lindbergia serrulatus*

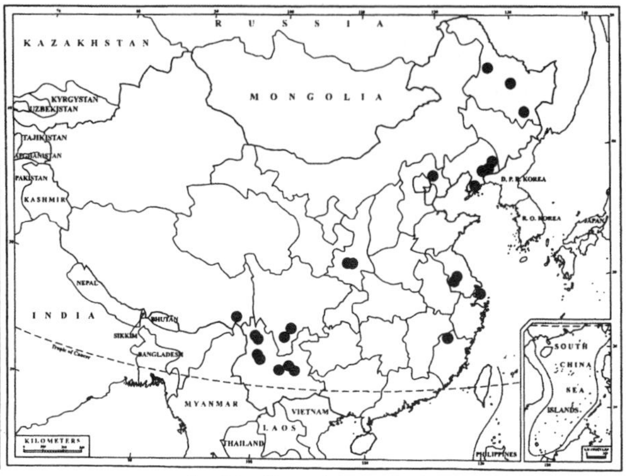

4. *Lindbergia sinensis*

3. ***Lindbergia serrulata*** C. Gao, T. Cao & W.-H. Wang in P.-C. Wu 齿边细枝藓 chi-bian xi-zhi xian
Fl. Bryophyt. Sin. 6: 130. 2001. Type. China: Yunnan, Kunming City, Xishan (Mt.), *X.-J. Li 3095* (holotype IFSBH; isotype HKAS).

Plants yellowish green, slightly shiny. *Primary stems* creeping, 2–3 cm long, lateral stems erect or ascending, long irregularly pinnately branched. *Stem leaves* broadly ovate, 1.1–1.2 mm × 0.4–0.6 mm, with a short apex; margins obviously serrulate above; branch leaves small, ovate, 0.6–1.0 mm × 0.35–0.40 mm; costa somewhat hyaline, slender, ending at middle of the leaf; upper and median leaf cells rhomboidal, 25–40 µm × 10–12 µm, thin-walled, smooth, projecting at upper ends; basal cells near margins flatly quadrate to subquadrate, 10–20 µm wide, basal cells along costa elongated, 25–40 µm × 10–12 µm wide. *Monoicous*. *Perichaetial leaves* somewhat sheathing at base, 1.6 mm × 0.5 mm, hyaline, with a long acute apex. *Setae* ca. 1 cm long, yellowish brown, erect, smooth; capsules ellipsoid, erect, symmetric, with small mouth; annuli differentiated, consisting of two rows of thick-walled cells; peristome single, not involute when dry, deeply inserted in inner side of capsule mouth, teeth narrowly lanceolate, 0.30–0.35 mm long, verrucose, often two teeth growing together; opercula conical, with obtuse rostrum. *Spores* brown, spherical, 20–25 µm in diameter, finely papillose.

Illustrations: Plate 436, figs. 1–11.
Habitat: on bases of tree trunks.
Distribution: endemic to China.

Chinese specimens examined: SICHUAN: Mu-li Co., *C. Gao 1996*, *20121* (IFSBH); Xi-chang City, *C. Gao 19887* (IFSBH). YUNNAN: Li-jiang Co., *M. Zang 1383*, *1390* (HKAS, IFSBH), *H. Li 1187*, *S.-M. Yu 111* (HKAS, IFSBH); Yu-long-xue Shan (Mt.), *Q.-W. Wang 3929* (IFSBH); Kun-ming City, *C. Gao 2126*, *T. Cao 21298* (IFSBH), *X.-J. Li 3038* (HKAS, IFSBH).

Lindbergia serrulatus is similar to *L. sinensis*, but differs from the latter by having plants irregularly and pinnately branched, leaves broadly ovate with short apices, upper leaf margins serrulate, leaf cells rhomboidal, with projections or papillae at upper ends, capsules ovoid, and annuli well developed, consisting of two rows of thick-walled cells.

4. ***Lindbergia sinensis*** (C. Müll.) Broth. 中华细枝藓 zhong-hua xi-zhi xian
Nat. Pflanzenfam. I(3): 993. 1907. *Schwetschkea sinensis* C. Müll., Nuovo Giorn. Bot. Ital., n. ser. 3: 111. 1896. Type. China: Shaanxi (Schen-si), Monte Si-ku-tsui-san, *Giraldi 908* (isotype H).
Lindbergia magniretis (C. Müll.) Broth., Nat. Pflanzenfam. I(3): 993. 1907. *Leskea magniretis* C. Müll., Nuovo Giorn. Bot. Ital., n. ser. 4: 274. 1897. Type. China: Shaanxi (Schen-si), in monte Lao-y-san, March 1896, *Giraldi s.n.*

Plants tiny, delicate, green to yellowish green, in small mats. *Primary stems* creeping, ca.1–3 cm long, irregularly branched; branches short, 0.5–1.0 cm long. *Leaves* densely imbricate when dry, spreading when moist; stem leaves broadly ovate or triangularly broad-ovate, 0.9–1.4 mm × 0.5–0.8 mm, cordate at base, acute to acuminate at apex; branch leaves ovate, 0.6–0.9 mm × 0.25–0.50 mm; margins plane, entire; costa single, reaching middle part of the leaf; median leaf cells polygonal to rounded quadrate, thick-walled, smooth, 12–17 µm × 7–12 µm; basal cells near margins flatly quadrate, 12–15 µm wide. *Autoicous*. *Perichaetial leaves* slightly involute, gradually acuminate upward from an oblong base, 1.2 mm × 0.3 mm, cells rectangular, pale-white. *Setae* yellowish brown, erect, 6–8 mm long; capsules erect, cylindrical, 2.0–2.4 mm long, more or less contracted at mouth when dry; peristome appearing single, exostome teeth shortly lanceolate, pale yellowish brown, verrucose, with blunt apex and lamella, reflexed when wet; endostome only with basal membrane left; opercula conical. *Spores* spherical, brown, 20–25 µm, papillose.

Illustrations: Plate 437, figs. 1–11.
Habitat: on tree trunks.
Distribution: endemic to China.

Selected Chinese specimens examined: FUJIAN: Mt. Wuyi, *M. Zang 301* (IFSBH). HEBEI: Xing-long Co., *F.-D. Chen s.n.* (IFSBH), *J.- X. Ji 818* (IFSBH). HEILONGJIANG: Ning-an Co., *C. Gao 6053* (IFSBH); Wu Da Lian Chi (River), *C. Gao 13215* (IFSBH); Xiao Xin An Ling (Lesser Khingan Mt.), Dai-ling, *C. Gao 258* (IFSBH). JIANGSU: Long Tan, *M. Zang 16* (HKAS, IFSBH); Nanjing City, *M. Zang 89* (HKAS, IFSBH). LIAONING: Feng-cheng Co., *X.-Y. Jia 865* (IFSBH); Heng-ren Co., *T. Cao 287* (IFSBH); Kuan-dian Co., *Buck 23737A*, *23605* (NY); Zhuang-he Co., *X.-Y. Jia 880214* (IFSBH). SHAANXI: Hu Co., *M. Wang s.n.* (HKAS, IFSBH); Mt. Sao-y-san, *Giraldi 163*, *2222* (BM, H, NY); In- kia-po, Oct, 1897, *Giraldi s.n* (as *Leskea magnireris*, H). SHANGHAI: Cong-

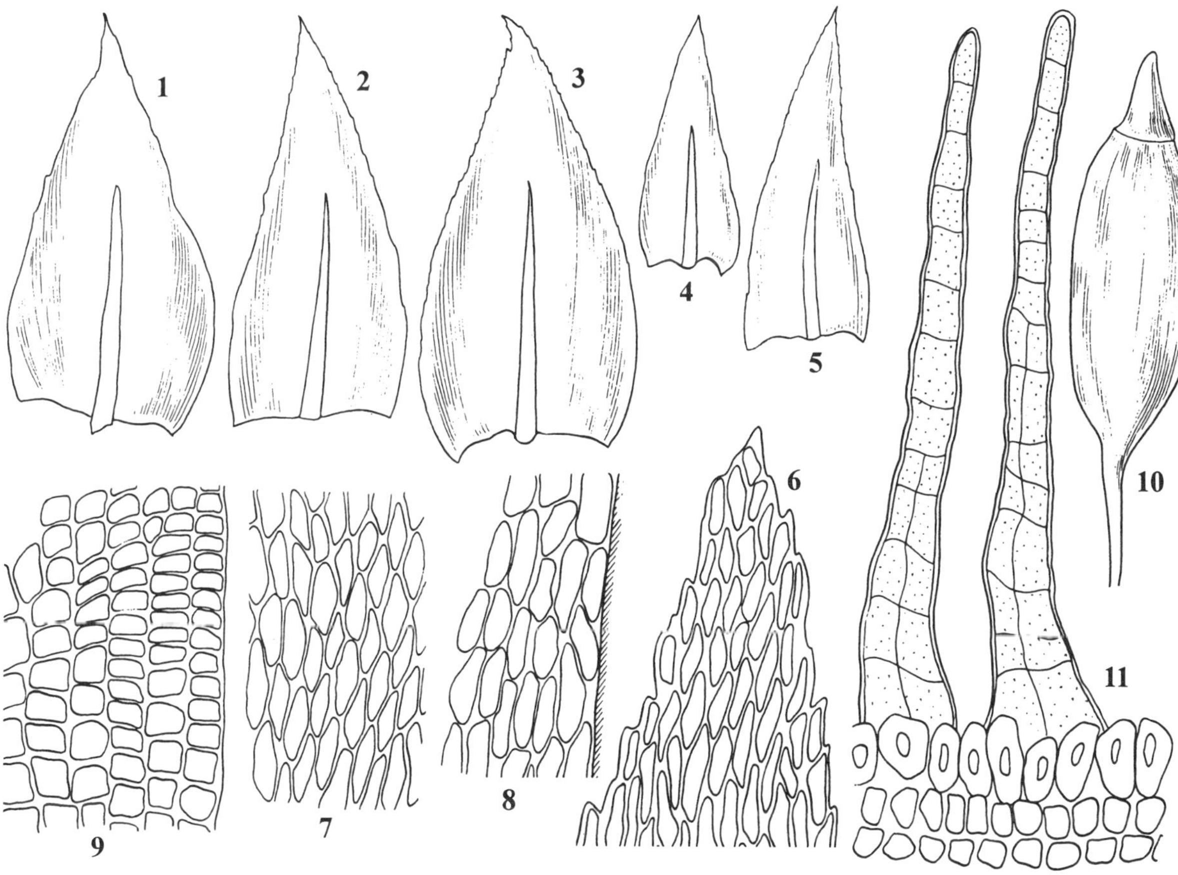

Plate 436. *Lindbergia serrulata*. 1–3, stem leaves (× 52). 4–5, branch leaves (× 52). 6, apical leaf cells (× 325). 7, median leaf cells (× 325). 8, juxtacostal leaf cells (× 325). 9, basal leaf cells (× 325). 10, capsule (× 52). 11, peristome teeth and annulus cells (× 325). (1–11, drawn from *X.-J. Li 3095*, HKAS.)

ming Co., *S.-X. Xu 309* (IFSBH). SICHUAN: Mei-gu Co., *G.-C. Zhang 212762* (IFSBH); Xi-chang City, *C. Gao 19743* (IFSBH); Yalung et Nganning-ho, *Handel-Mazzetti 2022* (H). XIZANG: Cha-yu Co., *G.-C. Zhang 14751, 14799* (IFSBH). YUNNAN: An-ning Co., Magill et al. 8043 (MO); Chu-xiong Co., *Redfearn et al. 1731* (MO); Dali City, *Redfearn 1071, 31072* (MO); Kunming City, *Handel-Mazzetti 403, 2002* (H); Xi Shan (Mt.), *T. Cao 21265* (IFSBH), *Redfearn et al. 2206* (MO), Hei Long Tan (Lake), *R.-L. Hu 63* (MO); Li-jiang Co., *R.-L. Hu & Z.-L. Liu 63* (ALTA); Lu-nan Co., *Redfearn et al. 226, 2268* (MO); Song-ming Co. *Redfearn et al. 1849* (MO); Yang-bi Co., *Redfearn et al. 146*; Wei-xi Co., *M. Zang 1888* (HKAS, IFSBH).

The distinctive features of this species are: 1) Stem leaves broadly ovate, usually cordate at the base and acute or acuminate at apex; 2) margins plane and entire; 3) leaf cells smooth; and 4) only exostome teeth exist.

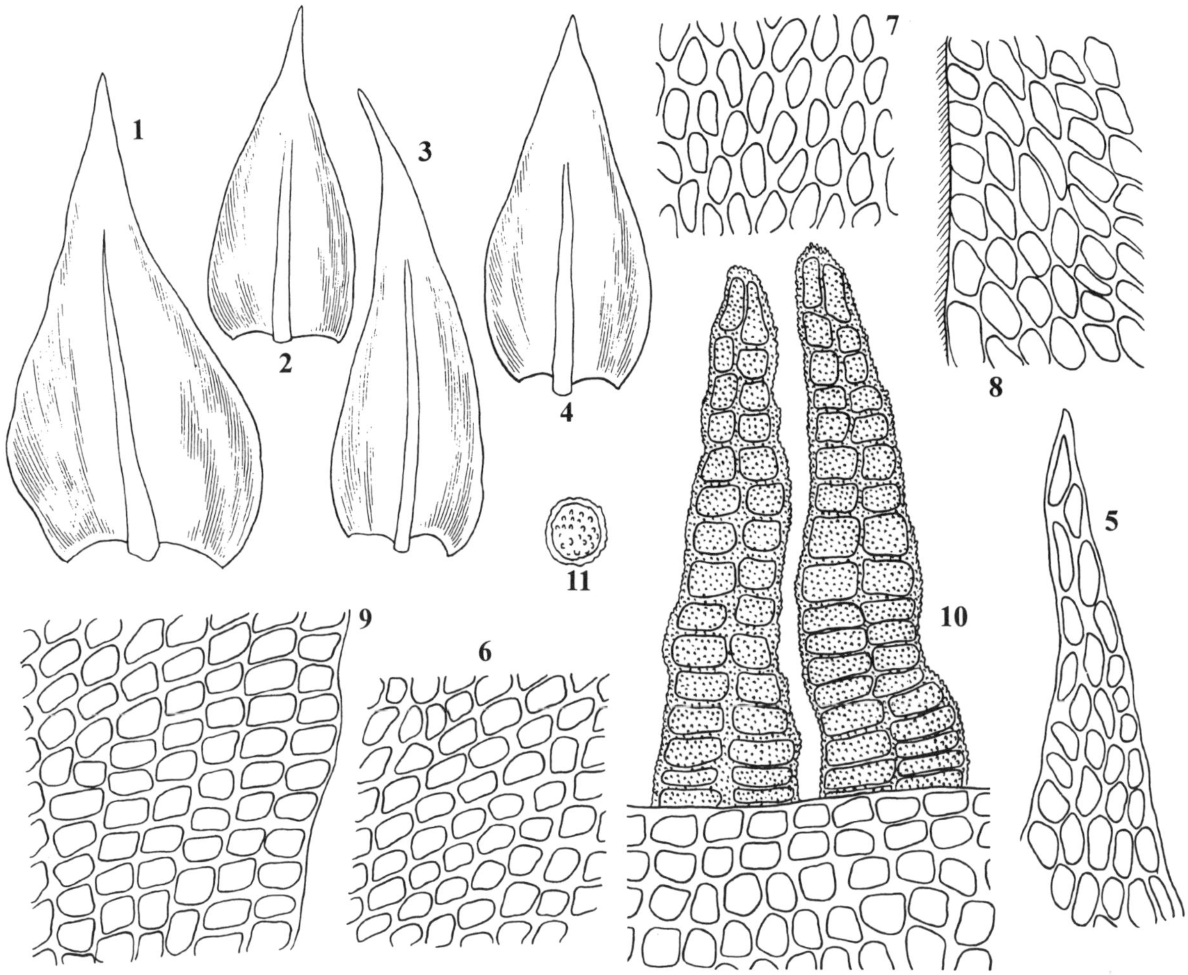

Plate 437. *Lindbergia sinensis*. 1–4, leaves (× 52). 5, apical leaf cells (× 325). 6, upper leaf cells (× 325). 7, median leaf cells (× 325). 8, juxtacostal leaf cells (× 325). 9, basal leaf cells (× 325). 10, peristome teeth (× 325). 11, spore (× 325). (1–11, drawn from *T. Cao & W.-H. Wang s.n.*, IFSBH.)

5. ***Okamuraea*** Broth. 褶藓属 zhe xian shu

Oefvers. Förh. Finska Vetensk.-Soc. 49(10): 2. 1906.

Rigodiopsis Dix. & Thér., Rev. Bryol., n. ser. 4: 155. 1933.

Plants green to yellowish green, slightly shiny, in mats. *Primary stems* creeping, with fascicular rhizoids; irregularly or fascicularly branched; branches rather dense, erect or curved, rounded, acuminate at tips, sometimes flagelliform. *Leaves* loosely appressed on stems when dry, spreading when moist, ovate-lanceolate, concave, slightly decurrent, plicate at base, with an acuminate apex; margins entire, recurved at base; costa single, slender to rather thick, reaching to the middle of leaf; leaf cells rhomboidal or oblong-rhomboidal, slightly thick-walled, smooth; alar cells usually subquadrate. *Dioicous. Inner perichaetial leaves* smaller than stem leaves, sheathing at base, suddenly narrowed to an acuminate apex. *Setae* elongate, erect, reddish, smooth; capsules erect or slightly pendent, oblong-ovoid, mouth not contracted; annuli not differentiated; peristome double; exostome teeth 16, narrowly lanceolate, usually fused at base, not striate on outer surfaces, borders differentiated at both sides, densely papillose; endostome segments reduced, hyaline, finely papillose; basal membrane high; cilia lacking; opercula short, rostrate. *Calyptrae* with sparse hairs. *Spores* ovoid, densely verrucose.

Three species and one variety of *Okamuraea* were recorded in China. Ando (1964) first reported *O. hakoniensis* var. *ussuriensis* (Broth.) Nog. from Jilin province. C. Gao (1977) reported *O. brachydictyon*, *O. hakoniensis*, *O. hakoniensis* var. *ussuriensis*, *O. hakoniensis* fo. *multiflagellifera* from northeastern China. Later, C. Gao (1985) reported *O. brachydictyon* from Xizang. Recently, S.-H. Lin & C.-K. Wang (1977) confirmed the distribution of *O. brevipes* Broth. in mainland China, although its distinction from *O. hakoniensis* is based only on the sporophyte characters, i.e. short setae (only 3–5 mm long) and subspherical capsules. Since no sporophyte of *O. brevipes* was seen by us, this species is not included in this treatment.

There are different views about the systematic status of this genus. When established, *Okamuraea* was compared to *Forsstroemia* Lindb. (Brotherus 1906) and was placed in the Brachytheciaceae of the Isobryales (Brotherus 1907). Later, Brotherus (1925) considered the genus to be close to *Lesquereuxia* and placed both genera in the Rhytidiaceae, which was followed by P.-C. Chen et al. (1978). Andrews (1954) disagreed with this later placement and thought that the genus should be placed in the Meteoriaceae because of the combinations of the following characters: the pilosus calyptrae, the leaf shape and cells, branching patterns, capsule shape and growth position. All these features pointed the genus to the Meteoriaceae. There are other opinions considering the genus to be close to *Homalothecium* and placing it in the Brachytheciaceae.

1. Plants with short branches; branch apices obtuse at apex and with gemmae or innovations; leaves ovate at base, acuminate at apex .... 1. *O. brachydictyon*
1. Plants with long branches; branches often flagelliform and attenuate at apex; leaves ovate-lanceolate at base, slenderly acuminate at apex .... 2. *O. hakoniensis*

1. ***Okamuraea brachydictyon*** (Card.) Nog. 短枝褶藓 duan-zhi zhe xian

J. Hattori Bot. Lab. 9: 10. 1953. *Brachythecium brachydictyon* Card., Beih. Bot. Centralbl. 17: 34. 1904. Syntypes. Korea: Ouen-San, *Faurie 134*; Fusan, *Faurie 145*.

Plants small, brownish green to yellowish green, dull or slightly shiny, in dense mats. *Primary stems* creeping, up to 3–4 cm long; crowded with short branches; rhizoids abundant; branches erect or pendent, covered with dense leaves, rounded, obtuse at apex, sometimes with branchlets bearing clustering gemmae; flagelliform branches absent. *Leaves* ovate to ovate-lanceolate, acuminate upward, somewhat plicate; margins plane, entire; costa slender to rather thick, generally ending at 2/3 the leaf length; median leaf cells oval, 12–19 µm × 8–10 µm; upper cells slightly elongate; basal cells short, 8.5–11.0 µm long, sometimes similar to alar cells; alar cells quadrate to subquadrate; branch leaves similar to stem leaves, but smaller. *Sporophytes* unknown.

Illustrations: Plate 438, figs. 1–6.
Habitat: on rocks or tree trunks.
Distribution: China, Korea, and Japan.

Chinese specimens examined: GUANGDONG: Le-chang Co., *L. Zhang 727* (SYS, IFSBH). HUBEI: Mt. Shen-nong-jia, *P.-C. Wu 267* (MO, PE). JILIN: Chang-bai Co., *C. Gao 1227* (IFSBH); Dun-hua City, *X.-Z. Lin 2745* (IFSBH); Jiao-he Co., *X.-G. Kang 1964* (IFSBH); Mt. Zhengfeng, *X.-G. Kang 7682, 2773* (IFSBH). LIAONING: Feng-cheng Co., *X.-Y. Jia 957* (IFSBH). SHANGHAI: Jia-ding Distr., *S.-M Xu 106, 240* (MO, PE). SICHUAN: Wen-chuan Co., *B. Allen 6718g* (MO). TAIWAN: Lushan (Mt.), *C.-C. Chuang B101458* (NY). ZHEJIANG: Ning-bo City, *Oldham s.n.* (H); De-qing Co., *P.-C. Wu 1213* (MO).

This species is different from *O. hakoniensis* (Mitt.) Broth. by its small plants, short branches without flagelliform branchlets, small and mostly plane leaves with obtuse apices, and shorter leaf cells (12–19 µm long vs. 20–30 µm long).

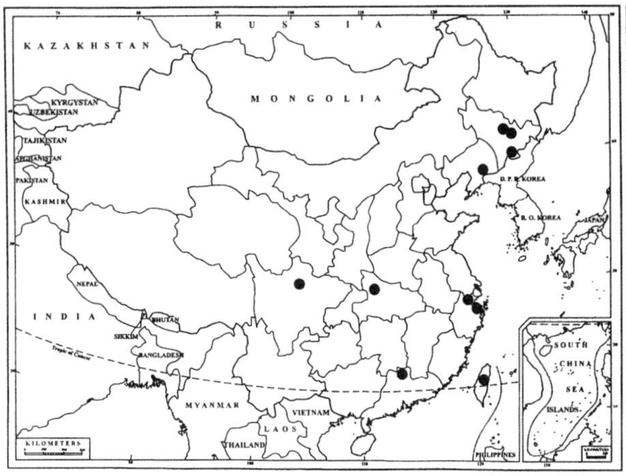

1. *Okamuraea brachydictyon*

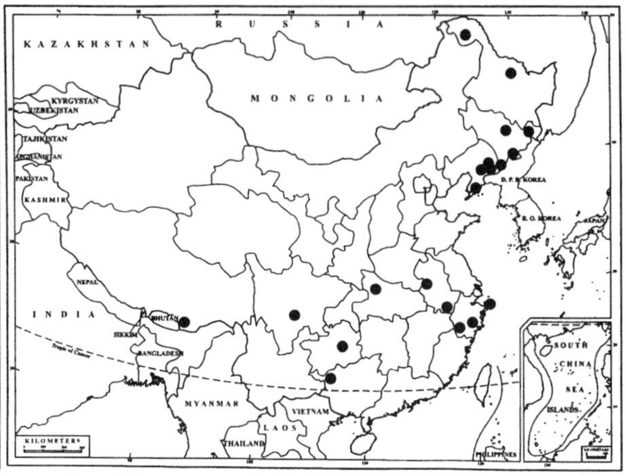

2a. *Okamuraea hakoniensis* var. *hakoniensis*

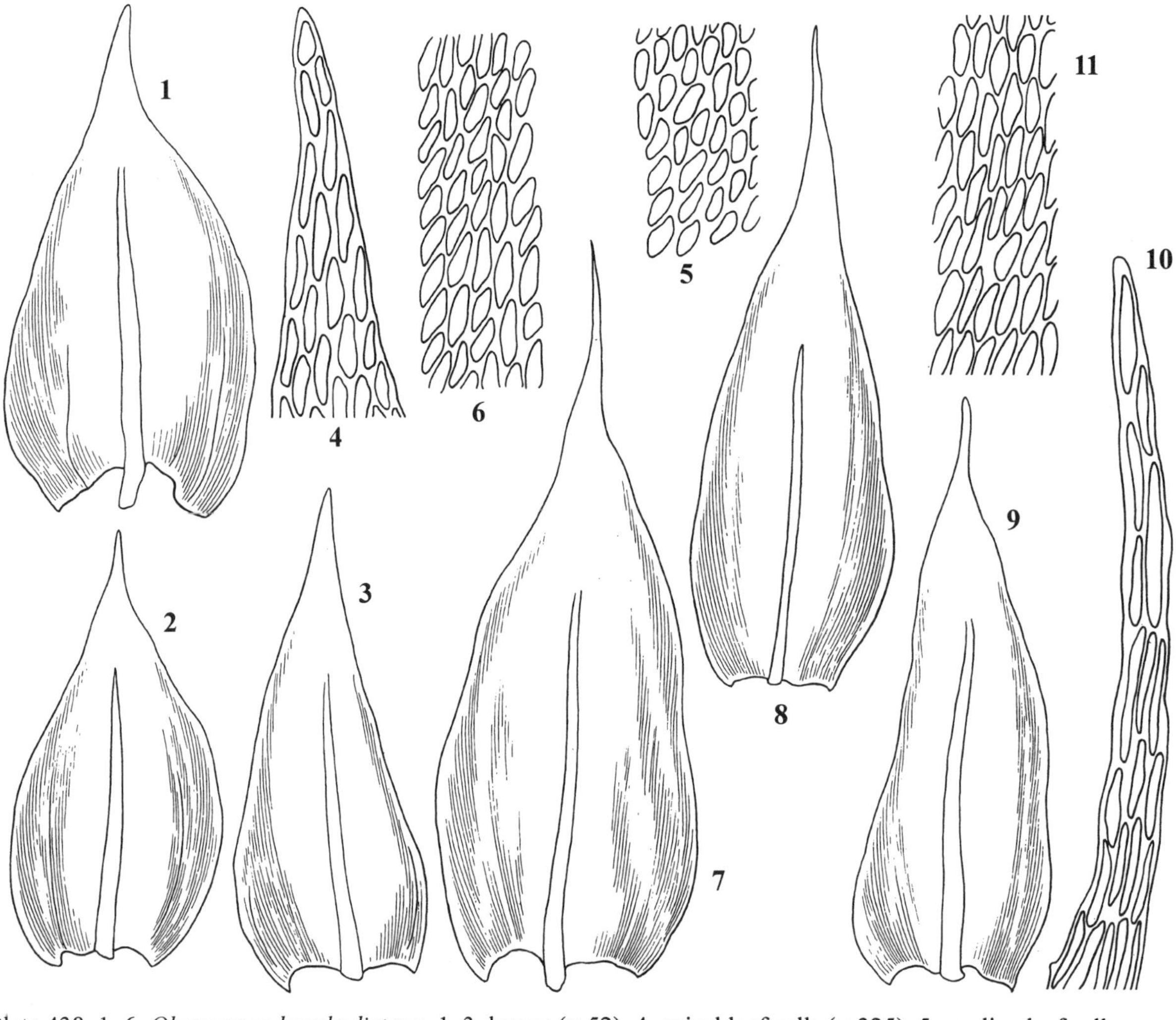

Plate 438. 1–6, *Okamuraea brachydictyon*. 1–3, leaves (× 52). 4, apical leaf cells (× 325). 5, median leaf cells (× 325). 6, basal leaf cells (× 325). 7–11, *Okamuraea hakoniensis* var. *hakoniensis*. 7–9, leaves (× 52). 10, apical leaf cells (× 325). 11, basal leaf cells (× 325). (1–2, 6, drawn from Guangdong, without collector and number; 3–5, *X.-G. Kang 1964*, IFSBH; 7–11, from *R.-L. Hong 575*, SHM.)

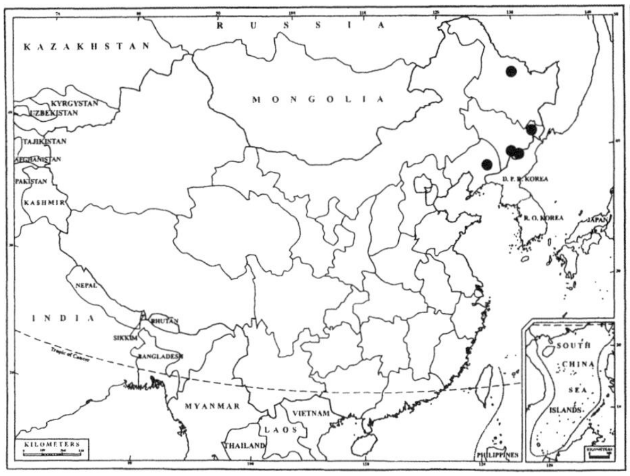

2b. *Okamuraea hakoniensis* fo. *multiflagellifera*

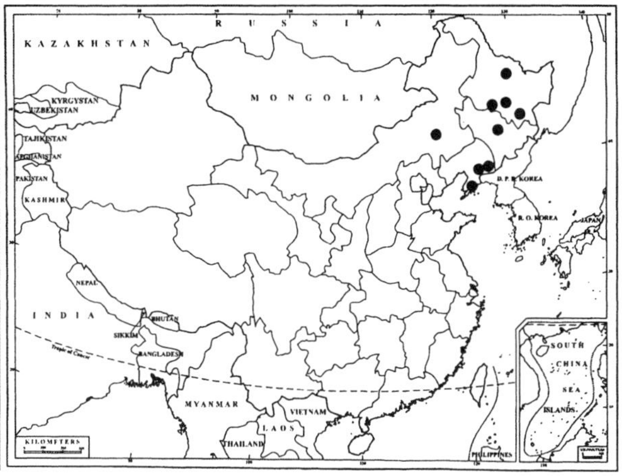

2c. *Okamuraea hakoniensis* var. *ussuriensis*

2. ***Okamuraea hakoniensis*** (Mitt.) Broth. 长枝褶藓 chang-zhi zhe xian
Nat. Pflanzenfam. I(3): 1133. 1908. *Hypnum hakoniense* Mitt., Trans. Linn. Soc. London, Bot. 3: 185. 1891. Type. Japan: Honsyu, Hakone Pass, October, *Bisset s.n.*

2a. ***Okamuraea hakoniensis*** var. ***hakoniensis*** 长枝褶藓原变种 chang-zhi zhe xian yuan bian-zhong

Plants robust, in dense tufts. *Stems* creeping, up to 7 cm long; branches 1–2 cm long, with few and short flagelliform branchlets. *Leaves* ovate to oblong-ovate, distinctly plicate, suddenly tapered to linear or lanceolate apex, 1.6–2.2 mm × 0.7–0.9 mm; margins plane, entire; costa rather slender, reaching half of the leaf length; median leaf cells oblong-oval, thick-walled, 20–30 µm × 5–8 µm; upper cells slightly elongate; basal cells short, oval to subquadrate, 13–20 µm × 6–10 µm, thick-walled; alar cells subquadrate. *Inner perichaetial leaves* linear-lanceolate, up to 2.2 mm long, with weaker costa. *Setae* erect, elongate, 13–25 mm long; capsules erect or pendent, cylindrical to oblong-ellipsoidal; exostome teeth lanceolate, with lamella, verrucose on surfaces; endostome segments reduced or absent; basal membrane high; cilia not differentiated; opercula obliquely long-rostrate. *Calyptrae* loosely covered with long hairs.

Illustrations: Plate 438, figs. 7–11.
Habitat: on tree trunks, sometimes on rocks.
Distribution: China, Japan, and Bhutan.

Chinese specimens examined: ANHUI: Huang Shan (Mt.), *C. Gao 30509* (IFSBH); Jin-zhai Co., *Z.-M. Zeng 323* (MO). GUANGXI: Tian-lin Co., *C. Gao 2636* (IFSBH). GUIZHOU: Lu Shan (Mt.), *B. C. Tan 91-696* (FH). HEILONGJIANG: Da-hai Forest Station, *C. Gao 9135* (IFSBH); Xiao Xin An Ling (Lesser Khingan Mt.) Wu-yi Distr., *C. Gao 3918* (IFSBH). HUBEI: Mt. Shen-nong-jia, P.-C. Wu 383b (MO), *Sino-American 1980 Team 656a* (MO, NY, PE). JILIN: Chang-bai Co., *C. Gao s.n.* (IFSBH); Jiao-he Co., *C. Gao 1675* (IFSBH); Ji-an Co., *C. Gao 7868* (IFSBH); Wang-qing Co., *C. Gao 8754* (IFSBH). LIAONING: Feng-cheng Co., *C. Gao 5900, 5978, 6888, 2448* (IFSBH); Heng-ren Co., *T. Cao 167, 263* (IFSBH); Kuan-dian Co., *X.-L. Chen 4606* (IFSBH); Zhuang-he Co., *X.-Y. Jia 425* (IFSBH). SICHUAN: Mt. Omei, *Allen 6555* (MO); *P.-C. Wu 3836* (MO, PE). XIZANG: Cuo-na Co., *S.-K. Chen 5107* (HKAS). ZHEJIANG: Mt. Kuo-chang, *C. Gao 601* (IFSBH), Mt. Jiu-long, *R.-L. Hong 757* (IFSBH); Putou Dao (Isl.), *P.-C. Wu 1343* (MO, PE).

2b. ***Okamuraea hakoniensis*** (Mitt.) Broth. fo. ***multiflagellifera*** (Okam.) Nog. 长枝褶藓鞭枝变型 chang-zhi zhe-xian bian-zhi bian-xing
J. Hattori Bot. Lab. 9: 9. 1953. *Okamuraea cristata* Broth. var. *multiflagellifera* Okam., J. Coll. Sci. Imp. Univ. Tokyo 36 (7): 42. 1915. Type. Japan: Hondo, Iyo Prov., Mt. Ishiduchi, *J. Shiraga s.n.*

Plants mostly epiphytic. *Leaves* ovate-lanceolate, with narrow base, long acuminate at upper part; flagelliform branchlets usually slender, covered with reddish brown fasciculate rhizoids; flagelliform leaves sparse, minute, ovate-lanceolate with long-acuminate apices.

Illustrations: Plate 439, figs. 1–6.
Habitat: on tree trunks.
Distribution: China and Japan.

Chinese specimens examined: HEILONGJIANG: Xiao Xin An Ling (Lesser Khingan Mt.), *C. Gao 641* (IFSBH). LIAONING: Ben-xi Co., *C. Gao 8408* (IFSBH). JILIN: Changbai Co., *C. Gao 1102* (IFSBH); Lin-jiang Co., *C. Gao 7757* (IFSBH); Wang-qing Co., *C. Gao 8932* (IFSBH).

2c. ***Okamuraea hakoniensis*** (Mitt.) Broth. var. ***ussuriensis*** (Broth.) Nog. 长枝褶藓乌苏里变种 chang-zhi zhe-xian wu-su-li bian-zhong

J. Hattori Bot. Lab. 9: 9. 1953. *Bryhnia ussuriensis* Broth., Turdy Troitskos.-Kyakhtinsk. Otd. Priamursk. Otd. Imp. Russk. Geogr. Obstch. 8 (3): 8. 1906. Type. Russia: Ussuri.

Plants rather robust. *Leaves* typically larger than other two varieties, 1.9–2.4 mm × 0.9–1.1 mm, usually piliferous at apex, obviously broad, not decurrent to slightly decurrent at base; leaf cells thick-walled. *Setae* (more than 25 mm long) and capsules longer than those of var. *hakoniensis*. This variety is widely distributed in northeastern China.

Illustrations: Plate 439, figs. 7–11.
Habitat: on rocks or trees.
Distribution: China, Korea, Japan, and Russia (Siberia).

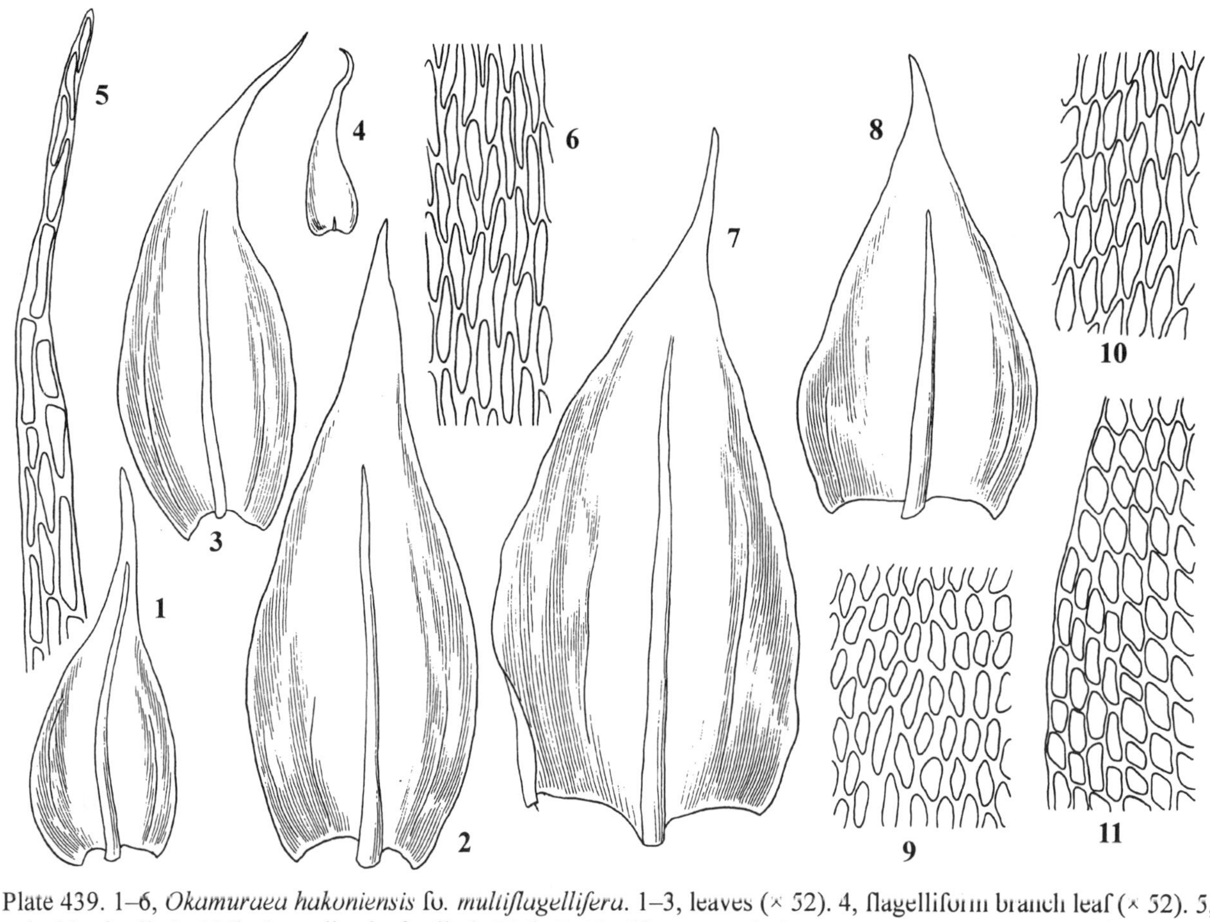

Plate 439. 1–6, *Okamuraea hakoniensis* fo. *multiflagellifera*. 1–3, leaves (× 52). 4, flagelliform branch leaf (× 52). 5, apical leaf cells (× 325). 6, median leaf cells (× 325). 7–11, *Okamuraea hakoniensis* var. *ussuriensis*. 7–8, leaves (× 52). 9, upper leaf cells (× 325). 10, median leaf cells (× 325), 11, basal leaf cells (× 325). (1–6, drawn from *C. Gao 1102*, IFSBH; 7–11, *C. Gao 3918*, IFSBH.)

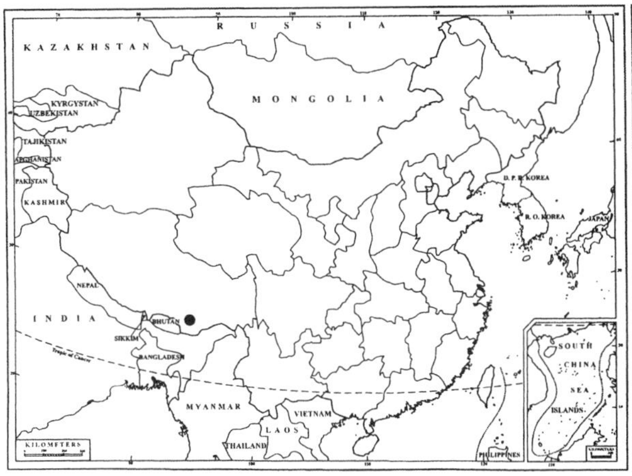

1. *Orthoamblystegium spurio-subtile*

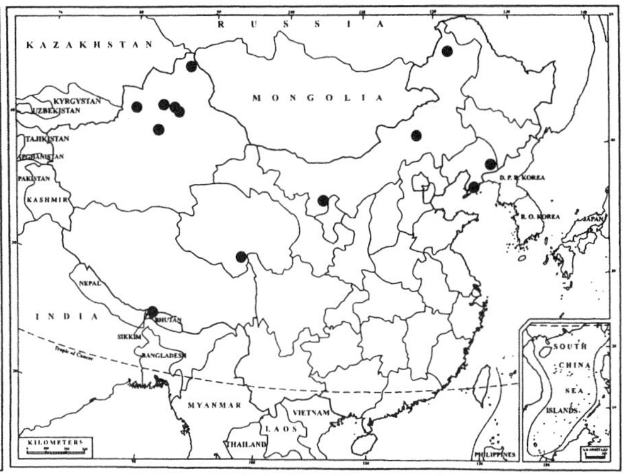

1. *Pseudoleskeella catenulata*

Chinese specimens examined: HEILONGJIANG: Lin-jiang Distr., *C. Gao 7745* (IFSBH); Ling-an Co., *C. Gao 6752*; *6163* (IFSBH); Shang-zhi Co., *Z.-G. Wang 1576* (IFSBH); Xiao Xing An Ling (Lesser Khingan Mt.), *C. Gao 3918* (IFSBH); Yi-chun Distr., *T.-Y. Ding 83003* (IFSBH). JILIN: Jiao-he Co., *X.-G. Kang 1964* (IFSBH); Lao-ye Ling (Mt.), *C. Gao 1590* (IFSBH). LIAONING: Feng-cheng Co., *C. Gao 6004* (IFSBH), *X.-Y. Jia 890* (IFSBH); Kuan-dian Co., *T. Cao et al. 38391*, *38306* (IFSBH), *Buck 23690* (IFSBH, NY); Zhuang-he Co., *X.-Y. Jia 381*, *74*, *162*, *408* (IFSBH). NEI MONGOL: Ba-lin You-qi (Co.), *X.-L. Bai 2423* (NY).

6. ***Orthoamblystegium*** Dix. & Sat. 拟柳叶藓属 ni liu-ye xian shu
Bot. Mag. (Tokyo) 50: 260. 1936.

Plants slender, dark green below, yellowish green above, in prostrate mats. *Primary stems* elongate, subpinnately branched; branches 5 mm long, tips curved when dry, with dense leaves; paraphyllia few, lanceolate. *Stem leaves* linear-lanceolate, gradually tapered from an ovate or triangular base to a lanceolate apex; margins slightly recurved; costa single, strong, yellowish, smooth, subpercurrent to percurrent; leaf cells long-rhomboidal or sublinear, smooth to projecting at upper ends; alar region consisting of few rectangular cells. *Dioicous*. *Perichaetia* on main stems. *Setae* smooth, brownish, up to 1 cm long; capsules erect, brownish, long-spherical to shortly cylindrical; annuli formed by thick-walled cells; peristome double; exostome teeth lanceolate, verrucose; endostome segments rudimentary; basal membrane low. *Spores* spherical to ellipsoidal, 14–18 µm in diameter.

This is a monotypic genus. The species was first reported in China by C. Gao et al. (1985) from Xizang province.

1. ***Orthoamblystegium spurio-subtile*** (Broth. & Par.) Kanda & Nog. 拟柳叶藓 ni liu-ye xian
Misc. Bryol. Lichenol. 9: 135. 1982. *Platydictya spurio-subtile* (Broth. & Par.) Schofield, J. Hattori Bot. Lab. 28: 37. 1965. *Amblystegium spurio-subtile* Broth. & Par., Rev. Bryol. 31: 94. 1904. Type. Japan: Honshu.
Orthoamblystegium nipponicum Dix. & Sak., Bot. Mag. (Tokyo) 50: 260. 1936.
Orthoamblystegium longinerve (Card.) Toy., Acta Phytotax. Geobot. 6: 105. 1937. *Schwetschkea longinervis* Card., Bull. Soc. Bot. Genève., ser. 2, 3: 279. 1911.

The description of this species is the same as that of the genus described above.

Illustrations: Plate 440, figs. 1–8.
Habitat: on tree trunks or rocks.
Distribution: China, Japan, and Korea.

Chinese specimens examined: XIZANG: Long-zi Co, Zhun-ba-da, *M. Zang 1035*, *1012* (HKAS, IFSBH).

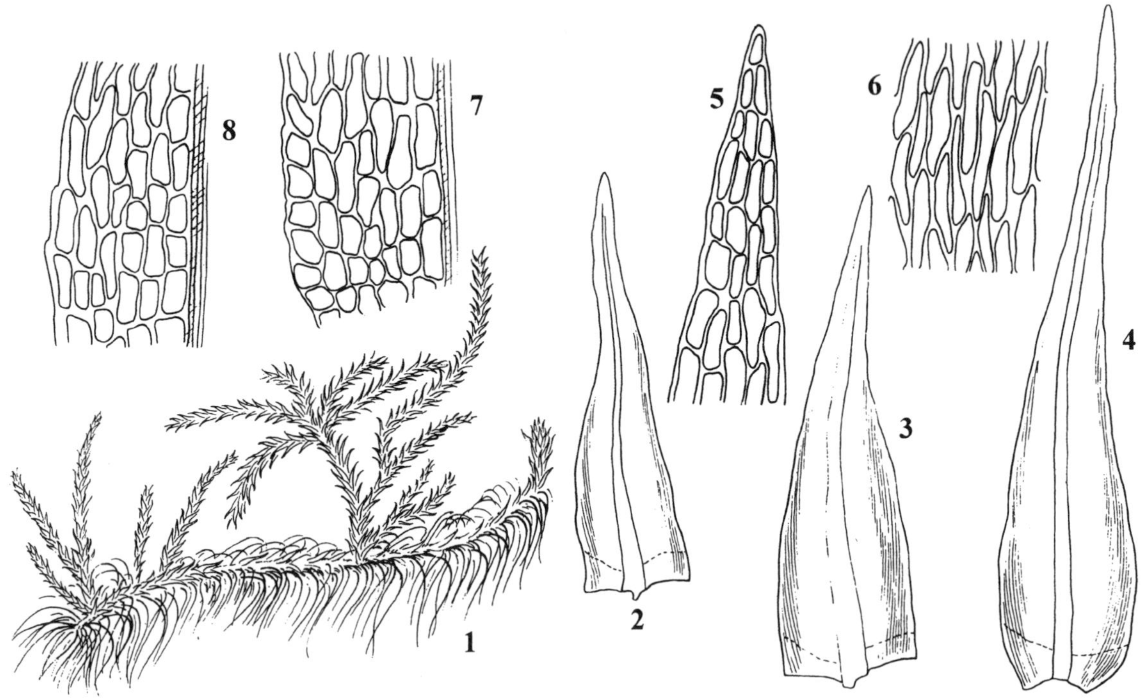

Plate 440. *Orthoamblystegium spurio-subtile*. 1, habit (× 5). 2–4, leaves (× 52). 5, apical leaf cells (× 325). 6, median leaf cells (× 325). 7–8, basal leaf cells (× 325). (1–8, drawn from *M. Zang 1035*, HKAS.)

7. ***Pseudoleskeella*** Kindb. 假细罗藓属 jia xi-luo xian shu
Gen. Eur. N.- Amer. Bryin. 20. 1897.

Plants slender, soft or rigid and fragile, dark green to brown, dull to slightly shiny. *Stems* thin, prostrate, regularly or irregularly pinnately branched; paraphyllia linear to lanceolate. *Stem leaves* imbricate when dry, spreading when moist, cordate, slightly decurrent at base, suddenly narrowed to a long acuminate apex; basal margins entire, plane or recurved, plicate below; costa single, up to middle of the leaf or unequally forked, sometimes very short or absent; upper and median leaf cells round or rhomboidal; basal cells near margins flatly quadrate, slightly thick-walled; branch leaves rather small, acuminate. *Dioicous*. *Inner perichaetial leaves* pale green, moderately sheathing at base, gradually narrowed to an acuminate apex; costa short or absent. *Setae* elongate, slender; capsules patent, oblong-cylindrical, curved, reddish brown, contracted at lower part of capsule mouth when dry; annuli differentiated; peristome double; exostome teeth narrowly lanceolate, yellowish, borders differentiated, densely transversely striate and lamellate on outer surfaces; endostome segments pale yellow, finely papillose, nearly as long as the teeth, thin, keeled; basal membrane high; two cilia present, shorter than teeth, sometimes rudimentary or absent; opercula conic, obliquely beaked. *Spores* small, yellowish green to green, nearly smooth.

There are about 6 species of *Pseudoleskeella* in the world, and three species are known in China. Potier de la Varde (1937) first reported *Pseudoleskeella catenulata* (Brid. ex Schrad.) Kindb. from Xinjiang. C. Gao (1977) reported *P. catenulata* and *P. tectorum* (Brid.) Kindb. from northeastern China.

1. Leaf cells mamillose on the back; costa short and slender, usually forked, rarely absent .... 2. *P. papillosa*
1. Leaf cells smooth; costa stout and short, single or forked, rarely double 2
   2. Costa up to the middle of leaf, single or forked above; branches short, curved at the apex; endostomal cilia well developed .... 1. *P. catenulata*
   2. Costa not reaching the middle of leaf, mostly forked at the base; branches sparse and long; endostomal cilia reduced .... 3. *P. tectorum*

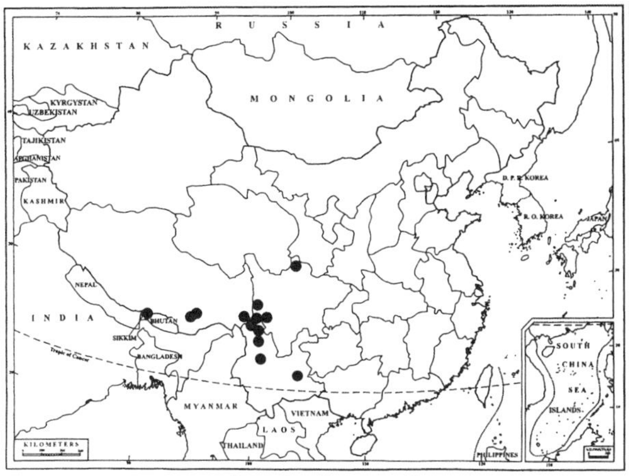

2. *Pseudoleskeella papillosa*

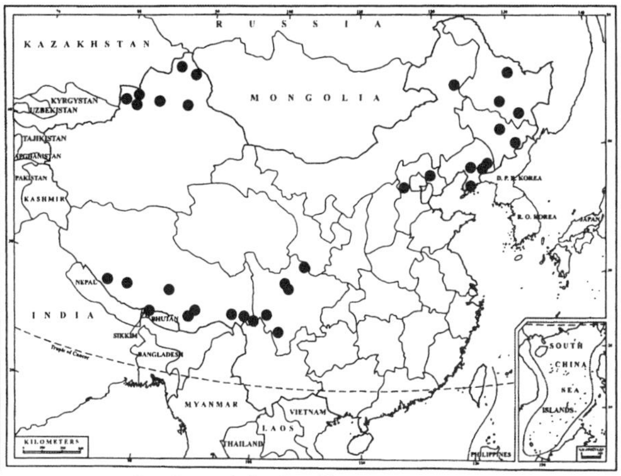

3. *Pseudoleskeella tectorum*

1. ***Pseudoleskeella catenulata*** (Brid. ex Schrad.) Kindb. 假细罗藓 jia xi-luo xian
Eur. N. Amer. Bryin. 1: 48. 1897. *Pterigynandrum catenulatum* Brid. ex. Schrad., J. Bot. (Schrader) 1801(1): 195. 1803. Type. Switzerland: *Bridel s.n.*

Plants soft, dark green to yellowish green, in loose or dense tufts. *Stems* creeping, irregularly pinnately branched; branches short, julaceous, curved at tips; pseudoparaphyllia small, linear-lanceolate. *Leaves* concave, cucullate, gradually tapered from cordate to ovate base to a mucronate apex, 0.6–0.8 mm × 0.30–0.45 mm; costa ending above middle of leaf, single or forked above; leaf cells smooth, short, small; median cells irregularly short-rhomboidal, 10–15 μm × 5–8 μm; basal cells near margins quadrate to flatly quadrate, 7.5–10.0 μm × 10.0-12.5 μm; basal cells along costa rectangular, 15–25 μm × 10 μm. *Dioicous*. *Setae* up to 1.5 cm long, yellowish brown; capsules pendent, recurved, yellowish brown or somewhat reddish; peristome double; exostome teeth narrowly lanceolate; cilia well developed. *Spores* spherical, 16–18 μm, finely papillose.

Illustrations: P.-C. Wu (ed.) 2001 (Pl. 61, figs. 1–8).
Habitat: on rocks, sometimes on rotten woods or tree trunks.
Distribution: China, Japan, Europe, Greenland, and North America.

Chinese specimens examined: LIAONING: Zhuang-he Co., *X.-Y. Jia 880253*, *880338* (IFSBH); Huan-ren Co., *T. Cao 142*, *143* (IFSBH). NEI MONGOL: Xi-lin-hao-ti City., *Z.-G. Gong 1054*, *1081* (IFSBH); Helan Shan (Mt.), *Z.-G. Tong 1330* (HIMC, IFSBH, MO); Da Xing An Ling (Greater Khingan Mt.), *C. Gao 3437* (IFSBH). QINGHAI: Cheng-duo Co., *B. C. Tan 95-1690* (FH, MO). XINJIANG: Fu-kang Co., *B.-C. Tan 93-963*, *93-998*, *93-1211* (FH, NY), *J.-C. Zhao 3214* (IFSBH), *C.-F. Zhao 2338* (IFSBH); He-jing Co., *S. He 951147* (MO); Ma-na-si Co., *X.-S. Wang 22* (MO, PE); Mt. Altai, *C.-F. Zhao 2457* (IFSBH); Mt. Bo-ge-da, *J.-C. Zhao 2791*, *2284* (IFSBH); Ni-le-ke Co., *S. He 951123* (MO); Urumqi Co., *S. He 951046* (MO). XIZANG: Ya-dong Co., *Xizang Team 7785* (MO, PE).

This species is characterized by having a slender costa that is either short, single or forked above, ending above the middle of leaf and by having cells rather small, short, and smooth. The distribution of this species in China is limited to northern areas. Many specimens previously identified as this species from southern provinces are *Pseudoleskeella tectorum*.

2. ***Pseudoleskeella papillosa*** (Lindb.) Kindb. 瘤叶假细罗藓 liu-ye jia xi-luo xian.
Eur. N. Amer. Bryin. 1: 49. 1897. *Heterocladium papillosum* (Lindb.) Lindb., Musci Scand. 37. 1879. *Pseudoleskea papillosa* (Lindb.) Jaeg., Ber. Thätigk. St. Gallischen Naturwiss. Ges. 1876–77: 243. 1878. *Leskea papillosa* Lindb., Bot. Not. 1872: 135. 1872. Type. Europe.

Plants slender, yellowish green, slightly shiny, in loose tufts. *Primary stems* creeping, 3–4 cm long, irregularly pinnately branched; paraphyllia few. *Leaves* appressed on stems, obviously secund when dry, spreading when moist, slightly concave, ovate at base, contracted to piliferous upward to the apex, usually secund, 0.7–0.9 mm × 0.5–0.6 mm; basal margins somewhat narrowly involute; costa slender, forked, ending below middle of the leaf; upper and median cells round-quadrate to elongate-hexagonal, to 10–15 µm long, obviously projecting at upper ends on back surfaces, more obvious on tender leaves; basal cells near margins flatly quadrate to quadrate; cells along costa rectangular. *Sporophytes* not seen.

Illustrations: P.-C. Wu (ed.) 2001 (Pl. 61, figs. 9–17).
Habitat: on rotten wood under forests.
Distribution: China, Europe, Greenland, and North America.

Chinese specimens examined: SICHUAN: Ba-tang Co., *S. He 31425* (MO); Dao-cheng Co., *S. He 31681* (MO), *J.-K. Yang 81-2763*, *81-2733* (HKAS, IFSBH); De-rong Co., *X.-J. Li 1525*, *81-2885* (HKAS, IFSBH); Mai-cheng Co., *L.-S. Wang 81-2556* (HKAS, IFSBH); Ruo-er-gai Co., *S. He 30608* (MO). XIZANG: Bi-tu Co., Ca Wa Long, *M. Zang 7117* (HKAS, IFSBH); Lang Co., *M. Zang 1809* (HKAS, IFSBH); Long-zi Co., *M. Zang 1091* (HKAS, IFSBH), Jia-yu Distr., *M. Zang 1208* (HKAS, IFSBH); Ri-dong Co., *M. Zang 5288* (HKAS, IFSBH); Ya-dong Co., *M. Zang 35* (HKAS, IFSBH). YUNNAN: Da-li Co., *Redfearn et al. 1524* (MO); Li-jiang Co., *M. Zang 1436* (HKAS, IFSBH); Lu-xi Co., *X.-Y. Jia 910113* (IFSBH, MO); Mei-li City, *M. Zang 8722* (HKAS, IFSBH); Mt. Shi-long, *X.-J. Li 956* (HKAS, IFSBH); Zhong-dian Co., *L.-S. Wang 1334*, *1337* (HKAS, IFSBH).

The particularly slender plants and the mammillae at the back surfaces of leaf cells make this species distinguishable from other species.

3. ***Pseudoleskeella tectorum*** (Brid.) Kindb. 瓦叶假细罗藓 wa-ye jia xi-luo xian
Eur. N. Amer. Bryin. 1: 48. 1897. *Leskeella tectorum* (Brid.) Hag., Kongel. Norske Vidensk. Selsk. Skr. (Trondheim) 1908 (9): 92. 1909. *Hypnum tectorum* Funck ex Brid., Bryol. Univ. 2: 582. 1827. Type. Germany: Karlsruhe.

Plants slender, rigid, green to yellowish green, brownish green when old, dull, in tufts. *Stems* creeping, varying greatly in length, irregularly branched; branches long to short, often flagelliform. *Leaves* cucullate, 0.7–1.2 mm × 0.4–0.6 mm, suddenly tapered from a broadly ovate base to an apiculate or blunt apex, indistinctly biplicate; costa short, forked from base or middle, ending below middle of leaf; margins plane, entire; leaf cells small, irregularly round-quadrate to shortly rhomboidal, smooth; median cells 20–25 µm × 6–8 µm; alar cells flatly quadrate, 7.5–12.5 µm × ca. 12.5 µm, in oblique rows upward. *Dioicous*. *Perichaetial leaves* abruptly narrowed from a semi-sheathing base to an apiculate apex upward, erect, 1.6 mm long. *Setae* 8–13 mm long; capsules erect or slightly pendent, cylindrical; annuli differentiated, consisting of 1–2 rows of cells; peristome double; exostome teeth long-lanceolate, reddish brown, 0.6–0.7 mm long, nearly smooth, lamellae present; endostome segments keeled; cilia absent. *Spores* 20–25 µm in diameter, verrucose.

Illustrations: P.-C. Wu (ed.) 2001 (Pl. 62, figs. 1–16).
Habitat: on rocks, rotten wood or tree trunks.
Distribution: China, Far East of Russia, Europe, and North America.

Chinese specimens examined: HEBEI: Wei Co., *C. Gao 30669* (IFSBH); Xing-long Co., *J.-X. Ji 983*, *B130*, *479* (IFSBH). HEILONGJIANG: Ning-an Co., *C. Gao 6770* (IFSBH); Shang-zhi Co., *Z.-W. Ao 85-7* (IFSBH); Xiao Xing An Ling (Lesser Khangan Mt.), *P.-C. Chen 206* (IFSBH); Yi-chun City, Wu-ying Distr., *C. Gao 5053* (IFSBH). JILIN: He-long Co., *X.-G. Kang 2991* (IESBH); Jiao-he Co., *X.-G. Kang 1889*, *1961* (IFSBH). LIAONING: Heng-ren Co., *T. Cao 465* (IFSBH); Kuan-dian Co., *T. Cao 38680* (IFSBH); Qian Shan (Mt.), *C. Gao 1943* (IFSBH); Zhuang-he Co., *X.-Y. Jia 511* (IFSBH). NEI MONGOL: Za-lan-tun City, *X.-L. Bai 2675*(NY). SICHUAN: Dao-cheng Co., *S. He 31707* (MO); Li Co., *Redfearn 34978* (MO); Ma-er-kang Co., *Redfearn 35056* (MO); Mai-cheng Co., *X.-J. Li 81-3038* (HKAS, IFSBH); Mu-li Co., *C. Gao 20289* (IFSBH), *M.-Z. Wang 23730* (IFSBH, PE); Nan-ping Co., *Redfearn 35364* (MO). XINJIANG: Bo-le Co., *B.-C. Tan 93-941*(FH, NY); Bur-qin Co., *B.-C. Tan 93-936* (NY); Huo-cheng Co., *B.-C. Tan 93-959* (FH, NY); Ni-le-ke Co., *S. He 951127a* (MO); Qing-he Co., *B.-C. Tan 93-894*, *93-890* (FH); Qi-tai Co., *B.-C. Tan 93885* (FH); Wu-su Co., *S. He 951098* (MO). XIZANG: Ang-ren Co., *J.-W. Zhang 7506* (IFSBH, PE); Bi-tu Co., Cha Wa Long, *M. Zang 7192* (HKAS, IFSBH); Cha-yu Co., *S.-G. Wu 1073* (IFSBH, PE); Long Co., *M. Zang 1663* (HKAS, IFSBH); Long-zi Co., Jia-yu Co., *M. Zang 1653* (HKAS, IFSBH), San An Qu Lin, *M. Zang 1604*, *1624* (HKAS, IFSBH), Zhun-ba Distr., *M. Zang 1104*, *1054* (HKAS, IFSBH); Ji-long Co., *S.- K. Chen 218*, *220* (IFSBH, PE); Lasa City, *Ecol. Dept. Tibet Group 7433* (PE, IFSBH); Ya-dong Co., *M. Zang 658*, *4955*a (HKAS, IFSBH), *S.- K. Chen 4953* (IFSBH, PE). YUNNAN: De-qin Co., *J.-K. Yang 2311* (HKAS, IFSBH).

8. ***Pseudoleskeopsis*** Broth. 拟草藓属 ni cao xian shu
Nat. Pflanzenfam. I(3): 1002. 1907.

Plants robust, green to yellowish green, sometimes brownish green, dull. *Stems* creeping, with sparse rhizoids; branches dense, short, blunt; paraphyllia sparse, narrowly lanceolate. *Leaves* loosely appressed when dry, slightly secund, erect to spreading when moist, ovate to oblong-ovate, slightly decurrent at base, often obtuse at apex; margins plane, crenate or serrate; costa strong, slightly flexuous, ending below apex; leaf cells small, oval to obliquely rhomboidal, smooth or mamillose; basal cells gradually becoming shortly rectangular; alar cells quadrate to flatly quadrate. *Monoicous*. *Inner perichaetial leaves* erect, pale green, narrowly lanceolate, costa extending up to leaf apex or percurrent, cells smooth. *Setae* elongate, slender; capsules pendent or horizontal, asymmetric, oblong-ovoid to long-cylindrical, slightly curved, with obvious apophysis; annuli differentiated; peristome double; exostome teeth narrowly lanceolate, yellowish, with dense transverse striae on outer surfaces, borders differentiated, lamellate on inner surfaces; endostome segments pale yellowish, finely papillose, folded, nearly as long as the exostome teeth, narrowly perforate or slit; basal membrane high; cilia 1–2, well developed; opercula conic, with short, blunt rostrum. *Calyptrae* cucullate. *Spores* small, finely papillose.

The first report of *Pseudoleskeopsis* in China was made by Thériot (1911), who described *P. serrulata* Card. & Thér. from Guizhou. Later, Brotherus (1929) and Bartram (1935) reported *P. decurvata* (Mitt.) Broth., *P. integrifolia* Broth., and *P. zippelii* (Dozy & Molk.) Broth., respectively. *Pseudoleskeopsis laticuspis* (Card.) Broth. was subsequently reported new to China by Dixon (1933). More recently, M.-J. Lai and J.-R. Wang-Yang (1976) reported *P. japonica* (Sull. & Lesq.) Iwats. from Taiwan. P.-C. Chen et al. (1978) listed 3 species and 1 variety for China, of which *P. orbiculata* (Mitt.) Broth. and *P. orbiculata* var. *laticuspis* (Card.) Thér. were reported new to the country. Most recently, R.-L. Hu & Y.-F. Wang (1981) reported *P. tosana* Card. from eastern China. Thus, a total of 8 species and 1 variety have since reported from China.

Thériot (1929) studied the genus taxonomically and recognized 5 species. Iwatsuki (1976) studied Asiatic species based on type specimens deposited in Helsinki and made numerous synonyms under *P. zippelii*. His treatment is useful in studying Chinese taxa. Recently, Crosby et al. (2000) recognized 8 species in the world, with 3 better-known species. In this flora two species are treated. The genus *Pseudoleskeopsis* is different from other genera in the Leskeaceae mainly by the structure of peristome.

1. Branch apices cuspidate; leaf cells longer, rectangular or nearly lineariform .... 1. *P. tosana*
1. Branch apices obtuse; leaf cells rounded-hexagonal or rhomboid.. 2. *P. zippelii*

1. ***Pseudoleskeopsis tosana*** Card. 尖叶拟草藓 jian-ye ni cao xian
Bull. Soc. Bot. Genève, ser. 2, 5: 317. 1913. Type. Japan.
Pseudoleskeopsis hattorii Nog., J. Hattori Bot. Lab. 2: 82. 1947.

Plants brownish green to yellowish green, in mats. *Primary stems* creeping, with numerous branches; branches short, erect, ca. 1 cm long. *Leaves* dense, curved when dry, keeled; stem leaves broadly lanceolate, gradually tapered to a long acuminate apex, 1.0–1.5 mm long; branch leaves ovate to oblong-ovate, acute, usually secund, asymmetric, 0.7–1.2 mm × 0.35–0.60 mm, shortly plicate; costa thick, strong, ending near apex, generally not percurrent; margins dentate, apical cells projecting and dentate at leaf tips; leaf cells smooth, thick-walled; median cells rectangular, 10–15 μm × 4–7 μm; basal cells shortly rectangular. *Inner perichaetial leaves* lanceolate, plicate, acuminate, up to 2.5 mm long; costa thin, slightly percurrent; cells rectangular, thin-walled. *Setae* 1.5–3.0 cm long, reddish brown; capsules pendent, 1.5–2.0 mm long, with apophysis, contracted at mouth when dry; annuli differentiated, consisting of one layer of cells; exostome teeth ca. 0.3 mm long, subulate, transversely striate below, finely papillose above; endostome segments as long as the teeth, folded; basal membrane high; cilia 2–3. *Spores* spherical, 13–15 μm in diameter.

Illustrations: Plate 441, figs. 1–6.

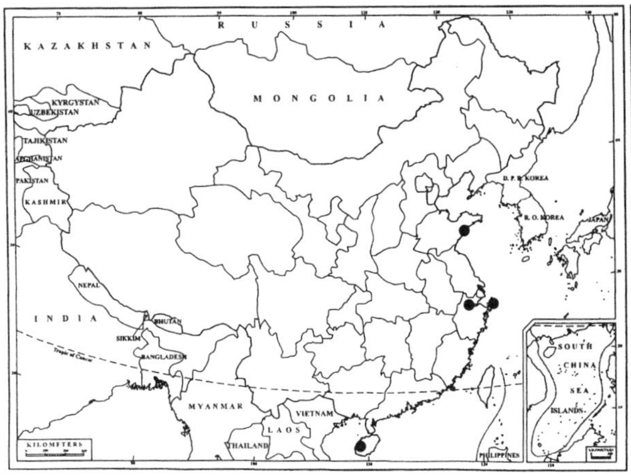

1. *Pseudoleskeopsis tosana*

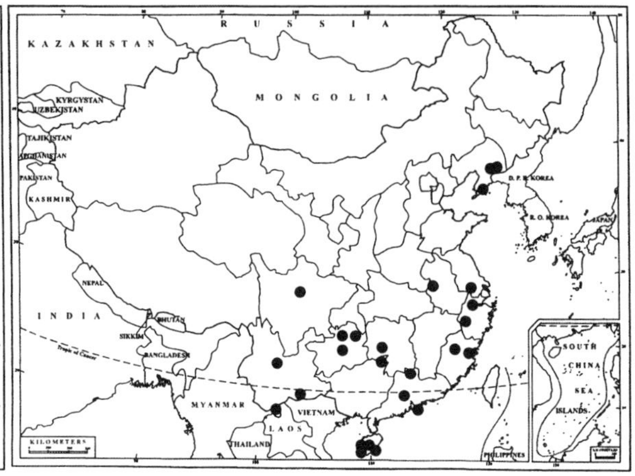

2. *Pseudoleskeopsis zippelii*

Habitat: on wet rocks.
Distribution: China and Japan.

Chinese specimens examined: HAINAN: Chang-jiang Co., *Redfearn 35725*, *35909*, *35914* (FH, MO). SHANDONG: Qing-dao City, Lao Shan Mt., *C. Gao 34534*, *34535*, *34576* (IFSBH). ZHEJIANG: Putuo Shan (Mt.), *D.-K. Li 7002* (IFSBH, SHM); Xitianmu Shan (Mt.), *R.-L. Hu 202* (ALTA).

This species is distinguished from *P. zippelii* by having acute leaf apices and rectangular to linear leaf cells. However, the Chinese specimens we examined sometimes have shorter leaf cells (only 10–15 μm × 4–7 μm).

2. ***Pseudoleskeopsis zippelii*** (Dozy & Molk.) Broth. 拟草藓 ni cao xian

Nat. Pflanzenfam. I(3): 1003. 1907. *Hypnum zippelii* Dozy & Molk., Ann. Sci. Nat. Bot., ser. 3, 2: 310. 1844. Type. Indonesia: Amboina, *Zippelinus s.n.*

Pseudoleskeopsis decurvata (Mitt.) Broth., Nat. Pflanzenfam. I(3): 1003. 1907. *Leskea decurvata* Mitt., J. Proc. Linn. Soc., Bot. 8: 154. 1864.

Pseudoleskeopsis integrifolia Broth., Symb. Sin. 4: 95. 1929. Type. China: Hunan,Yin-schan Wukang, *Handel-Mazzetti 12203* (holotype H).

Pseudoleskeopsis japonica (Sull.) Iwats., J. Hattori Bot. Lab. 29: 58. 1966. *Hypnum japonicum* Sull. in Perry, Narr. Exp. China Japan 2: 330. 1857.

Pseudoleskeopsis orbiculata (Mitt.) Broth., Nat. Pflanzenfam. I(3): 1003. 1907. *Hypnum orbiculatum* Mitt., J. Proc. Linn. Soc., Bot., Suppl. 1: 84. 1859.

Pseudoleskeopsis orbiculata var. *laticuspis* (Card.) Thér., Ann. Crypt. Exot. 2: 15. 1929. *Pseudoleskeopsis laticuspis* (Card.) Broth., Nat. Pflanzenfam. I(3): 1003. 1907. *Pseudoleskea laticuspis* Card., Beih. Bot. Centralbl. 17(1): 30. 1904.

Pseudoleskeopsis serrulata Card. & Thér., Bull. Acad. Int. Géogr. Bot. 21: 271. 1911. Type. China: Guizhou, Pingba, Chuts-shan, 10. 1. 1903, *Aortunat s.n.* (isotype H).

Plants rigid, dark green, yellowish green when young, in mats. *Stems* creeping, up to 3–4 cm long, irregularly branched; branches usually fascicular, erect, 0.5–1.0 cm long, slightly declinate when dry. *Stem leaves* loosely appressed on stems when dry, spreading when moist, triangularly ovate, 1.2–1.5 mm × 0.5–0.6 mm; upper margins crenate; costa thick, strong, ending below apex; branch leaves broadly ovate, involute or slightly secund when dry, spreading when moist, 1.0–1.4 mm × 0.5–0.8 mm, obtuse at apex; margins plane, usually dentate; costa ending before apex; upper and median leaf cells small, round hexagonal or rhomboid, 7–12 μm × 5–6 μm; basal leaf cells near margins subquadrate, along costa shortly rectangular. *Inner perichaetial leaves* lanceolate, 2–3 mm × 0.8 mm; cells linear, costa percurrent, broader below. *Setae* long 1.5–2.0 cm, reddish brown, sometimes two mature sporophytes borne in one perichaetium; capsules pendent, obovoid to oblong-ovoid, 1.5–2.0 mm long, with short, distinct apophysis, mouth contracted when dry; exostome and endostome equal in length, 0.4–0.6 mm long; exostome teeth lanceolate, nearly smooth, with lamella; endostome segments folded; basal membrane high; cilia 2. *Spores* spherical, 20–25 μm in diameter, verrucose.

Illustrations: Plate 441, figs. 7–14.
Habitat: often on moist rocks near streams or on tree trunks.
Distribution: China, Japan, Korea, Thailand, and the Philippines.

Chinese specimens examined: ANHUI: Jin-zhai Co., *Z.-M. Zeng 10168* (MO). FUJIAN: Fu-zhou City, Nov. 8, 1925, *H.-H. Chang s.n.* (H), Gushan (Mt.), *M. Zang 785* (HKAS, IFSBH); Jing-long Co., *M. Zang 334, 338* (HKAS, IFSBH); Nan-ping Co., *C. Gao 25562* (IFSBH). GUANGDONG: Mt. Ding-hu, *C. Gao 8226* (IFSBH); Ren-hua Co., *W.-T. Tsang 26192, 26062, 12212, 12178* (FH, NY). GUANGXI: Xing-an Co., *C. Gao 1176* (IFSBH). GUIZHOU: Fan-jing Shan (Mt.), *P.-C. Wu 23577* (FH); Li Shan (Mt.), *B.-C. Tan 91788* (FH); Sui-yang Co., *C. Gao 33657* (IFSBH). HAINAN: Bawang-ling (Mt.), *C. Gao 3279, 3357* (IFSBH); Chang-jiang Co., *Redfearn 33715* (H); Le-dong Co., *B.-J. Lin 4024* (H); Lin-shui Co., *Redfearn 36225* (H). HONG KONG: Tai-mo Shan (Mt.), *C.-H. Gao 26116* (IFSBH). JIANGSU: Ming Ling, *M. Zang 440* (HKAS, IFSBH); Yi-xing Co., *P.-C. Wu & Y.-X. Jin 2200* (MO). LIAONING: Feng-cheng Co., *Noda 469* (IFSBH); Kuan-dian Co., *Buck 23628* (NY), *T. Cao 38712* (IFSBH); Zhuang-he Co., *X.-Y. Jia 880026, 880131, 880386* (IFSBH). SICHUAN: Wo-long Co., *Y. Xian 62108* (HKAS, IFSBH). YUNNAN: Chu-xiong Co., *Redfearn et al. 17366* (MO); Meng-la Co., *Redfearn 33978* (MO); Ping-bien Co., *H.-T. Tsai 55294* (H). ZHEJIANG: Hang-zhou City, *Buck 23907* (NY), *R.-L. Hu 202* (MO); Mt. Jiu-long, *Z.-L. Liu 834* (IFSBH, SHM); Yuan Si, *M. Zang 229* (HKAS, IFSBH).

The distinct features of this species are: 1) plants rigid, with abundant irregular and dense branches; 2) leaves acuminate from broadly ovate base, with obtuse and dentate apices; 3) costa broader at leaf base, becoming thin and twisting upward; 4) leaf cells short and small, nearly parenchymatous; and 5) capsules pendent, peristome well developed, and cilia present.

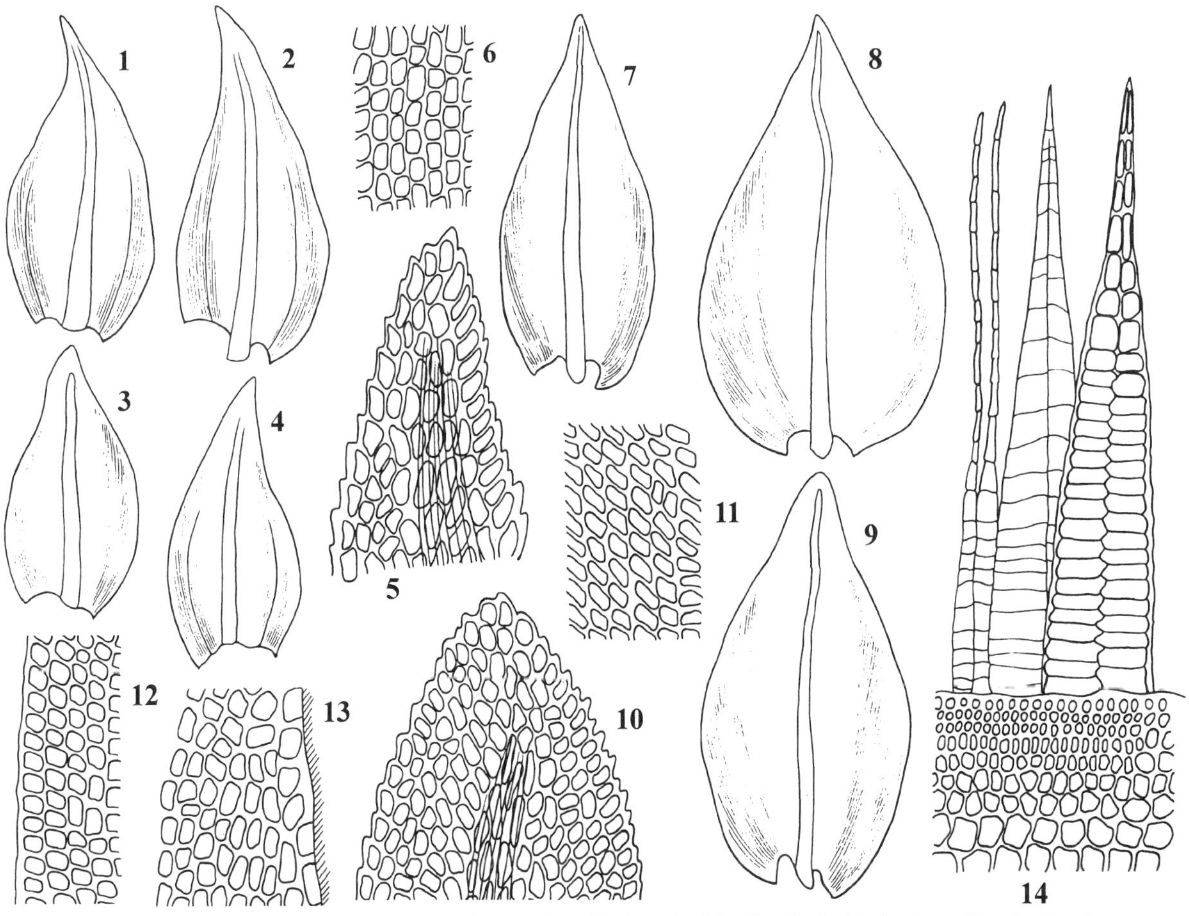

Plate 441. 1–6, *Pseudoleskeopsis tosana*. 1–4, leaves (× 52). 5, apical leaf cells (× 325). 6, median leaf cells (× 325). 7–14, *Pseudoleskeopsis zippelii*. 7–9, leaves (× 52). 10, apical leaf cells (× 325). 11, median leaf cells (× 325). 12, basal marginal leaf cells (× 325). 13, juxtacostal leaf cells (× 325). 14, peristome (× 325). (1–6, drawn from *C. Gao 34534*, IFSBH; 7–14, from *Z.-L. Liu 834*, SHM.)

9. ***Regmatodon*** Brid. 异齿藓属 yi-chi xian shu

Bryol. Univ. 2: 294. 1827.

Plants rigid when dry, dull or slightly shiny, green to yellowish green, in sparse to dense mats. *Stems* creeping, irregularly or pinnately branched; branches short, sometimes flagelliform, obtuse at tips; paraphyllia rare or lacking. *Stem leaves* small, mucronate; branch leaves imbricate when dry, erect-spreading when wet, oblong-ovate or ovate-lanceolate, slightly concave, recurved at apex; margins entire or sparsely dentate above; costa single, reaching middle part of leaf or slightly above; leaf cells thick-walled, rhomboidal; alar cells undifferentiated; several rows of basal cells near margins smaller, subquadrate. *Monoicous. Inner perichaetial leaves* erect, elongate, acuminate. *Setae* elongate, reddish brown, smooth or coarsely papillose; capsules erect to slightly pendent, ovoid to cylindrical; peristome double; exostome obviously shorter than endostome, the teeth shortly lanceolate, slightly blunt at tips, yellowish, smooth, transversely striate; endostome segments yellowish, smooth or papillose, 2–3 times longer than the teeth, acuminate, keeled, perforate at middle; basal membrane low; opercula conic, with obtuse tips. *Spores* spherical, coarsely papillose.

There are about 16 species of *Regmatodon* in the world. Most species are distributed around the tropical zones in a disconnected manner. Some bryologists placed the genus in a monotypic family, Regmatodontaceae, based on some unique features such as the peristome inserted below the capsule mouth and exostome teeth apparently shorter than endostome segments. Eight taxa of *Regmatodon* were recorded in the Chinese flora in the literature (Redfearn et al. 1996). Herzog (1925) first reported *R. orthostegius* Mont. from Yunnan province. Brotherus (1929) described two new taxa, *R. declinatus* var. *minor* Broth. from Guizhou and *R. handelii* Broth. (= *R. orthstegius* Mont.) from Yunnan as well as reporting *R. polycarpus* (Griff.) Mitt. (= *R. orthostegius*) from these two provinces. Bartram (1935) reported *R. declinatus* (Hook.) Brid. from Guangxi. P.-C. Chen et al. (1978) listed 4 species from China, with *R. schwabei* Herz. (= *R. declinatus*) reported new to the country. C. Gao (1981) published a new species, *R. longinervis*, from northeastern China and later C. Gao (1985) treated four species from Xizang with *R. serrulatus* (Dozy & Molk.) Bosch. & Sande Lac. newly reported to China. Four species of *Regmatodon* are recognized in this study.

1. Setae papillose 2
1. Setae smooth or nearly so 3
  2. Branch leaves oblong-ovate; margins entire; costa up to the middle of leaf or so .... 1. *R. declinatus*
  2. Branch leaves ovate-lanceolate; margins dentate; costa up to the upper leaf .... 4. *R. serrulatus*
3. Branch leaves ovate, acuminate above; leaf margin entire; costa up to the middle of leaf.......... 3. *R. orthostegius*
3. Branch leaves triangular; leaf margin with 1-celled sharp teeth; costa up to the apex of leaf.......... 2. *R. longinervis*

1. ***Regmatodon declinatus*** (Hook) Brid. 异齿藓 yi-chi xian

Bryol. Univ. 2: 294. 1827. *Pterogonium declinatum* Hook., Trans. Linn. Soc. London 9: 309. 1808. Type. Nepal: *Buchanan s.n.*

Regmatodon declinatus var. *minor* Broth., Symb. Sin. 4: 94. 1929. Type. China: Guizhou, Dschenning (Zhen-ning) and Hwangtsaoba (Xing-yi), *Handel-Mazzetti 10395* (holotype H).

Regmatodon schwabei Herz., J. Hattori Bot. Lab. 14: 66. 1955.

Plants small, brownish green, dull. *Primary stems* creeping, lateral stems erect or ascending, irregularly branched; branches unequal in length, curved, 1.0–1.5 cm long. *Leaves* oblong-ovate, 0.9–1.3 mm × 0.3–0.5 mm, acuminate, with short apices; leaf margins plane or slightly involute, entire; costa single, strong, reaching middle part of the leaf or slightly above; leaf cells thick-walled, irregularly rhomboidal or rectangular, 10–13 µm × 5–7 µm, with several rows of quadrate to subquadrate cells at basal margins. *Monoicous. Inner perichaetial leaves* sheathing at base, piliferous upwards, entire. *Setae* 5 mm long, erect, reddish brown, coarsely papillose; capsules erect, symmetric, cylindrical, ca. 2 mm long, mouth contracted; exostome teeth short, yellowish, ca. 0.2 mm long, blunt at apex, yellowish, with many lamellae; endostome segments yellowish, 0.40–0.45 mm long, papillose, perforate, with conduplicative processes; opercula with obtuse beaks. *Calyptrae* cucullate, ca. 2 mm long. *Spores* spherical, 22–25 µm in diameter, brownish, verrucose.

Illustrations: Plate 442, figs. 1–12.
Habitat: on trunks or rocks in forests.
Distribution: China, Nepal, and India.

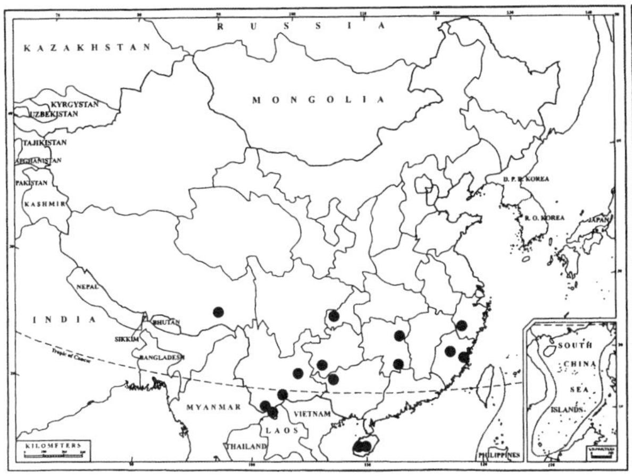

1. *Regmatodon declinatus*

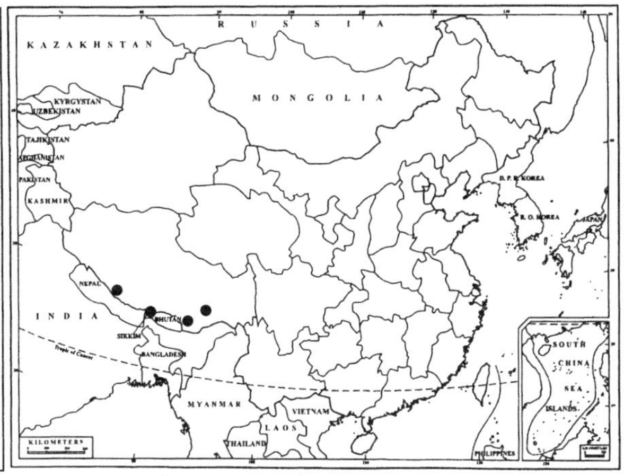

2. *Regmatodon longinervis*

Chinese specimens examined: CHONGQING: Chong-qing City, Ge-le Shan (Mt.), *B.-J. Chen 805* (IFSBH). FUJIAN: Fu-zhou City, *H.-H. Chang B173, B2445a* (FH); Nan-ping Co., *H.-H. Chung B173, B245a* (FH). GUANGXI: Ling-yun Co., Lao Shan, *S.-Y. Cheo 1963* (FH). GUIZHOU: *H. Chung B117* (FH). HAINAN: Bawangling (Mt.), *C. Gao 3208, 3248* (IFSBH); Li-mu Shan (Mt.), *Reese 17960* (MO, NY). HUNAN: Liu-yang Co., *C. Gao 336, 373* (IFSBH); Yi-zhang Co., Mang Shan (Mt.), *C. Gao 1320* (IFSBH). XIZANG: Mo-tuo Co., *Y.-G. Shu 3190-(1)* (HKAS, MO). YUNNAN: Gong Shan (Mt.), *Q.-W. Wang 7263* (PE); Lu-nan Co., *Redfearn et al. 2268* (MO); Meng-la Co., *Magill et al. 7689* (MO); Xishuangbanna, *Magill et al. 9789* (MO), *Redfearn et al. 2268* (MO); Yuan-yang Co., Feng-chun-ling (Mt.), *M. Zang 4839* (HKAS, IFSBH). ZHEJIANG: Mt. Jiulong, *Z.-L. Liu 885* (IFSBH, SHM).

This species is characterized by having oblong-ovate and acuminate leaves, smooth and entire leaf margins, and coarsely papillose setae.

2. ***Regmatodon longinervis*** C. Gao in C. Gao, G.-C. Zhang & T. Cao 长肋异齿藓 chang-lei yi-chi xian
Acta Bot. Yunnan. 3(4): 398. 1981. Type. China: Xizang, Yadong Co., *Y.-C. Yang 1b* (holotype ISFBH); Ji-long Co., *S.-K. Chen 283* (paratype IFSBH).

Plants slender, rigid, yellowish green above, greenish brown below, in dense tufts. *Primary stems* closely creeping on substratum, lateral stems erect or ascending, 1–2 times irregularly pinnately branched, flagelliform branches often present. *Stem leaves* lanceolate upward from broadly ovate base, 1.0–1.6 mm × 0.6–0.8 mm; branch leaves usually narrowly triangular, 1.0 mm × 0.5 mm; leaf margins plane or slightly involute, irregularly unicellarly serrulate above middle part; costa single, strong, ending below leaf apex; leaf cells thick-walled; upper cells irregularly rhomboidal, 7–12 μm × 3–4 μm, indistinctly projecting at the upper ends; several rows of cells near basal margins subquadrate; cells along costa somewhat elongated. *Monoicous. Perichaetial leaves* plane, smooth at apex. *Setae* smooth. *Mature sporophytes* not seen.

Illustrations: Plate 442, figs. 13–18.
Habitat: on rocks and bases of tree trunks in forests; alt. 2600–3100 m.
Distribution: endemic to China.

Chinese specimens examined: XIZANG:Ya-dong Co., *Y.-C. Yang 16* (IFSBH, PE); Mi-lin Co., *K.-Y. Lang 615* (IFSBH, PE); Cuo-na Co., *Y.-C. Yang 34* (IFSBH, PE).

The distinctive features of this species are: subpercurrent costa, irregularly and unicellularly dentate upper leaf margins, and smooth setae. We disagree with the treatment of this species by Eakin (1998) who synonymized it with *R. polycarpus* (Griff.) Mitt.

3. ***Regmatodon orthostegius*** Mont. 多蒴异齿藓 duo shuo yi chi xian
Ann. Sci. Nat. Bot., ser. 2, 17: 248. 1842. Type. India: Nilgiris, *Perrottet s.n.*

Regmatodon polycarpus (Griff.) Mitt., J. Proc. Linn. Soc., Bot., Suppl. 1: 127. 1859. *Anhyenium polycarpus* Griff., Calcutta J. Nat. Hist. 3: 275. Pl. 16. 1843 [1842].

Regmatodon handelii Broth., Akad. Wiss. Wien Sitzungsber., Math.-Naturwiss. Kl., Abt. 1, 133: 578. 1924. Type. China: Yunnan, Yunanfu (Kunming), *Handel-Mazzetti 260* (holotype H).

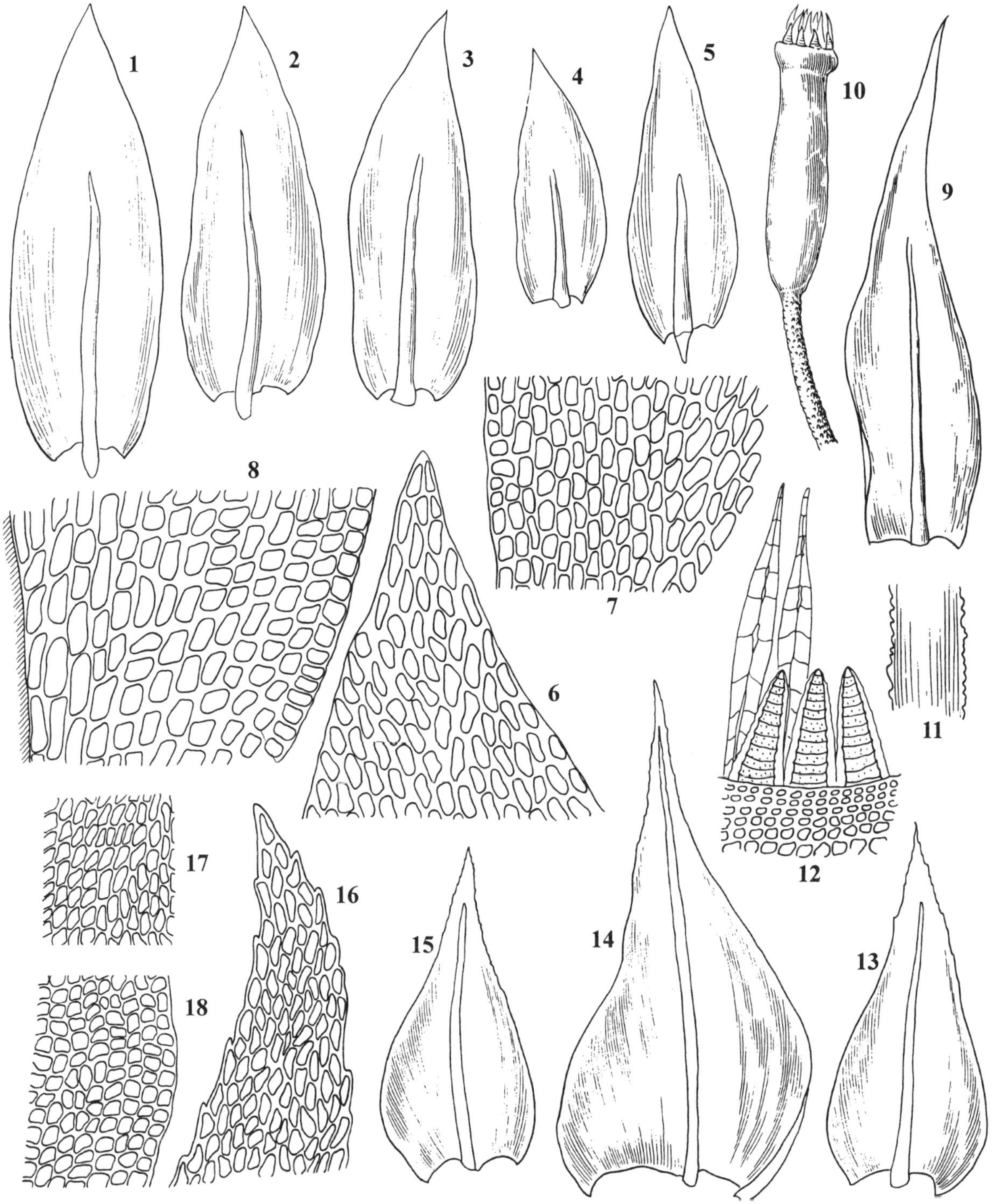

Plate 442. 1–12, *Regmatodon declinatus*. 1–3, stem leaves (× 52). 4–5, branch leaves (× 52). 6, apical leaf cells (× 325).7, median leaf cells (× 325). 8, basal leaf cells (× 325). 9, perichaetial leaf (× 52). 10, capsule (× 21). 11, portion of papillose seta (× 83). 12, peristome (× 100). 13–18, *Regmatodon longinervis*. 13–15, leaves (× 52). 16, apical leaf cells (× 325). 17, median leaf cells (× 325). 18, basal leaf cells (× 325). (1–12, drawn from *Handel-Mazzetti 10395*, H; 13–18, from *S.-K. Chen 283*, HKAS.)

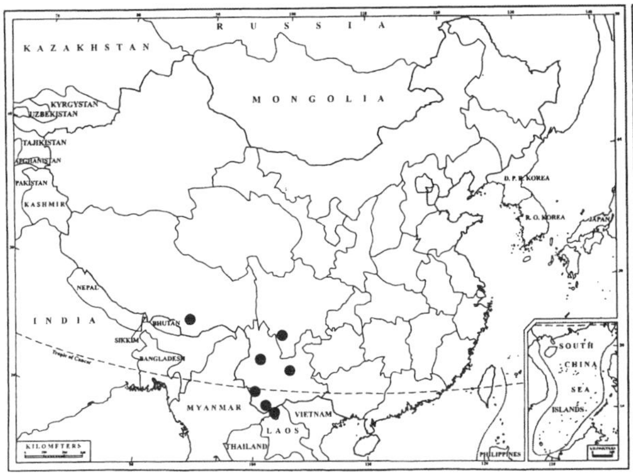

3. *Regmatodon orthostegius*

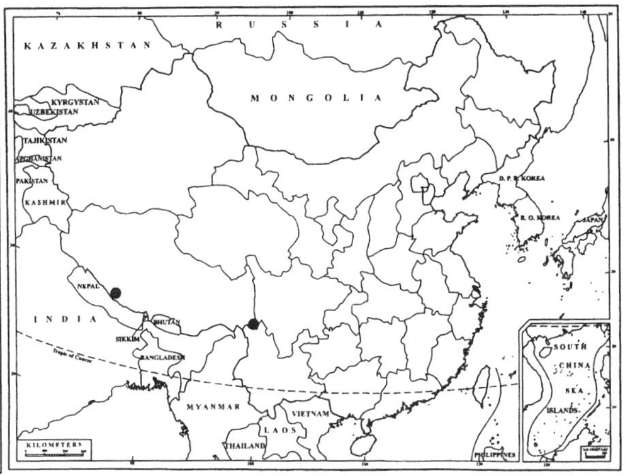

4. *Regmatodon serrulatus*

Plants rigid, about 2 cm long, yellowish green above, greenish brown below, in dense tufts. *Primary stems* creeping, irregularly branched; branches unequal in length, rounded, tips slightly curved. *Leaves* ovate, 0.8–1.1 mm × 0.3–0.4 mm, acuminate, with short apices; leaf margins slightly recurved, entire; costa single, strong, slightly short, ending at middle of leaf; leaf cells thick-walled; upper and median cells rhomboidal or round-quadrate, 13–15 µm × 5–7 µm; one to several rows of cells near basal margins quadrate; cells along costa slightly elongate. *Dioicous. Perichaetial leaves* sheathing at base, piliferous upward, ca. 2.0–2.5 mm long. *Setae* reddish brown, smooth, erect, ca. 1.0–1.2 cm long, sometimes 2–3 setae inserted in one perichaetium; capsules cylindrical, erect; exostome teeth shortly lanceolate, 0.15 mm long, shorter than endostome segments, transversely striate; endostome segments ca. 0.3 mm long; basal membrane absent; opercula conical, obtuse. *Spores* spherical, ca. 30 µm in diameter, coarsely papillose.

Illustrations: Plate 443, figs. 1–13.
Habitat: on bases of tree trunks.
Distribution: China, Sikkim, India, and Thailand.

Chinese specimens examined: SICHUAN: Mu-li Co., *C. Gao 19955, 20000, 920120, 20146* (IFSBH). XIZANG: Long-zi Co., *M. Zang 528* (HKAS, IFSBH). YUNNAN: Cang-yuan Co., *M. Zang 1436, 1423* (HKAS, IFSBH); Jing-dong Co., *J.-K. Zhang 1189* (HKAS, IFSBH); Kun-ming City, *Handel-Mazzetti 260* (H); Meng-la Co., *Redfearn 34216* (MO); Yang-bi Co., *Redfearn et al. 218, 340* (FH, MO, NY).

This species is distinguished from other species by having a short costa (only reaching the middle of leaves), smooth setae, and sometimes several capsules inserted in one perichaetium.

4. ***Regmatodon serrulatus*** (Dozy & Molk.) Bosch & Sande Lac. 齿边异齿藓 chi-bian yi-chi xian
Bryol. Jav. 2: 111. 1864. *Macrohymenium serrulatum* Dozy & Molk., Musci Frond. Ined. Archip. Ind. 6: 170. 56. 1848. Type. Indonesia: Java.

Plants rigid, yellowish green to greenish brown, slightly shiny, ca. 2–3 cm long. *Primary stems* creeping, lateral stems ascending, 1–2 times irregularly pinnately branched, branch apices taper-pointed. *Leaves* ovate to broadly ovate, acuminate or acute upward, 0.7–1.5 mm × 0.3–0.7 mm; leaf margins obviously unicellularly dentate above middle part, slightly involute on one side at base; costa single, thick, strong, ending at 1/2–2/3 the leaf length; leaf cells thick-walled; median and upper cells irregularly rhomboidal, 7–12 µm × 4–5 µm, several rows of cells at basal margins, subquadrate; basal cells near costa slightly elongate. *Monoicous. Setae* erect, coarsely papillose; capsules cylindrical; exostome teeth short; endostome segments well developed; opercula obtusely rostellate.

Illustrations: Plate 443, figs. 14–22.
Habitat: on rocks in forests.
Distribution: China and Indonesia (Java).

Chinese specimens examined: XIZANG: Ji-long Co., *S.-K. Chen 263* (HKAS, IFSBH). YUNNAN: De-qin Co., *J.-K. Yang 2279* (HKAS, IFSBH).

This species is easily distinguished by its ovate to broadly ovate leaves, obviously unicellularly dentate leaf margins, and coarsely papillose setae.

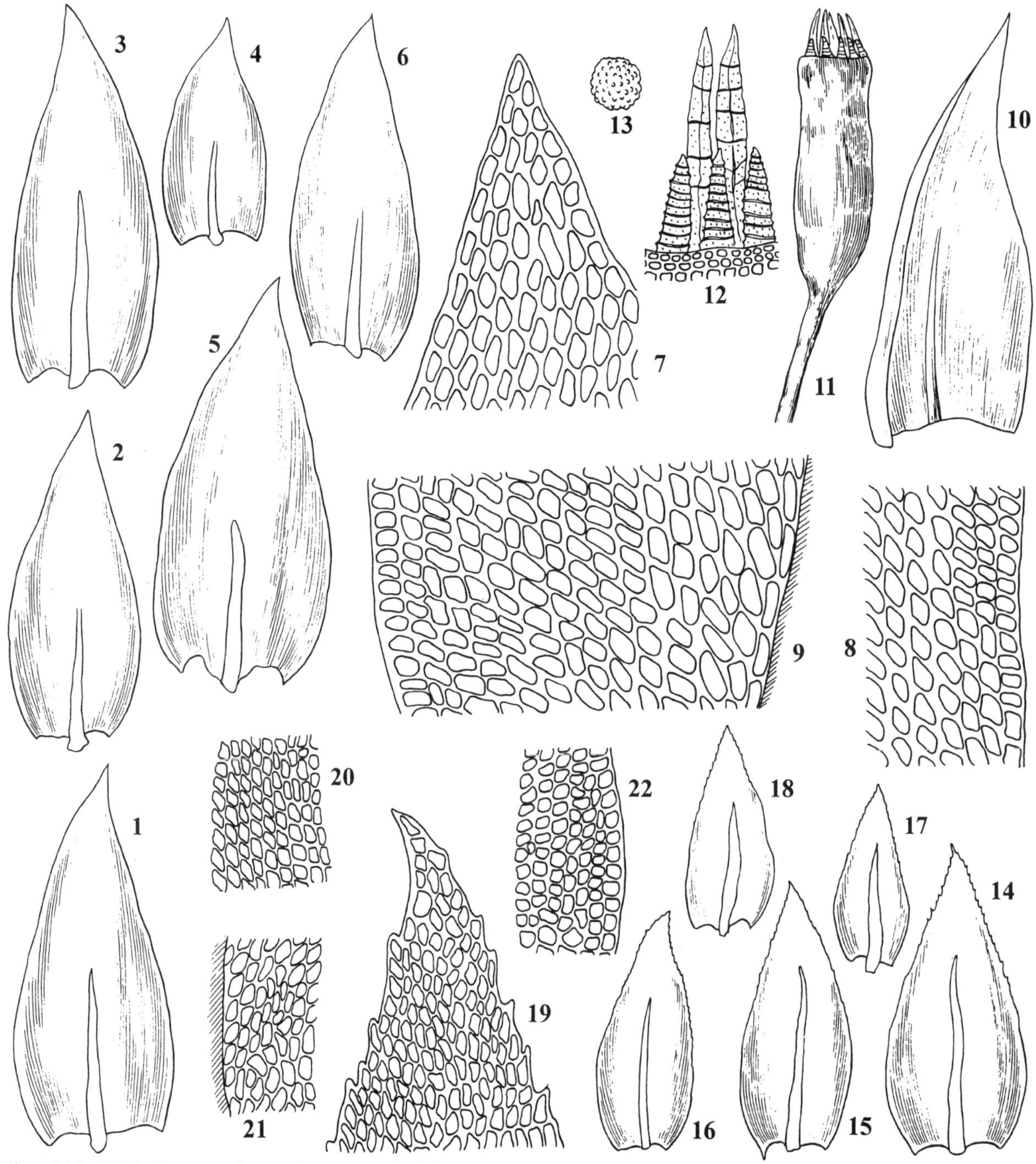

Plate 443. 1–13, *Regmatodon orthostegius*. 1–6, leaves (× 52). 7, apical leaf cells (× 325). 8, median marginal leaf cells (× 325). 9, basal leaf cells (× 325). 10, perichaetial leaf (× 52). 11, capsule (× 21). 12, peristome (× 83). 13, spore (× 325). 14–22, *Regmatodon serrulatus*. 14–18, leaves (× 52). 19, apical leaf cells (× 325). 20, median leaf cells (× 325). 21, juxtacostal leaf cells (× 325). 22, basal leaf cells (× 325). (1–13, drawn from *Handel-Mazzetti 260*, H; 14–22, from *S.-K. Chen 263*, HKAS.)

ANOMODONTACEAE

牛舌藓科 niu-she xian ke

by Wu Peng-cheng, Jia Yu, and Wang Mei-zhi

Plants small or robust, light green, yellowish green or brownish green, in loose or dense mats, rarely pendulous. *Stems* or primary stems creeping, secondary stems erect or ascending, irregularly pinnately or irregularly remotely branched, apices curved or caudate, often with asexual creeping branches; central strand present or absent; paraphyllia mostly lacking. *Stem leaves* and branch leaves similar or nearly identical, imbricate or erect-spreading when dry; stem leaf bases ovate or oblong ovate, acuminate or suddenly narrowed into a linear-lingulate or lanceolate apex, rarely transversally undulate or ovate to widely ovate, with acute apices; leaf margins plane or undulate, minutely serrate or minutely rounded-papillose, rarely remotely dentate or irregularly grossly serrate; costa single, up to the middle, rarely rigid, nearly reaching leaf apices; leaf cells rhomboidal, hexagonal or ovate-rhomboidal, thin-walled, multipapillose or thick-walled and smooth; basal median leaf cells ovate or oblong, hyaline at both sides of costa. *Branch leaves* smaller and narrower or slightly larger than stem leaves. *Monoicous* or dioicous. *Perichaetial leaves* usually large, sheathing at base, rarely small, gradually acuminate or lanceolate to a long subula. *Setae* slender, 5–20 mm long; capsules usually ovoid, rarely cylindrical or ovoid-cylindrical, smooth, stomata sparse or absent; annuli mostly differentiated; opercula conical, beaked; exostome teeth usually light yellow, rarely brownish yellow, papillose above, smooth or weakly striate at base; endostome segments usually short, rudimentary or absent or sometimes well developed, basal membrane low. *Calyptrae* cucullate, smooth, rarely papillose or ciliate. *Spores* nearly spherical, densely minutely papillose.

This family, elevated by Kindberg (1897) from subfamily Anomodontoideae of the Thuidiaceae, includes *Miyabea* Broth., *Haplohymenium* Dozy & Molk., *Anomodon* Hook. & Tayl., and *Herpetineuron* (C. Müll.) Card. However, some bryologists, e.g., Crum and Anderson (1981), placed the genera of the subfamily Anomodontoideae in the Leskeaceae. We consider the habitats and morphology of these genera to be different from those of the Leskeaceae; therefore, the establishment of Anomodontaceae as a separate family is warranted.

The distribution center of this family is mainly in East Asia and North America, with only *Herpetineuron* extending its range to the tropical and subtropical mountain regions of the world. *Miyabea* is endemic in East Asia, and it is mainly restricted in east coast mountain regions in China.

1. The branch apices of plants circinnate when dry; leaves triangular-lanceolate or narrowly ovate-lanceolate; costa strongly flexuous-sinuose above or nearly straight .... 3. *Herpetineuron*
1. The branch apices of plants not circinnate when dry; leaves mostly lingulate or widely ovate; costa straight throughout or only slightly curved 2
   2. Stem leaves and branch leaves widely ovate; leaf cells smooth .... 4. *Miyabea*
   2. Stem leaves and branch leaves lingulate; leaf cells densely papillose 3
3. Plants usually robust; leaf apices rounded .... 1. *Anomodon*
3. Plants slender; leaf apices rounded or obtusely acute .... 2. *Haplohymenium*

1. ***Anomodon*** Hook. & Tayl. 牛舌藓属 niu-she xian shu

Muscol. Brit. 79. 3. 1818.

Plants robust, rather rigid, dark green, yellowish green or brownish green, in lax or dense tufts or cushions. *Primary stems* creeping, secondary stems erect or ascending, remotely and irregularly pinnately branched; branch apices often curved, with leaves remotely arranged and revolute at apex when dry; commonly with stoloniferous branches; paraphyllia absent. *Stem leaves* similar to branch leaves, ovate or oblong-ovate at the base, rarely with small auricles, suddenly narrowed to a widely lingulate or oblong-lingulate acumen, rarely lanceolate, with a rounded apex, rarely serrate at the apex; leaf margins often undulate or rugose; costa single, mostly ending only a few cells below the apex; leaf cells isomorphic, hexagonal or rounded hexagonal, equally thick-walled, grossly multipapillose or sharply unipapillose; basal median cells usually hyaline along the costa. *Dioicous*. *Perichaetial leaves* lanceolate from a sheathing base. *Setae* slender, up to several centimeters long; capsules cylindrical, shortly cylindrical or ovoid, smooth, or strict when dry; stomata absent or rarely present; annuli mostly differentiated; opercula conical, sometimes beaked; exostome teeth light yellow or brownish yellow, narrowly lanceolate, densely papillose, rarely weakly striate at the base, nodulose or nodules absent; endostome segments pale, finely papillose, well developed or rudimentary, sometimes keeled; basal membrane low; cilia short and weak, rudimentary or lacking. *Calyptrae* smooth, rarely papillose above.

Species of *Anomodon* usually occur in subtropical and temperate regions. They often grow on tree trunks, rotten wood, or rock surfaces. Their morphological characters are quite variable due to the different habitats. The genus consists of more than 20 species in the world. Nine species and one variety were reported from China (Redfearn et al. 1996). The first record of *Anomodon* in China might be *A. sinensis* C. Müll. (C. Müller 1896) from Shaanxi. Later, Brotherus (1907, 1925, 1929), Reimers (1931), C.-Y. Yang (1936), and Potier de la Varde (1937) reported most of the species of the genus from various parts of China. C. Gao (1977) published a new species, *A. dentatus*, from northeastern China. P.-C. Chen et al. (1978) provided a detailed account of Chinese *Anomodon* with 2 subgenera and 9 species. Tan et al. (1996) added a new variety, *Anomodon solovjowii* Laz. var. *henanensis* B. C. Tan, Boufford & T.-S. Ying. In this study, 9 species, one subspecies, and one variety of *Anomodon* are treated. We cannot confirm the specimens of *Anomodon longifolius* (Brid.) Hartm. from Heilongjiang and Jilin reported by C. Gao et al. (1977). Its occurrence in China is questionable.

1. Plants large, up to 10 cm long; endostome segments ca. 2/3 the length of exostome teeth .... 9. *A. viticulosus*
1. Plants small to medium-sized, rarely larger; endostome segments mostly rudimentary, rarely ca. 1/2 the length of exostome teeth 2
   2. Each leaf cell with a single, setulose papilla...3
   2. Each leaf cell with multipapillae 4
3. Plants large; endostome segments short or rudimentary.. 1. *A. abbreviatus*
3. Plants medium-sized; endostome segments ca. 1/4 the length of exostome teeth ....7. *A. solovjovii*
   4. Leaf apices grossly dentate...2. *A. dentatus*
   4. Leaf apices entire to subentire...5
5. Stem leaves oblong-lingulate above; leaf cells minutely papillose.. 5. *A. perlinguatus*
5. Stem leaves lingulate above; leaf cells grossly papillose..6
   6. Leaf bases auriculate ....6. *A. rugelii*
   6. Leaf bases not auriculate 7
7. Branches attenuate at the apex; leaves widely ovate, shortly acute at the apex .... 3. *A. giraldii*
7. Branches not attenuate at the apex; leaves ovate-lingulate, rounded at the apex...8
   8. Stems without central strand; leaves not fragile, ligulate, not sheathing at base, erect or imbricate, hardly contorted when dry ....4. *A. minor*
   8. Stems with central strand; leaves fragile, lance-ligulate from an ovate, somewhat sheathing base, more or less incurved-contorted when dry .... 8. *A. thraustus*

1. ***Anomodon abbreviatus*** Mitt. 单疣牛舌藓 dan-you niu-she xian

Trans. Linn. Soc. London, Bot. 3: 187. 1891. Type. Japan: Joshim, *Bisset s.n.*

Anomodon asperifolius C. Müll., Nuovo Giorn. Bot. Ital., n. ser. 3: 117. 1896. Type. China: Shaanxi (Schen-si), *Giraldi s.n.*

Plants usually large, densely interwoven in lax cushions, erect spreading above, yellowish green to green, yellowish brown when old. *Primary stems* creeping, flagelliform; secondary stems erect or erect-spreading, up to 10 cm long, simple or sparsely branched; branches about 2–3 cm long; central strand undifferentiated. *Stem leaves* crisped when dry, erect-spreading when moist, gradually forming an oblong-lingulate lamina from an ovate base, about 2.5–5.0 mm long, rounded-obtuse or acute at the apex; margins papillose; costa vanishing near leaf apex; median leaf cells rounded quadrate to rounded hexagonal, ca. 10–20 μm wide, setulosely unipapillose, thick-walled, basal cells near costa oblong, smooth; branch leaves similar to stem leaves. *Dioicous*. *Perichaetial leaves* narrowly ovate-lanceolate. *Setae* yellowish brown to reddish brown, about 2–3 mm long; capsules ovoid, erect, reddish brown; exostome teeth yellowish brown, hyaline, papillose above, smooth at lower portion; endostome segments rudimentary; opercula cylindrical, shortly beaked. *Calyptrae* about 4 mm long, smooth. *Spores* up to 30 μm in diameter, papillose.

Illustrations: Granzow de la Cerda 1997 (Fig. 21, a–i).
Habitat: on tree trunks, rock surfaces or rotten logs; alt. 2780 m.
Distribution: China, Korea, and Japan.
Chinese specimens examined: XIZANG: Nei-dong Co., *S.-K. Chen 5139* (HKAS, PE).

2. ***Anomodon dentatus*** C. Gao 齿缘牛舌藓 chi-yuan niu-she xian
Fl. Musc. Chin. Boreali-Orient. 380. f. 165. 2, 170. 1977. Type. China: Jilin Prov., Mt. Changbai, *C. Gao 1231* (holotype IFSBH).

Plants large, rigid, dark green. *Primary stems* creeping, secondary stems erect or ascending, 4–8 cm long; central strand absent. *Leaves* crowded, crisped when dry, erect-spreading when moist, 1.5–3.0 mm long, gradually narrowed from a widely ovate base to an oblong-lingulate acumen, obtuse at the apex, grossly dentate; costa thick, up to leaf apex; leaf cells small, nearly quadrate, multipapillose, upper cells 5.0–7.5 µm, basal median cells narrowly oblong-oval or oval, hyaline, rarely remotely papillose, 7–10 × 30–50 µm. *Sporophytes* unknown.

Illustrations: Pl. 194, figs. 1–10; Granzow de la Cerda 1997 (Fig. 23, a–g).
Habitat: on tree trunks; alt. 1500–1600 m.
Distribution: endemic to China.

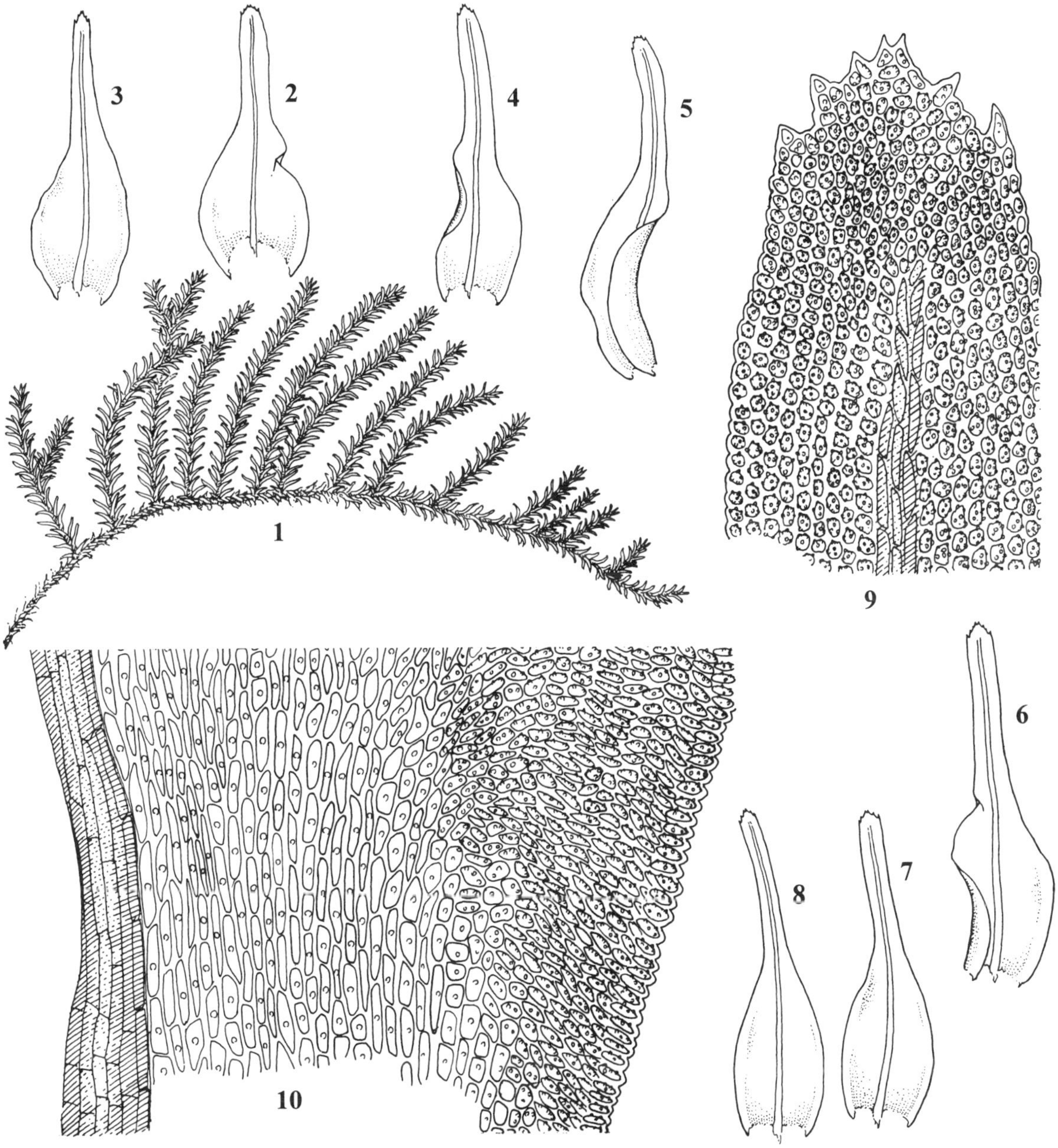

Plate 444. *Anomodon dentatus*. 1, habit (× 1). 2, stem leaf (× 28). 3–8, branch leaves (× 28). 9, apical leaf cells (× 330). 10, basal leaf cells (× 330). (1–10, drawn from *C. Gao 1231*, IFSBH.)

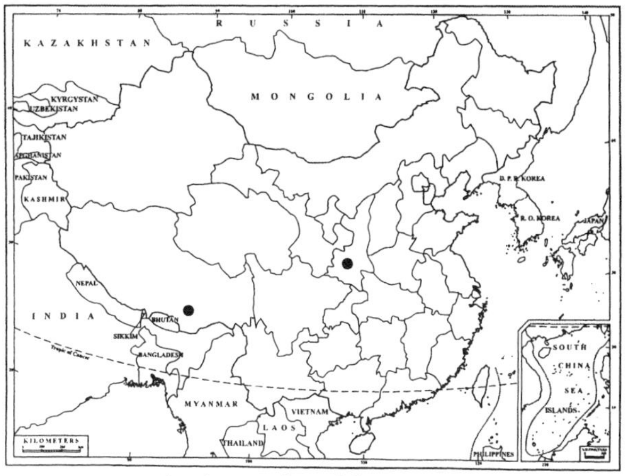

1. *Anomodon abbreviatus*

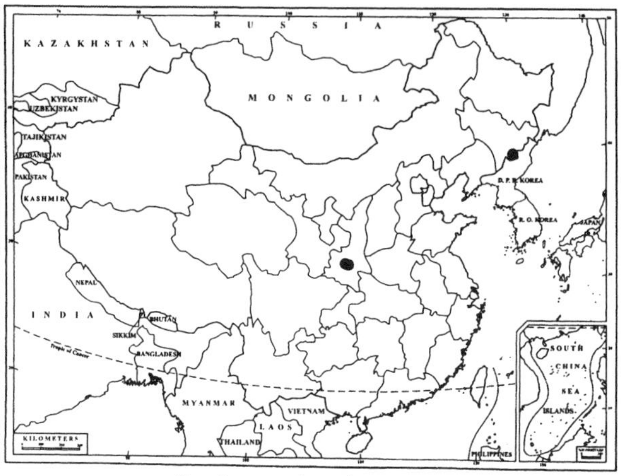

2. *Anomodon dentatus*

Chinese specimens examined: JILIN: Mt. Changbai, *C. Gao 22108* (IFSBH, MO). SHAANXI: Hu Co., *Z.-P. Wei 4340, 4343* (MO, WUK); Tai-bei Co., *Z.-P. Wei 5903* (MO, PE, WUK).

This species is characterized by the narrowly lingulate lamina with an obtuse and grossly dentate apex. It is only distributed in Northern China.

3. ***Anomodon giraldii*** C. Müll. 尖叶牛舌藓 jian-ye niu-she xian
Nuovo Giorn. Bot. Ital., n. ser. 3: 117. 1896. Type. China: Shaanxi (Schen-si), *Giraldi s.n.*

Plants usually tufted in large lax cushions, dark yellowish green to green, yellowish brown when old. *Primary stems* creeping, flagelliform, up to 7 cm long; secondary stems erect or ascending, ca. 3 cm long, irregularly pinnately branched, caudate; central strand absent. *Stem leaves* concave, imbricate when dry, widely ovate, shortly acuminate, ca. 1.8–2.5 mm long, opaque; leaf margins entire or with a few teeth near apex; costa nearly excurrent, smooth or papillose at the back; median leaf cells rhomboidal to oval, ca. 5–15 µm, thick-walled, with several gross papillae; basal leaf cells rather long and smooth near costa; branch leaves ca. 1/2 the stem leaf length. *Dioicous. Perichaetial leaves* ovate at base, gradually narrow to a lanceolate acumen; costa reaching 2/3 the leaf length. *Setae* yellowish brown or reddish brown, ca. 2 mm long; capsule cylindrical; opercula conical, obliquely beaked; peristome double; exostome teeth yellowish brown, hyaline above, finely papillose; endostome segments rudimentary; basal membrane ca. 1/2 the length of exostome teeth; cilia absent. *Calyptrae* smooth.

Illustrations: Plate 445, figs. 1–13; Granzow de la Cerda 1997 (Fig. 29, a–i).
Habitat: on rock surfaces or tree trunks at lower elevations, rarely on soil; alt. 800–1080 m.
Distribution: China, Korea, Japan, and Russia (Siberia).

Chinese specimens examined: ANHUI: Huang Shan (Mt.), *P.-C. Chen 6629* (PE), Qingliangta to Erdaoting, *P.-C. Chen 7257, 7363* (PE), Yungusi to Meimaofong, *P.-C. Chen 7549* (MO, PE). CHONGQING: Nan-chuan Co., Mt. Jinfu, *M.-Z. Wang 86-800* (PE). HUBEI: Mt. Shennongjia, *P.-C. Wu 346a* (MO, PE); Xing-shan Co., *J.-X. Luo 961213* (MO, PE). HUNAN: Xin-hua (Hsinhwa) Co., *Handel-Mazzetti 2793* (H, MO). SICHUAN: Mt. Omei, *Redfearn 34764b* (MO). ZHEJIANG: De-qing Co., *P.-C. Wu 1241* (MO, PE).

This species differs from other species of *Anomodon* by having irregularly pinnately branched plants, oblong-ovate leaf bases, and lanceolate leaf apices.

4. ***Anomodon minor*** (Hedw.) Lindb. 小牛舌藓 xiao niu-she xian
Bot. Not. 1865: 196. 1865. *Neckera viticulosa* Hedw. var. *minor* Hedw., Sp. Musc. Frond. 210. 48. f. 6–8. 1801. Type. U.S.A.: Pennsylvania, Lancaster, *Muhlenberg s.n.*
Anomodon grandiretis Broth., Akad. Wiss. Wien Sitzungsber., Math.-Naturwiss. Kl., Abt. 1, 133: 578. 1924, *syn. nov*. Type. China: Sichuan (Setschwan), Yuan-yan, 27°37′N, *Handel-Mazzetti 2079* (holotype H).

Anomodon leptodontoides C. Müll., Nuovo Giorn. Bot. Ital., n. ser. 4: 274. 1897. Type. China: Shaanxi (Schen-si), Schan-kio, Aug. 1895, *Giraldi s.n.*

Anomodon minor (Hedw.) Lindb. subsp. *integerrimus* (Mitt.) Iwats., J. Hattori Bot. Lab. 26: 41. 1963. *Anomodon integerrimus* Mitt., J. Linn. Soc. Bot. Suppl. 1: 126. 1859.

Anomodon planatus Mitt., J. Linn. Soc. Bot. Suppl. 1: 126. 1859.

Anomodon ramulosus Mitt., Trans. Linn. Soc. London, Bot. 3: 187. 1891. Type. China: Zhejiang (Chekiang), Ning-po City, *Oldham s.n.*

Anomodon sinensis C. Müll., Nuovo Giorn. Bot. Ital., n. ser. 3: 118. 1896. Type. China: Shaanxi (Schen-si), in monte Si-ku-tziu-san, July 1894, *Giraldi s.n.*

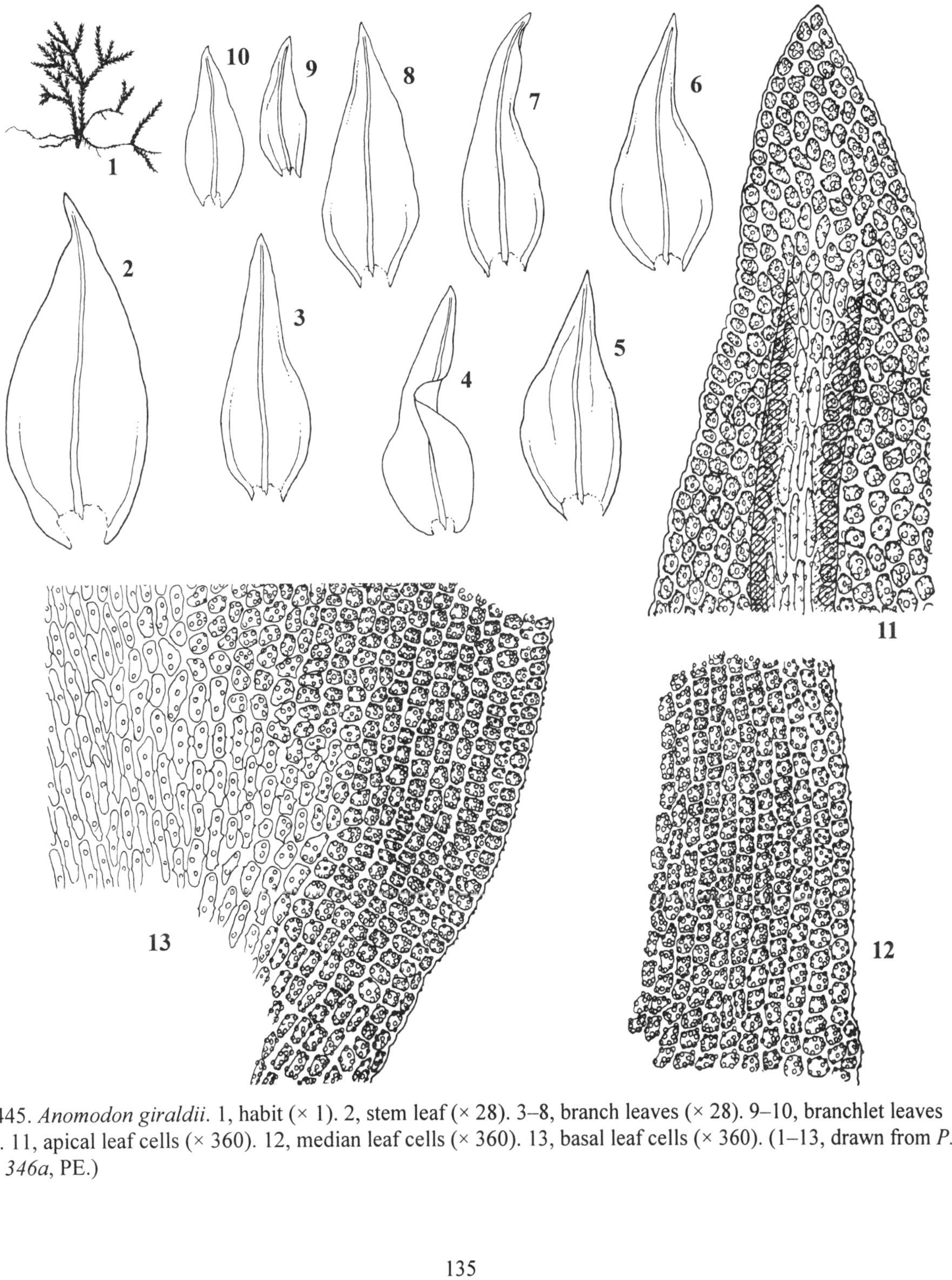

Plate 445. *Anomodon giraldii*. 1, habit (× 1). 2, stem leaf (× 28). 3–8, branch leaves (× 28). 9–10, branchlet leaves (× 28). 11, apical leaf cells (× 360). 12, median leaf cells (× 360). 13, basal leaf cells (× 360). (1–13, drawn from *P.-C. Wu 346a*, PE.)

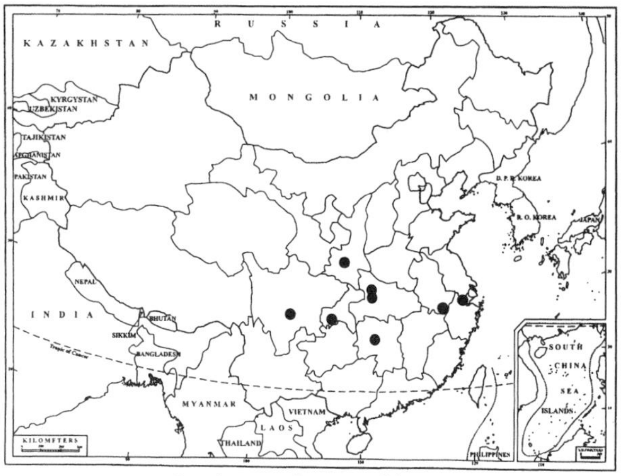

3. *Anomodon giraldii*

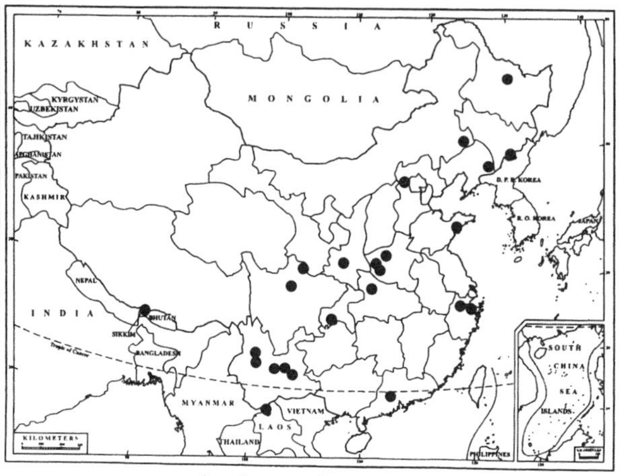

4. *Anomodon minor*

Plants usually slender, light green, brown when old, in loose tufted mats. *Primary stems* creeping, leaves mostly deciduous, secondary stems erect or ascending, 2–7 cm long, regularly or irregularly pinnately branched; branches ca. 5–7 mm long; central strand absent. *Stem leaves* crowded, appressed when dry, erect-spreading when moist, usually in 2-ranks, widely lingulate above, ca. 2 mm long, broadly rounded at apex; leaf margins slightly serrulate; costa ending below leaf apex, often forked above, hyaline; median leaf cells rounded quadrate or hexagonal, 5–15 µm wide, thick-walled, opaque, 3–8 papillae per cell; basal median cells oblong or rhomboidal. *Dioicous. Perichaetial leaves* ovate at base, gradually narrowed to a lanceolate acumen. *Setae* ca. 10 mm long; capsules oblong ovoid, ca. 2 mm long; exostome teeth yellowish brown, grossly papillose, hyaline above; endostome segments rudimentary; basal membrane rather low. *Calyptrae* ca. 3 mm long, smooth. *Spores* 20–30 µm in diameter, finely papillose.

Illustrations: P.-C. Wu (ed.) 2001 (Pl. 72, figs. 1–10).
Habitat: often growing on shady calcareous rock cliffs, rarely on trees; alt. 750–2700 m.
Distribution: China, Korea, Japan, India, Nepal, Bhutan, Russia, and North America.

Chinese examined specimens: CHONGQING: Nan-chuan Co., *X.-Y. Hu 0294* (MO, PE); Mt. Jingfu, *M.-Z. Wang 860517* (PE). GUANGDONG: Mt. Dinghu, *P.-H. Liang 46* (IBSC, MO). HEBEI: Mt. Xiao-wu-tai, *X.-W. Kong 1205* (MO, PE). HENAN: Nie-xiang Co., *J.-X. Luo 126* (MO, PE); Song Co., *J.-X. Luo 459* (MO, PE); Xi-xia Co., *J.-X. Luo 342* (MO, PE). HUBEI: Mt. Shennongjia, Sonbei, *P.-C. Wu 133* (PE), Yanrewan, *P.-C. Wu 188, 190* (MO, PE). HEILONGJIANG: Yi-chun City, Wu-ying, *Vitt 34907* (ALTA, MO). JILIN: Mt. Changbai, *Vitt 34748* (ALTA, MO). LIAONING: Heng-ren Co., *T. Cao & J.-Y. Feng 460* (IFSBH, MO). NEI MONGOL: Keerxinzuoyehouqi, Daqinggou, *X.-L. Bai 1315* (HIMC). SHAANXI: Hu Co., *Z.-P. Wei 4729* (MO); Mt. Tai-bai, *Z.-P. Wei 5886* (MO, WUK). SHANDONG: Qing-dao City, Laoshan (Mt.), *C. Gao & E.-Z. Bai 34523* (MO, PE). SICHUAN: Nan-ping Co., *S. He 30186* (MO); Wen-chuan Co., *Redfearn 35578* (MO). XIZANG: Ya-dong Co., *S. K. Chen 5019* (HKAS). YUNNAN: Chu-xiong Co., *Redfearn et al. 1725* (MO); Kun-ming City, *Handel-Mazzetti 689* (H); Lu-nan Co., *Redfearn & Y.-G. Su 2243* (MO); Meng-la Co., *Magill et al. 7638* (MO); Wei-shan Co., *Redfearn et al. 1204* (MO); Yang-bi Co., *Redfearn & Y.-G. Su 2055* (MO). ZHEJIANG: Mt. West Tian-mu, *R.-L. Hu 81-b* (HSNU, MO), *B.-C. Tan 93-555* (FH, MO).

5. ***Anomodon perlinguatus*** Broth. ex P.-C. Wu & Y. Jia 带叶牛舌藓 dai-ye niu-she xian
Acta Phytotax. Sin. 38: 260. f. 2. 2000. Type. China: Shaanxi, *P. Giraldi 1272* (holotype H).

Plants robust, in loose mats. *Stems* up to 3 cm long, irregularly pinnately branched; branches 1–2 cm long. *Leaves* appressed when dry, erect-spreading when moist, rather abruptly narrowed from a widely ovate base toward a lanceolate to lingulate lamina, up to 3 mm long, obtuse at apex; costa vanishing below leaf apex; leaf cells rounded hexagonal, median leaf cells 60–70 µm × 75–90 µm, finely multipapillose. *Sporophytes* unknown.

Illustrations: Plate 446, figs. 1–10.
Habitat: Unknown.

Distribution: endemic to China.
Specimen examined: see the type citation above.

Brotherus (1907, 1925), supplied this species name for the above-mentioned Chinese collection without Latin description. Based on the examination of the specimen, we described this new species (P.-C. Wu & Y. Jia 2000). Comparing it with other species of *Anomodon,* we considered the narrow lingulate stem leaves and multipapillose leaf cells to be distinguishable features in *Anomodon.*

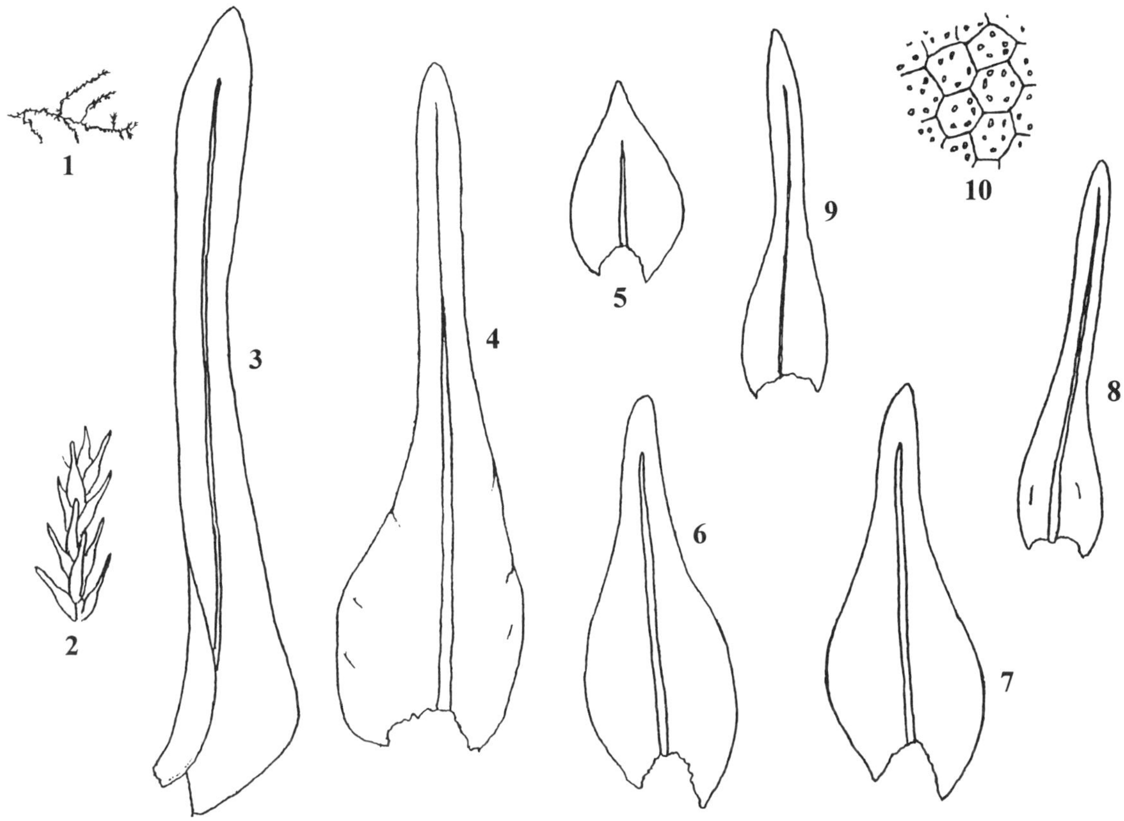

Plate 446. *Anomodon perlinguatus*. 1, habit (× 1). 2, portion of branch (× 7). 3–4, stem leaves (× 28). 5–7, basal stem leaves (× 28). 8–9, branch leaves (× 28). 10, median leaf cells (× 420). (1–10, drawn from *Giraldi 1272*, H.)

6. ***Anomodon rugelii*** (C. Müll.) Keissl. 皱叶牛舌藓 zhou-ye niu-she xian
Ann. Naturhist. Hofmns. 15: 214. 1900. *Hypnum rugelii* C. Müll., Syn. Musci. Frond. 2: 473. 1851. Type. U.S.A.: Tennessee, Smoky Mountains, *Rugel s.n.*

Plants small to medium-sized, yellowish green or yellowish brown, in densely tufted cushions. *Primary stems* creeping; secondary stems ascending, up to 4 cm long, irregularly pinnately branched; central strand absent. *Stem leaves* ca. 1.0–2.5 mm long, ovate or oblong ovate at base, usually auriculate at both angles; leaf margins entire or finely serrate above; costa slender, up to leaf apex, smooth or papillose on back surfaces; median leaf cells rounded quadrate or hexagonal, ca. 5–15 µm wide, 5–10 papillae per cell, thick-walled; basal cells near costa mostly oblong, hyaline; branch leaves narrowly ovate-lanceolate. *Dioicous*. *Perichaetial leaves* ovate-lanceolate, 2 mm long. *Setae* yellowish brown to reddish brown, 7–15 mm long; capsules erect; exostome teeth yellowish brown, papillose above, hyaline; endostome often short or absent. *Calyptrae* ca. 4 mm long, smooth. *Spores* 10–25 µm in diameter, papillose.

Illustrations: P.-C. Wu (ed.) 2001 (Pl. 73, figs. 1–10); Granzow de la Cerda 1997 (Fig. 17, a–b).
Habitat: alt. 260–1300 m.
Distribution: China, Korea, Japan, Nepal, India, Vietnam, Russia, Europe, and North America.

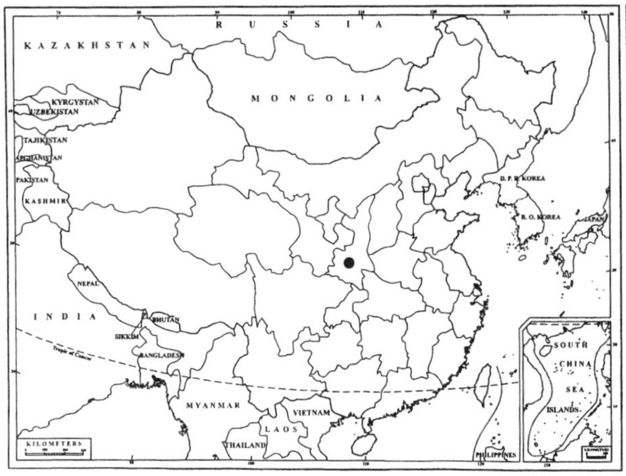

5. *Anomodon perlinguatus*

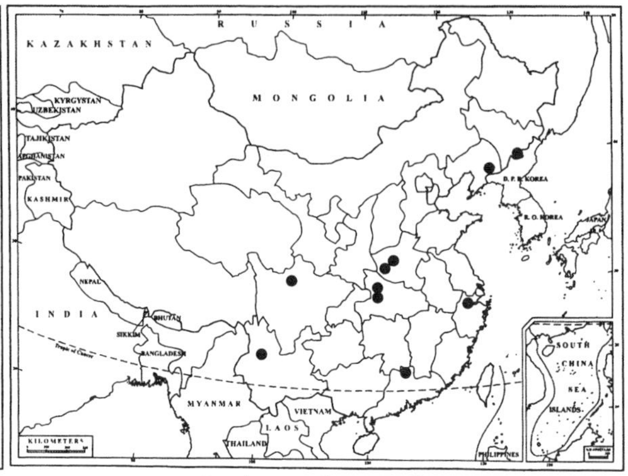

6. *Anomodon rugelii*

Chinese examined specimens: GUANGDONG: Le-chang Co., *Magill et al. 8087* (MO), *G.-Q. Zeng 544* (IBSC, PE). HENAN: Song Co., *J.-X. Luo 417* (MO, PE); Xi-xia Co., *J.-X. Luo 364* (MO, PE). HUBEI: Mt. Shennongjia, *1980 Sino-Amer. Exped. 295a* (MO, NY, PE); Xing-shan Co., *J.-X. Luo 961108* (MO, PE). JILIN: Mt. Changbai, *B.-D. Xin 0243* (MO, PE). LIAONING: Kuan-dian Co., *Buck 23776* (MO, NY). SICHUAN: Ma-er-kang Co., *Allen 6894* (MO). YUNNAN: Da-li Co., *Redfearn et al. 1608* (MO). ZHEJIANG: Mt. West Tian-mu, *R.-L. Hu 4* (HSNU, MO).

This species is characterized by the auriculate leaf bases and rather long leaf lamina.

7. ***Anomodon solovjovii*** Laz. 东亚牛舌藓 dong-ya niu-she xian
Rev. Bryol. Lichénol. 5: 45. 1933. Type. Russia: *Solovjov s.n.*
Anomodon solovjovii Laz. var. *henanensis* B. C. Tan, D. Boufford & T.-S. Ying, Acta Bot. Yunnan. 18(1): 69. 1996. Type. China: Henan Prov., Nei-xiang Co., Bao-tian Man Natural Reserve, *D. Boufford et al. 26458b* (holotype FH, isotypes H, PE).

Plants medium-sized, yellowish green, brown when old, in dense mats. *Secondary stems* 3–6 cm long, sparsely branched, branches ca. 2 cm long; central strand present. *Stem leaves* crowded, appressed when dry, erect-spreading when moist, abruptly narrowed from an ovate base toward a lingulate acumen, ca. 1.2–2.5 mm long, rounded obtuse at apex, rarely acute; leaf margins papillosely serrate; costa stout, vanishing below the apex, papillose on back surfaces; leaf cells rounded quadrate to hexagonal, 7–15 µm wide, with a single setulose papilla over each cell lumen, thick-walled; basal leaf cells oblong along costa, smooth; branch leaves similar to the stem leaves, but smaller. *Dioicous*. *Perichaetial leaves* narrowly lanceolate. *Setae* rather short, less than 1 mm long, yellowish brown; capsules erect, ca. 2 mm long, ovoid, reddish brown; annuli persistent; exostome teeth yellowish brown, papillose above, smooth at lower portion; endostome segments only 1/4 the length of exostome, finely papillose; cilia lacking. *Calyptrae* papillose above. *Spores* 16–30 µm in diameter, finally papillose.

Illustrations: B. C. Tan et al. 1996 (Figs. 1–4, as *Anomodon solovjovii* var. *henanensis*).
Habitat: usually growing on trees.
Distribution: China, Korea, and Russia (Siberia).

Chinese specimens examined: SHAANXI: Qinling Mts., *Z.-P. Wei 4729*, *6503*, *6574* (WUK, PE).

This species is very close to *Anomodon abbreviatus*, but differs by its smaller plants, shorter leaves with lingulate acumens, and well developed endostome segments. B. C. Tan, D. Boufford, and T.-S. Ying (1996) described *Anomodon solovjovii* var. *henanensis* from Henan province based on a very weak distinction from *Anomodon solovjovii* var. *solovjovii*. Granzow de la Cerda (1997) synonymized *A. solovjovii* var. *henanensis* under this species. Therefore, this species is newly reported to China.

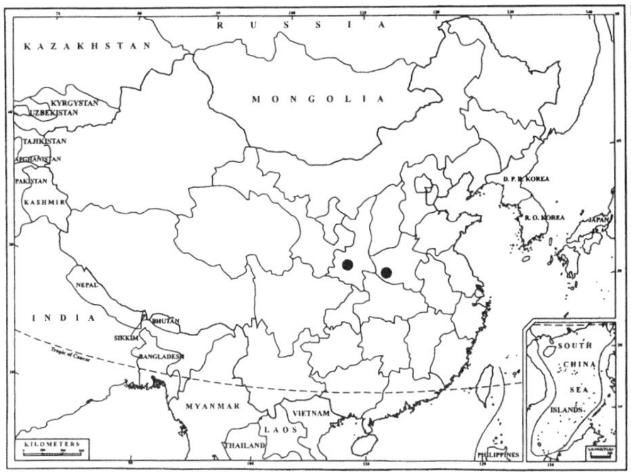

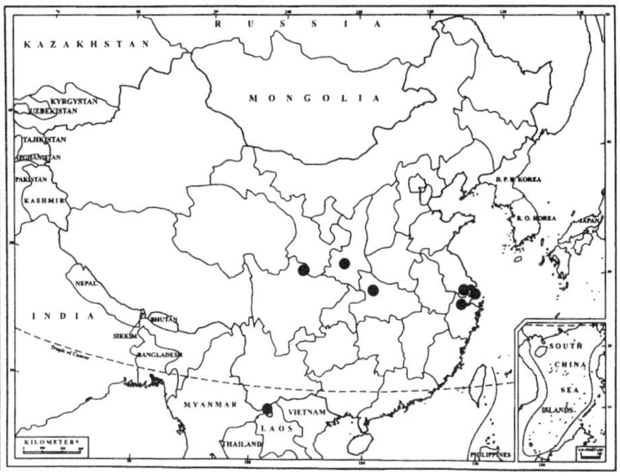

7. *Anomodon solovjovii* 8. *Anomodon thraustus*

8. ***Anomodon thraustus*** C. Müll. 碎叶牛叶藓 sui-ye niu-ye xian

Nuovo Giorn. Bot. Ital., n. ser. 5: 207. 1898. Type. China: Shaanxi (Schen-si), in monte Tui-kio-san, Sept. 1896, *Giraldi s.n.*

?*Anomodon rotundatus* Par. & Broth., Rev. Bryol. 36: 11. 1909. Type. China: Shanghai, Xujiahui, *Courtois & Henry s.n.*

Plants rather slender, dark green when dry, bright green when moist, in loose mats. *Stems* creeping, elongate, irregularly pinnately branched; branches erect to ascending, somewhat julaceous; central strand differentiated. *Leaves* lingulate, 1.2–1.8(–2.0) mm × 0.5–0.6 mm, abruptly narrowed at the middle from an ovate base to an obtuse to rounded apex, delicate and fragile, somewhat crisped, secund when dry, reflexed when moist; leaf margins plane, entire; costa single, 1/2–4/5 the leaf length, often indistinct or sometimes forked above the middle, papillose on the back; median and upper cells mostly rounded hexagonal, ca. 8–15 μm wide, multipapillose, thick-walled; basal juxtacostal cells elongate, 10–20 μm × 5–9 μm, thick-walled, smooth or unipapillose. *Dioicous. Perichaetial leaves* shortly ovate-lanceolate. *Setae* reddish brown, 5–13 mm long; capsules cylindrical, erect, symmetrical; opercula conic, shortly obliquely beaked; annuli well differentiated; peristome double; exostome teeth lanceolate, papillose at base, hyaline above; endostome very reduced or rudimentary; basal membrane very low or absent; cilia absent. *Calyptrae* cucullate, smooth or slightly glabrous. *Spores* spherical, 10–18 μm in diameter, smooth to finely papillose.

Illustrations: Granzow de la Cerda 1997 (Fig. 5, a–f); Watanabe 1972 (Pl. 47, figs. 1–17).

Habitat: on dead logs, soil, or moist rocks.

Distribution: China, Korea, Japan, Nepal, India, Russia, and Mexico.

Chinese specimens examined: HUBEI: Shennongjia, *P.-C. Wu* 67, 69, *133* (MO, PE). JIANGSU: Wu Co., Dongshan, *P.-P. Ling 622* (NAS), *P.-C. Wu & Y.-X. Jing 1647* (NAS). SICHUAN: Nan-ping Co., *S. He 30378, 30019* (MO). YUNNAN: Xishuangbanna, *W.-X. Xu 11374* (PE). ZHEJIANG: Lin-an Co., *P.-C. Wu 7616* (MO, PE).

9. ***Anomodon viticulosus*** (Hedw.) Hook. & Tayl. 牛舌藓 niu-she xian

Muscol. Brit. 79. 1818. *Neckera viticulosa* Hedw., Sp. Musc. Frond. 209. 1801. Type. Germany: Giessen.

Anomodon subintegerrimus Broth. & Par., Rev. Bryol. 27: 77. 1900.

Plants robust, dark yellowish green to pale yellowish green, brown or yellowish brown when old, caespitose, in loosely tufted mats. *Primary stems* creeping, nudicaulous, secondary stems erect, 4–10 cm long; branches remote, 1–3 cm long; central strand present. *Stem leaves* crisped when dry, erect-spreading when moist, crowded, ovate or oblong, sheathing at base, gradually narrowed to elongate lingulate, rounded obtuse at the apex, 2.5–6.0 mm long; leaf margins papillose; costa single, nearly excurrent; median leaf cells rounded hexagonal or hexagonal, 10–22 μm wide, densely papillose; branch leaves narrowly ovate-lingulate, slightly smaller than the stem leaves. *Dioicous. Perichaetial leaves* ovate, lanceolate toward the apex, ca. 1.5 mm long. *Setae* slender, yellowish brown, smooth; capsules ovoid, 2–3 mm long; opercula conic with beaks; exostome teeth yellowish brown, ca. 0.5

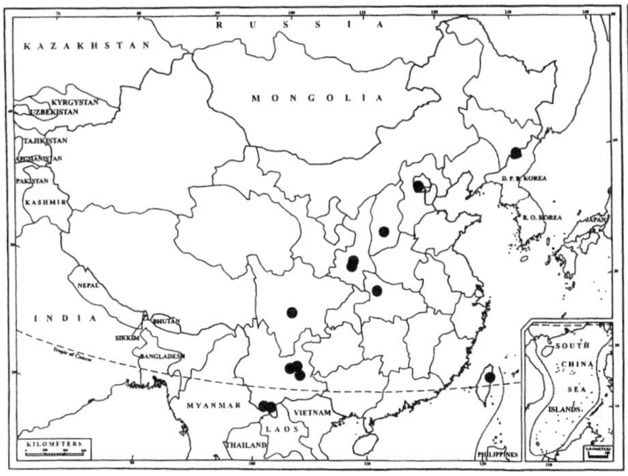

9. *Anomodon viticulosus*

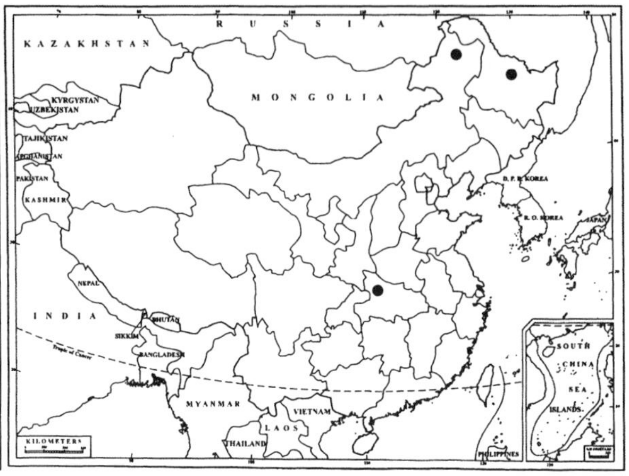

1. *Haplohymenium flagelliforme*

mm long, papillose above; endostome segments ca. 2/3 the length of exostome teeth; basal membrane very low. *Calyptrae* smooth. *Spores* ca. 15 µm in diameter.

Illustrations: P.-C. Wu (ed.) 2001 (Pl. 73, figs. 11–18); Granzow de la Cerda 1997 (Fig. 11, a–e).
Habitat: frequently growing on calcareous rock surfaces, rarely on other rocks.
Distribution: China, Korea, Japan, India, Pakistan, Myanmar, Vietnam, Russia (Siberia), Europe, North America, and North Africa.

Chinese specimens examined: BEIJING: Fang-shan Co., *B.-C. Chen 362* (MO, PE). HUBEI: Mt. Shennongjia, *P.-C. Wu 223, 254b* (MO, PE). JILIN: Mt. Changbai, *C.-J. Lin 2953* (PE). SHAANXI: Mt. West Tai-bai, *Z.-P. Wei 6603* (MO, WUK); Zhou-zhi Co., *M.-X. Zhang 446* (MO, WUK). SHANXI: Qinyuan Co., Mt. Lingkongshan, *Y.-J. Ling 85072, 85076* (SXU, PE). SICHUAN: Tian-quan Co., *S. He 32053* (MO). TAIWAN: Hua-lien Co., *Shevock 15071* (MO). YUNNAN: Kun-ming City, *Redfearn 1853* (MO); Lu-nan Co., *Redfearn et al. 2251* (MO); Meng-hai Co., *Magill et al. 7867* (MO); Meng-la Co., *Redfearn et al. 33957* (MO); Song-ming Co., *Redfearn et al. 1853* (MO); Xishuangbanna, *P. C. Wu 21836* (PE), *W.-X. Xu 6267, 6336* (PE).

2. ***Haplohymenium*** Dozy & Molk. 多枝藓属 duo-zhi xian shu
Musc. Frond. Ined. Archip. Ind.: 127, f. 40. 1846.

Plants delicate, usually yellowish green to brownish green, loosely pendulous from substrata. *Stems* mostly irregularly branched or remotely branched; paraphyllia absent. *Stem leaves* and branch leaves similar, imbricate when dry, erect-spreading or squarrose when moist, plicae absent, ovate or oblong-ovate at the base, tapering into narrow, lingulate or lanceolate, mostly rounded, rarely acute apices; leaf margins plane, finely serrulate or rounded papillose, rarely serrate at the apex; costa single, up to the middle of leaf, rarely stout and reaching to the leaf apex; leaf cells hexagonal or rounded hexagonal, thin-walled, roughly papillose; basal median cells hyaline, without papilla. *Monoicous. Perichaetial leaves* rather large, erect, oblong-ovate at base, acuminate above. *Setae* 1.5–5.0 mm long; *capsules* ovoid to oblong-ovoid, symmetrical, brown; annuli narrow, well developed; opercula conical with beaks; exostome teeth light yellow, narrowly lanceolate, papillose, perforate; endostome segments reduced; basal membrane very low. *Calyptrae* cucullate, sometimes lobed at base, granulate at apex, rarely with erect hairs.

Haplohymenium is distinguished from *Anomodon* by its smaller and slender plants, ovoid to oblong-ovoid capsules, grossly papillose and irregularly thickened peristome teeth, reduced endostomal development, and remotely pilose calyptrae. The first report of *Haplohymenium* in China was made by Brotherus (1925). In an earlier time, Cardot (1905) treated *H. sieboldii* (Dozy & Molk.) Dozy & Molk. as *Anomodon submicrophyllus*. Later, Brotherus (1928) published a new species, *H. pellucens* Broth. (= *H. sieboldii*) from Taiwan. P.-C. Chen et al. (1978) listed six species of *Haplohymenium* from China, and five species are recognized in this study. There are about 8 species of *Haplohymenium* in the world, but most of them are dealt with under the genus *Anomodon* according to Granzow de la Cerda (1997).

1. Plants with flagelliform branches .... 1. *H. flagelliforme*
1. Plants without flagelliform branches 2
   2. Stem leaves and branch leaves ovate, gradually tapering to an acute acumen .... 4. *H. sieboldii*
   2. Stem leaves and branch leaves ovate or widely ovate, suddenly narrowed to a lanceolate acumen....................3
3. Stems and branches without central strands; leaf cells spinosely unipapillose; costa extending to near the leaf apex .... 2. *H. longinerve*
3. Stems and branches with central strands; leaf cells multipapillose; costa extending above the middle of leaf........4
   4. Branch leaves usually less than 0.5 mm long; basal median leaf cells short .... 3. *H. pseudo-triste*
   4. Branch leaves ca. 1 mm long; basal median leaf cells elongate.. 5. *H. triste*

1. ***Haplohymenium flagelliforme*** Savicz-Ljubitskaya 鞭枝多枝藓 bian-zhi duo-zhi xian
   Bot. Mater. Inst. Sporov. Rast. Glavn. Bot. Sada RSFSR 1(7): 98. 1922. *Anomodon flagelliformis* (Savicz-Ljubitskaya) Granzow de la Cerda, Contr. Univ. Michigan Herb. 21: 245. 1997. Type. Russia: Siberia, southern Ussuri, *Boulavkina s.n.*
   Haplohymenium cristatum Nog., J. Jap. Bot. 20: 146. 1945.

Plants slender, yellowish green or dull green, dark brown when old, in small loose mats. *Stems* creeping, spreading, up to 5 cm long, irregularly branched; branches sparse and remote, pendulous, flagelliform; central strand absent. *Stem leaves* cordate, concave, tapering to a short acute apex, up to 1 mm long; leaf margins reflexed, remotely teethed; costa up to 3/4 the leaf length, opaque; median leaf cells rounded hexagonal, 7–15 µm in diameter, thick-walled, grossly multipapillose; branch leaves squarrose when moist, other features similar to those of stem leaves. *Sporophytes* not seen.

Illustrations: Plate 447, figs. 1–9.
Habitat: on trees or calcareous rock surfaces in mountain regions.
Distribution: China, Japan, and Russia (Siberia).

Chinese specimens examined: HEILONGJIANG: Yi-chun City, *Kozlov 35* (H, MO). HUBEI: Mt. Shennongjia, *P.-C. Wu 98* (PE), *P.-C. Wu 359* (MO). NEI MONGOL: northern Daxinganling, *X.-L. Bai 240* (NMU).

This species is characterized by having flagelliform branches and grossly serrate leaf apices.

2. ***Haplohymenium longinerve*** (Broth.) Broth. 长肋多枝藓 chang-le duo-zhi xian
   Nat. Pflanzenfam. I(3): 986. 1907. *Anomodon longinerve* Broth., Hedwigia 39: 243. 1899. Type. Japan: Shikoku, Kochi, *Makino s.n.*
   Haplohymenium longinerve fo. *piliferum* (Broth. & Yas) Nog., Kumamoto J. Sci., Sect. 2, Biol. 3: 30. 1957. *Haplohymenium piliferum* Broth. & Yas., Öfver. Förh. Finska Vetensk.-Soc. 62A(9): 33. 1921.
   Haplohymenium spinosum Nog., J. Jap. Bot. 13: 789. 3. 1937.

Plants slender, yellowish green or dark green, becoming yellowish brown or brown when old, in loose mats. *Stems* creeping, up to more than 5 cm long, irregularly pinnately branched; branches sparse, flagelliform; central strand absent. *Stem leaves* and branch leaves differentiated; stem leaves widely ovate, suddenly narrowed to an acuminate subula, ca. 0.3–0.6 mm long; margins plane; costa ending only a few cells below the apex; median leaf cells rounded hexagonal, ca. 8–15 µm wide, thick-walled, with a single sharp papilla at the middle of each cell; basal median leaf cells oblong, smooth; branch leaves widely ovate, gradually narrowed into a wide acumen; costa ending above the leaf middle; leaf cells rounded hexagonal, with a single setulose papilla or multipapillose. *Dioicous*. *Perichaetial leaves* ovate lanceolate, up to 2 mm long. *Setae* yellowish brown or reddish brown, 1.5–2.0 mm long; capsules and peristome similar to those of *H. triste*. *Calyptrae* cucullate, with a few erect hairs. *Spores* ca. 20–30 µm in diameter, densely minutely papillose.

Illustrations: Watanabe 1972 (Pl. 44, figs. 1–18).
Habitat: mostly on trees and rock cliffs in the forests of subalpine mountain regions.
Distribution: China, Korea, and Japan.

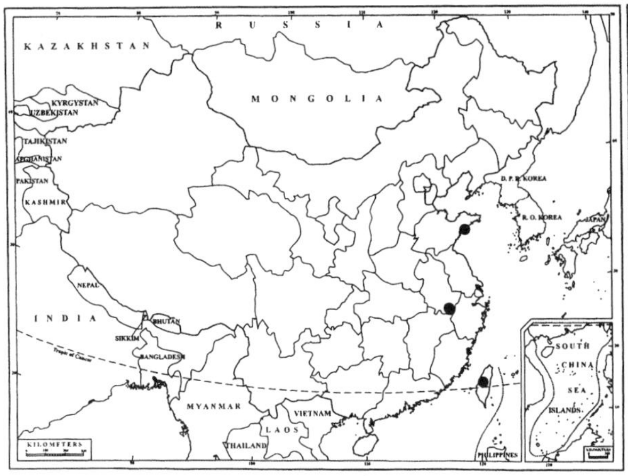

2. *Haplohymenium longinerve*

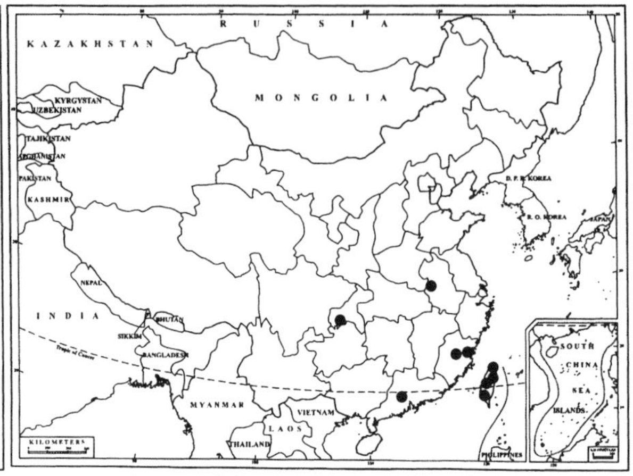

3. *Haplohymenium pseudo-triste*

Chinese specimens examined: ANHUI: Mt. Huangshan, *P.-C. Chen et al. 6937*, *7000* (PE), *6951* (MO, PE). SHANDONG: Qing-hai City, Laoshan (Mt.), *C. Gao & E.-Z. Bai 34488* (IFSBH, MO). TAIWAN: Nan-tou Co., *Nakn 13863* (Watanabe 1972).

The setulose papillae of the leaf cells are characteristic of this unique, East Asiatic species.

3. ***Haplohymenium pseudo-triste*** (C. Müll.) Broth. 拟多枝藓 ni-duo-zhi xian

Nat. Pflanzenfam. I(3): 986. 1907. *Anomodon pseudo-tristis* (C. Müll.) Granzow de la Cerda, Contr. Univ. Michigan Herb. 21: 243. 1997. *Hypnum pseudo-triste* C. Müll., Bot. Zeit. 13: 786. 1855. Type. South Africa: Cape of Good Hope, *Ecklon s.n.*

Haplohymenium fasciculare Nog., J. Jap. Bot. 13: 791. 1939.

Haplohymenium filiforme (Thwait. & Mitt.) Broth., Nat. Pflanzenfam. I(3): 1236. 1909. *Anomodon filiformis* Thwait. & Mitt., J. Linn. Soc., Bot. 13: 308. 1873, *hom. illeg.*

Haplohymenium mithouardii var. *viride* Thér., Ann. Cryptog. Exot. 5: 179. 1932. Type. China: Fujian (Fukien) Prov., Yenping (Nan-ping) Co., *C.-C. Chung B101.*

Haplohymenium pellucens var. *obtusifolium* Broth., Ann. Bryol. 1: 21. 1928. Type. China: Taiwan, Taihoku, *Sasaoka 2864.*

Haplohymenium submicrophyllus (Card.) Nat. Pflanzenfam. I(3): 986. 1907. *Anomodon submicrophyllus* Card., Beih. Bot. Centralbl. 19: 128. 1905.

Haplohymenium tenerrimum Broth., Rev. Bryol. 2: 10. 1920.

Plants very slender, yellowish green or brownish green, in small loose mats. *Stems* creeping, up to 3 cm long, regularly pinnately branched or irregularly pinnately branched; branches 3 mm long; central strand present. *Stem leaves* and branch leaves similar; slightly concave, imbricate on stems when dry, erect-spreading when moist, widely ovate at base, with short lanceolate apices, ca. 0.3–0.6 mm long; leaf margins plane, densely papillose; costa slender, ending at 3/4 the leaf length; median leaf cells rounded hexagonal or hexagonal, 6–12 µm wide, 2- to multipapillose, thin-walled; basal median leaf cells oblong, smooth. *Dioicous*. *Perichaetial leaves* ovate-lanceolate. *Setae* yellowish brown to reddish brown, up to 2 mm long; capsules ovoid; exostome teeth very similar to those of *H. triste*. *Calyptrae* covered with sparse, erect hairs. *Spores* densely and minutely papillose.

Illustrations: P.-C. Wu (ed.) 2001 (Pl. 71, figs. 10–17).

Habitat: mostly on trees, rarely on rock surfaces; alt. 200–500 m.

Distribution: China, Korea, Japan, Thailand, Vietnam, Sri Lanka, New Zealand, U.S.A. (Hawaii), and South Africa.

Chinese specimens examined: ANHUI: Jin-zhai Co., *Z.-M. Zeng T0127* (MO). CHONGQING: Nan-chuan Co., Mt. Jingfu, *P.-C. Wu 21155* (PE), *X.-Y. Hu 687* (MO, PE). FUJIAN: Ning-de City, *J. Zhu 1289*, *1326* (HSNU). GUANGDONG: Mt. Ding-hu, *P.-J. Lin 148* (IBSC, MO). TAIWAN: Nan-tou Co., *S.-H. Lin s.n.* [Bryophytes of Taiwan Exsiccatae ser.5, no. 219] (MO); Ping-tung Co., *Shevock 15055* (MO); Tai-pei Co., *Z.-Y. Jiang 5485* (TAI).

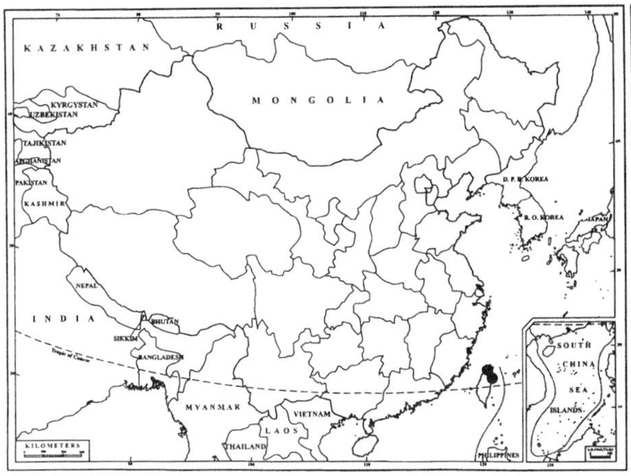

4. *Haplohymenium sieboldii*

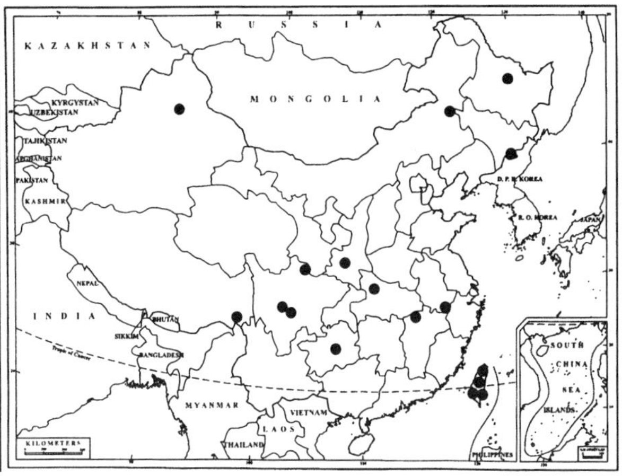

5. *Haplohymenium triste*

4. ***Haplohymenium sieboldii*** (Dozy & Molk.) Dozy & Molk. 多枝藓 duo-zhi xian

Musc. Frond. Ined. Archip. Ind. 4: 127. T. 40. 1846. *Anomodon sieboldii* (Dozy & Molk.) Granzow de la Cerda, Contr. Univ. Michigan Herb. 21: 243. 1997. *Neckera sieboldii* (Dozy & Molk.) C. Müll., Syn. Musc. Frond. 2: 81. 1850. *Leptohymenium sieboldii* Dozy & Molk., Ann. Sci. Nat. Bot., ser. 3, 2: 310. 1844. Type. Japan: *Siebold s.n.*

Haplohymenium pellucens Broth., Ann. Bryol. 1: 21. 1928. Type. China: Taiwan, Taihoku, *Sasaoka s.n.*

Haplohymenium pinnatum Nog., J. Jap. Bot. 13: 410. 1937.

Plants slender, yellowish green or brownish green, in lax mats. *Stems* creeping, ca. 1–3 cm long, irregularly pinnately branched; branches 3–5 mm long; central strand absent. *Stem leaves* rather small, appressed when dry, erect-spreading when moist, slightly concave, ovate, gradually narrowed into a short acute apex, ca. 0.6 mm long; leaf margins papillose; costa covering ca. 2/3 the leaf length; median leaf cells rounded quadrate or hexagonal, thin-walled, multipapillose; basal median cells oblong or oblong-ovate, smooth; branch leaves slightly larger than stem leaves, ovate or oblong ovate, isomorphic. *Dioicous*. *Perichaetial leaves* oblong, with narrow lanceolate apex. *Setae* yellowish brown, ca. 5 mm long; capsules oblong ovoid; peristome teeth similar to those of *H. triste*. *Calyptrae* minutely papillose and rarely erect ciliated. *Spores* nearly spherical, papillose.

Illustrations: Watanabe 1972 (Pl. 43, figs. 1–19).
Habitat: usually growing on tree trunks, rarely on rock surfaces.
Distribution: China, Korea, and Japan.

Chinese specimens examined: TAIWAN: Tai-pei Co., *Saso 1729* (TNS).

5. ***Haplohymenium triste*** (Cesati in De Notaris) Kindb. 暗绿多枝藓 an-lv duo-zhi xian

Rev. Bryol. 26: 25. 1899. *Anomodon tristis* (Cesati in De Notaris) Sull. & Lesq., Musci Hep. U.S. (repr.) 241 [Schedae 52]. 1856. *Hypnum triste* (Cesati in De Notaris) C. Müll., Syn. 2: 478. 1851. *Leskea tristis* Cesati in De Notaris, Syllab. Musc. 67. 1838. Type. Italy: Lake Maggiore, *Cesati s.n.*

Haplohymenium sinensi-triste (C. Müll.) Broth., Nat. Pflanzenfam. I(3): 986. 1907. *Anomodon sinensi-tristis* C. Müll., Nuovo Giorn. Bot. Ital., n. ser. 3: 118. 1896. Type. China: Shaanxi (Schen-si), *Giraldi s.n.*

Haplohymenium formosanum Nog., Trans. Nat. Hist. Soc. Taiwan 26: 43. 3 f. 14–17. 1936. Type. China: Taiwan, Tai-nan Co., Mt. Kodama, *Noguchi 5867*. Synonymized by Granzow de la Cerda (1997).

Haplohymenium longiglossum P.-C. Chen, Repert. Spec. Nov. Regni Veg. 58: 31. 1955. Type. China: Sichuan (Szetschwan) Prov., Mt. Omei, *P.-C. Chen 5353* (holotype PE). Synonymized by Granzow de la Cerda (1997).

Haplohymenium microphyllum (Broth. & Par.) Broth., Nat. Pflanzenfam. I(3): 986. 1907, *hom. illeg. Anomodon microphyllum* Broth. & Par., Rev. Bryol. 31: 56. 1904.

Hypnum fragile (Wils.) C. Müll., Syn. 2: 471. 1851. *Anomodon fragilis* Wils., London J. Bot. 7: 275. 1848. Type. China: Chusan, *Anderson s.n.* Synonymized by Granzow de la Cerda (1997).

Plants rather slender, yellowish green to brownish green, in loose mats. *Stems* creeping, up to 5 cm long, usually irregularly pinnately branched; branches 3–5 mm long; central strand present. *Stem leaves* imbricate when dry, erect-spreading when moist, ovate or widely ovate at base, abruptly narrowed to a lanceolate or lingulate apex; leaf margins densely papillose; costa slender, up to the leaf middle or above; median leaf cells rounded quadrate, ca. 10 µm wide, densely multipapillose; basal median cells elliptic, smooth; branch leaves similar to stem leaves, lingulate above. *Dioicous*. *Perichaetial leaves* widely oblong-ovate at base, narrowly lanceolate at apex; costa extending up to leaf apex. *Setae* slender, yellowish brown, ca. 1.5 cm long; capsules ovoid, erect; opercula conical; annuli large; exostome lightly yellowish brown, divided above into 2 lanceolate teeth, densely papillose. *Spores* large, up to 10 µm in diameter, finely papillose.

Illustrations: P.-C. Wu (ed.) 2001 (Pl. 70, figs. 10–17).
Habitat: usually on trees, rarely on moist shady rock surfaces; alt. 500–1000 m.
Distribution: China, Korea, Japan, Russia (Siberia), Europe, Central and North America (including Hawaii).

Chinese specimens examined: ANHUI: Huang Shan (Mt.), *P.-C. Chen 6534* (PE). GUIZHOU: Gui-yang City, *Gui-yang Teacher's Coll. 86* (MO). HEILONGJIANG: Yi-chun City, Wu-ying, *Vitt 34863* (ALTA, MO). HUBEI: Mt. Shennongjia, Tong-cha-gou, *P.-C. Wu 80* (MO, PE), Yangriwan, *P.-C. Wu 260* (H, MO, NY, PE). JIANGXI: Lushan (Mt.), *P.-C. Chen et al. 225* (MO, PE). JILIN: Mt. Changbai, *C. Gao 22126* (IFSBH, MO). NEI MONGOL: Keerginyouyiqiang, Mt. Beilangjiguan, *X.-L. Bai 1560* (HIMC). SICHUAN: Nan-ping Co., *S. He 30157a* (MO); Tian-quan Co., *S. He 32096* (MO). TAIWAN: Mt. Alishan, *Z.-Y. Jiang* (MO, TAI); Tai-nan Co., *Noguchi 2475* (NICH), *5867* (TNS); Tai-pei Co., *S.-H. Lin s.n.* [Bryophytes of Taiwan Exsiccatae ser. 2, no. 69] (MO); Tai-tung Co., *S.-H. Lin s.n.* [Bryophytes of Taiwan Exsiccatae ser. 4, no. 169] (MO). XINJIANG: Eastern Tianshan Mts., *J.-C. Zhao 86499* (J.-C. Zhao 1993). XIZANG: Cha-yu Co., without exact specimen mumber [X.- J. Li (ed.) 1985].

This widely distributed species is the most common *Haplohymenium*. Its habit and size are variable. The leaves are very similar to those of *Anomodon*; however, the slender appearance of the plants and the imbricate leaves separate it from the species of *Anomodon*. *Haplohymenium pseudo-triste* is very close to *H. triste* morphologically, except that the plants of the former species are much more slender.

3. ***Herpetineuron*** (C. Müll.) Card. 羊角藓属 yang-jiao xian shu

Beih. Bot. Centralbl. 19 (2): 127. 1905. *Anomodon* sect. *Herpetineuron* C. Müll., Flora 73: 496. 1890. *Bryonorrisia* Stark & Buck, Brittonia 38: 52. 1986.

Plants medium-sized to robust, rigid, yellowish green to brownish green in lax tufts. *Primary stems* creeping, often with asexual stoloniferous branches, secondary stems erect or ascending, irregularly to sparsely branched, circinnate at apices when dry, caudate. *Stem leaves* and branch leaves similar, ovate-lanceolate or widely lanceolate, often undulate, concave and conspicuously keeled, acuminate above; leaf margins entire at base, irregularly grossly dentate above; costa strong, not percurrent, gradually slender and obviously zigzag above; leaf cells nearly isomorphic, hexagonal, opaque, smooth, thick-walled; branch leaves rather small and narrow. *Dioicous*. *Perichaetial leaves* light yellow, from a widely sheathing base narrowed into a lanceolate acumen. *Setae* reddish brown; capsules ovoid to cylindrical, brown, symmetrical, stomata present; annuli well developed; opercula conical; exostome teeth light in color, lanceolate, fenestrate, densely papillose; endostome segments linear, pale, densely papillose; basal membrane well developed. *Calyptrae* smooth.

Bryonorrisia acutifolia (Mitt.) Enroth was reported from Sichuan, Taiwan, and Yunnan provinces (Redfearn et al. 1996). Granzow de la Cerda (1989) recently treated *Bryonorrisia* Stark & Buck as a synonym of *Herpetineuron* and made the new combination of *H. acutifolium* (Mitt.) Granzow de la Cerda.

1. Leaves triangular-lanceolate, costa ca. 2/3 the leaf length, ending well below the leaf apex, almost straight throughout; leaf margins only serrulate at the apex; leaf cells sometimes projecting on the dorsal side of leaves...... .... 1. *H. acutifolium*
1. Leaves narrowly ovate-lanceolate, costa ending near or at the apex, strongly flexuous-sinuose above; leaf margins strongly serrate near the apex; leaf cells smooth .... 2. *H. toccoae*

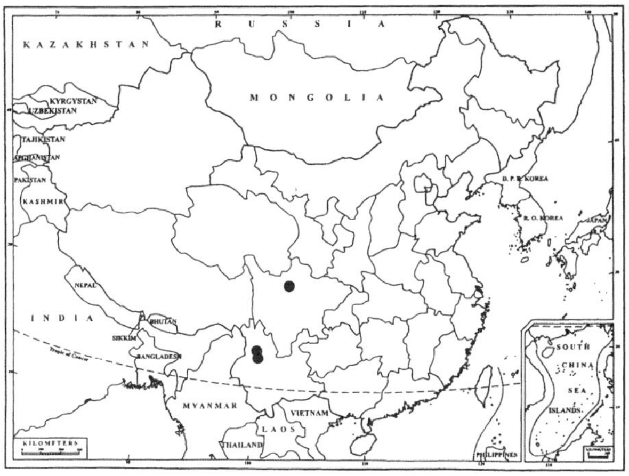

1. *Herpetineuron acutifolium*

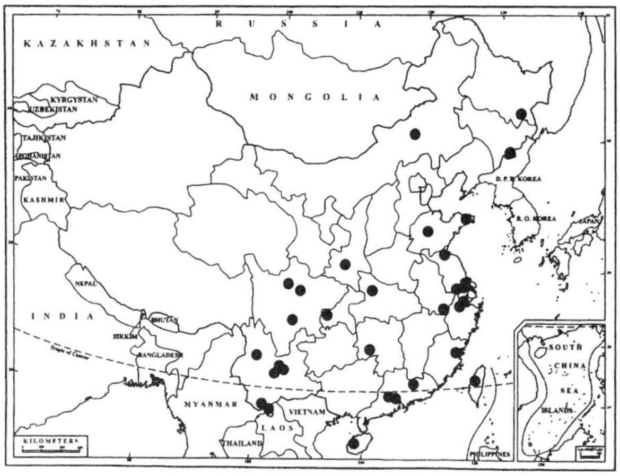

2. *Herpetineuron toccoae*

1. ***Herpetineuron acutifolium*** (Mitt.) Granzow de la Cerda 尖叶羊角藓 jian-ye yang-jiao xian
Bryologist 92: 385. 1989. *Bryonorrisia acutifolia* (Mitt.) Enroth, J. Bryol. 16: 407. 1991. *Anomodon acutifolius* Mitt., J. Proc. Linn. Soc., Bot., Suppl. 1: 126. 1859. Type. India: Kumaon.

Plants medium-sized, yellowish green to green above, brownish below, in dense tufts. *Primary stems* prostrate, secondary stems irregularly pinnately branched; branches erect to more or less arcuate, usually attenuate and flagelliform; central strand present. *Leaves* of secondary stems and branches triangular-lanceolate, 1.2–1.8 mm × 0.7–0.8 mm, abruptly acuminate near the apex, somewhat concave, appressed when dry, erect-spreading when moist; margins plane, mostly entire, only serrulate at the apex; costa extending just beyond 2/3 the leaf length, almost straight throughout, only slightly sinuose above; leaf cells shortly rectangular to rhomboidal, smooth, sometimes projecting on the dorsal side of leaves; alar cells not differentiated. *Dioicous*. *Perichaetial leaves* not sheathing at base, shortly lanceolate with acute apex. *Setae* elongate, 9–10 mm long, reddish brown; capsules erect, symmetric, cylindrical, 2–3 mm long; opercula rostrate; annuli not differentiated; exostome teeth lanceolate, papillose and cross striate at base; endostome segments linear, nearly as long as the teeth, smooth; basal membrane low. *Calyptrae* not seen. *Spores* spherical, ca. 20 µm in diameter, finely papillose.

Illustrations: Granzow de la Cerda 1997 (Fig. 31, a–i).
Habitat: on rocks or bark.
Distribution: China, Japan, Nepal, India, and Mexico.

Chinese specimens examined: SICHUAN: Ma-er-kang Co., *Allen 6948* (MO). YUNNAN: Da-li Co., Diancang Shan (Mt.), *Redfearn & Y.-G. Su 1520* (MO, PE); Yang-bi Co., *Redfearn et al. 298* (MO, PE), *Redfearn & Y.-G. Su 1003* (MO, PE).

2. ***Herpetineuron toccoae*** (Sull. & Lesq.) Card. 羊角藓 yang-jiao xian
Beih. Bot. Centralbl. 19 (2): 127. 1905. *Anomodon toccoae* Sull. & Lesq., Musci Bor. Amer.: 52. 1856. Type. U.S.A.: Georgia, Toccoa Falls, *C. L. Lesquereux*, Musci Bor. Amer. exsicc. 1, 240.
Anomodon devolutus Mitt., J. Linn. Soc., Bot., Suppl. 1: 127. 1959.
Anomodon flagelliferus C. Müll., Nuovo Giorn. Bot. Ital., n. ser. 4: 273. 1897. Type. China: Shaanxi (Schen-si), Hu-schien (Co.), Sche-kin-tsuen, Dec. 1895, *Giraldi s.n.*
Herpetineuron attenuatum Okam., J. Coll. Sci. Imp. Univ. Tokyo 38 (4): 54. 24. 1916.
Herpetineuron formosicum Broth., Ann. Bryol. 1: 22. 1928. Type. China: Taiwan, Taichu Hori, *Suzuki 1734*.
Herpetineuron serratinerve Sak., Bot. Mag. (Tokyo) 62: 107. 10. 1949.

Plants medium-sized, yellowish green to green above, dark green at base, caespitose in large loose mats. *Stems* irregularly branched; branches circinnate when dry. The other characters are the same as those of the genus.

This species is very common and nearly cosmopolitan. The plant size, leaf shape and length, and the marginal teeth are variable.

Illustration: P.-C. Wu (ed.) 2001 (Pl. 77, figs. 1–8).

Habitat: frequently on shady calcareous rocks, sometimes growing on tree trunks in deciduous forests; alt. 200–1000 m.

Distribution: China, Japan, Korea, India, Sri Lanka, Thailand, Indonesia, the Philippines, New Caledonia, South and North America.

Chinese specimens examined: ANHUI: Huang Shan (Mt.), *P.-C. Chen 6523* (PE). CHONGQING: Jiang-bei Co., *T.-Y. Li & L.-Q. Xia 54* (PE). FUJIAN: near Fuzhou City, *H.-H. Chung 239A* (MO). GUANGDONG: Guangzhou City, *Levine 1843* (MO); Mt. Dinghu, *Magill et al. 8073* (MO), *P.-J. Lin 153* (IBSC, MO); Mt. Lofu, *Magill et al. 8133* (MO); Mt. Sha-lo, *Taam 1056A* (MO); Zhao-qing City, *Redfearn 34386* (MO). HAINAN: Chang-jiang Co., *Reese 17505* (LAF, MO). HEILONGJIANG: Ning-an Co., *X.-L. Zhu 13* (PE). HUBEI: Mt. Shennongjia, *P.-C. Wu 186* (MO, PE). HUNAN: Cheng-bu Co. *T.-R. Cao 91310* (PE). JIANGSU: Chang-shou City, *P.-P. Ling & Q. Wang 213* (NAS); Lian-yun-gang City, *P.-P. Ling & Q. Wang 3004* (NAS); Nan-ton City, *P.-P. Ling & Q. Wang 2865, 2876* (NAS); Wu-xian Co., *P.-C. Wu & Y.-X. Jin 1946* (MO, PE); Yi-xing Co., *T. Cao 38771a* (MO). JILIN: Mt. Changbai, *Vitt 34750* (ALTA, MO). NEI MONGOL: Xi-lin-hao-te Qi (Co.), *Z.-G. Tong 1062* (HIMC). SHANDONG: Mt. Kunyu, *C. Gao & E.-Z. Bai 34302* (MO); Tai-shan (Mt.), *C. Gao & E.-Z. Bai 34079* (MO). SICHUAN: Ma-er-kang Co., *Allen 6911* (MO); Mt. Omei, *Redfearn 34804* (MO); Wen-chuan Co., *Allen 6685* (MO). TAIWAN: Nan-tou Co., *Z.-Y. Jiang 765* (PE, TAI), *S.-H. Lin s.n.* [Bryophytes of Taiwan Exsiccatae ser. 3, no. 121] (MO). YUNNAN: An-ning Co., *Magill et al. 8048* (MO), *Redfearn et al. 34315* (MO), *Koponen 37956* (H, MO); Chu-xiong Co., *Redfearn et al. 1723* (MO); Jing-hong Co., *Redfearn 33691* (MO); Kun-ming City, *Redfearn & S. He 2103* (MO); Lu-feng Co., *Redfearn & S. He 1799* (MO); Meng-hai Co., *Magill et al. 7889* (MO); Wei-shan Co., *Redfearn et al. 1232* (MO); Xishuangbanna, *W.-X. Xu 6307, 6318, 6437* (PE), *M.-Z. Wang 4253* (PE), *P.-C. Wu 29106* (PE); Yang-bi Co., *Redfearn et al. 38* (MO). ZHEJIANG: De-qing Co., *P.-C. Wu 1289* (MO, PE); Hang-zhou City, *B.-C. Tan 93-504* (FH, MO); Lin-an Co., *P.-C. Wu 24872* (MO, PE).

4. ***Miyabea*** Broth. 瓦叶藓属 wa-ye xian shu

Nat. Pflanzenfam. I(3): 984. 1907.

Plants slender, rigid, dark green, in loosely compact mats. *Primary stems* creeping, secondary stems erect, ascendant or dendroid, irregularly branched, densely foliolate; central strand undifferentiated. *Stem leaves* and branch leaves similar, imbricate when dry, erect-spreading when moist, concave, ovate or widely ovate, acute above; leaf margins nearly entire, only remotely dentate at the apex; costa single, up to the middle or sometimes to leaf apex. *Setae* slender, exserted; capsules erect, ovoid, symmetrical; opercula with long beaks; exostome teeth 16, well developed, smooth or papillose; endostome absent. *Calyptrae* cucullate. *Spores* nearly spherical, densely papillose.

The genus *Miyabea*, containing 3 species, only occurs in East Asia. In China, the genus was first collected in Huang Shan (Mt.), Anhui province. It was later discovered in Mt. West Tianmo and northeastern China. The distribution of this genus is mainly restricted in the areas of east coastal mountains, north of 30°N and east of 110°E.

Systematically, *Miyabea* is very close to *Haplohymenium*. Both genera have imbricate leaves when dry, usually grow on tree trunks, and lack or have reduced endostomal development. However, the absence of lamellae and longitudinal division of the peristome teeth in *Haplohymenium* differentiates it from *Miyabea*.

1. Plants regularly pinnately branched .... 3. *M. thuidioides*
1. Plants mostly irregularly pinnately branched 2
   2. Leaves widely ovate or ovate .... 1. *M. fruticella*
   2. Leaves cordate or widely ovate .... 2. *M. rotundifolia*

1. ***Miyabea fruticella*** (Mitt.) Broth. 瓦叶藓 wa-ye xian

Nat. Pflanzenfam. I(3): 984. 1907. *Lasia fruticella* Mitt., Trans. Linn. Soc. London, Bot. 3: 173. 1891. Type. Japan: Tochigi Pref., *Bisset s.n.*

Plants slender, yellowish green or yellowish brown, not glossy, forming loosely tufted cushions. *Stems* creeping, pendulous, up to more than 5 cm long, regularly or irregularly pinnately branched, the apices of branches

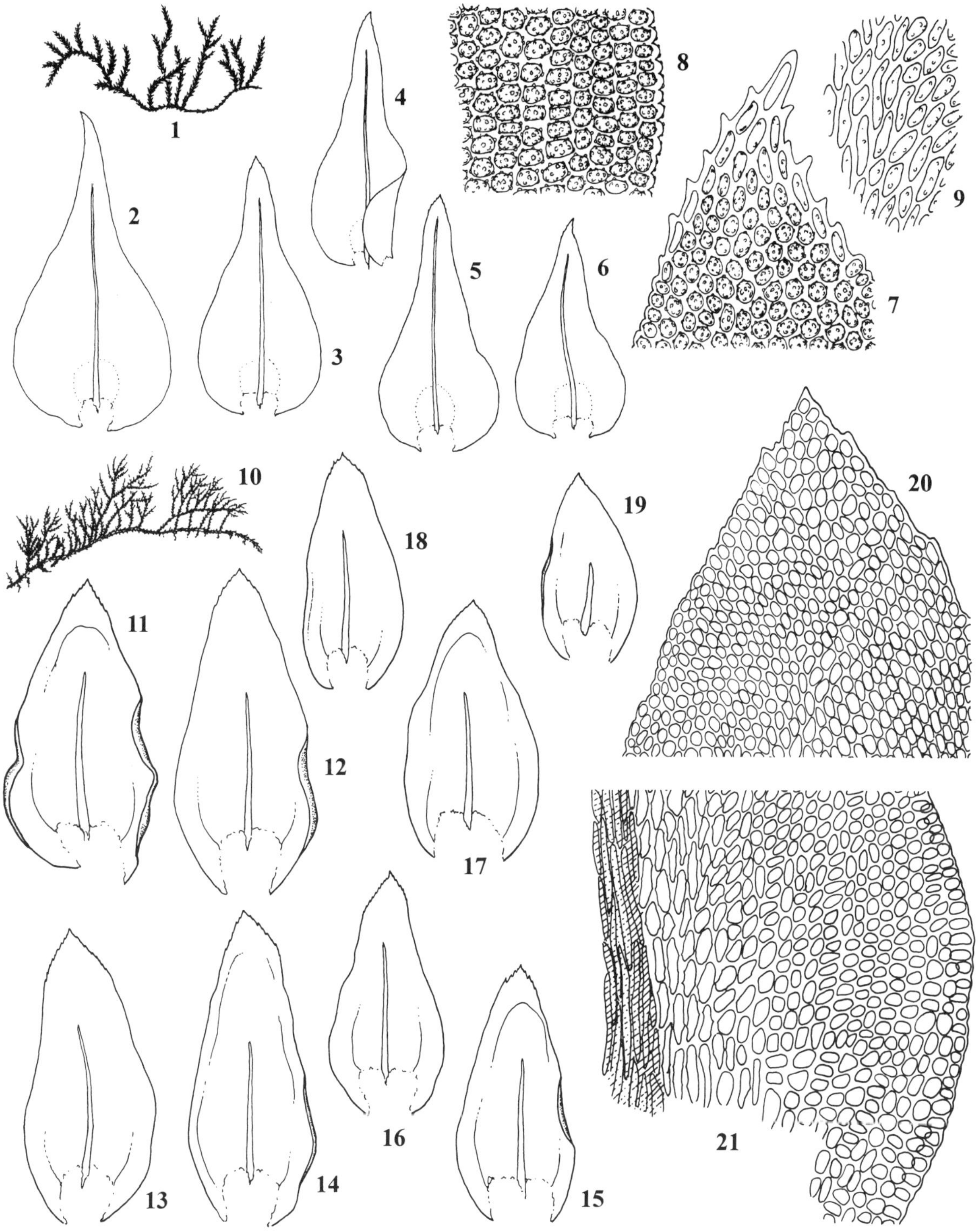

Plate 447. 1–9, *Haplohymenium flagelliforme*. 1, habit (× 1). 2, stem leaf (× 40). 3–6, branch leaves (× 40). 7, apical leaf cells (× 360). 8, median marginal leaf cells (× 360). 9, juxtacostal leaf cells (× 360). 10–21, *Miyabea fruticella*. 10, habit (× 1). 11–14, stem leaves (× 45). 15–19, branch leaves (× 45). 20, apical leaf cells (× 260). 21, basal leaf cells (× 260). (1–9, drawn from *P.-C. Wu 98*, PE; 10–21, from *Luteyn 1148*, PE.)

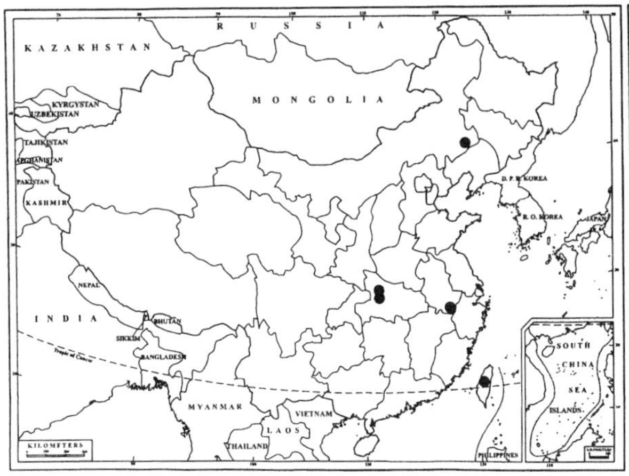

1. *Miyabea fruticella*

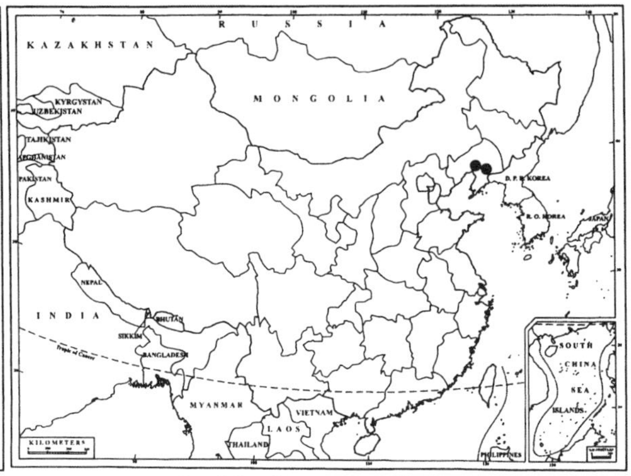

2. *Miyabea rotundifolia*

often secund to one side; central strand undifferentiated. *Stem leaves* and branch leaves nearly isomorphic; stem leaves imbricate when dry, erect-spreading when moist, widely ovate, with acute apices, ca. 1.0 mm × 0.5–0.8 mm; leaf margins nearly entire, only remotely serrulate; costa single, up to the middle of leaves; leaf cells quadrate, rhomboidal or hexagonal, ca. 4–10 μm in diameter, equally thick-walled; basal cells nearly hexagonal, thick-walled; branch leaves slightly smaller, ovate. *Dioicous*. *Perichaetial leaves* narrowly ovate to lanceolate, sheathing at base, up to 1.5 mm long; costa reaching to the apex. *Setae* smooth, ca. 1 mm long; capsules ovoid, reddish brown, with a short neck; exothecial cells rhomboidal to hexagonal; exostome teeth yellowish brown, lanceolate, with transversal lamellae; endostome absent; opercula sharply beaked. *Calyptrae* cucullate, smooth. *Spores* variable in size, 25–40 μm in diameter, minutely papillose.

Illustrations: Plate 447, figs. 10–21.
Habitat: mostly growing on trees in mountain regions, rarely on rock surfaces.
Distribution: China, Korea, and Japan.

Chinese specimens examined: ANHUI: Huang Shan (Mt.), Baieling, *P.-C. Chen et al. 7414*, *7419* (MO, PE), near Yungu Temple, *P.-C. Chen et al. 7515* (PE), Qingliangtai, *P.-C. Chen et al. 7347* (PE). HUBEI: Mt. Shennongjia, *1980 Sino-Amer. Exped. No. 1148b* (BM, MO, NY, PE); Xing-shan Co., *J.-X. Luo 961243* (MO, PE). NEI MONGOL: Liaohe Plain, Daqinggou, *X.-L. Bai 1315* (HIMC). TAIWAN: Nan-tou Co., collector unknown (PE).

This species is typically an East Asiatic taxon. In China, the first record of this species was collected from Huangshan (P.-C. Chen & P.-C. Wu 1965). It was later found in Mt. Xiaoxinganling, Mt. Changbai (C. Gao 1977), and Mt. West Tianmu (R.-L. Hu & Y.-F. Wang 1981).

2. ***Miyabea rotundifolia*** Card. 圆叶瓦叶藓 yuan-ye wa-ye xian
Bull. Soc. Bot. Genève, ser. 2, 1: 132. 1909. Type. Korea: Yjyang-Tjyen, *Faurie 321*.

Plants medium-sized, yellowish or lightly yellowish brown when young, yellowish brown when old, loosely compact. *Stems* up to 5 cm long, irregularly branched or regularly pinnately branched; central strand undifferentiated. *Stem leaves* and branch leaves similar, approximate when dry, erect-spreading when moist, cordate or broadly ovate, ca. 1 mm long, slightly concave; leaf margins entire below, minutely serrulate near apex; costa single, reaching 2/3 the leaf length; leaf cells oval or rhomboidal, 8–15 μm × 4–10 μm, thick-walled; basal cells elliptical, up to 40–60 μm long. *Setae* reddish brown, 5 mm long or more; capsules ovoid; peristome teeth well developed, yellowish brown. *Spores* polymorphic, finely papillose.

Illustrations: Watanabe 1972 (Pl. 39, figs. 1–16).
Habitat: on rock surfaces by the edge of forests.
Distribution: China, Korea, and Japan.

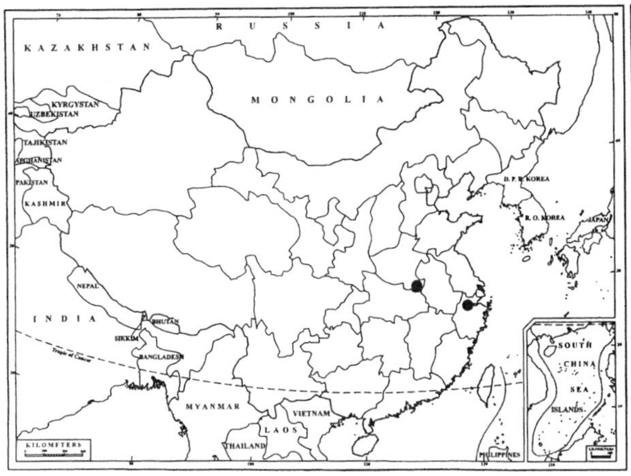

3. *Miyabea thuidioides*

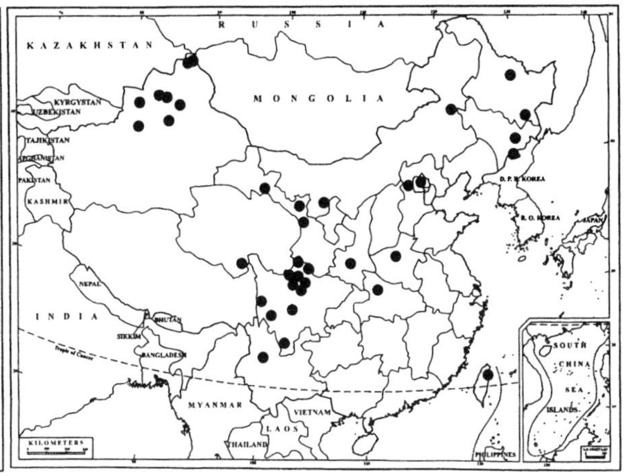

1. *Abietinella abietina*

Chinese specimens examined: LIAONING; Mt. Fenghuang, *Z.-F. Fang s.n.* (IFP G), *C. Gao 6893* (IFP); Anshan City, Qianshan (Mt.), *S.-E. Liu 6859* (IFP, PE).

The widely cordate leaves of this species are obviously different from those of *Miyabea fruticella* and *M. thuidioides*.

3. ***Miyabea thuidioides*** Broth. 羽形瓦叶藓 yu-xing wa-ye xian
Oefvers. Förh. Finska Vetensk.-Soc. 62: 32. 1921. Type. Japan: Kochi Pref., Mt. Kuishi, *Nakanishiki 3-1*.

Plants rather large, yellowish green, lightly yellowish brown or brown when old, in loosely aggregated mats. *Stems* up to 5 cm long, irregularly branched to regularly pinnately branched; central strand undifferentiated. *Stem leaves* and branch leaves similar, leaves approximate when dry, erect-spreading when moist, ovate to widely ovate, ca. 1.5 mm long, concave; leaf margins entire, only remotely serrate above; costa reaching 1/2–2/3 the leaf length; median leaf cells rhomboidal or rounded quadrate, 8–10 µm × 10–12 µm, strongly thick-walled; basal leaf cells oblong-ovate to narrowly ovate, smooth, rarely slightly papillose. *Sporophytes* unknown.

Illustrations: Watanabe 1972 (Pl. 40, figs. 1–16).
Habitat: growing on tree trunks and branches in forests.
Distribution: China and Japan.

Chinese specimens examined: HENAN: Shang-chang Co., Mt. Huangbo, *Xin Xian Normal College 2289* (PE). ZHEJIANG: Mt. West Tianmu, near Laodian Temple, *P. C. Wu 8094* (PE).

This species is uncommon both in China and in southern Japan. It is similar to *Miyabea fruticella*, but the latter species differs by its smaller plants, erect-spreading branches, and small median leaf cells.

THUIDIACEAE

羽藓科 yu-xian ke

by Wu Peng-cheng, Wang Mei-zhi, and Zhong Ben-gu

Plants small to robust, mostly dark green, yellowish green or brownish green, dull, rigid when dry, often in small mats or thick mats after years. *Stems* creeping or ascending above, irregularly branched or regularly pinnately, bipinnately or tripinnately branched; central strand present or absent; paraphyllia often present, single or branched, usually numerous. *Stem leaves* and branch leaves often differentiated; leaves arranged in several rows, appressed or usually imbricate when dry, erect-spreading when moist, ovate, widely ovate or ovate-triangular, acuminate or filiform above; leaf margin entire, finely serrulate or papillose; costa mostly single, usually reaching 2/3 the leaf length or excurrent, rarely short, weak, forked or inconspicuous; leaf cells mostly hexagonal or rounded hexagonal, thick-walled, mostly papillose; basal cells longer and pitted; marginal leaf cells nearly quadrate. *Autoicous* or dioicous. *Perichaetia* laterally growing on stems; perichaetial leaves usually ovate-lanceolate. *Setae* slender, light reddish when old, smooth or densely papillose; capsules declinate, asymmetrical, smooth; stomata rare, at the base of capsules or lacking; annuli usually present; opercula conical, mostly rostrate; peristome double; exostome teeth 16, lanceolate, light yellowish or yellowish brown, pale above, usually transversely striolate in the low parts, papillose in the upper parts, borders differentiated; endostome segments narrowly lanceolate, pale, papillose; basal membrane high; cilia often present. *Calyptrae* cucullate, smooth, rarely with hairs or papillae. *Spores* spherical.

The Thuidiaceae, one of the large moss families, are widely distributed on every continent except Antarctica from tropical rain forests to temperate coniferous forests. Some of the genera in the Thuidiaceae are East Asiatic endemics, such as *Hylocomiopsis* endemic to Japan and *Thuidiopsis* to Japan and Korea. Worldwide, most of the Thuidiaceae occur in Asia, Central and South America, and Oceania. The species of this family grow in various places and frequently occur under grasses or on rock surfaces, rotten logs, tree trunks, and forest ground in large mats with other mosses. They can be found from sea level to above 4000 m. Many species of the family play an important role in protecting against water and soil erosion in alpine coniferous forests.

The distinct features of the Thuidiaceae include the *Hypnum*-type peristome, the pinnately branched stems often bearing paraphyllia, and papillose leaf cells. Brotherus' (1925) classification and systematic arrangement of the family between the Leskeaceae and Amblystegiaceae are generally accepted among bryologists. However, some bryologists have different opinions on the classification. Crum and Anderson (1981) removed the primitive genera *Haplohymenium*, *Anomodon*, and *Herpetineuron* from Thuidiaceae to Leskeaceae and transferred *Pseudoleskeella* and *Pseudoleskea* of the Leskeaceae and *Myurella* of the Theliaceae to Thuidiaceae. Generally, the evolutionary relationships among genera of the Thuidiaceae are well established; nonetheless, there are some genetic differences in this family that reflect its systematic relationship with related families.

Brotherus (1925) divided the Thuidiaceae into four subfamilies: Heterocladioideae, Anomodontoideae, Thuidioideae, and Helodioideae. This classification reflects not only infra-family relationships, but also systematic relationships with related families such as the Leskeaceae. We move *Haplocladium* from the Anomodontoideae to Thuidioideae considering their appearance, branching forms, branch apices, leaf cells, and paraphyllia types. The remaining genera of the Anomodontoideae, with *Anomodon*, *Haplohymenium*, *Herpetineuron*, and *Miyabea*, are moved from the Thuidiaceae to the Anomodontaceae. *Abietinella* is transferred to Helodioideae. Both Crum and Anderson (1981) and Noguchi (1991) treated *Abietinella* and *Rauiella* in *Thuidium*; however, we maintain them separate from *Thuidium* based on the differences in habit and leaf morphology. We agree with Buck and Crum (1990) in elevating the subgenus *Microthuidium* Limpr. of *Thuidium* to generic rank as *Cyrto-hypnum*. In this study, 14 genera in the Thuidiaceae are treated. Touw (2001a) redefined the Thuidiaceae and presented a realignment of taxa traditionally accommodated in *Thuidium*; however, his opinions appeared too late to be taken into consideration for our treatment.

The systematic arrangement of Chinese Thuidiaceae:

Subfamily I. Heterocladioideae: *Heterocladium* Schimp. in B. S. G., *Leptocladium* Broth., and *Leptopterigynandrum* C. Müll.

Subfamily II. Thuidioideae: *Boulaya* Card., *Claopodium* (Lesq. & James) Ren. & Card., *Cyrto-hypnum* (Hampe) Hampe, *Haplocladium* (C. Müll.) C. Müll., *Pelekium* Mitt., *Rauiella* Reim., and *Thuidium* B. S. G.

Subfamily III. Helodioideae: *Abietinella* C. Müll., *Actinothuidium* (Besch.) Broth., *Bryonoguchia* Iwats. & Inoue, and *Helodium* (Sull.) Warnst.

Key to the subfamilies

1. Plants delicate; leaf costae double, short or absent .... Subfamily 1. Heterocladioideae
1. Plants rather robust; leaf costa single, usually ending above the middle of leaf, rarely excurrent 2
   2. Plants usually simple or pinnately branched, rather robust; mostly growing in swampy or moist habitats .... Subfamily 3. Helodioideae
   2. Plants 1–3 pinnately branched, rather small to medium-sized; often growing in moist, sometimes dry habitats .... Subfamily 2. Thuidioideae

The history of studying Chinese Thuidiaceae started at the beginning of this century with most of the publications involving materials from one mountain or a localized region. Cardot and Thériot (1904) examined the specimens collected by Elm Bodinier and L. Martin in northern Guizhou, and published *Thuidium japonicum* Dozy & Molk. and *T. pycnothallum* C. Müll. Later, Par. (1909) published a new species, *Anomodon rotundatus* Par. & Broth., and reported *Thuidium aciculum* Broth. based on Courtois and Henry's collection from Qiudong, Jiangsu province. Brotherus (1929) reported 10 genera and 28 species of Chinese Thuidiaceae from southwestern China. Among them *Leptocladium* Broth. was described as a new genus; six new species of *Leptopterigynandrum* C. Müll., *Claopodium* (Lesq. & Jam.) Ren. & Card., and *Haplocladium* (C. Müll.) C. Müll. were described. This was the first comprehensive treatment concerning Chinese Thuidiaceae. Reimers (1931) published the book entitled "Beiträge zur Mossflora China, I" after examining the collections of R.-C. Ching, S.-S. Sin, K.-K. Wang, and P. Klautke et al. from Mt. Magan, Zhejiang province, and Mt. Yao, Guangxi province. Six species of *Thuidium* were reported. Bartram (1935) reported *Claopodium assurgens* (Sull. & Lesq.) Card. and *Haplocladium capillatum* (Mitt.) Reim. based on S.-Y. Cheo's collections from Guangxi and H. H. Chung's collections from Fujian (Fukien) provinces. Although C.-W. Wang (1935) published a list of 1064 Chinese mosses from 18 provinces, he included only a few genera and species of the Thuidiaceae. *Hylocomiopsis ovicarpa* (Besch.) Card. was reported from Jilin province by Koponen et al. (1983). We are unable to confirm the specimen at this time. Touw (2001b) appeared too late to be considered for our study.

1. Plants slender; costae double, short, weak or absent 2
1. Plants rather robust, rarely slender; costa single, mostly up to the upper part of leaf, rarely excurrent 4
   2. Leaf cells unipapillose or multipapillose .... 9. *Heterocladium*
   2. Leaf cells smooth 3
3. Leaves oblong-ovate; leaf cells nearly linear .... 10. *Leptocladium*
3. Leaves ovate to broadly ovate; leaf cells rhombic to rhomboidal .... 11. *Leptopterigynandrum*
   4. Plants usually robust; growing in swamps, alpine forests or sunny coniferous forests 5
   4. Plants rather slender; growing in moist forest ground, grasslands, decaying logs or rock surfaces 8
5. Plants regularly and pinnately branched, silver-shining; leaves multiplicate; leaf cells smooth .... 2. *Actinothuidium*
5. Plants remotely and pinnately branched, without shining color; leaves little-plicate; leaf cells papillose...............6
   6. Plants often growing on sunny and sometimes dry forest ground; leaf cell walls strongly thickened, with conspicuous pores .... 1. *Abietinella*
   6. Plants usually growing on moist forest ground or swamps; leaf cell walls not strongly thickened....................7
7. Leaf bases widely cordate; leaf cells grossly spinulosely papillose .... 4. *Bryonoguchia*
7. Leaf bases ovate; leaf cells with a single rounded papilla .... 8. *Helodium*
   8. Plants 2–3 pinnately branched .... 14. *Thuidium*
   8. Plants irregularly pinnately branched or simple pinnately branched 9
9. Dioicous or autoicous; setae setulose; calyptrae campanulate, covered with long or short hairs, lobed at base......... .... 12. *Pelekium*
9. Mostly autoicous; setae smooth or papillose; calyptrae usually cucullate, smooth, unlobed at base........................ .... 6. *Cyrto-hypnum*
   10. Plants usually brownish green, rigid when dry; paraphyllia abundant on stems 11
   10. Plants usually yellowish green or dark green, delicate when dry; paraphyllia few on stems 12
11. Leaf cells unipapillose; cilia absent .... 3. *Boulaya*
11. Leaf cells unipapillose or multipapillose; cilia 2–3 .... 13. *Rauiella*
   12. Leaf cells opaque, unipapillose or multipapillose .... 5. *Claopodium*
   12. Leaf cells rather hyaline, unipapillose .... 7. *Haplocladium*

1. ***Abietinella*** C. Müll. 山羽藓属 shan-yu xian shu
Nuovo Giorn. Bot. Ital., n. ser. 3: 115. 1896.

Plants variable in size, depending on habitats, usually rather robust, up to 10 cm long, stout, yellowish green, not shining, in tufted lax cushions. *Stems* ascending, distinctly pinnately branched; paraphyllia dense, lanceolate or deeply lobed to filamentous; branches 1 cm long, apices usually obtuse, sometimes caudate; central strand present. *Stem leaves* and branch leaves differentiated; stem leaves cordate-ovate at base, gradually narrowed to a long lanceolate apex, multiplicate; margins grossly dentate; costa stout, vanishing at leaf apex; leaf cells oval-hexagonal to rhomboidal, coarsely unipapillose; branch leaves rather small, oblong-ovate, apices short. *Dioicous. Inner perichaetial leaves* ovate-lanceolate, deeply multiplicate. *Setae* slender, smooth; capsules cylindrical, arched; annuli present; opercula conic; peristome double; exostome teeth lanceolate, trabeculate, bordered; endostome segments nearly as long as the teeth; cilia absent. *Calyptrae* cucullate. *Spores* finely papillose.

Species of this genus usually grow in calcareous mountain regions under sunny, moist ground often in large patches or cushions. Based on its pinnately branched plants and peristome features, Crum and Anderson (1981) and Noguchi (1991) treated the species of *Abietinella* under *Thuidium*. However, in our opinion, the thick cushion habits, thick-walled leaf cells, and the absence of cilia separate *Abietinella* from *Thuidium*. It is close to *Actinothuidium* systematically.

The genus consists of two species. C.-Y. Yang (1936) first reported *Abietinella abietina* (Hedw.) Fleisch. from China followed by Potier de la Varde (1937) and C. Gao (1977). P.-C. Chen et al. (1978) reported *A. hystricosa* (Mitt.) Broth., from Guangdong Province, but we are unable to confirm this record.

1. Leaves broadly ovate, with a short lanceolate apex; costa papillose on back .... 1. *A. abietina*
1. Leaves ovate-lanceolate, with slender lanceolate apex; costa smooth on back .... 2. *A. hystricosa*

1. ***Abietinella abietina*** (Hedw.) Fleisch. 山羽藓 shan-yu xian
Musci Buitenzorg 4: 1497. 1922. *Thuidium abietinum* (Hedw.) Schimp. in B.S.G., Bryol. Eur. 5: 165. 485 (fasc. 49–51 Mon. 9. 5). 1852. *Hypnum abietinum* Hedw., Sp. Musc. Frond. 353. 1801. Type. Europe.
Abietinella giraldii C. Müll., Nuovo Giorn. Bot. Ital. n. ser. 3: 115. 1896. Syntypes. China: Shaanxi (Schen-si), Huo-gia-zien, 20 Sept. 1890, *Giraldi s.n.*; Lun-san, Jun. 1891, *Giraldi s.n.*; Sche-liu-san, Aug. 1893, *Giraldi s.n.*; Si-ku-tziu-san, July 1894, *Giraldi 919* (syntypes BM).

Plants robust, up to 10 cm long, rigid, yellowish or dark green above, brownish at base, in densely or laxly interwoven cushions. *Stems* regularly pinnately branched, ascending or erect; central strand present; paraphyllia numerous, foliose or filamentous, branched, apical cells multipapillose. *Stem leaves* broadly ovate at base, gradually narrowed to a lanceolate apex, deeply plicate, ca. 1–5 mm × 1 mm; leaf margins nearly plane or slightly revolute, serrate; costa stout, reaching 3/4 the leaf length, papillose at back; median leaf cells linear, mostly curved, 10–12 μm × 6–8 μm, thick-walled, unipapillose; apical leaf cells linear or oblong-oval, smooth; branch leaves concave, appressed when dry, erect-patent when moist, lanceolate to widely ovate, with a short acumen; margins entire or finely serrulate; costa reaching the upper portion of leaves; leaf cells oval to nearly quadrate, unipapillose. *Dioicous. Perichaetial leaves* ovate or widely ovate, gradually or suddenly narrowed to a lanceolate apex, costa absent. *Capsules* not seen.

Illustrations: P.-C. Wu (ed.) 2001 (Pl. 106, figs. 1–10).
Habitat: on moist, nutritious coniferous forest floors or on dry calcareous rock surfaces.
Distribution: China, Japan, Russia, and North America.

Chinese specimens examined: BEIJING: *Licent 218* (BM). GANSU: Huo-shan (Mt.), *Meyer s.n.* (NY); Tianzhu Co., *S.-Z. Hu 148* (MO); Wu-wei Co., *S.-Z. Hu 2096a* (MO). HEBEI: Mt. Dongling, *C.-Y. Yang 206* (BM); Mt. Xiaowutai, *Chao C6* (BM); Mt. Xiling, *Q.-W. Wang 1418* (MO, PE). HEILONGJIANG: Ning-an Co., *Z.-Z. Lin 44* (MO); Yi-chun City, Dai-ling, *Z.-H. Zhang 4143* (NY). HENAN: Song shan (Mt.), *J.-X. Lou s.n.* (MO, PE). HUBEI: Mt. Shennongjia, *P.-C. Wu 508* (MO, PE). JILIN: An-tu Co., *Koponen 36830* (NY); Mt. Changbai, *B.-D. Xin 00221* (IFSBH, MO). NEI MONGOL: Mt. Daqing, *Z.-G. Tong 802* (MO); Mt. Helan, *Z.-G. Tong 1285* (MO). QINGHAI: Yushu Co., *B.-C. Tan 951641* (MO), *B.-C. Tan 952018* (NY). SHAANXI: In-kio-po, *Giraldi 1775* (BM); Mt. Taibai, *X.-*

J. Li 640 (MO). SICHUAN: A-ba Co., *S. He 30990* (MO); Ba-tang Co., *S. He 31326* (MO); Dao-cheng Co., *S. He 31715a* (MO); Hong-yuan Co., *S. He 30902* (MO); Li Co., *Allen 6756A* (MO); Lu-huo Co., *Y. Jia J02911* (MO, PE); Ma-er-kang Co., *S. He 31272* (MO); Mao-wen Co., *Allen 7352* (MO); Nan-ping Co., *S. He 30058* (MO, NY); Ruo-er-gai Co., *S. He 30871b* (MO); Song-pan Co., *S. He 30578* (MO); Tian-quan Co., *C. Gao 18248* (IFSBH, MO); Yan-yuan (Yenyuen) Co., *Handel-Mazzetti 2200* (BM). TAIWAN: Ilan Co., *C.-I. Peng 83-26* (MO). XINJIANG: Mt. Altai, *R.-C. Qin 2419* (MO, PE); Bu-er-jin Co., *R.-C. Qin 2142* (MO, PE); He-jing Co., *S. He 951153* (MO); Ku-che Co., *S. He 951213* (MO); Ni-le-ke Co., *S. He 951124* (MO); Sha-wan Co., *R.-C. Qin 3671* (MO, PE); Urumqi Co., *S. He 951064* (MO); Wu-su Co., *S. He 951119* (MO). YUNNAN: Da-li Co., *Redfearn et al. 1639* (MO, NY, PC).

This species is widely distributed in northern and southwestern China, but rare in central China. In the field, it is very much like *Actinothuidium hookeri* (Mitt.) Broth. due to their similar branching patterns.

2. ***Abietinella hystricosa*** (Mitt.) Broth. 美丽山羽藓 mei-li shan-yu xian
Nat. Pflanzenfam. (ed. 2), 11: 327. 1925. *Thuidium hystricosum* Mitt. in Seemann, J. Bot. 1: 356. 1863. Type. England.

Plants usually fairly robust to robust, up to 10 cm long, rarely slender, yellowish green or dark green, brownish at lower portion, in densely tufted cushions. *Stems* ascending or erect, 0.65 mm in diameter, regularly pinnately branched; branches rigid, apex obtuse, 7–15 mm long; central strand present; paraphyllia numerous, foliose or filamentous, branched, smooth. *Stem leaves* ovate-oblong, gradually narrowed to a lanceolate acumen, weakly plicate, often secund, ca. 3.0 mm × 1.0–1.5 mm; leaf margins finely serrulate; costa up to 3/4 the leaf length, coarsely papillose on back; leaf cells narrowly oval, smooth or unipapillose; branch leaves narrowly ovate, concave, slightly plicate, sometimes indistinctly multipapillose; median cells hexagonal or rhomboidal, smooth, rarely minutely unipapillose. *Sporophytes* not seen.

Illustrations: Watanabe 1972 (Pl. 28, figs. 19–27).
Habitat: on base of trees, rocks or soil.
Distribution: China, Japan, and Europe.

Chinese specimens examined: GANSU (Kansu): Lanzhou (Lan Tcheon), *Licent 252* (BM, MO). HEBEI: Wei-xian Co., Mt. Xiaowutai, *H.-B. Cui 2* (MO, PE); She-xian Co., *H.-B. Cui 231* (MO, PE). SHAANXI: Mt. Lun-san, *Giraldi s.n.* (BM); Mt. Qinling, *Hugh 111* (BM). SICHUAN: De-ge Co., *J.-L. 2043* (MO, PE); Song-pan Co., *Allen 7290* (MO). XINJIANG: Mt. Tianshan, *K.-J. Guan 1492*, *1857* (MO, PE).

2. ***Actinothuidium*** (Besch.) Broth. 锦丝藓属 jin-si xian shu
Nat. Pflanzenfam. I(3): 1019. 1908. *Thuidium* sect. *Actinothuidium* Besch., Ann. Sci. Nat., Bot., ser. 7, 15: 82. 1891.

Plants very stout, robust, rarely slender, yellowish green to brownish green, somewhat shining, in tufted or large cushions or mats. *Primary stems* erect, simple, sometimes forked; secondary stems densely pinnately branched; paraphyllia multibranched, dense on stems and branches; rhizoids reddish brown at lower portion of stems, rarely at the apex of branches. *Stem leaves* and branch leaves more or less polymorphic; stem leaves appressed when dry, widely cordate, gradually narrowed to a lanceolate acumen, deeply plicate, secund above; margins slightly revolute above, remotely dentate; costa thick, vanishing near leaf apex; leaf cells oblong-rhomboidal or oblong-hexagonal, slightly papillose at upper corners; thin-walled, hyaline; branch leaves ovate, with short lanceolate apex. *Dioicous*. *Setae* slender, up to 5 cm long; capsules cylindrical, inclined or horizontal, slightly curved; annuli differentiated, caducous when mature; opercula conic; peristome double; exostome teeth narrowly lanceolate, trabeculate below; endostome segments nearly as long as the teeth, lanceolate, pale yellowish, perforated, smooth; basal membrane high; cilia 3, shorter than the segments, smooth, nodulose. *Calyptrae* not seen. *Spores* not seen.

Actinothuidium has the largest plants in the Thuidiaceae. It is characterized by having erect plants, regularly pinnately branched stems, widely ovate or triangularly ovate leaf bases, papillose branch leaf cells with papillae at upper corners, and the cylindrical capsules. The genus is very close to *Hylocomiopsis* Card.

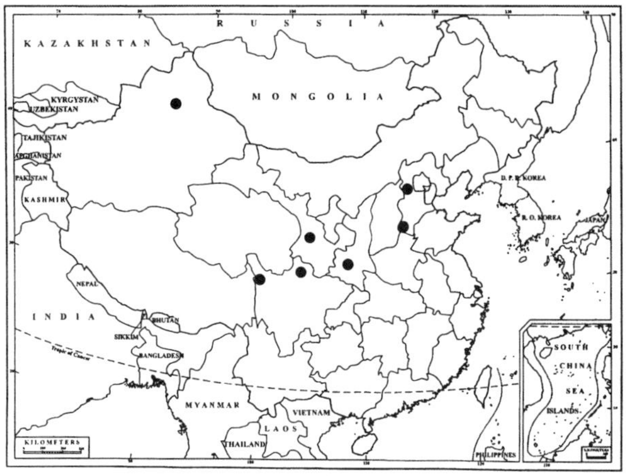

2. *Abietinella hystricosa*

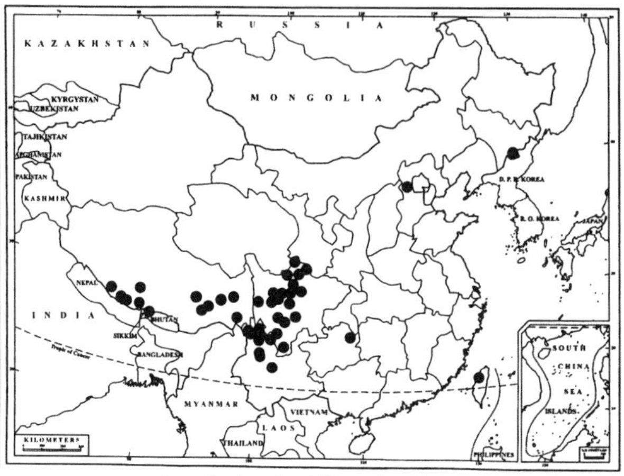

1. *Actinothuidium hookeri*

Actinothuidium is mainly distributed in Himalayan mountain regions and found in various places of China. The genus used to include two species with *Actinothuidium sikkimense* Warnst. endemic to Himalayan regions. Recently, Watanabe (1991) reduced the latter species to synonymy of *A. hookeri* (Mitt.) Broth. This species occurs in northeastern, northwestern and southwestern China as well as Taiwan alpine forest regions.

1. ***Actinothuidium hookeri*** (Mitt.) Broth. 锦丝藓 jin-si xian

Nat. Pflanzenfam. I(3): 1019. 1908. *Thuidium hookeri* (Mitt.) Jaeg., Ber. Thätigk. St. Gallischen. Naturwiss. Ges. 1876–77: 264. 1878. *Leskea hookeri* Mitt., J. Linn. Soc., Bot., Suppl. 1: 132. 1859. Syntypes. Nepal: *Wallich s.n.*; India: Sikkim, *J. D. Hooker 1104, 1105*.

Actinothuidium sikkimense Warnst., Hedwigia 57: 112. 1915.

Plants robust to very robust, up to 15 cm long, yellowish green above, light brown at base, in large interwoven cushions. *Stems* erect or ascendant, regularly pinnately branched; branches 2.0–2.5 cm long; central strand absent; paraphyllia dense on stems, few on branches, foliose or branched. *Stem leaves* gradually narrowed from a broadly ovate to triangularly ovate base to a lanceolate apex, sometimes deeply plicate, ca. 3.0 mm × 1.5 mm; margins grossly serrate, rarely entire; costa stout, ending at leaf middle, base with a few paraphyllia; median leaf cells narrowly rhomboidal, ca. 25–35 µm × 8 µm, thin-walled, smooth or with 1–2 papillae or unipapillose at front corners; apical leaf cells linear, smooth; basal leaf cells strongly pitted; branch leaves densely appressed when dry, erect-spreading when moist, ovate to widely ovate; margins dentate; costa reaching 4/5 the leaf length. *Sporophytic features* the same as those of the genus.

Illustrations: Plate 448, figs. 1–11.

Habitat: usually on forest ground or rock surfaces in alpine mountain regions and in loose cushions covering large pieces of forest ground; alt. 800–4300 m.

Distribution: China, Bhutan, Nepal, India, and Myanmar.

Chinese specimens examined: GUIZHOU: Fanjingshan (Mt.), *Crosby 15833* (MO). HEBEI: Mt. Xiao-wu-tai, *Q.-W. Wang 1042* (PE). JILIN: Mt. Chang-bai., *Invest. Group of the Ministry of Forestry s.n.* (PE). SICHUAN: A-ba Co., *S. He 31025* (MO); Ba-tang Co., *S. He 31602* (MO); Da-jin Co., *Z.-T. Guan 4021* (PE); Dao-fu Co., *Z.-T. Guan 620* (PE); Jiu-long Co., *Z.-T. Guan 552* (PE); Li Co., *Allen 6756* (MO); Lu-ding Co., *Y. Jia 1827* (PE); Ma-er-kang Co., *S. He 31288* (MO), *Y. Jia 2922* (PE); Mao-wen Co., *Allen 7335* (MO); Mi-yi Co., *Y.-D. Zhu 20508* (PE); Mian-ning Co., *S.-F. Zhu 20401* (MO); Mu-li Co., *Forest Ministry Team 23* (PE); Nan-ping Co., *S. He 30127* (MO); Mt. Omei, *Allen 6475* (MO); Ruo-er-gai Co., *S. He 30760* (MO); Song-pan Co., *S. He 30532b* (MO); Xiao-jin Co., *P.-X. Li 38* (PE); Ya-jiang Co., *S. He 31985* (MO). TAIWAN: Miao-li Co., *Shevock 18089* (MO). XIZANG: Ba-su Co., *Y. Jia 5421* (PE); Bo-mi Co., *Y. Jia 5748* (PE); Cha-yu Co., *Z.-C. Ni 65a* (PE); Ding-jie Co., *Z.-C. Ni 102* (MO, PE); Ji-long Co., *S.-K. Chen 310* (HKAS, PE); Lin-zhi Co., *Y. Jia 5957* (PE); Mi-lin Co., *K.-Y. Lang 611* (MO); Mt. Everest, *X.-G. Wang 29* (PE); Nie-la-mu Co., *K.-Y. Lang 1171* (PE); Ri-ge-ce Co., *W.-L. Chen 766* (PE); Ya-dong Co., *Xizang Team 7726* (MO). YUNNAN: Between Landsang-djiang and Lu-djiang, *Handel-Mazzetti 8446* (BM); Chu-xiong Co., *Z.-T. Guan 7388* (PE); Da-li Co., *Redfearn et al. 1193* (MO); Gong-shan Co., *Q.-W. Wang 6725* (PE); Li-jiang Co., *Q.-W. Wang 4016* (PE); Mt. Jade Dragon, *Croat 77627* (MO); Ning-lang Co., *Z.-T. Liu 4948b* (PE); Yang-bi Co., *Redfearn & Y.-G. Su 765* (MO, NY, PC); Zhong-dian Co., *H.-F. Xie 1940* (PE).

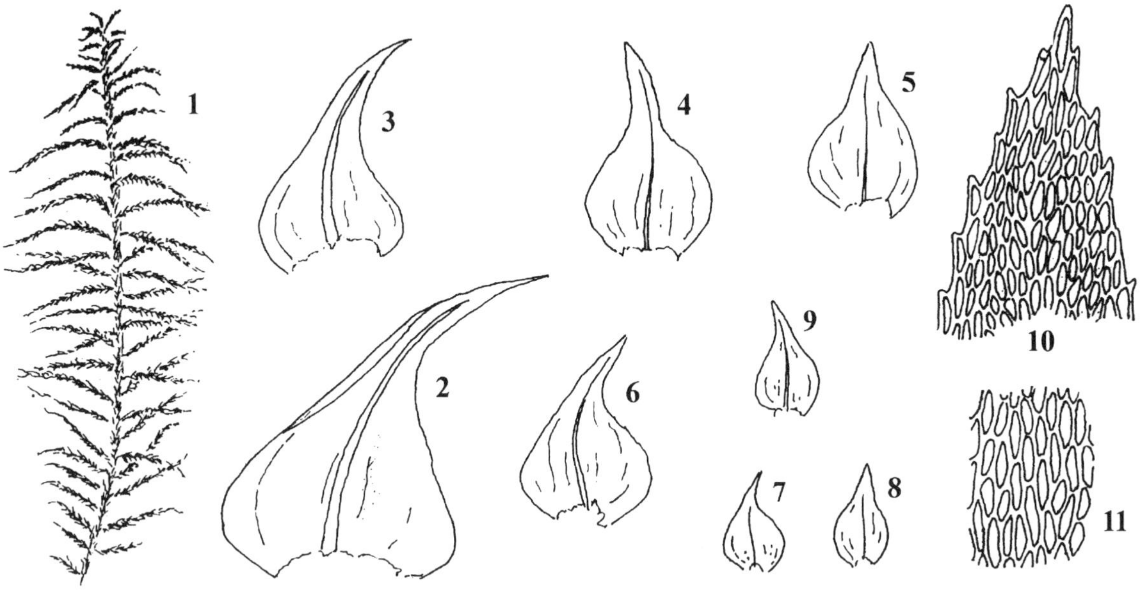

Plate 448. *Actinothuidium hookeri*. 1, habit (× 1). 2, stem leaf (× 22). 3–6, branch leaves (× 22). 7–9, branchlet leaves (× 22). 10, apical leaf cells (× 210). 11, basal juxtacostal leaf cells (× 210). (1–11, drawn from *E. E. Maire s.n.*, PE.)

3. ***Boulaya*** Card. 虫毛藓属 chong-mao xian shu
Rev. Bryol. 39: 1. 1912.

Plants rather robust, green to yellowish green, brownish green when old, in loose mats. *Stems* creeping, up to 10 cm long, regularly pinnately branched; branches ca. 5 mm long; paraphyllia numerous, foliose to filamentous, branched and papillose; central strand absent. *Stem leaves* and branch leaves differentiated; stem leaves appressed when dry, erect-spreading when moist, widely cordate, toward a lanceolate, secund apex; leaf margins serrate; costa single, excurrent; median leaf cells rhomboidal to ovate, ca. 6–8 µm × 15.0 µm, thick-walled, multipapillose or smooth; branch leaves ovate to oblong-ovate, gradually tapering to a short apex; margins entire or serrulate; costa extending about 3/4 the leaf length. *Dioicous*. *Perichaetial leaves* sheathing at base, with a narrowly lanceolate apex. *Setae* slender, reddish brown to yellowish brown, smooth; capsules ovoid, erect, mostly reddish brown; opercula beaked; exostome teeth lanceolate, yellowish brown, hyaline above, finely papillose; endostome segment pale yellow, about 1/2 the exostome length; cilia absent. *Calyptrae* cucullate. *Spores* ovoid to spherical, finely papillose.

Boulaya is endemic to East Asia with only one species. *Boulaya mittenii* (Broth.) Card. was first reported from Mt. Changbai, Jilin province (C. Gao 1977). We recently found it from Xizang. It shows a discontinued distribution from Mt. Changbai to the Himalayan region. The genus *Boulaya* is closely related to *Abietinella*, but differs from the latter by the weakly plicate stem leaves, multipapillose median leaf cells, and the erect capsules.

1. ***Boulaya mittenii*** (Broth.) Card. 虫毛藓 chong-mao xian
Rev. Bryol. 39: 2. 1912. *Thuidium mittenii* Broth., Hedwigia 38: 246. 1899. Type. Japan. Hokkaido, *Faurie 115*.
Boulaya latifolia Okam., J. Coll. Sci. Imp. Univ. Tokyo 36: 26. 1915.
Boulaya mittenii (Broth.) Card. var. *attenuata* Nog., J. Jap. Bot. 1: 412. 1932.

The species description is the same as that of the genus.

Illustrations: Plate 449, figs. 1–12.
Habitat: commonly on tree trunks and rock surfaces in forests; alt. 1000–3100 m.
Distribution: China, Japan, Korea, and Russian Far East.

Chinese specimens examined: HEILONGJIANG: Ning-an Co., *X.-L. Zhu 254* (PE). NEI MONGOL: Ya-ke-shi City, *X.-L. Bai 2445* (HIMC). XIZANG: Ya-dong Co., *Z.-X. Tan 21* (IFSBH, PE).

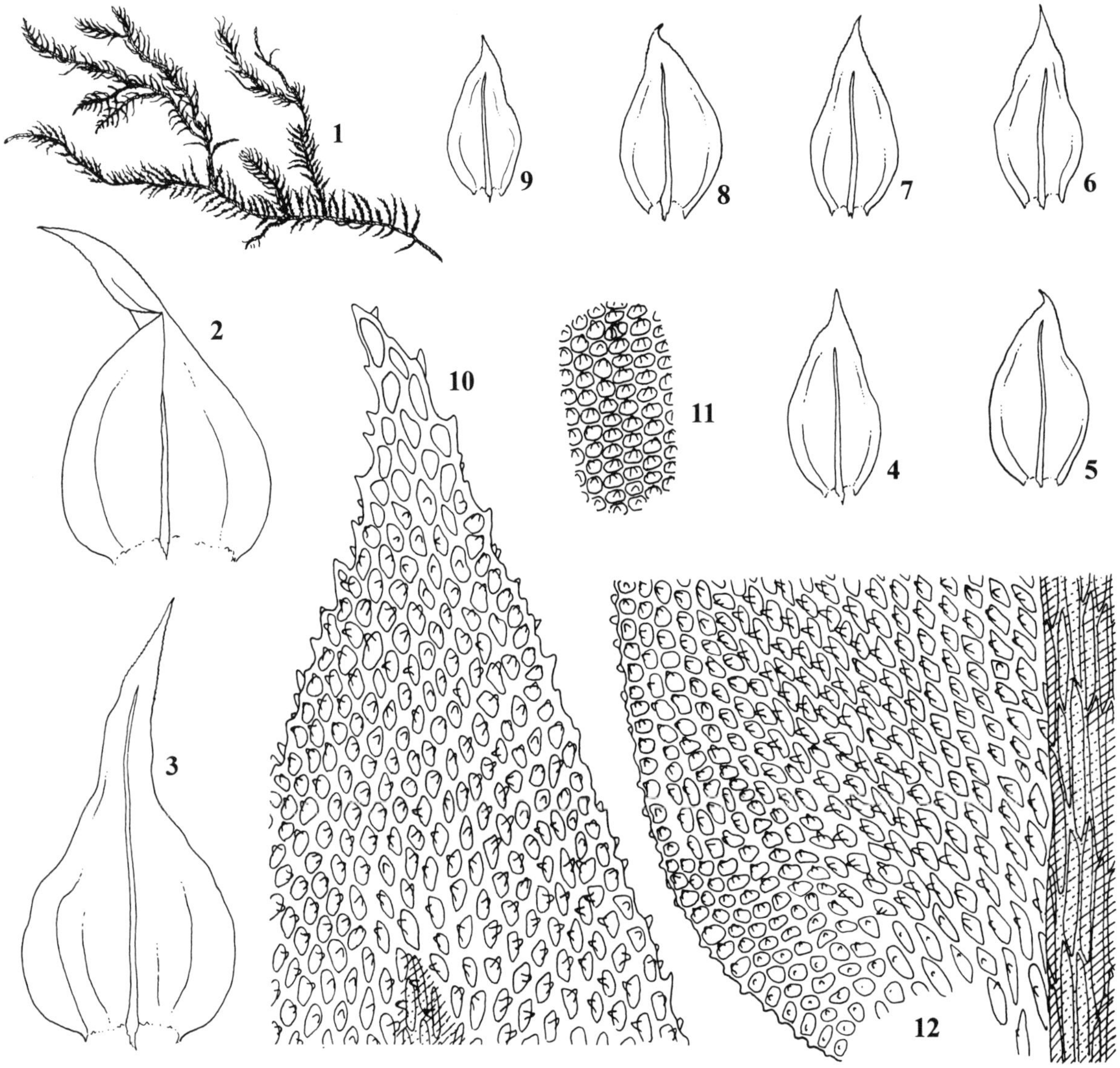

Plate 449. *Boulaya mittenii.* 1, habit (× 1). 2–3, stem leaves (× 24). 4–8, branch leaves (× 24). 9, branchlet leaf (× 24). 10, apical leaf cells (× 365). 11, median leaf cells (× 365). 12, basal leaf cells (× 365). (1–12, drawn from *Z.-X. Tan 21*, PE.)

4. ***Bryonoguchia*** Iwats. & Inoue 毛羽藓属 mao-yu xian shu
Misc. Bryol. Lichenol. 5(7): 107. 1970.
Bryochenea C. Gao & G.-C. Chang, Bull. Bot. Res. Harbin 2 (4): 113. 1982.

Plants robust, 5–10 cm long, yellowish green to light green above, yellowish brown at low portion, in loosely interwoven patches. *Stems* creeping, ca. 0.7 mm in diameter, regularly bipinnately to tripinnately

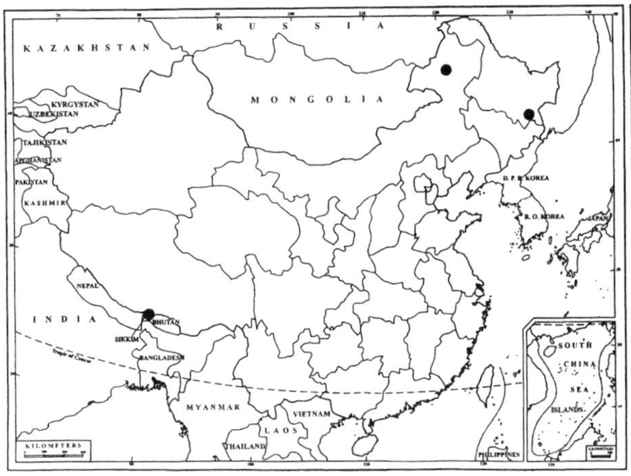

1. *Boulaya mittenii*

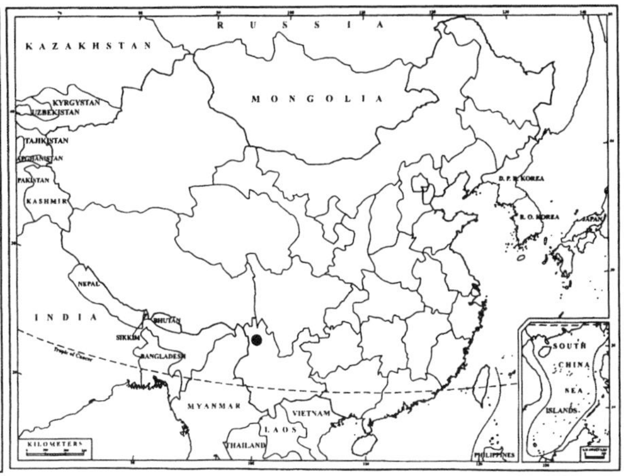

1. *Bryonoguchia brevifolia*

branched; branches flagelliform, with dense rhizoids; central strand present; paraphyllia foliose or filamentous, mostly branched, densely on stems and branches. *Stem leaves* and branch leaves differentiated; stem leaves flexuous when dry, erect-patent when moist, ovate or cordate-ovate, gradually narrowed to a lanceolate apex, 1.5–2.0 mm long; margins coarsely dentate, papillose at base; costa stout, ending at upper portion of leaves, rarely excurrent, with a few papillae or paraphyllia on back; median leaf cells rhomboidal or rectangular, 5–8 µm × 15–25 µm, thick-walled, unipapillose; basal median cells smooth; branch leaves concave, triangularly ovate, with short lanceolate apex, 0.6–0.9 mm long; margins grossly serrate; costa ending at the middle of leaf; median leaf cells usually rhomboidal, setulosely unipapillose; apical cells 1–3 papillae. *Dioicous*. *Perichaetial leaves* oblong at base, suddenly narrowed to a lanceolate acumen, upper margins long ciliate. *Setae* reddish brown to yellowish brown, 3–5 cm long, smooth; capsules cylindrical, horizontal, reddish brown to yellowish brown, 3–4 mm long; opercula conic; exostome teeth lanceolate 0.5–0.6 mm long, yellowish brown to reddish brown, trabeculate at base, finely papillose above; endostome segments yellowish brown, as long as the teeth; membrane 1/2 the segment height, minutely papillose; cilia hyaline, 2–3. *Calyptrae* cucullate, smooth. *Spores* rounded, 10–15 µm in diameter, minutely papillose.

Bryonoguchia contains 2 species in the world, commonly on rock surfaces, humus, decaying logs or tree trunks in high altitude forests. *Bryonoguchia molkenboeri* (Sande Lac.) Iwats. & Inoue was first reported from northeastern China as *Tetracladium molkenboeri* (Sande Lac.) Fleisch. (C. Gao et al. 1977). C. Gao and G.-C. Zhang (1982) published the genus, *Bryochenea* with *B. ciliata* as its type species and included another species, *B. sachalinensis*. Most recently, P.-C. Wu and Y. Jia (2000) reduced *Bryochenea ciliata* to synonymy of *Bryonoguchia molkenboeri* and transferred *Bryochenea sachalinensis* to *Helodium*. The genus *Bryochenea*, therefore, is treated as a synonym of *Bryonoguchia*. In this study, we recognize two species, *Bryonoguchia molkenboeri* and a recently published new species *B. brevifolia* S.-Y. Zeng (S.-Y. Zeng 1991).

1. Stem leaves broadly cordate, with short apex, multiplicate; margins irregularly grossly dentate or setulosely serrate; leaf cells mostly linear, smooth .... 1. *B. brevifolia*
1. Stem leaves broadly ovate, with long lanceolate apex, somewhat plicate; margins entire, papillose; leaf cells mostly oblong-oval, setulosely unipapillose .... 2. *B. molkenboeri*

1. ***Bryonoguchia brevifolia*** S.-Y. Zeng 短叶毛羽藓 duan-ye mao-yu xian
Acta Bot. Yunnan. 13 (4): 377. 1991. Type. China: Yunnan, Wei-xi Co., Weidong, alt. 3000 m, *D.-C. Zhang 743a* (holotype HKAS).

Plants large, up to 8 cm long, rather rigid, yellowish green, not shining, in tufted and in interwoven patches. *Stems* often forkedly pinnately branched; branches 1.0–1.2 cm long, apex caudate; paraphyllia irregularly forkedly branched, dense on stems. *Stem leaves* suddenly narrowed from widely cordate base to a short apex, multiplicate;

margins irregularly grossly serrate; costa stout, reaching leaf apex; leaf cells linear, usually smooth; branch leaves ovate to triangularly ovate, other characters similar to those of stem leaves. *Sporophytes* not seen.

Illustrations: Plate 450, figs. 1–11.
Habitat: on wet rock surfaces.
Distribution: endemic to China.
Specimens examined: see the type citation above.

This species is different from *B. molkenboeri* by having multiplicate and broad stem leaves, grossly serrate leaf margins, and linear and smooth leaf cells. These characters make this species look like Hylocomiaceae.

2. ***Bryonoguchia molkenboeri*** (Sande Lac.) Iwats. & Inoue 毛羽藓 mao-yu xian
Musc. Bryol. Lichénol. 5(7): 107. 1970. *Thuidium molkenboeri* Sande Lac., Ann. Mus. Bot. Lugduno-Batavum 2: 298. 1866. Type. Japan.
Bryochenea ciliata C. Gao & G.-C. Chang, Bull. Bot. Res. 2(4): 15. 1982. Type. China: Hubei, Wu-feng Co., *Y.-S. Fu 5023* (holotype IFSBH).
Thuidium komarovii Savicz-Ljubitskaya, Trudy Imp. S.-Peterburgsk. Bot. Sada 39: 153. 1923.

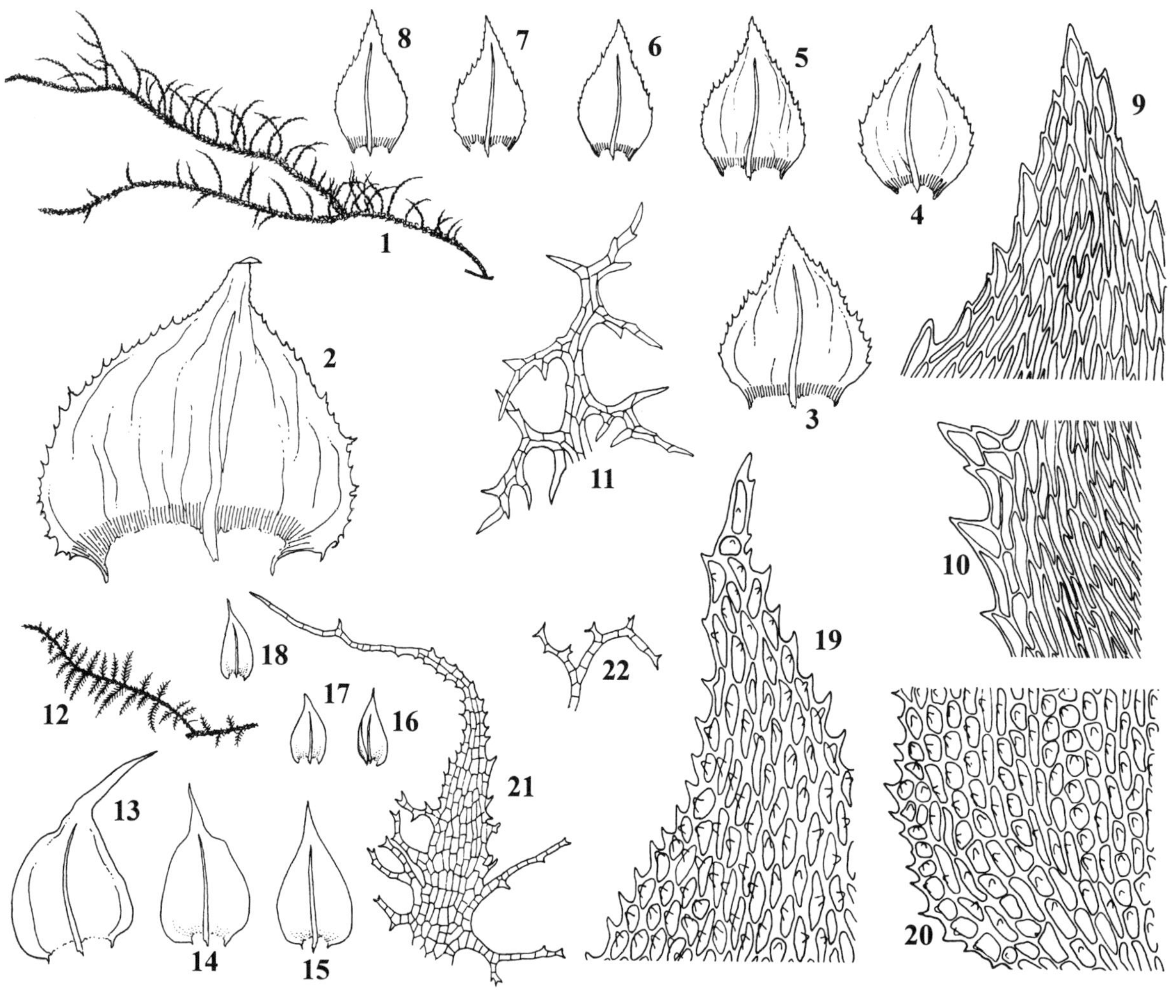

Plate 450. 1–11, *Bryonoguchia brevifolia*. 1, habit (× 1). 2, stem leaf (× 38). 3–5, branch leaves (× 38). 6–8. branchlet leaves (× 38). 9, apical leaf cells (× 365). 10, basal marginal leaf cells (× 365). 11, paraphyllia (× 185). 12–22, *Bryonoguchia molkenboeri*. 12, habit (× 1). 13, stem leaf (× 28). 14–15, branch leaves (× 28). 16–18, branchlet leaves (× 28). 19, apical leaf cells (× 365). 20, basal leaf cells (× 365). 21–22, paraphyllia (× 140). (1–11, drawn from *D.-C. Zhang 743a*, HKAS; 12–22, from *B.-D. Xin 375*, PE.)

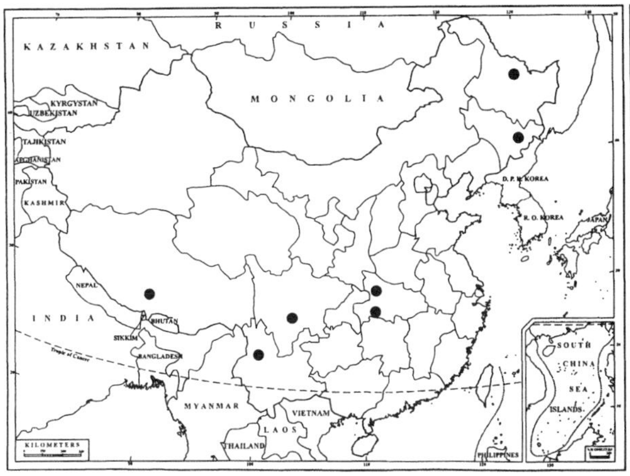

2. *Bryonoguchia molkenboeri*

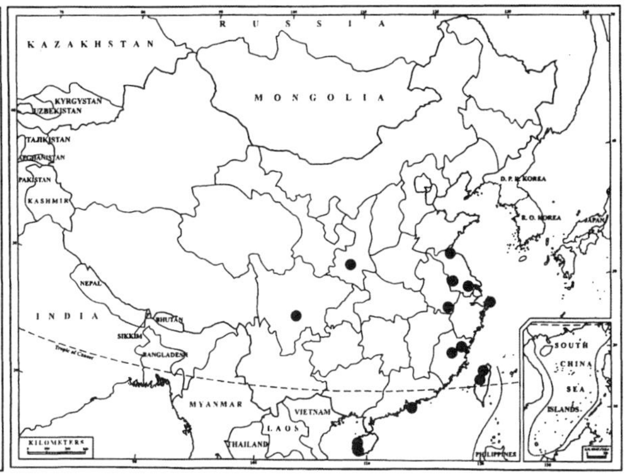

1. *Claopodium aciculum*

Plants rather robust, up to 10 cm long, yellowish green to light green above, yellowish brown below. *Stems* regularly bi- to tripinnately branched; branches up to 10 mm long, flagelliform, with a few rhizoids; central strand present; paraphyllia foliose or filamentous, branched. *Stem leaves* incurved or crisped when dry; erect-patent when moist, from widely ovate base or cordate base suddenly narrowed to a lanceolate apex, ca. 2 mm long, usually lightly longitudinal plicate; margins serrate; costa thick, reaching near leaf apex, rarely excurrent, slightly papillose on back; median leaf cells rhomboidal to rectangular, ca. 20 µm × 6 µm, unipapillose; branch leaves triangularly ovate, with a short apex, ca. 0.5–0.8 mm long; margins serrate; costa vanishing above leaf middle; median leaf cells rhomboidal, setulosely unipapillose, apical cells 1–3, papillose. *Dioicous*. *Perichaetial leaves* lanceolate, from oblong-ovate base suddenly narrowed to a lanceolate apex. *Setae* reddish to yellowish brown, 3–5 cm long; capsules cylindrical, horizontal; opercula long beaked; peristome double; exostome teeth yellowish to reddish brown, ca. 0.5 mm long; endostome segments nearly as long as exostome teeth, lightly yellowish, papillose; cilia 2–3, hyaline, finely papillose. *Calyptrae* cucullate, smooth. *Spores* nearly rounded, ca. 15 µm in diameter, papillose.

Illustrations: Plate 450, figs. 12–22.
Habitat: on rotten logs, rock surfaces or moist soil in high elevation forests; alt. 3300 m.
Distribution: China, Japan, Korea, and Russia.

Chinese specimens examined: HUBEI: Mt. Shennongjia, *Luteyn 682* (MO, NY, PE). HEILONGJIANG: Yi-chun City, Wu-ying, *Vitt 34873* (ALTA, MO). JILIN: An-tu Co., *B.-D. Xin 229, 245, 262, 375* (PE), *Koponen 37274, 37283* (H, PE). SICHUAN: E-mei Co., *Allen 6487* (MO). XIZANG: Rekaze (X.-J. Li et al. 1985). YUNNAN: Yang-bi Co., *Redfearn & Y.-G. Su 569* (MO).

5. ***Claopodium*** (Lesq. & Jam.) Ren. & Card. 麻羽藓属 ma-yu xian shu
Rev. Bryol. 20: 16. 1893. *Hypnum* subg. *Claopodium* Lesq. & Jam., Man. Mosses N. America 317, 327. 1884.

Plants small to medium-sized, sometimes delicate light green or yellowish green, brownish green when old, not shining, forming lax interwoven mats. *Stems* creeping or ascending, pinnately or irregularly branched, smooth or rarely papillose; central strand absent; paraphyllia mostly lacking. *Stem leaves* and branch leaves slightly to clearly differentiated; stem leaves loosely arranged, usually circinate when dry; ascending when moist, prolonged into a lanceolate or filiform acumen from an ovate or triangularly ovate base, rarely flexuous; leaf margins erect or slightly involute, serrate or serrulate; costa rigid, excurrent or vanishing below leaf apex, smooth or papillose on back; leaf cells rhomboidal, hexagonal or oblong ovate, opaque, unipapillose or finely multipapillose; marginal cells rather long and smooth; basal cells long and hyaline; branch leaves rather small, short, rarely revolute or flexuous above. *Dioicous*. *Inner perichaetial leaves* ovate-lanceolate prolonged into a narrow acumen. *Setae* slender, smooth or roughened; capsules oblong ovoid, brown, inclined to pendulous; annuli differentiated; opercula conic with beaks; peristome double; exostome teeth lightly yellow, bordered, striate and trabeculate; endostome smooth or finely papillose, the segments as long as the teeth, lanceolate, keeled; cilia 2–3, nodulose. *Calyptrae* cucullate, aciliate.

Claopodium contains about 8 species in the world mostly occuring in East Asia, with two species from North America, and one from Europe. Species usually grow on tree trunks or rocks, rarely on soil. Seven species are treated for China in this study. *Claopodium crispifolium* (Hook.) Ren. & Card. reported from Zhejiang province by R.-L. Hong & R.-L. Hu (1984) and Y.-M. Fang & Koponen (2001) cannot be confirmed.

1. Stems and branches papillose; leaf cells multipapillose .... 5. *C. pellucinerve*
1. Stems and branches not papillose; leaf cells unipapillose 2
   2. Leaf cells papillose at the upper corner .... 7. *C. rugulosifolium*
   2. Leaf cells papillose in the center 3
3. Stem leaves and branch leaves clearly differentiated 4
3. Stem leaves and branch leaves only slightly differentiated 5
   4. Leaf margins mostly recurved .... 2. *C. assurgens*
   4. Leaf margins usually plane .... 4. *C. leptopteris*
5. Costa slender, ending at 2/3 the leaf length .... 3. *C. gracillimum*
5. Costa stout, ending below just below the leaf apex 6
   6. Leaf cells longer, ovate or rhomboid .... 1. *C. aciculum*
   6. Leaf cells short, hexagonal or nearly rectangular .... 6. *C. prionophyllum*

1. ***Claopodium aciculum*** (Broth.) Broth. 狭叶麻羽藓 xia-ye ma-yu xian

Nat. Pflanzenfam. I(3): 1009. 1908. *Thuidium aciculum* Broth., Hedwigia 30: 245. 1899. Type. Japan: Nagasaki Pref., *Wichura 1413c*.

Claopodium aciculum var. *brevifolium* Card., Bull. Soc. Bot. Genève, ser. 2, 4: 378. 1912.

Claopodium gracillescens Dix., Hong Kong Naturalist His. Suppl. 2: 23. 1933. Type. China: Hong Kong, Amoy Is. *Herklots B3* (holotype BM).

Claopodium prionophylloides Broth. ex Sasaoka, Bot. Mag. (Tokyo) 35: 68. 1921. Type. China: Taiwan, Shinchiku Prov., Hantoso, *Sasoke 671*.

Claopodium sinicum Broth. & Par., Rev. Bryol. 38: 57. 1911. Type. China: Anhui (Ngan Hoei), Niang-kia-kiao, *Courtois 164* (isotype PC).

Claopodium viridulum Card., Bull. Soc. Bot. Genève, ser. 2, 3: 292. 1911.

Haplocladium minutifolium Thér., Ann. Crypt. Exot. 3(2–3): 70. 1930. Syntypes. China: Fujian (Fukien), Buong Kang, Yenping (Nanping), *Ching 76*; Korea: Hallaisan, *Taquet 468*.

Plants slender, green to yellowish green, brownish at base, forming loosely interwoven small patches. *Stems* creeping, 1–2 cm long, irregularly pinnately branched; branches 3–4 mm long; central strand present; paraphyllia absent. *Stem leaves* and branch leaves not clearly differentiated; leaves crowded at stem tips, appressed when dry, erect-spreading when moist, lanceolate to oblong-ovate, acuminate above, ca. 0.6 mm long; leaf margins serrate; costa narrow, nearly excurrent, smooth; leaf cells oblong-rhomboidal to rhomboidal, ca. 12.0 µm × 5.0 µm, thin-walled, with single centrally located rounded papilla. *Dioicous*. *Perichaetial leaves* ovate-lanceolate, costa inconspicuous, yellowish brown above, reddish brown below. *Peristome teeth* not observed. *Spores* ca. 10–15 µm in diameter, finely papillose.

Illustrations: Plate 451, figs. 1–11.
Habitat: on rock surfaces covered with soil and shady moist slopes in lower mountain regions.
Distribution: China, Korea, Japan, and Laos.

Chinese specimens examined: FUJIAN (Fukien): Buongkang, Yengping, *H.-H. Chung s.n.* (PE); Ning-de City, *J. Zhu 49, 260b* (HSNU). HAINAN: Chang-jiang Co., *Redfearn 35885* (MO, NY); Le-dong Co., *Redfearn 36047* (MO, NY). HONG KONG: Amoy Is., *Ah Nin 12A* (BM), *Herklots 17C* (BM); Da-pu, *P.-P. H. But & C.-H. Gao 26343b* (CUHK). JIANGSU: Chang-shu City, *D.-K. Li 17604* (SHM); Lian-yun-gong City, *P.-P. Lin & Q. Wang 3206, 3395* (NAS); Nan-jing City, *Courtois 1299* (PE). SHAANXI: Hu Co., *Z.-P. Wei 4393* (MO, PE). SICHUAN: Mt. Omei, *Allen 6499* (MO). TAIWAN: Miao-li Co., *S.-S. Lin s.n.* [Bryophytes of Taiwan Exsiccatae ser. 3, no. 110] (MO); Xin-zhu Co., *H.-H. Chung* 44144 (TAI). ZHEJIANG: Puduo Isl., *P.-C. Wu 1322* (MO).

This species is very close to *Claopodium prionophyllum* (C. Müll.) Broth., but differs by having smaller plants and elongate-oval to rhomboidal leaf cells and by its restricted distribution in East Asia.

2. ***Claopodium assurgens*** (Sull. & Lesq.) Card. 大麻羽藓 da ma-yu xian

Bull. Soc. Bot. Genève, ser. 2, 3: 283. 1911. *Hypnum assurgens* Sull. & Lesq., Proc. Amer. Acad. Arts Sci. 4: 279. 1859. Type. Japan: Kagoshima Pref., *Wright s.n.*

Claopodium assurgens (Sull. & Lesq.) Card. var. *attenuatum* Nog., J. Jap. Bot. 14 : 29. 1938. Type. China: Taiwan, *Suzuki s.n.*

Claopodium crispulum (Bosch & Sande Lac.) Broth., Nat. Pflanzenfam. I(3): 1009. 1908. *Pseudoleskea crispula* Bosch & Sande Lac., Bryol. Jav. 2: 125, tab. 228. 1865.

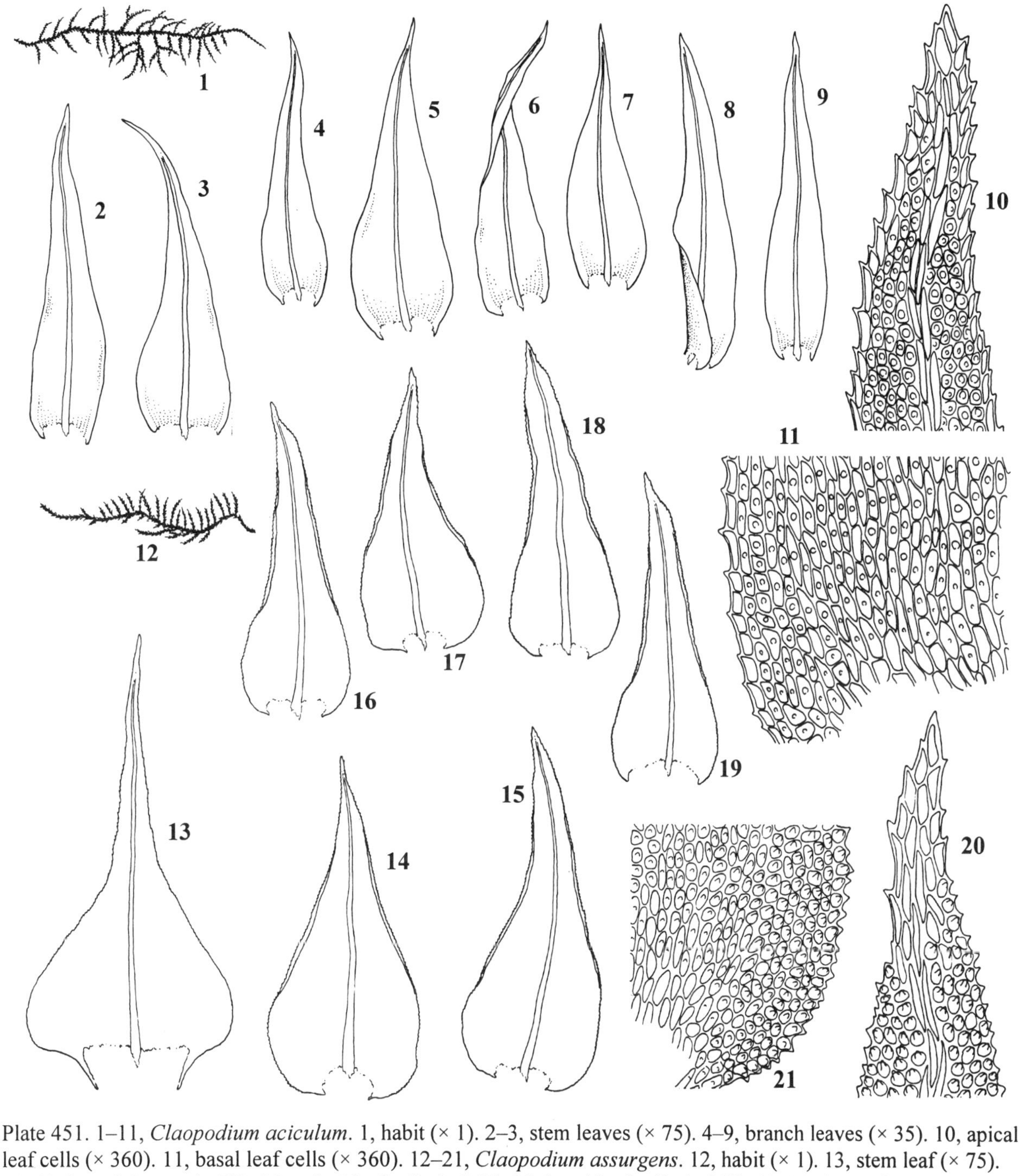

Plate 451. 1–11, *Claopodium aciculum*. 1, habit (× 1). 2–3, stem leaves (× 75). 4–9, branch leaves (× 35). 10, apical leaf cells (× 360). 11, basal leaf cells (× 360). 12–21, *Claopodium assurgens*. 12, habit (× 1). 13, stem leaf (× 75). 14–19, branch leaves (× 75). 20, apical leaf cells (× 360). 21, basal leaf cells (× 360). (1–11, drawn from *P.-C. Wu 85266*, PE; 12–21, from *P.-C. Wu 85132*, PE.)

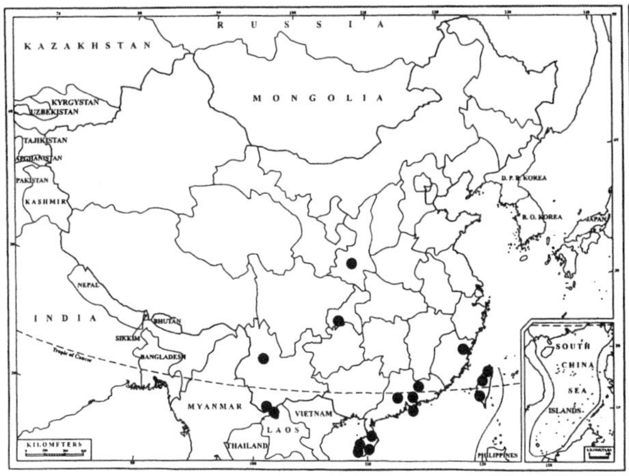

2. *Claopodium assurgens*

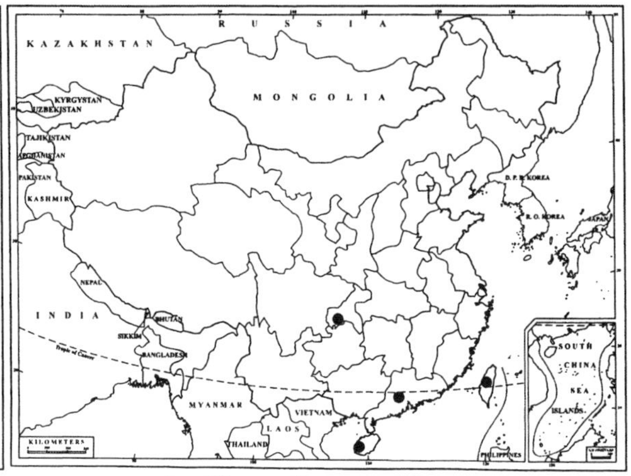

3. *Claopodium gracillimum*

Plants rather large, soft, yellowish green to olive green, brownish green when old, in loose interwoven mats. *Stems* up to 5 cm long, often irregularly pinnately branched; branches 5–8 mm long; central strand differentiated; paraphyllia absent. *Stem leaves* different from the branch leaves; stem leaves mostly crisped when dry, erect-spreading when moist, ovate to ovate-triangular from a wide base toward a narrow acumen; leaf margins serrate, revolute above the leaf middle; costa percurrent or slightly excurrent, smooth on back; median leaf cells oval to rounded quadrate, 6–12 µm × 5–6 µm, grossly papillose; apical leaf cells oblong-ovate to oblong rhomboidal; branch leaves about 1/2 the stem leaf length. *Dioicous*. *Dwarf male plants* growing on the leaves of female plants. *Perichaetial leaves* narrowly elongate-lanceolate, suddenly narrowed into a setulose apex, margins serrate above; costa weak and short. *Setae* reddish brown, about 1.0–1.5 cm long, minutely papillose; capsules cylindrical; opercula long beaked; peristome double; exostome teeth lanceolate, striate at lower portion, finely papillose above; endostome segments nearly as long as the exostome teeth; basal membrane about 2/5 the height of segments; cilia mostly 2. *Spores* ca. 12–17 µm in diameter, finely papillose.

Illustrations: Plate 451, figs. 12–21.
Habitat: under grasses, on tree trunks or shady moist rock surfaces in low mountain regions; alt. 300–580 m.
Distribution: China, India (Assam), and Indonesia.

Chinese specimens examined: CHONGQING: Jin-fou-shan (Mt.), *M. Z. Wang 860567* (PE), *X. Y. Hu 71, 503* (PE). FUJIAN: Ning-de City, *J. Zhu 620, 1509* (HSNU). GUANGDONG: Bo-luo Co., *Redfearn et al. 34427* (MO, NY, PC); Mt. Dinghu, *Magill et al. 8090* (M), *H. Wu B242* (MO); Mt. Lofu, *Magill et al. 8124* (MO). HAINAN: Chang-jiang Co., *Redfearn 35726* (MO, NY); Hai-kou City, *Redfearn 35685* (MO); Mt. Jianfengling, *P.-C. Chen et al. 16* (MO, PE); Le-dong Co., *Redfearn 35941* (MO); Ling-shui Co., *Redfearn 36068* (MO, NY). HONG KONG: Amoy Is., *Herklots 4C* (BM); Mt. Dayushan, Fengwangjing, *C.-H. Cao & P. P. H. But 90283d* (CUHK). SHAANXI: Mt. Qin-ling, *P.-C. Chen 22a* (MO); Wei-tze-ping, *Licent 88a* (BM). TAIWAN: Miao-li Co., *S.-S. Lin s.n.* [Bryophytes of Taiwan Exsiccatae ser. 2, no. 59] (MO); Ping-tung Co., *Shevock 15051* (MO); Tai-pei Co., *C. C. Chuang 5329* (NY, PE, TAI). YUNNAN: Da-li Co., *Redfearn et al. 1264* (MO, NY); Jing-hong Co., *Magill et al. 7846* (MO); Meng-la Co., *Magill et al. 7789* (MO); Meng-lun Co., *L. Zhang 849* (IBSC, PE).

This species with revolute leaf margins is easily distinguished from other species of *Claopodium* in China.

3. ***Claopodium gracillimum*** (Card. & Thér) Nog. 细麻羽藓 xi ma-yu xian
J. Hattori Bot. Lab. 27: 33. 1964. *Diaphanodon gracillimus* Card. & Thér., Bull. Acad. Int. Géogr. Bot. 18: 2. 1908. Type. Japan: Kagoshima, Amami-oshima Is., *Ferrie s.n.*

Plants very slender, green to yellowish green, in loose interwoven mats. *Stems* creeping, spreading, ca. 1.5 cm long, irregularly pinnately branched; branches remote, ca. 3 mm long; central strand differentiated; paraphyllia sparse, foliose. *Stem leaves* and branch leaves similar; stem leaves appressed when dry, erect-spreading when moist, ovate to ovate-triangular, acuminate above, ca. 0.3–0.4 mm long; leaf margins finely serrate; costa weak, reaching

2/3 leaf length; leaf cells hexagonal to rhomboidal, 12–20 µm × 4–8 µm, thin-walled, with a central papilla; branch leaves ovate to oblong-ovate, concave. *Sporophytes* unknown.

Illustrations: Plate 452, figs. 1–10.
Habitat: on rocks.
Distribution: China and Japan.

Chinese specimens examined: CHONGQING: Nan-chuan Co., *M. Z. Wang 860128*, *X.-Y. Hu 829* (PE). GUANGDONG: *Z. H. Li 85206* (PE, SYS). HAINAN: Chang-jiang Co., *Redfearn 35701* (MO, NY). TAIWAN: Nan-tou Co., *C. C. Chung* & *W. B. Schofield 600-15* (NY, PE, UBC).

This species is close to *Claopodium aciculum*, but differs from the latter by having smaller plants, a short and weak leaf costa, and short and papillose apical leaf cells.

4. ***Claopodium leptopteris*** (C. Müll.) P.-C. Wu & M.-Z. Wang 卵叶麻羽藓 luan-ye ma-yu xian
Acta Phytotax. Sin. 38: 261. 2000. *Haplocladium leptopteris* C. Müll., Nuovo Giorn. Bot. Ital., n. ser. 3: 116. 1896. Type. China: Shaanxi (Schen-si), *Giraldi s.n.*

Plants delicate, ca. 1–2 cm long, in loose mats. *Stems* irregularly sparsely branched; paraphyllia absent. *Stem leaves* gradually tapering to a lanceolate acumen from an ovate leaf base, 0.7–0.9 mm long, basal stem leaves ovate, sharply acute; costa slender, up to 2/3 the leaf length or vanishing at upper leaf portion; leaf margins entire or somewhat serrulate; branch leaves ovate-lanceolate to narrowly lanceolate; leaf cells with small single papilla, marginal cells mostly smooth; upper leaf cells smaller, oblong-hexagonal, 4–9 µm × 5–14 µm, thin-walled; median marginal leaf cells hexagonal to oblong multipolygonal, 5–10 µm × 6–15 µm; basal leaf cells oblong-hexagonal to narrowly polygonal; thin-walled. *Male inflorescence* and female inflorescence not seen.

Illustrations: Plate 452, figs. 11–19.

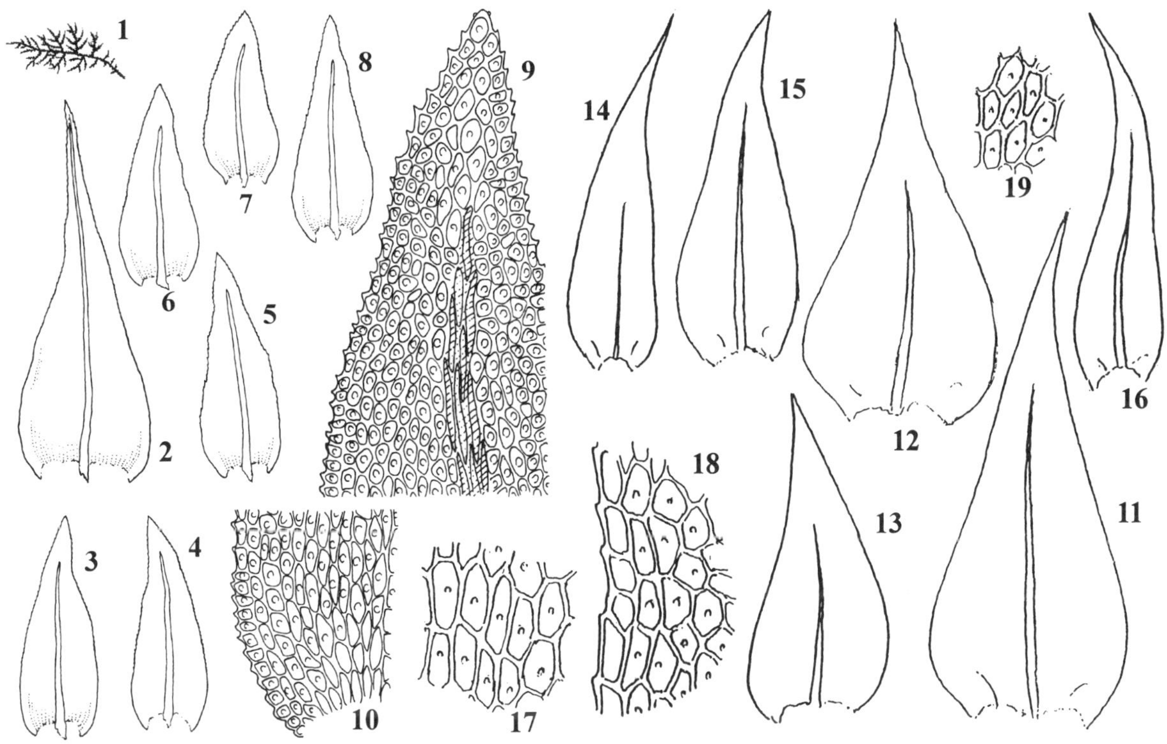

Plate 452. 1–10, *Claopodium gracillimum*. 1, habit (× 1). 2, stem leaf (× 75). 3–8, branch leaves (× 75). 9, apical leaf cells (× 360). 10, basal leaf cells (× 360). 11–19, *Claopodium leptopteris*. 11–12, stem leaves (× 75). 13, basal stem leaf (× 75). 14–16, branch leaves (× 75). 17, apical leaf cells (× 460). 18, median marginal leaf cells (× 460). 19, basal leaf cells (× 460). (1–10, drawn from *Z.-H. Li 85206*, SYS; 11–19, from *Giraldi 939*, H.)

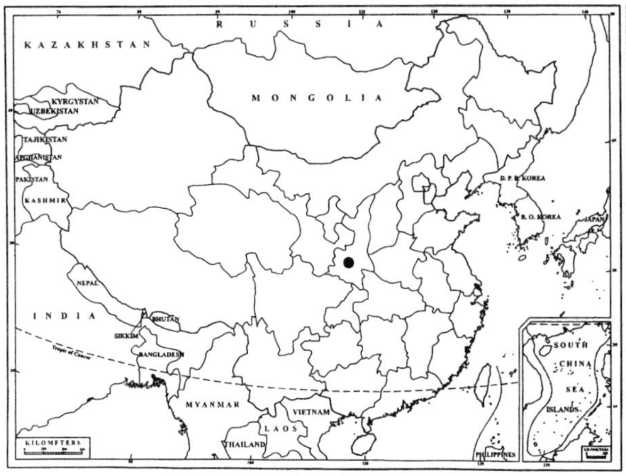

4. *Claopodium leptopteris*

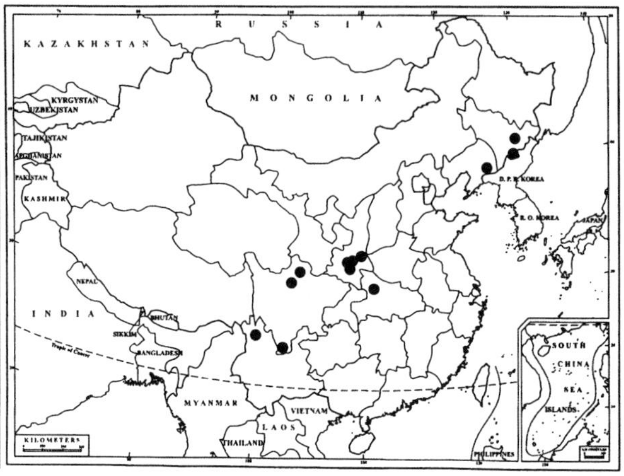

5. *Claopodium pellucinerve*

Habitat: unknown.
Distribution: endemic to China.

Chinese specimen examined: SHAANXI: Mt. Taibai? (without exact locality), *Giraldi 939* (H).

This species is transferred from *Haplocladium* based on its sharing of the characteristics of *Claopodium* such as stems without paraphyllia, stem leaves without a wide base, and branch leaves narrowed.

5. ***Claopodium pellucinerve*** (Mitt.) Best 多疣麻羽藓 duo-you ma-yu xian

Bryologist 3: 19. 1900. *Leskea pellucinerve* Mitt., J. Linn. Soc., Bot., Suppl. 1: 130. 1859. Syntypes. India: Northwest Himalayas, Simla, *Thomson 1125, 1132, 1146.*

Claopodium papillicaule (Broth.) Broth., Nat. Pflanzenfam. I(3): 1009. 1908. *Thuidium papillicaule* Broth., Hedwigia 38: 245. 1899.

Claopodium piliferum Broth., Akad. Wiss. Wien Sitzungsber., Math.-Naturwiss. Kl., Abt. 1. 133: 578. 1924. Type. China: Sichuan (Setschwan), Yen-yuen (Yan-yuan), *Handel-Mazzetti 2186* (isotype BM).

Claopodium pugionifolium (Broth.) Broth., Nat. Pflanzenfam. I(3): 1009. 1908. *Thuidium pugionifolium* Broth. & Par., Rev. Bryol. 31: 59. 1904.

Claopodium subpiliferum (Lindb. & Arn.) Broth., Nat. Pflanzenfam. I(3): 1009. 1908. *Anomodon subpilifer* Lindb. & Arn., Kongl. Svenska Vetenskapsakad. Handl. 23(10): 111. 1890.

Claopodium tenuissimum Dix., Rev. Bryol., n.s. 1: 187. 1928. Type. China: Shaanxi (Schen-si), Teou-mou-kong & Fang-yang-soeur, *Licent 133b* (holotype BM).

Plants small to medium-sized, sometimes slender, yellowish green to brownish green, forming interwoven mats. *Stems* creeping, up to ca. 5 cm long, irregularly pinnately branched; central strand present; stems and branches densely minutely papillose. *Stem leaves* and branch leaves differentiated; stem leaves ovate-triangular, gradually narrowed to a filiform acumen from a wide base, 1.0–1.5 mm long; leaf margins papillose; costa vanishing near leaf apex, papillose on back; median leaf cells rhomboidal to elliptical, ca. 20 µm × 10 µm, thin-walled, densely multipapillose; apical leaf cells narrow, hyaline, smooth; branch leaves oblong, apex caudate, costa reaching above midleaf; basal median cells hyaline, smooth. *Dioicous*. *Perichaetial leaves* oblong, abruptly narrowed toward an elongate acumen, costa distinct, leaf cells smooth. *Setae* elongate; capsules ovoid; peristome double; typical in the genus. *Spores* spherical, ca. 15–20 µm in diameter, finely papillose.

Illustrations: Plate 453, figs. 1–16.
Habitat: mostly on shady moist rocks or soil, rarely on rotten logs or tree trunks; alt. 1900–3400 m.
Distribution: China, Korea, Japan, Pakistan, and India.

Chinese specimens examined: HUBEI: Mt. Shennongjia, *Luteyn 1847c* (MO, NY, PE), *P. C. Wu 87* (PE). JILIN: An-tu Co., *Koponen 37223* (NY); Mt. Changbai, *B.-D. Xin 0264* (MO). LIAONING: Kuan-ding Co., *Buck 23641* (NY). NE China: Zhanghekou, *H. Sasaoka 3945* (PE), *P. V. Sinzev s.n.* (PE). SHAANXI: Ao-wo Co., *M-X.*

Zhang 496 (MO); Huashan (Mt.), *P.-C. Chen et al. 5* (MO, PE); Hu Co., *Z.-P. Wei 4479* (MO); Mt. Qinling, *P.-C. Chen et al. 29* (MO, PE); Mt. Taibai, *M.-X. Zhang 61* (MO); Ning-shan Co., *P.-C. Chen et al. 82* (MO, PE); Zhou-zhi Co., *M.-X. Zhang 507* (MO). SICHUAN: Ma-er-kang Co., *Redfearn 35022* (as *Claopodium aciculum*, MO, NY); Song-pan Co., *S. He 30463a* (MO), *Redfearn 35212* (MO, NY); Yanyuan to Yalung, *Handel-Mazzetti 2186* (PE). YUNNAN: Wei-xi Co., *J.-X. Xi 510* (PE).

This species is easily distinguished from other species of *Claopodium* by its rhomboidal and densely multipapillose leaf cells.

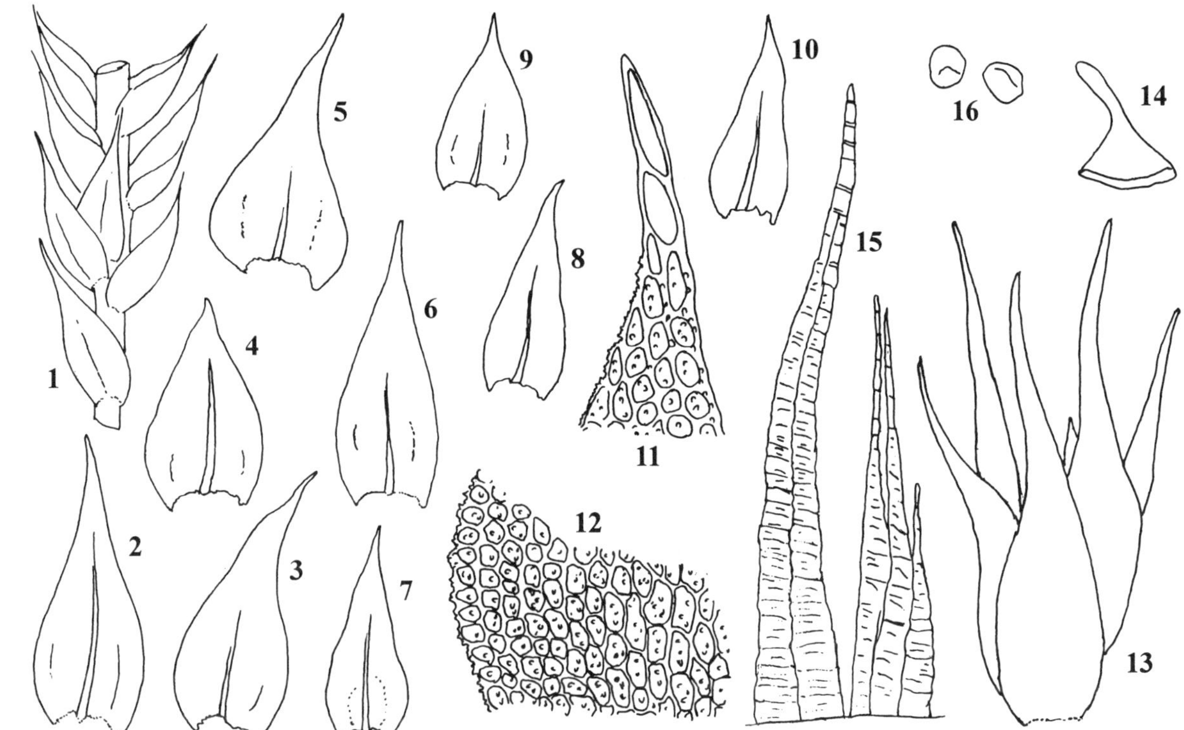

Plate 453. *Claopodium pellucinerve*. 1, branch (× 5). 2–6, stem leaves (× 50). 7–10, branch leaves (× 50). 11, apical leaf cells (× 315). 12, basal leaf cells (× 315). 13, perichaetium (× 50). 14, operculum (× 20). 15, peristome (× 200). 16, spores (× 690). (1–16, drawn from *P. V. Sinzev s.n.*, PE.)

6. ***Claopodium prionophyllum*** (C. Müll.) Broth. 齿叶麻羽藓 chi-ye ma-yu xian

Nat. Pflanzenfam. I(3): 1009. 1908. *Hypnum prionophyllum* C. Müll., Syn. Musc. Frond. 2: 481. 1851. Type. Nepal: *Wallich s.n.*

Claopodium amblystegioides Dix., Proc. Linn. Soc. New South Wales 55: 281. 1930.

Claopodium nervosum Fleisch., Musci Buitenzorg 4: 1504. 1923. *Hypnum nervosum* Harv. in Hook., Icon. Pl. 1: 24, f. 3. 1836, *hom. illeg.*

Plants medium-sized, up to 10 cm long, dark green to yellowish green, brownish at base when old, in spreading mats. *Stems* delicate, creeping, irregularly pinnately branched; branches 3–5 mm long; central strand present; paraphyllia foliose, sparse. *Stem leaves* and branch leaves similar; stem leaves appressed or slightly revolute when dry, erect-spreading when moist, ovate-lanceolate, up to 1 mm long; margins serrate; costa single, percurrent; leaf cells hexagonal, unipapillose, median cells 5.0–7.5 µm × 5.0 µm, thin-walled. *Dioicous. Dwarf male plants* on the leaves of female plants. *Perichaetial leaves* ovate-lanceolate, serrate at the apex, costa short, inconspicuous. *Setae* slender, reddish brown to yellowish brown, 1–2 cm long, smooth; capsules ovoid, horizontal or cernuous, exothecium cells rectangular to hexagonal; peristome double; exostome teeth yellowish brown, more than 0.5 mm long, papillose above; endostome segment nearly as long as the teeth, lightly yellowish, finely papillose; basal membrane about 1/2 the height of segments, cilia usually 2. *Spores* about 15 µm in diameter, finely papillose.

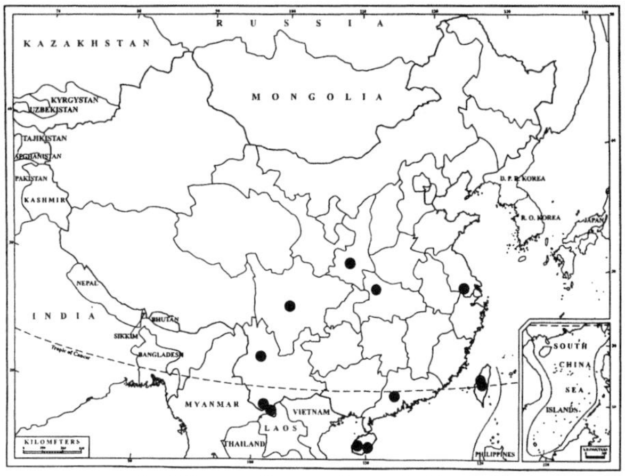

6. *Claopodium prionophyllum*

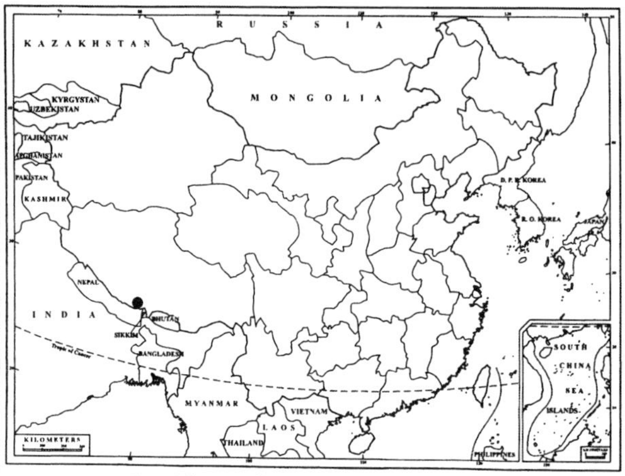

7. *Claopodium rugulosifolium*

Illustrations: Watanabe 1972 (Pl. 32, 1–15).
Habitat: mostly on calcareous rock surfaces, rarely on soil.
Distribution: China, Korea, Japan, U.S.A. (Hawaii), and Fiji.

Chinese specimens examined: GUANGDONG: Shao-qing City, *Redfearn et al. 34383* (MO, PC). HAINAN: Ling-shui Co., *Redfearn 36077* (MO, NY). HUBEI: Mt. Shennongjia, *P.-C. Wu 237d* (PE). JIANGSU: Yi-xinq City, *P.-C. Wu & Y.-X. Jing 2282* (NAS, PE). SHAANXI: Qin-ling Mts, *P.-C. Chen et al. 340122* (PE); Mt. Taibai *M.-X. Zhang 262* (MO). SICHUAN: Lu-ding Co., *Y. Jia J01875* (MO, PE). TAIWAN: Alishan (Mt.), *U. Ara 14532* (TAI); Tai-chung Co., *Shevock 18143* (MO). YUNNAN: Jing-hong Co., *Redfearn et al. 33682* (as *Claopodium aciculum*, MO, NY); Meng-hai Co., *Redfearn et al. 34092* (MO, NY); Meng-la Co., *Magill et al. 7793* (MO); Xishungbanna, *P.-C. Wu 21663*, *21675*, *22001* (PE); Yang-bi Co., *Redfearn et al. 230* (MO, NY).

This species is very close to *Claopodium aciculum*, but differs by having larger plants and leaves.

7. ***Claopodium rugulosifolium*** S.-Y. Zeng 皱叶麻羽藓 zhou-ye ma-yu xian
Acta Bot. Yunnan. 3(2): 263. 1981. Type. China: Xizang, Ding-jie Co., alt. 2800 m, *S.-K. Chen 2064b* (holotype HKAS).

Plants minute, delicate, yellowish green or brownish green, brown when old, in densely interwoven mats. *Stems* 1–2-pinnately branched; paraphyllia absent. *Leaves* rugulose when dry, erect-spreading when moist; stem leaves oblong-rounded to widely ovate, gradually tapering to a lanceolate acumen; leaf margins inconspicuously differentiated, serrate above; costa single, smooth, vanishing near leaf apex; leaf cells rhomboidal to polygonal, unipapillose at the front corner, thin-walled; branch leaves smaller than stem leaves, with short apices. *Dioicous*? *Setae* slender, pale reddish brown, smooth; capsules oblong, horizontal or erect-patent; peristome double, well developed; exostome teeth lanceolate, with stripes and lamellae, pale yellowish brown; endostome segments irregularly papillose, with high basal membrane; cilia 2. *Calyptrae* not seen. *Spores* not seen.

Illustrations: Plate 454, figs. 1–15.
Habitat: on shady moist soil; alt. 2800 m.
Distribution: endemic to China.

Chinese specimens examined: XIZANG: Ding-jie Co., *S.-K. Chen 2064b* (HKAS).

This species is very close to *Haplocladium* in having similar leaf shapes and leaf cells that are unipapillose at front corners. We tentatively accept this species in *Claopodium*. Its systematic position needs to be further considered in the future.

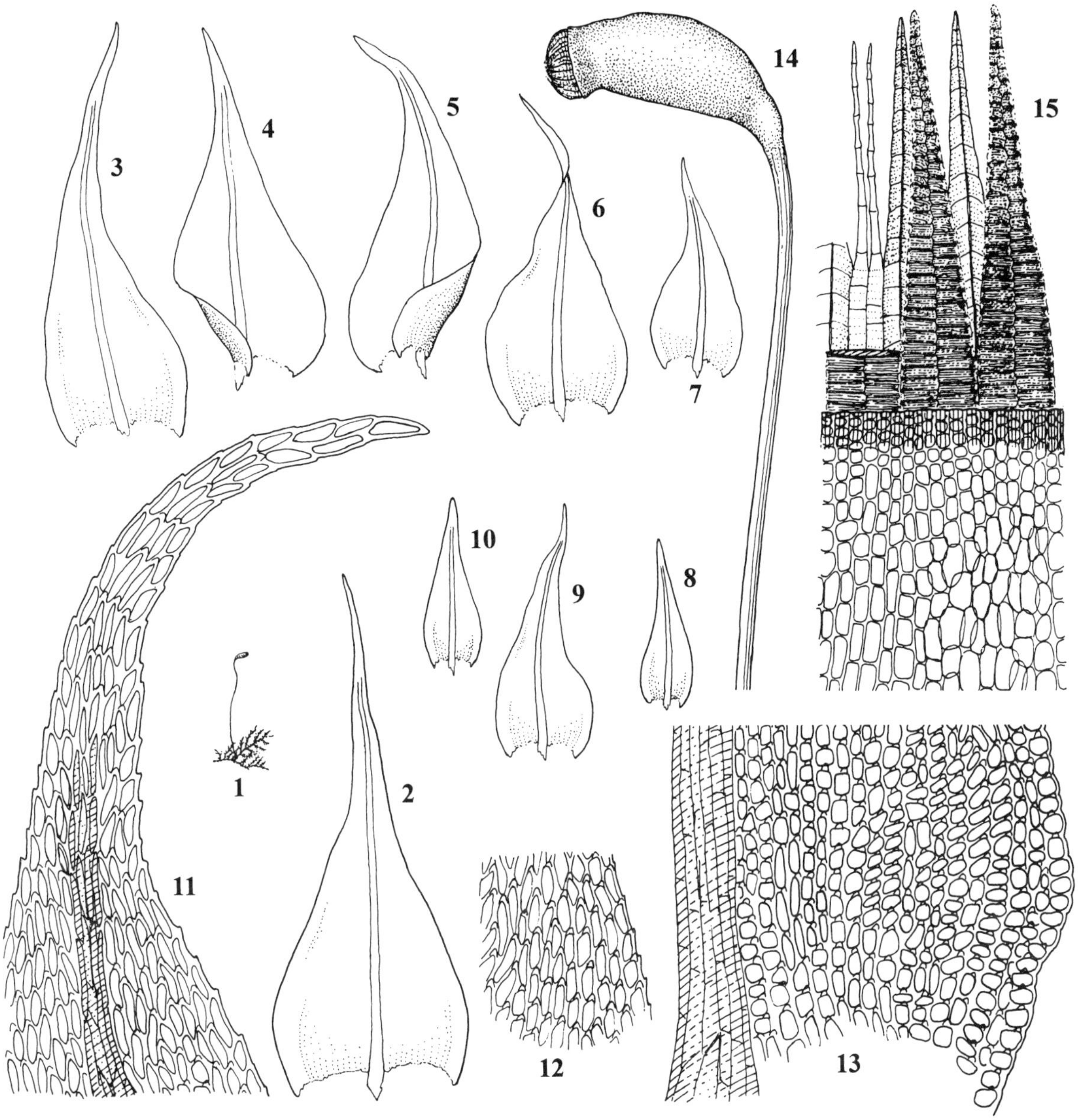

Plate 454. *Claopodium rugulosifolium*. 1, habit (× 1). 2, stem leaf (× 75). 3–6, branch leaves (× 75). 7–10, branchlet leaves (× 75). 11, apical stem leaf cells (× 443). 12, median stem leaf cells (× 443). 13, basal stem leaf cells (× 443). 14, capsule (× 14). 15, peristome (× 140). (1–15, drawn from *S.-K. Chen 2064b*, HKAS.)

6. ***Cyrto-hypnum*** Hampe & Lor. 细羽藓属 xi-yu xian shu
Bot. Zeitung (Berlin) 27: 455. 1869.

Plants slender, soft, yellowish green, olive green or dark green, rarely pale green, usually in lax patches. *Stems* creeping, ca. 2–3 cm long, rarely up to 5 cm long, irregularly pinnately to bipinnately branched; central strand usually present, rarely absent; paraphyllia densely or loosely growing on stems, sometimes also at the base of branches, rarely densely on branches, usually filamentous, ca. 1–5 cells long, rarely 10(–20) cells long, sometimes branched, apex acute, smooth or truncate, 2 or multipapillose; branches 3–5 mm long. *Stem leaves* remote, ovate, widely ovate to widely cordate or triangular at base, suddenly narrowed forming a lanceolate or slender apex, sometimes squarrose above, often slightly plicate; margins revolute at base or mostly revolute, serrulate or coarsely dentate, rarely entire; costa rigid, vanishing near leaf apex or 3/4 the leaf length, rarely rather weak, sometimes papillose on back, rarely hyaline; branch leaves ovate or widely ovate, rarely widely cordate to ovate-triangular, with short obtuse apex, rarely acute; leaf cells hexagonal, rounded hexagonal or rhomboidal; basal central cells mostly rectangular, thin- or slightly thick-walled, with single conical or fine papilla or multipapillae per cell. *Autoicous* or dioicous? *Perichaetial leaves* oblong, with long lanceolate apex, sometimes weakly plicate; margins entire or serrulate, rarely with several long or short hairs at the shoulder; costa sometimes excurrent forming a long hair. *Setae* orange red or orange yellow, slender, ca. 1.0–2.5 cm long, usually smooth or sometimes densely papillose or mammillose, rarely rough above; capsules ovoid or oblong-ovoid or shortly cylindrical, nearly horizontal to inclined; opercula conic, 0.5–1.0 mm long; peristome *Hypnum*-type; exostome teeth 16, mostly lightly yellowish brown, long lanceolate, hyaline above, densely papillose, transversally striate at base; endostome segments as long as the teeth, lightly yellowish, keeled, perforated; basal membrane ca. 1/2 the length of segments; cilia 1–3, nodulose, rarely rudimentary. *Calyptrae* cucullate, roughened at the apex. *Spores* 10–20 μm in diameter, smooth or finely papillose.

Cyrto-hypnum is mostly distributed in tropical and southern subtropical forests. *Cyrto-hypnum minutulum* (Hedw.) Buck & Crum is excluded from China. The previous report of this species in China (Redfearn et al. 1996) appears to be based on incorrectly identified specimens. This species is primarily a European and North American species (Crum & Anderson 1981). Ten species are recognized in this study, with three species newly reported from the country. Species distribution of the genus is concentrated in Yunnan, Taiwan, and Fujian provinces.

1. Cortical cells of stems and branches papillose; leaf cells with 3–8 minute papillae .... 7. *C. pygmaeum*
1. Cortical cells of stems and branches smooth; leaf cells mostly with a single, sharp papilla, rarely with multipapillae 2
    2. Paraphyllia densely growing on stems and branches, up to 10(–20) cells long; endostomal cilia mostly not developed (rudimentary) .... 10. *C. vestitissimum*
    2. Paraphyllia densely growing on stems and the bases of main branches, 2–8 cells long; endostomal cilia usually well developed 3
3. The apical cells of paraphyllia acute; stem leaves usually shortly acuminate 4
3. The apical cells of paraphyllia usually truncate; stem leaves acuminate or lanceolate above 5
    4. Stem leaves mostly broadly ovate; setae smooth .... 5. *C. microphyllum*
    4. Stem leaves broadly ovate or triangular; setae rough above .... 6. *C. minusculum*
5. Leaf cells multipapillose 6
5. Leaf cells unipapillose 7
    6. Margins of inner perichaetial leaves entire; setae smooth .... 8. *C. tamariscellum*
    6. Margins of inner perichaetial leaves somewhat ciliate; setae densely papillose .... 4. *C. gratum*
7. Stem leaf apices squarrose; setae only rough above .... 2. *C. contortulum*
7. Stem leaf apices not squarrose; setae smooth or papillose 8
    8. Pseudoparaphyllia absent .... 3. *C. fuscatum*
    8. Pseudoparaphyllia growing at the base of branch 9
9. Stem leaves mostly cordate-triangular or broadly triangular; inner perichaetial leaves not ciliate .... 9. *C. versicolor*
9. Stem leaves oblong-ovate, lanceolate toward apex; inner perichaetial leaves long ciliate above .... 1. *C. bonianum*

1. ***Cyrto-hypnum bonianum*** (Besch.) Buck & Crum 纤枝细羽藓 qian-zhi xi-yu xian

Contr. Univ. Michigan Herb. 17: 65. 1990. *Thuidium bonianum* Besch., Bull. Soc. Bot. France 34: 98. 1887. Type. Vietnam: Ha-noi, Nam-cong, *R. P. Bon 2413*.

Thuidium lejeuneoides Nog., J. Jap. Bot. 13: 88. 1937. Type. China: Taiwan, Tai-pei Co., Urai, *Suzuki 495*.

Plants minute, yellowish green, dark green when old, in loose patches. *Stems* creeping, ca. 2–3 cm long, irregularly bipinnately branched; branches ca. 3 mm long; central strand undifferentiated; paraphyllia densely growing on stems, filiform, 3–6 cells long, rarely branched, papillose, truncate or rarely acute at apex; pseudoparaphyllia lanceolate or foliose, margins papillose, often present at the base of branches. *Stem leaves* slightly concave, appressed when dry, erect-spreading when moist, oblong-ovate at base, suddenly narrowed to a shortly lanceolate apex, ca. 0.3–0.4 mm long; margins serrate; costa vanishing at 3/4 the leaf length, minutely papillose at the back; median leaf cells hexagonal or rounded hexagonal, ca. 5–15 µm long, thin-walled, with a single, small papilla; apical leaf cells oblong or hexagonal, smooth; branch leaves concave, imbricate, usually ovate, with a short apex, arranged in two ranks. *Autoicous*. *Inner perichaetial leaves* with several flexuous cilia above, irregularly serrate along lower margins. *Setae* very slender, reddish brown, ca. 1 cm long, densely papillose; exothecial cells inconspicuously mammillose; peristome similar to that of *Cyrto-hypnum pygmaeum* (Schimp.) Buck & Crum; cilia 1. *Spores* smooth.

Illustrations: Plate 455, figs. 1–14.
Habitat: on rotten logs.
Distribution: China, Japan, Thailand, and Vietnam.

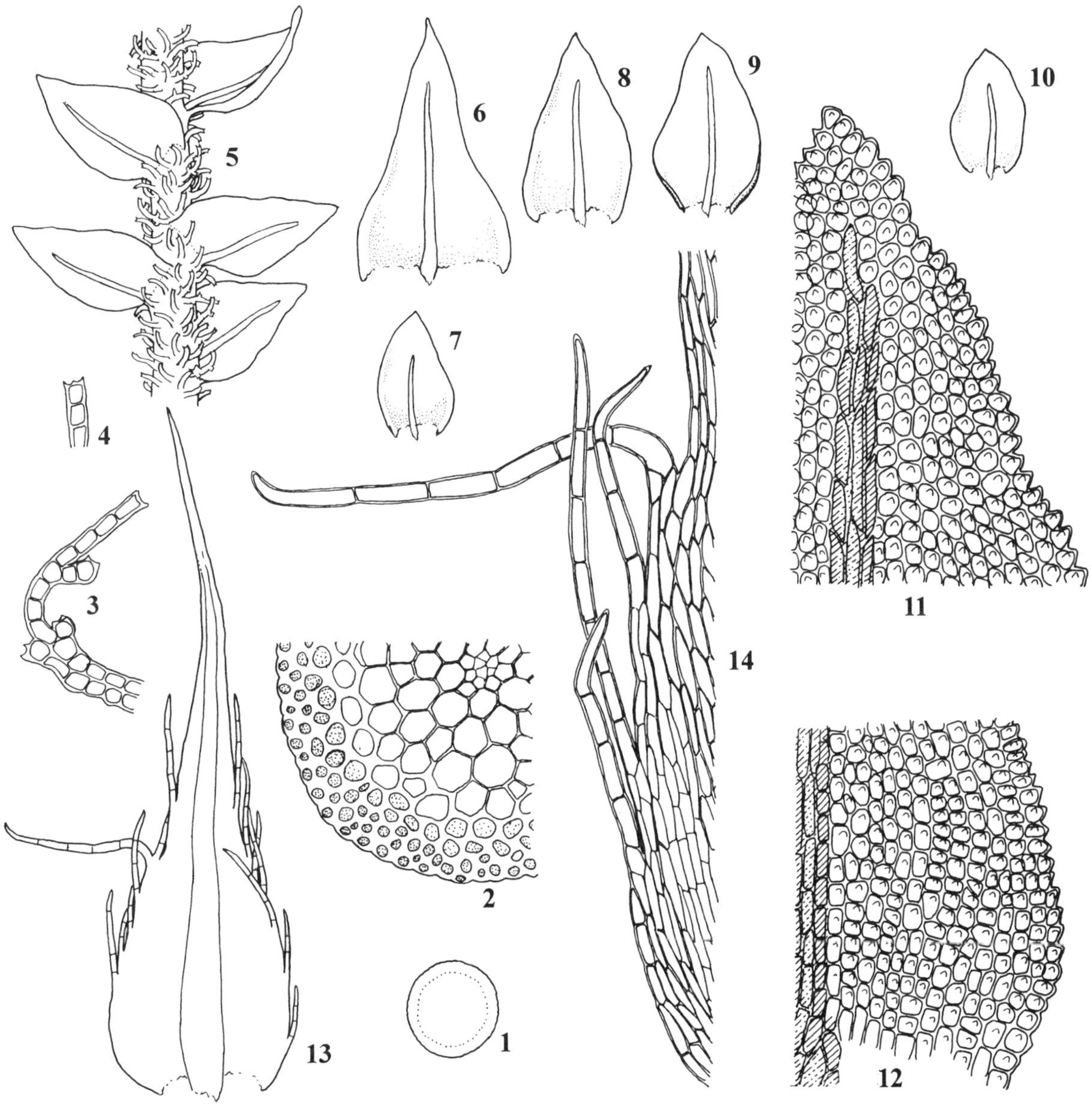

Plate 455. *Cyrto-hypnum bonianum*. 1, stem cross section (× 78). 2, portion of stem cross section (× 366). 3–4, paraphyllia (× 366). 5, portion of primary branch (× 78). 6, stem leaf (× 90). 7–10, branch leaves (× 90). 11, apical leaf cells (× 366). 12, basal leaf cells (× 366). 13, inner perichaetial leaf (× 90). 14, marginal cells of inner perichaetial leaf (× 258). (1–14, drawn from *Redfearn et al. 33681*, MO.)

Chinese specimens examined: YUNNAN: Jing-hong Co., *Redfearn et al. 33681*, *33715* (MO, NY), *33716* (MO, NY, PC); Meng-la Co., *Redfearn et al. 33875*, *33962* (MO, NY), *L. Zhang 378* (MO); Xishuangbanna, *W.-S. Hsu 6500*, *6755*, *6855* (PE), *M.-Z. Wang 4073* (PE), *P.-C. Wu 21683* (PE).

2. ***Cyrto-hypnum contortulum*** (Mitt.) P.-C. Wu, Crosby & S. He 美丽细羽藓 mei-li xi-yu xian
Chenia 6: 6. 1999. *Thuidium contortulum* (Mitt.) Jaeg., Ber. Thätigk. St. Gallischen. Naturwiss. Ges. 1876–77: 256. 1878. *Leskea contortula* Mitt., J. Linn. Soc., Bot., Suppl. 1: 134. 1859. Type. India: Eastern Himalayas, Kurseong, *J. D. Hooker 1124*.
Thuidium venustulum Besch., Ann. Sci., Nat. Bot., ser. 7, 15: 78. 1892. Type. China: Yunnan, Tsin-choui-ho, *Delavay 4635* (holotype BM).

Plants small, yellowish green, in small loose patches. *Stems* prostrate, bipinnately branched; paraphyllia minute, only growing on stems, papillose. *Stem leaves* remote, ca. 0.5 mm long, shortly acuminate; margins serrate;

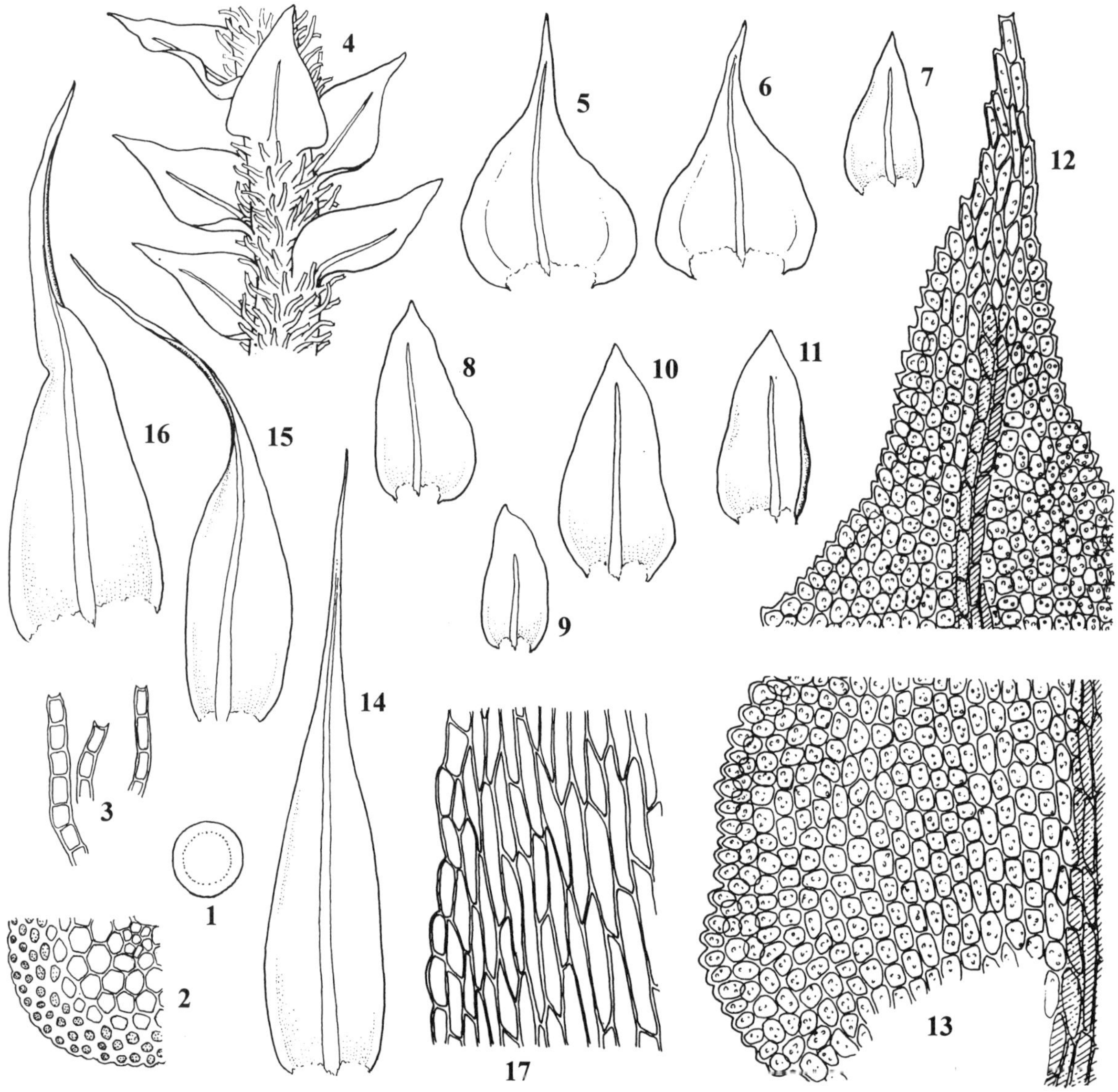

Plate 456. *Cyrto-hypnum contortulum*. 1, stem cross section (× 52). 2, portion of stem cross section (× 258). 3, paraphyllia (× 366). 4, primary branch (× 73). 5–6, stem leaves (× 90). 7–11, branch leaves (× 90). 12, apical leaf cells (× 366). 13, basal leaf cells (× 366). 14–16, perichaetial leaves (× 90). 17, marginal cells of perichaetial leaf (× 366). (1–17, drawn from *Redfearn et al. 33882*, MO.)

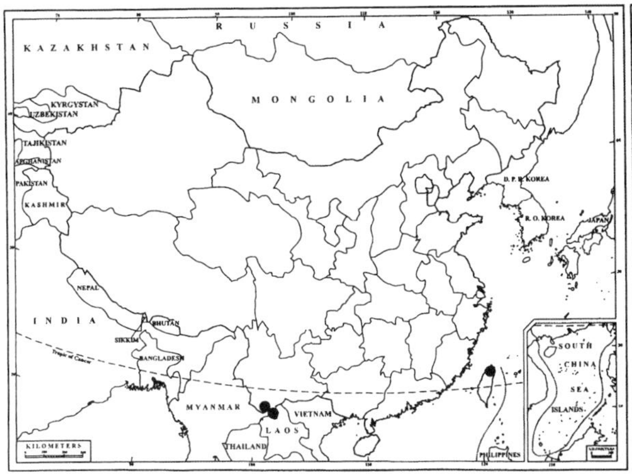

1. *Cyrto-hypnum bonianum*

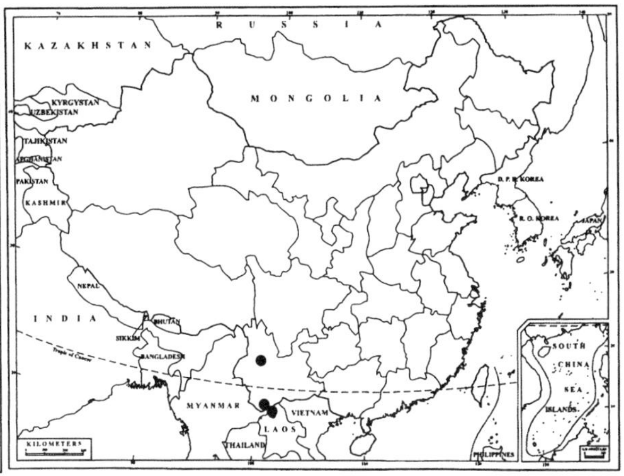

2. *Cyrto-hypnum contortulum*

costa ending 2/3 the leaf length; median leaf cells rounded rectangular, ca. 7 µm wide, centrally unipapillose; branch leaves oblong-ovate, ca. 0.3 mm long. *Monoicous*. *Sporophytes* on main stems. *Perichaetial leaves* oblong-lanceolate, with narrow apices. *Setae* erect, rough above, ca. 1 cm long; capsules erect, ovoid; peristome *Hypnum*-type; exostome teeth and endostome segments in equal length; segments keeled, perforated; basal membrane ca. half the height of the segment; cilia 2, nodulose.

Illustrations: Plate 456, 1–17.
Habitat: on tree trunks; alt. 1200–1260 m.
Distribution: China, Japan, India, Pakistan, the Philippines, and New Guinea.

Chinese specimens examined: YUNNAN: Jing-hong Co., *Redfearn et al. 34253* (as *Thuidium lejeuneoides*, NY); Meng-hai Co., *Redfearn et al. 34237* (MO, NY); Meng-la Co., *P. L. Redfearn 33882*, *34854* (MO, NY, SMSU).

In China, this species was previously treated or named as *Thuidium tamariscellum* (C. Müll.) Bosch & Sande Lac. or *T. sparsifolium* (Mitt.) Jaeg.

3. ***Cyrto-hypnum fuscatum*** (Besch.) P.-C. Wu, Crosby & S. He 尖毛细羽藓 jian-mao xi-yu xian
Chenia 6: 10. 1999. *Thuidium fuscatum* Besch., Ann. Sci. Nat. Bot., ser. 7, 15: 78. 1892. Type. China: Yunnan, Santchang-Kiou, *Delavay 4768* (holotype BM; isotype H).
Cyrto-hypnum talongense (Besch.) Buck & Crum, Contr. Univ. Michigan Herb. 17: 67. 1990. *Thuidium talongense* Besch., Ann. Sci. Nat. Bot. ser. 7, 15: 81. 1892. Type. China: Yunnan, Talong-tan, *Delavay 4849d* (holotype BM).
Thuidium burmense Bartr., Rev. Bryol. Lichénol. 23: 252. 1954.
Thuidium koelzii Robins., Bryologist 71: 92. 1968.

Plants slender, yellowish green, in interwoven mats. *Stems* creeping or ascending, regularly pinnately branched; central strand present; paraphyllia densely on stems, apical cells sharp, 3–6 cells long. *Stem leaves* large, remote, gradually narrowed from an ovate base to a lanceolate apex, ca. 1 mm long; margins serrate; costa ending below leaf apex; leaf cells hexagonal, unipapillose; branch leaves dense, revolute when dry, erect-patent when moist, ca. 0.35 mm long, oblong-ovate; margins serrate; costa reaching near leaf apex. *Autoicous*. *Perichaetial leaves* oblong-lanceolate, suddenly narrowed to a lanceolate apex, up to 2 mm long. *Setae* erect, ca. 2.5 cm long, smooth; capsules inclined to horizontal, short cylindrical; opercula conical, beaked; peristome teeth *Hypnum*-type; endostome segments shorter than the teeth; cilia single, short or rudimentary. *Spores* very small.

Illustrations: Plate 457, figs. 1–13.
Habitat: on rotten logs, soil or tree trunks.
Distribution: China and India.

Chinese specimens examined: YUNNAN: He-kou Co., *W.-M. Zhu 41* (PE); Song-min Co., *Koponen 37779* (H, NY), *37826* (H, MO); Xishuangbanna, *P.-C. Wu 21901* (PE).

This species is very close to *C. minusculum* (Mitt.) Buck & Crum, but is characterized by having short and acute paraphyllia, elongate stem leaf apices, narrow and sharp branch leaf apices, and erect or slightly arcuate capsules.

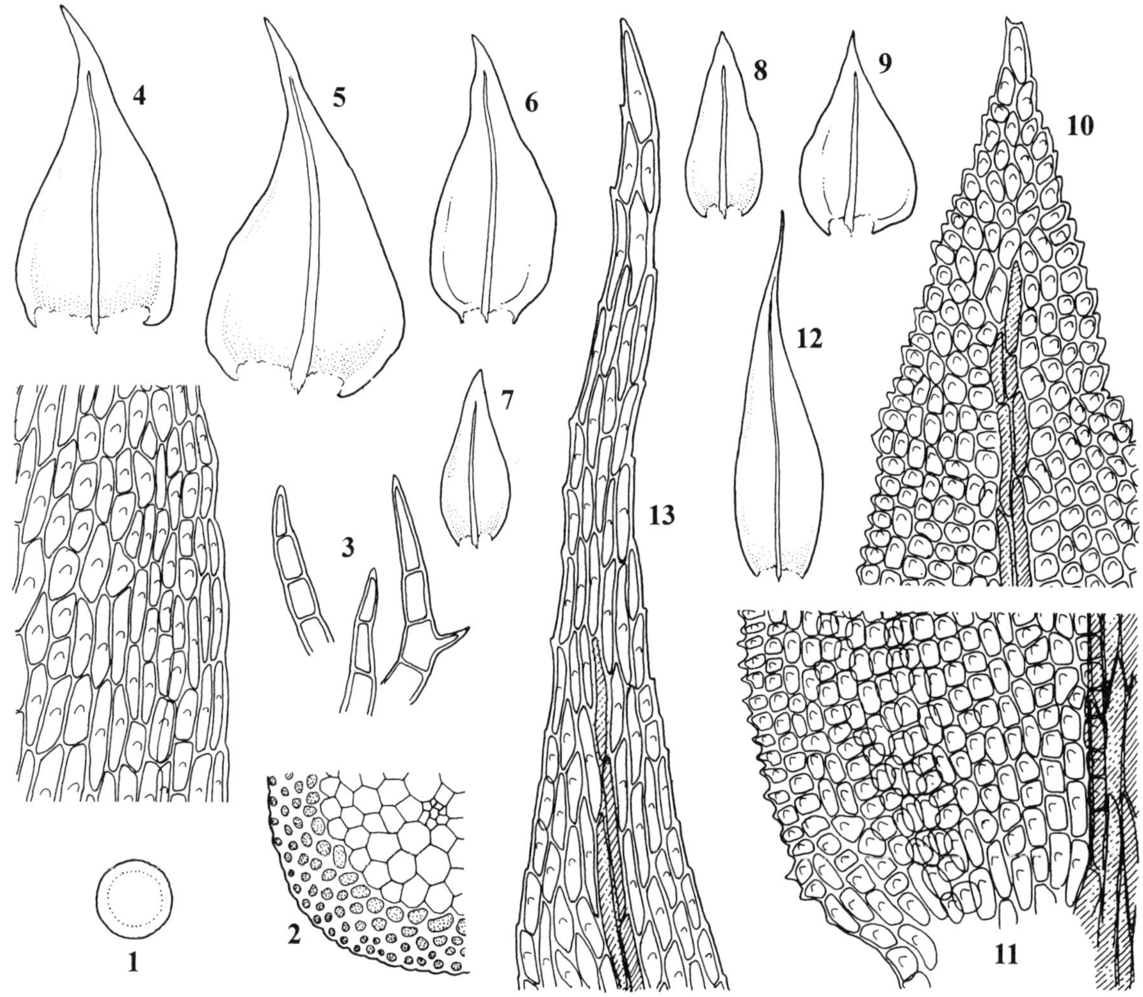

Plate 457. *Cyrto-hypnum fuscatum*. 1, stem cross section (× 52). 2, portion of stem cross section (× 258). 3, paraphyllia (× 443). 4–5, stem leaves (× 63). 6–9, branch leaves (× 63). 10, apical leaf cells (× 443). 11, basal leaf cells (× 443). 12, perichaetial leaf (× 63). 13, apical perichaetial leaf cells (× 443). (1–13, drawn from *Koponen 37779*, H.)

4. ***Cyrto-hypnum gratum*** (P. Beauv.) Buck & Crum 密毛细羽藓 mi-mao xi-yu xian

Contr. Univ. Michigan Herb. 17: 65. 1990. *Thuidium gratum* (P. Beauv.) Jaeg., Ber. Thätigk. St. Gallischen Naturwiss. Ges. 1876–77: 256. 1878. *Hypnum gratum* P. Beauv., Prodr. Aethéogam. 64. 1805. Type. West Africa: Benin.

Cyrto-hypnum pelekinoides (P.-C. Chen) Buck & Crum, Contr. Univ. Michigan Herb. 17: 67. 1990. *Thuidium pelekinoides* P.-C. Chen, Sunyatsenia 6: 190. f. 23. 1941. Types. China: Hainan, Pepu (Bei-bo), *C. Ho 1130b*, *1131* (syntypes PE).

Plants delicate, medium-sized, up to 4 cm long, grayish green to yellowish green or dark green, becoming brownish when old, in small patches. *Stems* creeping, bipinnately branched; cortical cells 3–5 layers, central strand present; paraphyllia densely growing on stems and the base of branches, filamentous, rarely foliose, simple or

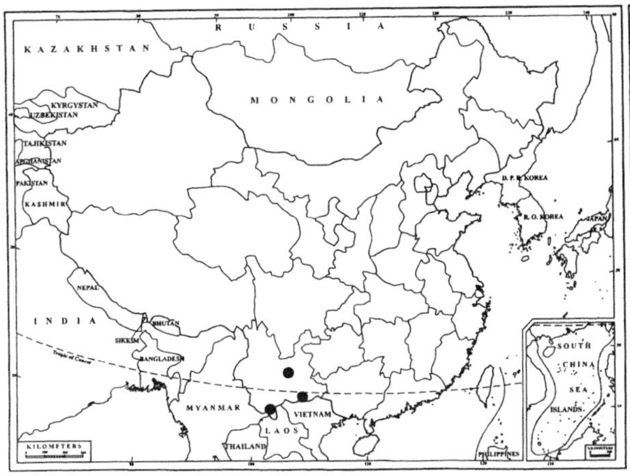

3. *Cyrto-hypnum fuscatum*

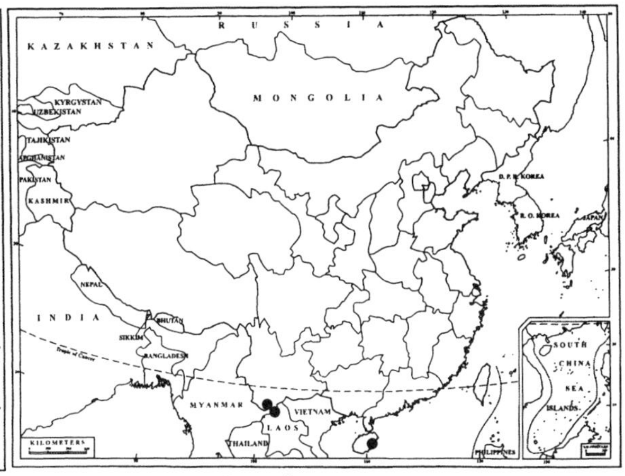

4. *Cyrto-hypnum gratum*

branched, up to 8 cells long, apical cell truncate, pluripapillose. *Stem leaves* remote, suddenly narrowed from an ovate or triangular base to a lanceolate or narrowly lanceolate apex, 0.8–1.0 mm × ca. 0.6 mm; margins revolute or recurved, serrate or cellulose; costa single, consisting of 3–4 rows of narrow and thick-walled cells, papillose on back; leaf cells 7.5–10.0 µm, hexagonal to irregularly rectangular, with single acute or 2–4 fine papillae, walls equally thickened; branch leaves dense, appressed when dry, widely cordate to ovate-triangular, with a short, obtuse apex, rarely acute; marginal cells projected, round-toothed, finely papillose; costa reaching 2/3–4/5 the leaf length, papillose on back. *Autoicous*. *Female inflorescences* on stems, with 7–8 perichaetial leaves; inner perichaetial leaves oblong-ovate; costa broad, excurrent into a long awn; margins remotely crenulated, sometimes with 8–10 single celled, hyaline, cilia above. *Setae* ca. 1.0–1.5 cm long, lightly brownish, reddish when old, densely granulate; capsules ovoid to shortly cylindrical, ca. 0.5–1.2 mm long, horizontal or inclined; opercula 1 mm long; peristome teeth well developed, cilia 1–3, isolated or connected, sometimes rudimentary. *Calyptrae* cucullate, coarse at the apex. *Spores* 10–20 µm in diameter.

Illustrations: P.-C. Wu (ed.) 2001 (Pl. 96, figs. 1–13).
Habitat: on moist soil, calcareous rocks, rotten logs or tree trunks in forests; alt. 800–1200 m.
Distribution: China, Nepal, Burma, Thailand, Indonesia, Malaysia, Fiji, Samoa, Central and West Africa, and Madagascar.

Chinese specimens examined: YUNNAN: Xishuangbanna, *W.-X. Xu 6556* (PE); Meng-la Co., *Redfearn et al. 33795* (MO, NY, SMSU); Jing-hong Co., *Redfearn 34294* (MO, NY, SMSU).

This species is characterized by having long ciliate stem leaf apices and by the papillosity of leaf cells, paraphyllia, and setae.

5. ***Cyrto-hypnum microphyllum*** (Schwaegr.) P.-C. Wu, Crosby & S. He 小叶细羽藓 xiao-ye xi-yu xian
Chenia 6: 7. 1999. *Leptohymenium microphyllum* (Schwaegr.) Schwaegr., Sp. Musc. Frond. Suppl. 3, 2(2): Index. 1830. *Haplohymenium microphyllum* Schwaegr., Sp. Musc. Frond. Suppl. 3, 2(1): [unpaged]. Pl. 271. 1829. Type. India: *Wallich s.n.*
Cyrto-hypnum haplohymenium (Harv.) Buck & Crum, Contr. Univ. Michigan Herb. 17: 66. 1990. *Thuidium haplohymenium* (Harv.) Jaeg., Ber. Thätigk. St. Gallischen. Naturwiss. Ges. 1876–77: 251. 1878. *Leskea haplohymenium* (Harv.) Mitt., J. Proc. Linn. Soc., Suppl. 1: 135. 1859. *Hypnum haplomenium* Harv., London J. Bot. 2: 21. 1843.
Cyrto-hypnum squarrosulum (Ren. & Card.) Buck & Crum, Contr. Univ. Michigan Herb.17: 67. 1990. *Thuidium squarrosulum* Ren. & Card., Bull. Soc. Roy. Bot. Belgique 38 (1): 31. 1900.
Thuidium brachymenium Herz., Hedwigia 65: 164. 1925.

Plants delicate, yellowish green, in small patches. *Stems* 1.0–1.5 cm long, creeping and spreading, irregularly branched to 1- or 2-pinnately branched; paraphyllia filamentous, densely on stems, apical cells sharp. *Stem leaves* remote, widely cordate or widely ovate, apex short, 0.5 mm long; margins serrate; costa ending below

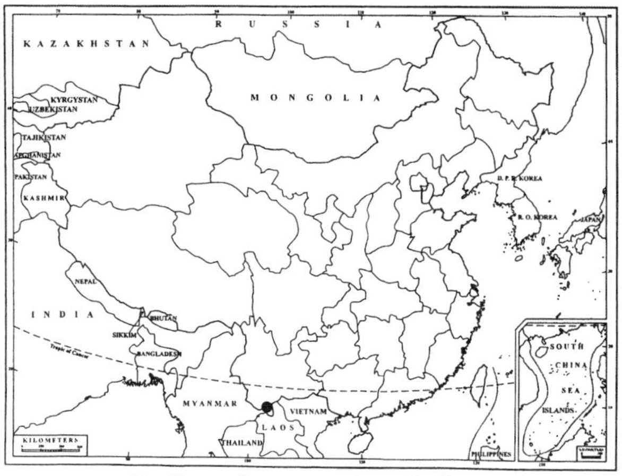

5. *Cyrto-hypnum microphyllum*

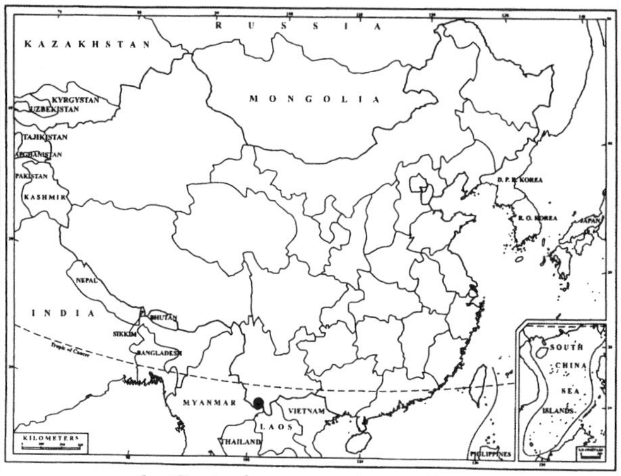

6. *Cyrto-hypnum minusculum*

leaf apex; leaf cells rounded hexagonal, unipapillose; branch leaves widely ovate, with a short apex, squarrose when moist. *Autoicous*. *Sporophytes* on main stems. *Perigonial leaves* concave, widely ovate, suddenly narrowed to a short lanceolate apex. *Perichaetial leaves* oblong-lanceolate, with a long acuminate apex; margins shortly ciliate above, ca. 1 mm long. *Setae* erect, curved above, 2 cm long; capsules oblong-ovoid, horizontal or declined, ca. 2–5 cm long; peristome double; exostome teeth lanceolate, hyaline near apex, densely papillose, trabeculate; endostome segments lightly yellowish; basal membrane ca. 2/5 the segment length; cilia 2, weak, short, nodulose. *Spores* large, 18 µm in diameter, papillose.

Illustrations: Gangulee 1978 (Fig. 816).
Habitat: alt. 3000 m
Distribution: China, Himalayan region, India, and Thailand.

Chinese specimens examined: YUNNAN: Xishuangbanna, *J.-X. Luo 86218a* (PE); Baitonling, *S. Ten 110* (PE).

This species is characterized by its stems densely covered with paraphyllia, branches circinate when dry, and leaves squarrose when moist.

6. ***Cyrto-hypnum minusculum*** (Mitt.) Buck & Crum 糙柄细羽藓 cao-bing xi-yu xian

Contr. Univ. Michigan Herb. 17: 66. 1990. *Thuidium minusculum* (Mitt.) Jaeg., Ber. Thätigk. St. Gallischen Naturwiss. Ges. 1876–77: 256. 1878. *Leskea minuscula* Mitt., J. Linn. Soc., Bot., Suppl. 1: 134. 1859. Syntypes. India: Khasia, *J. D. Hooker & T. Thomson 1071, 1072, 1092.*

Thuidium asperulisetum Ren. & Card., Bull. Soc. Roy. Bot. Belgique 38(1) 32. 1900.

Plants slender, yellowish green to brownish green, in densely interwoven mats. *Stems* creeping, irregularly pinnately branched, rarely bipinnately branched; branches usually flexuous, ca. 3 mm long; central strand differentiated, rarely absent; paraphyllia densely growing on stems, filamentous, 1–5 cells long, apical cells acute; pseudoparaphyllia foliose, at the base of branches. *Stem leaves* remote, widely cordate, squarrose above, with acute or short obtuse apex; margins undulate, serrulate, revolute at base; costa rather wide, ending below leaf apex; branches widely ovate, slightly concave, apex widely obtuse; margins serrate; costa vanishing 5–9 cells below leaf apex; leaf cells rectangular to hexagonal, 7–10 µm wide, unipapillose on both sides, walls equally thickened. *Autoicous*. *Female inflorescences* on main branches. *Perichaetial leaves* suddenly narrowed from an oblong-ovate base to a long lanceolate apex, serrate above. *Setae* erect, 1.5–2.0 cm long, flexuous, coarse above, smooth below; capsules cylindrical, inclined to horizontal; peristome teeth *Hypnum*-type; endostome segments nearly as long as exostome teeth; basal membrane moderately high, cilia 2. *Spores* not seen.

Illustrations: Plate 458, figs. 1–11.
Habitat: growing on tree trunks; alt. 1250–1300 m.
Distribution: China, Nepal, and India.

Chinese specimens examined: YUNNAN: Xishuangbanna, Meng-hai Co., *Redfearn 34167* (MO, NY, SMSU).

This species is characterized by having stem and branch leaves with broad bases, stems densely covered with acute paraphyllia, and setae coarsely papillose above.

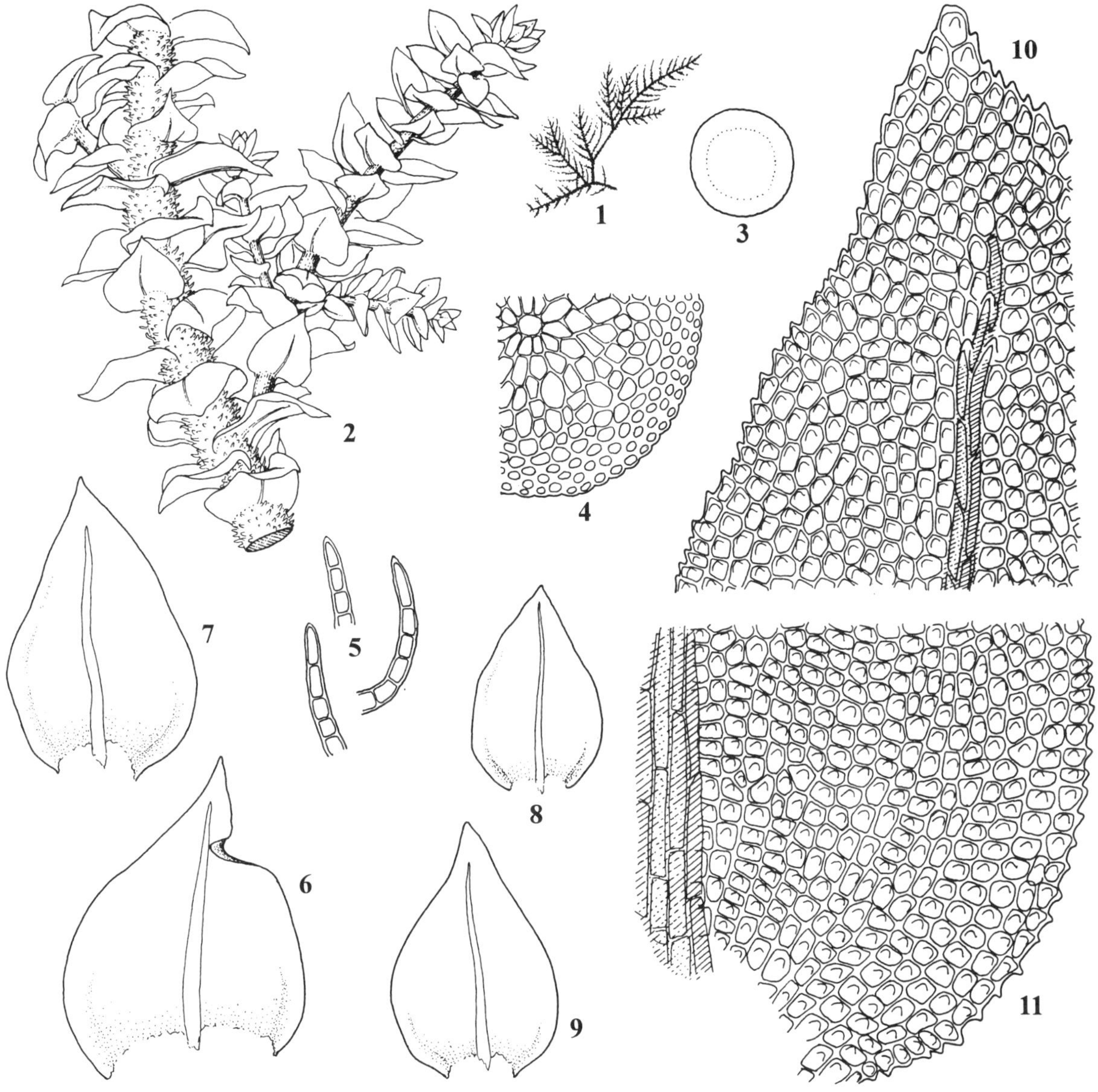

Plate 458. *Cyrto-hypnum minusculum*. 1, habit (× 2). 2, stem and branch (× 36). 3, stem cross section (× 75). 4, portion of stem cross section (× 260). 5, paraphyllia (× 266). 6–7, stem leaves (× 90). 8–9, branch leaves (× 90). 10, apical leaf cells (× 442). 11, basal leaf cells (× 442). (1–11, drawn from *Redfearn 34167*, MO.)

7. ***Cyrto-hypnum pygmaeum*** (Schimp.) Buck & Crum 多疣细羽藓 duo-you xi-yu xian

Contr. Univ. Michigan Herb. 17: 67. 1990. *Thuidium pygmaeum* Schimp. in B. S. G., Bryol. Eur. 5: 162. 1852. Type. U.S.A.: Ohio, Columbus.

Thuidium perpapillosum R. Watanabe, J. Jap. Bot. 34: 280. 1959.

Plants slender, small to medium-sized, 1–3 cm long, dark green to yellowish green, brownish when old, in small, interwoven mats. *Stems* creeping, regularly bipinnately branched; central strand present; paraphyllia filamentous, only on stems, apical cells 2–4 long, papillose; surfaces of both stems and branches papillose. *Stem*

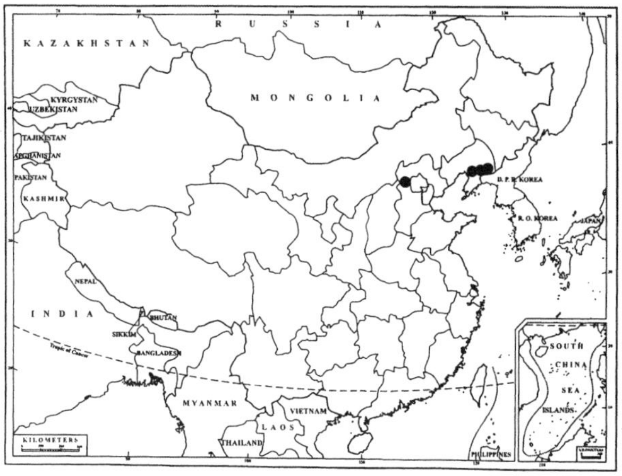

7. *Cyrto-hypnum pygmaeum*

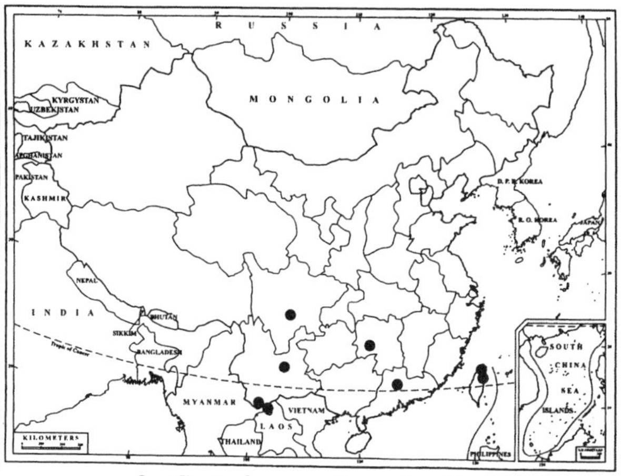

8. *Cyrto-hypnum tamariscellum*

leaves remote, appressed when dry, triangularly ovate, 0.3 mm long, acuminate at apex; margins serrate, revolute; costa hyaline, ending at upper portion of leaves, smooth on back; median leaf cells quadrate to polygonal, 5–12 µm wide, thin walls, 3–8 papillae per cell, apical cells 1–3 long, papillose or smooth; branch leaves concave, incurved when dry, narrowly ovate to triangular, apex short, 0.2 mm long. *Autoicous. Perichaetial leaves* oblong-ovate, acuminate to lanceolate toward apex; margins serrate above. *Setae* slender, yellowish brown to reddish brown, ca. 1 cm long, smooth; capsules oblong-ovate, curved, reddish brown; exostome teeth yellowish brown; endostome segments lightly yellowish or hyaline, finely papillose; basal membrane 1/2 the height of endostome segments; cilia 2. *Spores* ca. 10 µm in diameter, minutely papillose.

Illustrations: P.-C. Wu (ed.) 2001 (Pl. 93, figs. 1–10).
Habitat: mostly on calcareous rocks, rarely on moist soil.
Distribution: China, Korea, Japan, and North America.

Chinese specimens examined: HEBEI: Zhuo-lu Co., without collector's name 28 (PE). LIAONING: Kuan-dian Co., *Buck 23709, 23767* (NY); Mt. Koô (Mt. Fenghuang), *Kobayasi 3892* (BM); Qianshan (Mt.), *C. Gao 1993* (IFSBH, PE).

8. ***Cyrto-hypnum tamariscellum*** (C. Müll.) Buck & Crum 密枝细羽藓 mi-zhi xi-yu xian

Contr. Univ. Michigan Herb. 17: 68. 1990. *Thuidium tamariscellum* (C. Müll.) Bosch & Sande Lac., Bryol. Jav. 2: 120. 1865. *Hypnum tamariscellum* C. Müll., Bot. Zeitung (Berlin) 12: 573. 1854. Type. India: Nilghiri.

Cyrto-hypnum rubiginosum (Besch.) Buck & Crum, Contr. Univ. Michigan Herb. 17: 67. 1990. *Thuidium rubiginosum* Besch., Ann. Sci. Nat. Bot. ser. 7, 15: 80. 1892. Type. China: Yunnan: Ta-long-tan, *Delavay 4849* (holotype BM; isotype PC).

Cyrto-hypnum sparsifolium (Mitt.) Buck & Crum, Contr. Univ. Michigan Herb. 17: 67. 1990. *Thuidium sparsifolium* (Mitt.) Jaeg., Ber. Thätigk. St. Gallischen. Naturwiss. Ges. 1876–77: 248. 1878. *Leskea sparsifolia* Mitt., J. Proc. Linn. Soc., Bot., Suppl. 1: 135. 1859.

Thuidium bipinnatulum Mitt., Trans. Linn. Soc. London, Bot. 3: 190. 1891.

Thuidium micropteris Besch., Ann. Sci. Nat. Bot., ser. 7, 17: 367. 1893.

Plants soft, dark green to yellowish green, brownish when old, in thin, interwoven mats. *Stems* 2–4 cm long, regularly bipinnately branched; branches dense, 1–3 mm long; central strand present; paraphyllia numerous on stems, rarely on branches, filamentous, 1–3 cells long, apical cells 2–4, papillose. *Stem leaves* remote, slightly concave, broadly triangular to triangularly ovate, 0.3–0.4 mm, rarely 0.7 mm long, acuminate at apex; margins serrulate below, entire near apex; costa stout, ending near leaf apex, smooth or toothed on back; median leaf cells rhomboidal to oval, ca. 7–12 µm wide, multipapillose; branch leaves ovate to oblong-ovate, apex short, ca. 0.2 mm long, costa vanishing at 1/2–2/3 the leaf length; median leaf cells rounded quadrate to hexagonal, multipapillose, apical cells 2–4 long, papillose. *Autoicous. Perichaetial leaves* widely ovate to oblong-ovate at base, suddenly narrowed to a lanceolate apex, serrate. *Setae* yellowish brown to reddish brown, 1.0–1.5 cm long, smooth below, roughened above;

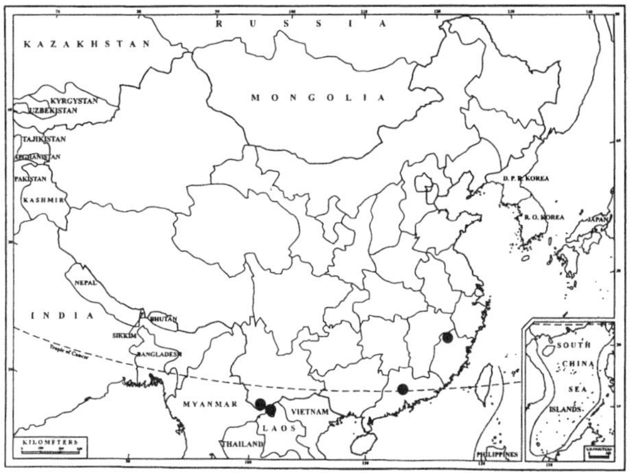

9. *Cyrto-hypnum versicolor*

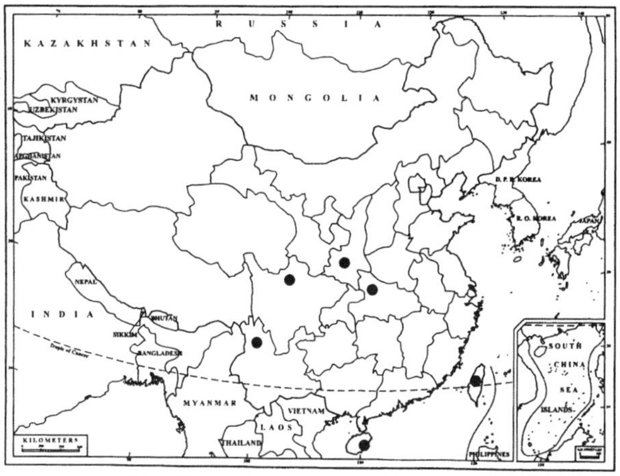

10. *Cyrto-hypnum vestitissimum*

capsules ca. 2 mm long, ovoid; peristome teeth similar to those of *C. pygmaeum*; endostome segments as long as the teeth; basal membrane 1/2 the height of endostome segments, keeled, perforate; cilia 2, nodulose. *Spores* ca. 15 µm in diameter, smooth.

Illustrations: Watanabe 1972 (Pl. 2, figs. 1–18, as *Thuidium tamariscellum*).
Habitat: on forest ground, humus, tree trunks or moist soil; alt. 500–2500 m.
Distribution: China, Korea, Japan, India, Myanmar, and Thailand.

Chinese specimens examined: GUANGDONG: Mt. Lung T'au, *Nat. Geogr. Soc. Cent. Exp. 12216* (as *Thuidium rubiginosum*, BM). HUNAN: Wu-gang City (Wukang), Yun-shan, *Handel-Mazzetti 11192* (BM). SICHUAN: Mt. Omei, *P.-C. Chen 187a* (PE), *K.-Y. Lang 303b* (PE). TAIWAN: Hua-lien Co., *C.-I. Peng 83-71* (MO); Ilan Co., *C.-I. Peng 83-47* (MO); Tai-pei Co., *C.-C. Chuang 5426* (PE, UBC), *M.-J. Lai 7330* (MO). YUNNAN: Jing-hong Co., *Redfearn et al. 34005* (MO); Kun-ming City, *Handel-Mazzetti 516* (BM), *Koponen 37524* (H); Meng-la Co., *L. Zhang 492* (IBSC, PE).

9. ***Cyrto-hypnum versicolor*** (C. Müll.) Buck & Crum 红毛细羽藓 hong-mao xi-yu xian
Contr. Univ. Michigan Herb. 17: 68. 1990. *Thuidium versicolor* (C. Müll.) Jaeg. Ber. Thätigk. St. Gallischen Naturwiss. Ges. 1876–77: 249. 1878. *Hypnum versicolor* Hornsch. ex. C. Müll., Syn. Musc. Frond. 2: 494. 1851. Type. South Africa: Uitenhage.

Plants slender, small to medium-sized, up to 4 cm long, yellowish green to dark green or brownish, creeping to ascending, in small, lax mats. *Stems* remotely or densely bipinnately branched, sometimes with strong branches; rhizoids dense; central strand present; paraphyllia densely growing on stems, usually simple, 2–6 cells long, apical cells truncate and papillose, rarely acute and branched. *Stem leaves* remote, cordate-triangular to widely cordate-triangular, 0.25–0.50 µm long, gradually or suddenly narrowed to a lanceolate apex, usually secund above; margins narrow at base or widely revolute, papillose or entire; costa stout, vanishing below leaf apex; branch leaves dense, ovate to triangularly ovate; leaf cells polygonal to rectangular, 3–6 µm × 6–12 µm, both antical and ventral sides with sharp papillae. *Autoicous*. *Inner perichaetial leaves* oblong-ovate at base, long lanceolate above, sometimes longitudinally plicate; margins serrate, rarely ciliate. *Setae* ca. 2 cm long, orange red to orange yellow, finely papillose above; capsules oblong, brown to orange, inclined to horizontal; opercula 0.5–1.0 mm long; cilia 1–3, mostly connected at base. *Spores* 10–17 µm in diameter.

Illustrations: Plate 459, figs. 1–19.
Habitat: on rotten logs or tree trunks in wet forests, also on rocks, caves or soil; alt. 1100–1150 m.
Distribution: China, Madagascar, and Central and South Africa.

Chinese specimens examined: FUJIAN: Mt. Wuyi, *P.-J. Lin 386* (IBSC, NY). GUANGDONG: *K.-P. To et al.* s.n. (BM, NY). YUNNAN: Jing-hong Co., *Redfearn et al. 34005* (MO, NY, SMSU); Meng-hai Co., *Redfearn et al. 34141* (MO, NY); Meng-la Co., *Redfearn et al. 33900a* (MO, NY).

This species is characterized by having abundant reddish rhizoids mingling with stem leaves, apical cells of paraphyllia either truncate or acute, and the perichaetial leaves ciliate at margins.

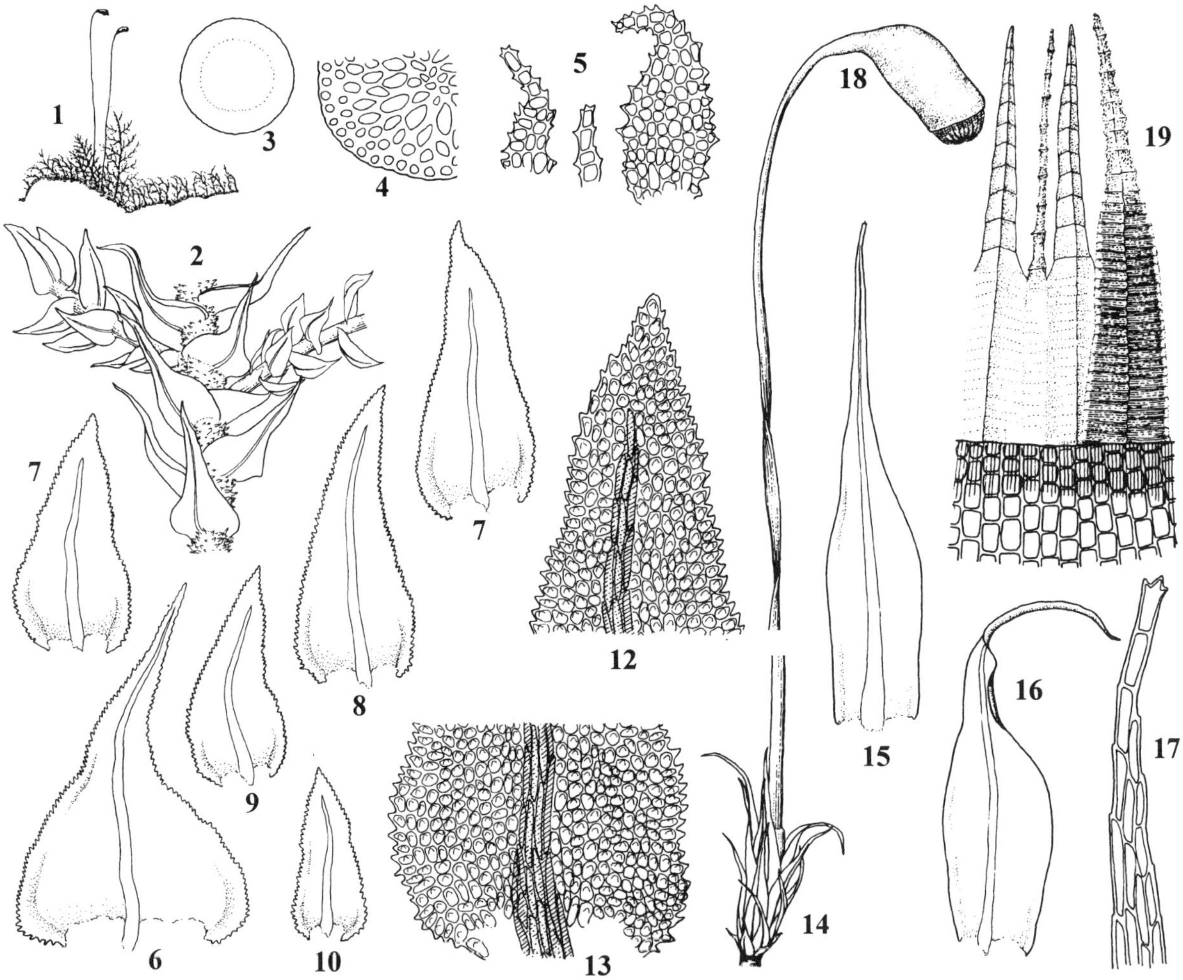

Plate 459. *Cyrto-hypnum versicolor*. 1, habit (× 1.5). 2, portion of stem and branch (× 39). 3, stem cross section (× 142). 4, portion of stem cross section (× 275). 5, paraphyllia (× 275). 6, stem leaf (× 103). 7–9, branch leaves (× 103). 10–11, branchlet leaves (× 103). 12, apical leaf cells (× 275). 13, basal leaf cells (× 275). 14, perichaetium (× 24). 15–16, perichaetial leaves (× 55). 17, apical perichaetial leaf cells (× 275). 18, capsule (× 24). 19, peristome (× 194). (1–19, drawn from *S.-P. Du et al. 12211*, PE.)

10. ***Cyrto-hypnum vestitissimum*** (Besch.) Buck & Crum 多毛细羽藓 duo-mao xi-yu xian
Contr. Univ. Michigan Herb. 17: 67. 1990. *Thuidium vestitissimum* Besch., Ann. Sci. Nat. Bot., ser. 7, 15: 79. 1892. Type. China: Yunnan, Santchang-kiou (Hokin), *Delavay 4768d* (holotype BM; isotype H).
Thuidium lepidoziaceum Sak., Bot. Mag. (Tokyo) 69: 88. 1947.

Plants delicate, medium-sized, ca. 3–5 cm long, dark green to yellowish green, brownish when old, in thin, interwoven matches. *Stems* regularly bipinnately branched; branches 5 cm long; central strand differentiated; paraphyllia densely growing on stems and branches, filamentous, branched, up to 10 cells long, apical cells multipapillose. *Stem leaves* incurved when dry, erect-patent when moist, ca. 0.6 mm long, gradually narrowed from a triangular base to a lanceolate apex, secund toward one side; margins serrate; costa hyaline, up to 4/5 the leaf length, mammillose on back; median leaf cells opaque, rhomboidal to rectangular, 8–20 µm wide, thin-walled, each cells with a single fine papilla; apical cells oblong, smooth; branch leaves appressed when dry, triangular to triangularly ovate; costa up to or near leaf apex. *Autoicous*. *Perichaetial leaves* erect, oblong-lanceolate, serrate. *Seta* reddish brown, ca. 2 cm long, smooth; capsules oblong-ovoid or short-cylindrical, ca. 1.5 mm long, horizontal or declinate; opercula conic, long beaked; peristome teeth *Hypnum*-type; cilia undeveloped. *Spores* ca. 10–15 µm in diameter, finely papillose.

Illustrations: Plate 460, figs. 1–14.
Habitat: on rock cliffs, limestone rocks or humus in forests; alt. 2200–2500 m.
Distribution: China, Japan, India, and Russia (Siberia).

Chinese specimens examined: HAINAN: Baipo, *G. He 1131* (PE). HUBEI: Shennongjia Mt., *Luteyn 1462* (MO, NY, PC, PE), *P.-C. Wu 386* (MO, PE). SHAANXI: Wei-tze-ping, *Licent 57* (BM). SICHUAN: Ma-er-kang Co., *Y. Jia 3547*, *3548* (PE). TAIWAN: Nan-tuo Co., *Nakn 13506* (TAI).

This species is characterized by having abundant filamentous paraphyllia on branches. The paraphyllia are papillose and ca. 10 cells long.

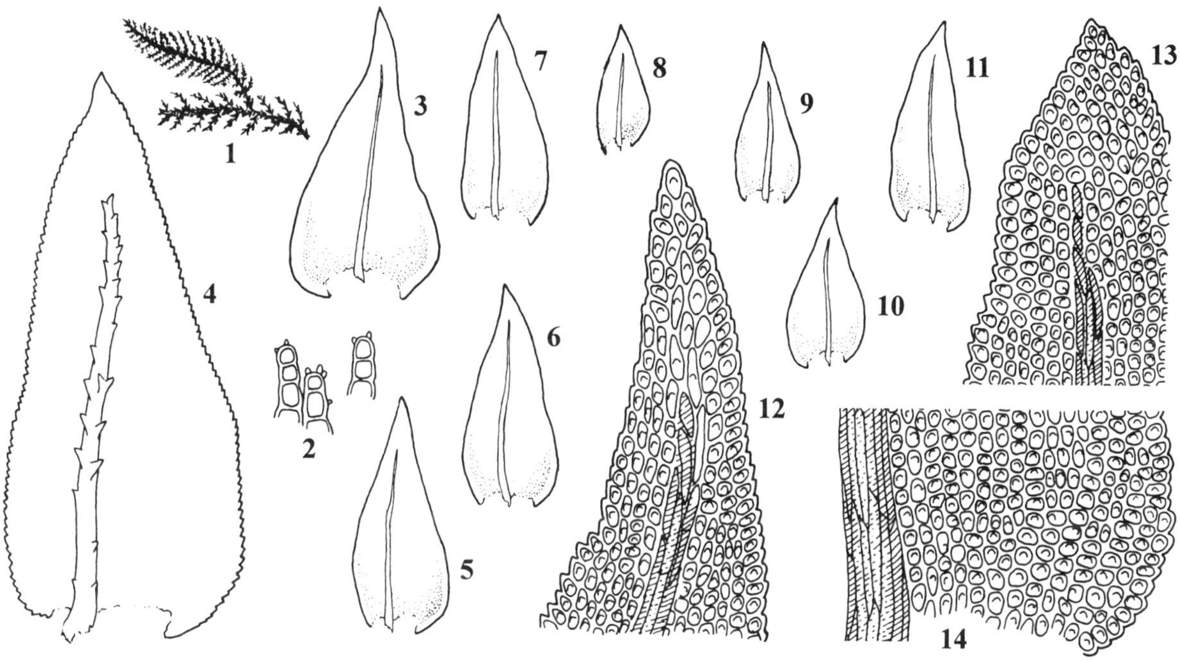

Plate 460. *Cyrto-hypnum vestitissimum*. 1, habit (× 1). 2, paraphyllia (× 385). 3, stem leaf (× 78). 4, branch leaf, showing spinulose costa (× 180). 5–7, branch leaves (× 180). 8–11, branchlet leaves (× 180). 12, apical stem leaf cells (× 385). 13, apical branch leaf cells (× 385). 14, basal branch leaf cells (× 385). (1–14, drawn from *Y.-G. Su 2593*, HKAS.)

7. ***Haplocladium*** C. Müll. 小羽藓属 xiao-yu xian shu

Hedwigia 38: 149. 1899.
Bryohaplocladium R. Watan. & Z. Iwats., J. Jap. Bot. 56: 259. 1981.

Plants rather slender to medium-sized, yellowish green, brownish green or light green, in loose interwoven mats or mixed with other mosses. *Stems* creeping, nearly pinnately or irregularly branched; paraphyllia sparse or numerous, variable in shape; central strand present. *Stem leaves* and branch leaves somewhat differentiated, revolute or slightly secund when dry, erect-spreading when moist; stem leaves plicate, ovate, toward a lanceolate apex; margins plane or slightly revolute; costa single, percurrent to excurrent; leaf cells irregularly quadrate or rhomboidal, walls equally thickened, unipapillose or projecting at upper ends; branch leaves rather small and narrow. *Autoicous*. *Inner perichaetial leaves* oblong-ovate, with a narrow lanceolate apex, pale green. *Setae* slender, smooth, reddish brown; capsules ovoid to cylindrical, horizontal or inclined, contracted below the mouth, stomata developed; annuli present; opercula conical, shortly beaked; exostome teeth yellowish or yellowish brown; endostome segments well developed; cilia 2–3, nodulose. *Calyptrae* smooth. *Spores* medium-sized, papillose.

Haplocladium is closely related to *Thuidium*, but differs from the latter by its small, simple pinnately branched stems with fewer paraphyllia, somewhat differentiated stem and branch leaves, and unipapillose or projecting leaf cells.

The species of *Haplocladium* often occur under shady, moist grasses, rocks, rotten logs, and tree trunks. The genus is usually confined to low elevation near 1000 m. There are about 15 species of *Haplocladium* in the world, and five species are found in China. Recently, Y.-M. Fang and Koponen (2001) reported *H. perparaphyllum* Watan. from China, but we have not had an opportunity to study the specimens.

1. Stem leaves broadly ovate at the base, gradually tapering to a lanceolate apex 2
1. Stem leaves ovate at the base, widely and shortly lanceolate at apex 3
    2. Leaf cells projecting at the upper ends .... 1. *H. angustifolium*
    2. Leaf cells unipapillose .... 4. *H. microphyllum*
3. Paraphyllia numerous on stems and branches; the apex of branch leaves short and acute .... 5. *H. strictulum*
3. Paraphyllia sparse on stems and branches; the apex of branch leaves usually lanceolate 4
    4. Paraphyllia mostly filamentous .... 2. *H. discolor*
    4. Paraphyllia foliose, often branched .... 3. *H. larminatii*

1. ***Haplocladium angustifolium*** (Hampe & C. Müll.) Broth. 狭叶小羽藓 xia-ye xiao-yu xian
Nat. Pflanzenfam. I(3): 1008. 1917. *Bryohaplocladium angustifolium* (Hampe & C. Müll.) Watanabe & Iwats., J. Jap. Bot. 56: 259. 1981. *Hypnum angustifolium* Hampe & C. Müll., Bot. Zeitung (Berlin) 13: 88. 1855. Type. South Africa: Krakakamma, *Ecklon s.n.*

Haplocladium incurvum Broth., Symb. Sin. 4: 99. 1929. Type. China: Hunan, Changsha (Tschangscha) City, *Handel-Mazzetti 12912* (isotype BM).

Haplocladium lutescens (Card.) Broth. ex Thér., Ann. Cryptog. Exot. 3: 82. 1930. *Pseudoleskea lutescens* Card., Bull. Soc. Bot. Genève, ser. 2, 3: 284. 1911. Type. China: Taiwan, Takao, *Faurie 2399.*

Haplocladium subulaceum (Mitt.) Broth., Nat. Pflanzenfam. I(3): 1008. 1907. *Leskea subulacea* Mitt., J. Proc. Linn. Soc., Bot., Suppl. 1: 131. 1859.

Haplocladium subulaceum (Mitt.) Broth. var. *amblystegioides* (Broth. & Par.) Thér., Ann. Cryptog. Exot. 3: 93. 1930. *Haplocladium amblystegioides* (Broth. & Par.) Broth., Nat. Pflanzenfam. I(3): 1008. 1917. *Thuidium amblystegioides* Broth. & Par., Rev. Bryol. 31: 57. 1904.

Haplocladium subulaceum (Mitt.) Broth. var. *fuscissimum* (C. Müll.) Thér., Ann. Cryptog. Exot. 3: 93. 1930. *Thuidium fuscissimum* (C. Müll.) Par., Ind. Bryol. 1281. 1898. *Haplocladium fuscissimum* C. Müll., Nuovo Giorn. Bot. Ital., n. ser. 4: 275. 1897. Type. China: Shaanxi (Schen-si), in monte Lao-y-san, March 1896, *Giraldi s.n.*

Haplocladium subulaceum (Mitt.) Broth. var. *macropilum* (C. Müll.) Thér., Ann. Cryptog. Exot. 3: 91. 1930. *Pseudoleskea macropilum* (C. Müll.) Salmon., J. Linn. Soc. Bot. 34: 471. 1900. *Thuidium macropilum* (C. Müll.) Par., Ind. Bryol. 1285. 1898. *Haplocladium macropilum* C. Müll., Nuovo Giorn. Bot. Ital., n. ser. 3: 116. 1896. Syntypes. China: Shaanxi (Schen-si), Hua-tzo-pin, 20 June 1894, *Giraldi s.n.*; Si-ku-tziu-san, July 1894, *Giraldi s.n.*

Haplocladium subulaceum (Mitt.) Broth. var. *subulatum* (Card.) Thér., Ann. Cryptog. Exot. 3: 91. 1930. *Haplocladium subulatum* Card., Bull. Soc. Bot. Genève, ser. 2, 3: 282. 1911.

Thuidium rubicundulum (C. Müll.) Par., Ind. Bryol. Suppl. 1: 321, 1900. *Haplocladium rubicundulum* C. Müll., Nuovo Giorn. Bot. Ital., n. ser. 5: 208. 1898. Syntypes. China: Shaanxi (Schen-si), Sche-kin-tsuen, April 1896, *Giraldi 2211* (syntype BM); Huo-kia-zaea, March 1897, *Giraldi s.n.*; Schan-gen-ze, March 1897, *Giraldi s.n.*

Plants small to medium-sized, green to yellowish green, brown when old, forming interwoven patches. *Stems* creeping, up to 10 mm long, regularly pinnately branched, branches ca. 5 mm long; central strand present; paraphyllia lanceolate, mostly on stems and rarely on branches. *Stem leaves* concave, gradually tapering from ovate to widely ovate base to a slender lanceolate apex; leaf margins revolute or plane, serrate; costa stout, often excurrent; median leaf cells quadrate to rhomboidal, 5–25 µm wide, thick-walled, projecting at upper ends; apical leaf cells narrow, rhomboidal, smooth; branch leaves ovate to narrowly ovate at base, with a short or long lanceolate apex. *Dioicous. Perichaetial leaves* entire at the margin. *Setae* 3–4 cm long, orange-red; opercula conical; peristome similar to *H. microphyllum* (Hedw.) Broth.

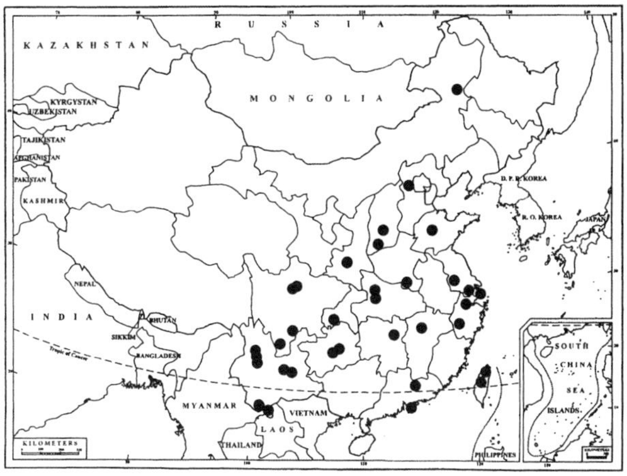

1. *Haplocladium angustifolium*

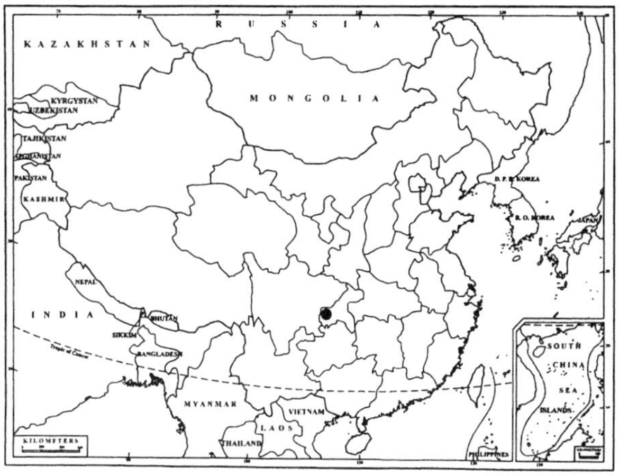

2. *Haplocladium discolor*

Illustrations: P.-C. Wu (ed.) 2001 (Pl. 88, figs. 1–17).
Habitat: mostly on rocks, rarely on the bases of trees or rotten logs; alt. 1500–1700 m.
Distribution: China, Korea, Japan, Vietnam, Burma, India, Pakistan, Russia (Siberia), Europe, and Africa.

Chinese specimens examined: CHONGQIAN: Mt. Jinfou, *M.-Z. Wang 21146* (PE). GUANGDONG: Mt. Luofu, *Magill et al. 8139* (MO). GUIZHOU: Gui-yang City, *Gui-yang Teacher's Coll. 139* (MO, PE); Ping-ba (Pin-fa) Co., *Cavalerie 1737* (PC). HEBEI: Mt. Dong-ling, *L.-Q. Yu 3* (MO, PE). HENAN: Xin-yang City, *J.-X. Luo 75* (MO, PE). HONG KONG: *Bowing s.n.* (NY). HUBEI: Mt. Shennongjia, *P.-C. Wu 60, 61, 103, 112a* (MO, PE); Xing-shan Co., *J.-X. Luo 961270* (MO, PE). JIANGSU: Nan-jing (Nanking) City, *Steward 943* (BM, NY); Wu Co., *P.-C. Wu & Y.-X. Jin 1631* (MO, PE). JIANGXI: Nan-chang City, Hb. Mitten, *Collector unknown* (NY). NEI MONGOL: Zha-lan-dun City, Ba-lin, *X.-L. Bai 2504* (NY). SHAANXI: Mt. Qinling, *P.-C. Chen et al. 22* (MO, PE); Mt. Taibai, *Z.-P. Wei 5233* (MO); Tui Kio San, *Giraldi 2204* (BM, PC). SHANDONG: Tai-shan (Mt.), *C. Gao & E.-Z. Bai 34029a* (MO). SHANGHAI: Zaukadoo, *Bailey s.n.* (BM, NY). SHANXI: Yuan-qu Co., *Y.-J. Ling 85021* (PE); Xin-yuan Co., *Y.-J. Ling 85069, 85070, 85071* (PE, SXU). SICHUAN: Lei-bo Co., *P.-C. Chen 5891* (MO, PE); Li Co., *X.-J. Li 915* (MO, PE); Mt. Er-lang, *C. Gao 18252* (MO); Yenyuen (Yan-yuan) Co., *Handel-Mazzetti 2598* (BM). TAIWAN: Nang-tou Co., *C.-C. Chung & W. B. Schofield 807* (BCU, NY, PE); Tai-pei Co., *M.-J. Lai 10053* (as *Haplocladium microphyllum*, MO, NY). YUNNAN: An-ning Co., *Magill 8062* (MO); Da-li Co., *X.-Y. Jia 910075* (MO); Lu-nan Co., *Redfearn et al. 2290* (MO); Meng-hai Co., *Redfearn 34107* (MO); Meng-la Co., *Magill et al. 7877* (MO); Rui-li Co., *X.-Y. Jia 910063* (MO); Song-ming Co., *Koponen 37815* (H, MO); Wei-shan Co., *Redfearn et al. 1212* (MO, NY); Yang-bi Co., *Redfearn & Y.-G. Su 144* (MO). ZHEJIANG: Hang-zhou City, *Buck 23853* (NY); Sui-chang Co., *Z.-L. Liu & R.-L. Hong 35* (HSNU, MO).

This species is characterized by having projecting leaf cells and long lanceolate leaf apices.

2. **Haplocladium discolor** (Par. & Broth.) Broth. 退色小羽藓 tui-se xiao-yu xian
Nat. Pflanzenfam. I(3): 1007. 1907. *Bryohaplocladium discolor* (Par. & Broth.) Watanabe & Iwats., J. Jap. Bot. 56: 259. 1981. *Thuidium discolor* Par. & Broth., Rev. Bryol. 31: 58. 1904. Type. Japan: Aomori Pref., *Faurie 1875*.
Haplocladium miser (Par. & Broth.) Broth., Nat. Pflanzenfam. I(3): 1007. 1907. *Thuidium miser* Par. & Broth., Rev. Bryol. 31: 59. 1904.

Plants small, soft, dark green, brownish when old. *Stems* procumbent, ca. 3 cm long, irregularly branched; central strand present; paraphyllia sparse, filamentous. *Stem leaves* appressed when dry, spreading when moist, ovate to oblong-ovate, with a widely lanceolate apex, ca. 1 mm long; margins serrate; costa reaching 3/4 the leaf length; median leaf cells quadrate to hexagonal, 5–15 µm wide, thick-walled, unipapillose; branch leaves oblong-ovate, with a short lanceolate apex, rarely obtuse. *Perichaetial leaves* up to 2 mm long, widely lingulate. *Setae* yellowish brown, 8–10 mm long; capsules erect to nearly erect or pendulous, oblong-cylindrical; peristome features the same as those of *H. microphyllum* (Hedw.) Broth. *Spores* papillose.

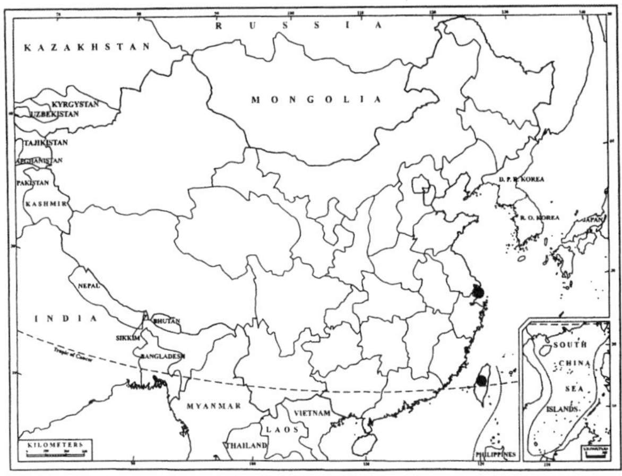

3. *Haplocladium larminatii*

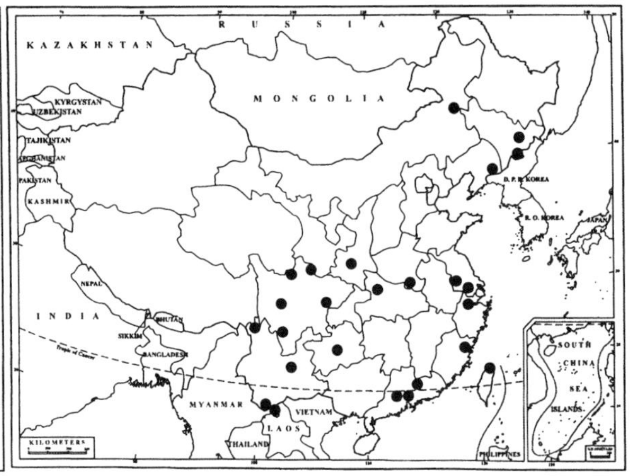

4. *Haplocladium microphyllum*

Illustrations: Watanabe 1972 (Pl. 19, figs. 1–15).
Habitat: on shady and moist soil.
Distribution: China, Korea, and Japan.

Chinese specimens examined: CHONGQING: Chong-qing City, *P.-C. Chen 41* (PE).

This species is characterized by its oblong-ovate leaves and rhomboidal leaf cells.

3. ***Haplocladium larminatii*** (Broth. & Par.) Broth. 瓦叶小羽藓 wa-ye xiao-yu xian
Nat. Pflanzenfam. (ed. 2) 11: 320. 1925. *Bryohaplocladium larminatii* (Broth. & Par.) Watanabe & Iwats., J. Jap. Bot. 56: 259. 1981. *Pseudoleskea larminatii* Broth. & Par., Rev. Bryol. 31: 57. 1904. Type. Vietnam: Hanoi.
?*Claopodium leskeoides* (Broth. & Par.) Broth., Nat. Pflanzenfam. (ed. 2) 11: 318. 1925. *Ctenidium leskeoides* Broth. & Par., Rev. Bryol. 37: 3. 1910. Type. China: Shanghai, Zi Ka Wei, 1908, *Courtois s.n.*
Haplocladium imbricatum Broth., Ann. Bryol. 1: 22. 1928. Syntypes. China: Taiwan, Taizhong, *T. Suzuki 2826, 2851, 2855.*
Haplocladium leskeoides Card., Bull. Soc. Bot. Genève, ser. 2, 3: 282. 1911.

Plants small, green to yellowish green, yellowish brown when old, loosely spreading, in interwoven patches. *Stems* procumbent, 3 cm long, regularly pinnately branched; branches ca. 1 cm long; paraphyllia sparse, lanceolate, often branched. *Stem leaves* variable, appressed when dry, broadly ovate, cordate to ovate-lanceolate, ca. 0.5–1.0 mm long; margins plane or partly revolute, serrate; costa reaching near leaf apex, smooth on back; median leaf cells pentagonal to hexagonal, thick-walled, uni- to mostly multipapillose; apical cells triangular to rhomboidal, smooth or finely papillose; branch leaves widely ovate-lanceolate, costa vanishing below the leaf apex. *Perichaetial leaves* entire. *Setae* reddish brown; capsules cylindrical; peristome features the same as those of *H. microphyllum* (Hedw.) Broth.

Illustrations: Watanabe 1972 (Pl. 21, figs. 1–14).
Habitat: unknown.
Distribution: China, Japan, and Vietnam.

Chinese specimen examined: JIANGSU: Without locality, *Courtois* 117 (PE).

This species is characterized by its ovate-lanceolate branch leaves with undulate leaf margins.

4. ***Haplocladium microphyllum*** (Hedw.) Broth. 细叶小羽藓 xi-ye xiao-yu xian
Nat. Pflanzenfam. I(3): 1007. 1907. *Bryohaplocladium microphyllum* (Hedw.) Watanabe & Iwats., J. Jap. Bot. 56: 260. 1981. *Haplocladium microphyllum* (Hedw.) Broth. subsp. *eumicrophyllum* Reim., Hedwigia 227. 1937, *nom. illeg. Hypnum microphyllum* Hedw., Sp. Musc. Frond. 269. Pl. 69: f. 1–4. 1801. Type. Jamaica: *Swartz s.n.*

Claopodium fulvellum Herz., Hedwigia 65: 163. 1925. Type. China: Yunnan, Beiyun Temple, alt. 3000 m, *S. Ten 45* (isotype PE).
Claopodium integrum P.-C. Chen in S.-J. Li, Investig. Alp. For. W. Sichuan. 423. 1963, *num nud.*
Haplocladium capillatum (Mitt.) Broth. fo. *robustum* Dix. in Yang, Sci. Rep. Natl. Tsing Hua Univ., Ser. B, Biol. Sci. 2(2): 127. 1936, *nom. nud.*
Haplocladium capillatum (Mitt.) Broth. fo. *sublaecifolium* Dix. in Yang, Sci. Rep. Natl. Tsing Hua Univ., Ser. B, Biol. Sci. 2(2): 127. 1936, *nom. nud.*
Haplocladium capillatum (Mitt.) Broth. var. *mittenii* Thér., Ann. Cryptog. Exot. 3: 87. 1930.
Haplocladium capillatum (Mitt.) Broth. var. *subcapillatum* (Broth. & Par.) Thér., Ann. Cryptog. Exot. 3: 89. 1930. *Thuidium subcapillatum* Broth. & Par., Bull. Herb. Boissier, ser. 2, 2: 929. 1902.
Haplocladium microphyllum (Hedw.) Broth. subsp. *capillatum* (Mitt.) Reimers, Hedwigia 76: 227. 1937. *Haplocladium capillatum* (Mitt.) Broth., Nat. Pflanzenfam. I(3): 1008. 1907. *Amblystegium capillatum* (Mitt.) Mitt., Trans. Linn. Soc. London, Bot. 3: 186. 1891. *Pseudoleskea capillata* (Mitt.) Sauerbeck in Jaeger, Ber. Thätigk. St. Gallischen Naturwiss. Ges., 1877–78: 475. 1880. *Thuidium capillatum* (Mitt.) Jaeg., Ber. Thätigk. St. Gallischen. Naturwiss. Ges., 1876–77: 252. 1878. *Leskea capillata* Mitt., J. Proc. Linn. Soc., Bot., Suppl. 1: 130. 1859.
Haplocladium microphyllum (Hedw.) Broth. var. *papillariaceum* (C. Müll.) Redfearn, B. C. Tan & S. He, J. Hattori Bot. Lab. 79: 233. 1996, *comb. inval. Haplocladium capillatum* (Mitt.) Broth. var. *papillariaceum* (C. Müll.) Thér., Ann. Cryptog. Exot. 3: 86. 1930. *Thuidium papillariaceum* (C. Müll.) Par., Ind. Bryol. 1287. 1898. *Haplocladium papillariaceum* C. Müll., Nuovo Giorn. Bot. Ital., n. ser. 4: 275. 1897. Type. China: Shaanxi (Schen-si), Pan-ko-tschien, *Giraldi s.n.*
Haplocladium microphyllum (Hedw.) Broth. var. *latifolium* (Sande Lac.) Thér., Ann. Cryptog. Exot. 3: 77. 1930. *Haplocladium latifolium* (Sande Lac.) Broth., Nat. Pflanzenfam. I(3): 1008. 1907. *Pseudoleskea latifolia* Sande Lac., Ann. Mus. Bot. Lugduno-Batavum 2: 297. 1866.
Haplocladium paraphylliferum Broth., Symb. Sin. 4: 99. 1929. Syntypes. China: Yunnan, Landsangjiang, *Handel-Mazzetti 8194*; Sichuan, Muli, *Handel-Mazzetti 7531* (syntype BM).
Haplocladium spurio-capillatum Broth., Nat. Pflanzenfam. I(3): 1008. 1907.
Thuidium gracile Schimp. in B.S.G., Bryol. Eur. 5: 161 (fasc. 49–51 Mon. 5). 1852.
Thuidium occultissimum (C. Müll.) Par., Ind. Bryol. Suppl. 1: 321. 1900. *Haplocladium occultissimum* C. Müll., Nuovo Giorn. Bot. Ital., n. ser. 5: 208. 1898. Type. China: Shaanxi (Schen-si), Huo-kia-zaez, April 1896, *Giraldi 2082* (isotype BM).

Plants medium-sized, green or yellowish green, yellowish brown when old, usually in interwoven patches. *Stems* prostrate, up to 5 cm long, regularly pinnately branched; central strand present; paraphyllia lineariform to lanceolate, densely on stems, sparsely or absent on branches. *Stem leaves* appressed when dry, gradually prolonged to a slender apex from a widely ovate base, up to 1.2 mm long; leaf margins plane or partly revolute, serrate; costa excurrent, rarely percurrent or extending below leaf apex; median leaf cells triangular to hexagonal, about 5–20 µm wide, thin-walled, unipapillose; basal median leaf cells rectangular, smooth; branch leaves widely ovate, with a short lanceolate apex; costa ending below leaf apex or slightly excurrent. *Perichaetial leaves* oblong-ovate at base, costa excurrent, forming a piliform apex. *Setae* slender, about 2 cm long, reddish brown, smooth; capsule inclined, ca. 2 mm long, constricted below mouth; peristome double; exostome teeth yellowish brown, broadly lanceolate, striate below, papillose above; endostome segments nearly as long as exostome teeth, yellowish, finely papillose; basal membrane about 1/2 the length of endostome segments; cilia 2–3. *Spores* about 15 µm in diameter, papillose.

Illustrations: P.-C. Wu (ed.) 2001 (Pl. 89, figs. 19).
Habitat: mostly on rotten logs in low mountain regions, sometimes also on soil or rocks.
Distribution: China, Korea, Japan, Sikkim, India, Russia, Europe, and North America.

Chinese specimens examined: FUJIAN: Ning-de City, *J. Zhu 9* (HSNU). GUANGDONG: Guang-zhou City, *P.-J. Lin 5* (IBSC); Mt. Luofu, *Redfearn 34420* (MO); Zhao-qing City, Qixingyan, *Magill et al. 8105* (MO). GUIZHOU: Ping-ba (ping-fa) Co., *Cavalerie 1759* (PC). HENAN: Xin-yang City, *J.-X. Luo 12* (MO, PE). HUBEI: Mt. Shennongjia, *P.-C. Wu 102, 377* (MO, NY, PE). JIANGSU: Nan-jing City, *Z.-P. Wei 4257, 4259* (MO, PE); Wu-ci City, *P.-C. Wu & Y.-X. Jin 1987* (MO, PE). JILIN: An-tu Co., *Koponen 37217* (NY); Bai-shan (Mt.), *C. Gao 22237* (MO); Mt. Changbai, *C. Gao 22106* (NY). LIAONING: Kuan-dian Co., *Buck 23607, 23670* (NY), *23755* (as *Haplocladium perparaphyllum*, NY). NEI MONGOL: Keerxinyouyi Qian Qi (Co.), *X.-L. Bai 1443, 1753* (HIMC). SHAANXI: Hu Co., *B.-Z. Guo & P.-Y. Li 4733* (MO, PE); Qinling Mts, *P.-C. Chen 110, 468* (PE), *P.-C. Chen et al.*

387 (MO, NY, PE). SICHUAN: A-ba Co., *S. He 30989a* (MO); Nan-chong Co., *J.-X. Luo 26* (MO, PE); Nan-ping Co., *S. He 30094* (MO); Ya-jiang Co., *S. He 32000* (MO). TAIWAN: Tai-pei Co., *M.-J. Lai 7992* (MO). YUNNAN: Meng-hai Co., *Magill 7884* (MO); Meng-la Co., *Redfearn et al. 33983* (MO, NY). ZHEJIANG: Mt. West Tian-mu, *R.-L. Liang 29* (HSNU, MO).

5. ***Haplocladium strictulum*** (Card.) Reimers 东亚小羽藓 dong-ya xiao-yu xian

Hedwigia 76: 199. 1937. *Bryohaplocladium strictulum* (Card.) Watanabe & Iwats., J. Jap. Bot. 56: 260. 1981. *Thuidium strictulum* Card., Beih. Bot. Centralbl. 17: 29, f. 18. 1904. Type. Korea: Ouen-San, *Faurie 33*.

Haplocladium fauriei Watanabe, J. Jap. Bot. 38: 31. 1963. *Thuidium fauriei* Broth. & Par., Bull. Herb. Boiss., ser. 2, 2: 928. 1902, *nom. illeg.*

Haplocladium himalayanum Bartr., Bull. Torrey Bot. Club 82: 27. 1955.

Thuidium substrictulum Dix., Rev. Bryol. Lichénol. 7: 112. 1934. Type. China: Liaoning (Ryonei), Mt. Koô (Fenghuang), *Kobayashi 3986* (holotype BM).

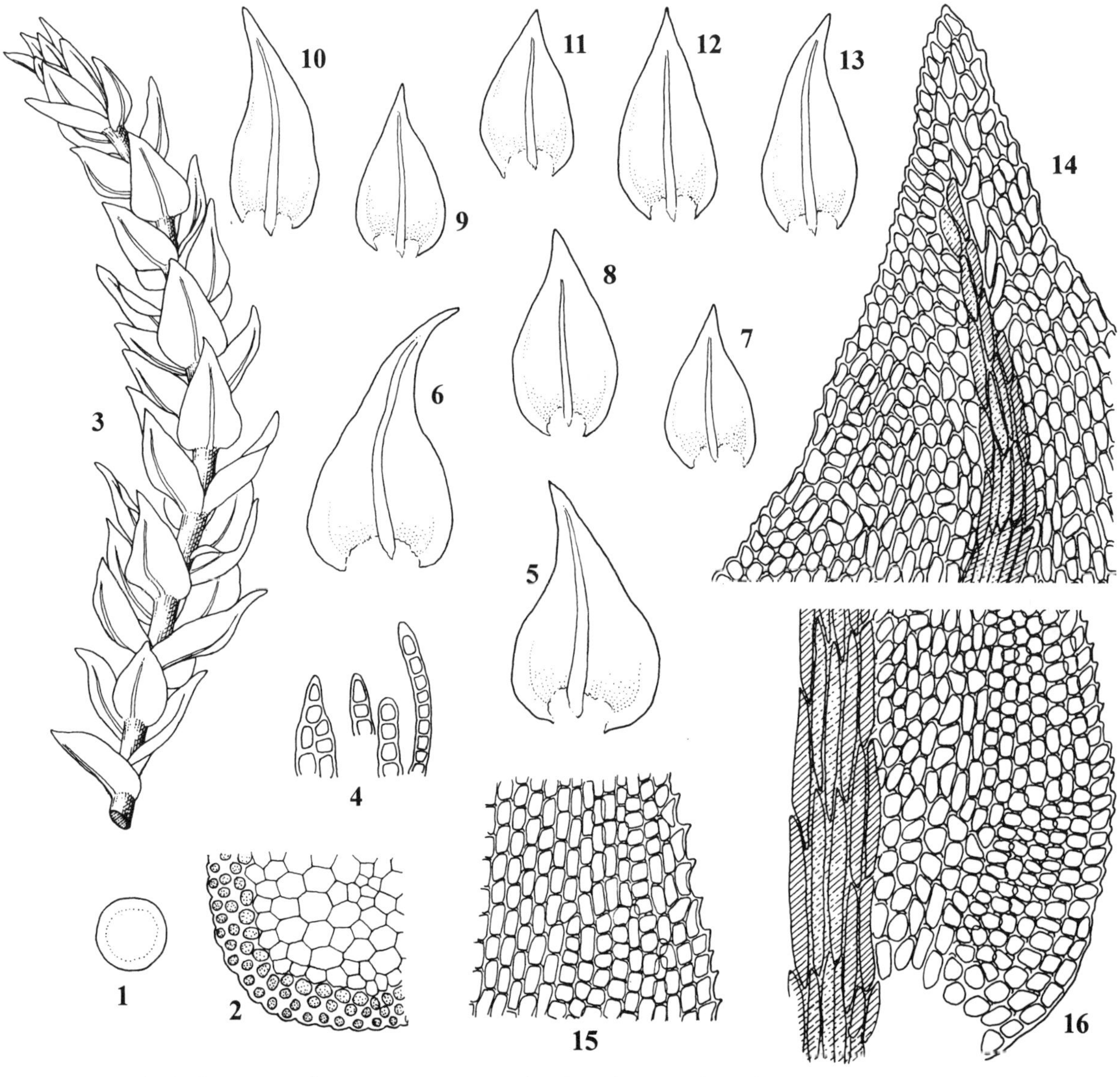

Plate 461. *Haplocladium strictulum.* 1, stem cross section (× 56). 2, portion of stem cross section (× 256). 3, branch (× 42). 4, paraphyllia (× 365). 5–6, stem leaves (× 64). 7–13, branch leaves (× 64). 14, apical leaf cells (× 365). 15, upper marginal leaf cells (× 365). 16, basal leaf cells (× 365). (1–16, drawn from *X.-L. Bai 2581*, HIMC.)

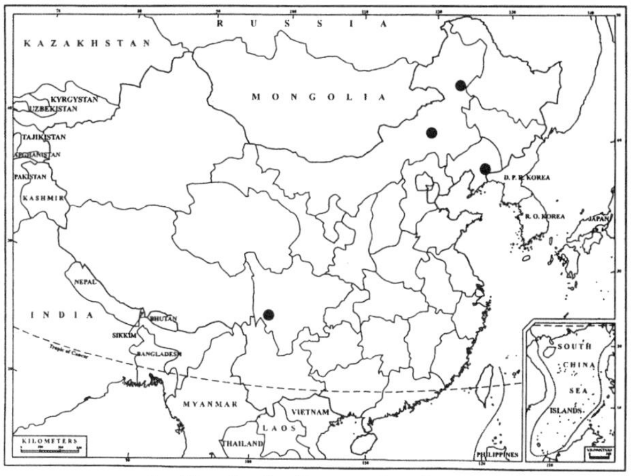

5. *Haplocladium strictulum*

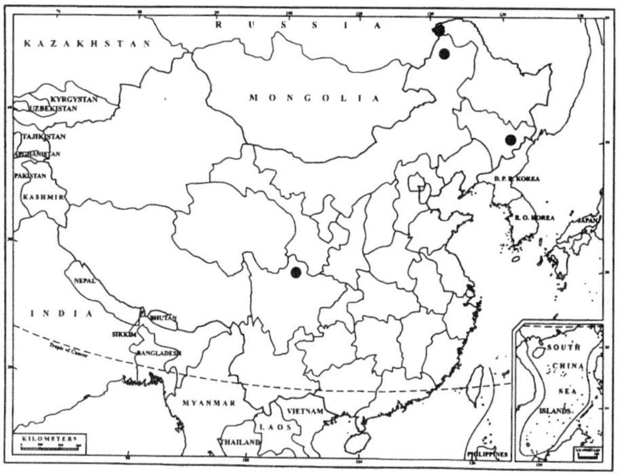

1. *Helodium paludosum*

Plants small to medium-sized, dark green to slightly brownish, in loose patches. *Stems* procumbent, up to 7 cm long, usually irregularly pinnately branched; branches 2–5 cm long; central strand differentiated; paraphyllia numerous, lanceolate, apical cells truncate. *Stem leaves* incurved when dry, up to or more than 1 mm long, ovate, widely ovate to triangular, lanceolate, delicately somewhat plicate; margins serrate; costa percurrent or slightly excurrent, slightly setulosely papillose; median leaf cells rhomboidal to oblong, thick-walled, unipapillose or strongly projecting at the upper ends, ca. 5–15 µm wide; branch leaves ovate to oblong-ovate, with triangular apex; costa 2/3 the leaf length. *Perichaetial leaves* ovate-lanceolate, multiplicate; margins entire. *Setae* elongate; capsules cylindrical, slightly arcuate; peristome features the same as those of *H. microphyllum*. *Spores* nearly spherical, finely papillose.

Illustrations: Plate 461, figs. 1–16.
Habitat: on soil over rocks in forests.
Distribution: China, Korea, and Japan.

Chinese specimens examined: NEI MONGOL: Balinzuoqi, Beiyinyula forest farm, *X.-L. Bai 2581* (HIMC, NY, PE); Zha-lan-duan Çity, *X.-L. Bai 2625* (HIMC). SICHUAN: Dao-cheng Co., *S. He 31690a* (MO).

8. ***Helodium*** Warnst. 沼羽藓属 zhao-yu xian shu
Krypt.-Fl. Brandenburg, Laubm. 2: 675, 692. 1905, *nom. cons.*
Thuidium sect. *Elodium* (Sull.) Braithw., Brit. Moss. Fl. 3: 8. 1896. *Hypnum* subg. *Elodium* (Sull.) Lesq. & Jam., Man. N. Amer. Moss. 329. 1884. *Thuidium* subg. *Elodium* (Sull.) Lindb., Musc. Scand. 31. 1879. *Elodium* (Sull.) Aust., Musci Appalach. 52. 1870. *Hypnum* sect. *Elodium* Sull. in Gray, Manual (ed. 2), 668. 1856.

Plants rather robust, green or yellowish green, rarely brownish, not shinning or slightly shinning, in tufts or cushions. *Stems* erect or ascending, simple or forkedly branched, but usually regularly pinnately branched; paraphyllia dense, filamentous, simple or branched, pale, brownish when old; central strand absent. *Stem leaves* and branch leaves similar, gradually narrowed from an ovate-lanceolate or cordate base to a lanceolate apex, longitudinally plicate, restricted at base; margins entire or serrate, revolute above, long ciliate or smooth at base; costa rather rigid, reaching upper portion of leaf apex; leaf cells hyaline, elongate-hexagonal or oval-hexagonal, basal cells rectangular. *Autoicous* or dioicous. *Inner perichaetial* leaves erect, lanceolate, deeply plicate, pale; costa ending below leaf apex. *Setae* slender, smooth; capsules ovoid or cylindrical, strongly recurved or horizontal, brownish; annuli present, stomata present at hypophysis; opercula conic; exostome teeth yellowish brown; cilia weak, usually 3. *Calyptrae* smooth. *Spores* not seen.

Helodium consists of three species, and all of them were found in China (C. Gao et al. 1977). However, only two species were accepted by Crosby et al. (2000) because *Helodium sachaliense* was placed in the genus *Bryonoguchia*. Although *Helodium blandowii* (Web. & Mohr) Warnst. was reported from northeastern China (C. Gao et al. 1977), we could not confirm the report. Two species are treated in this study.

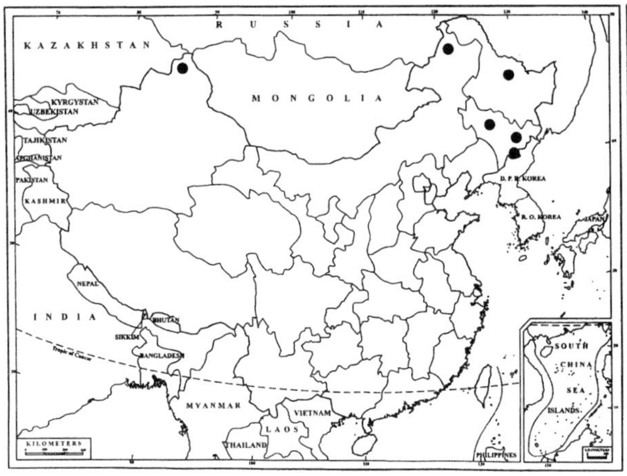

2. *Helodium sachalinense*

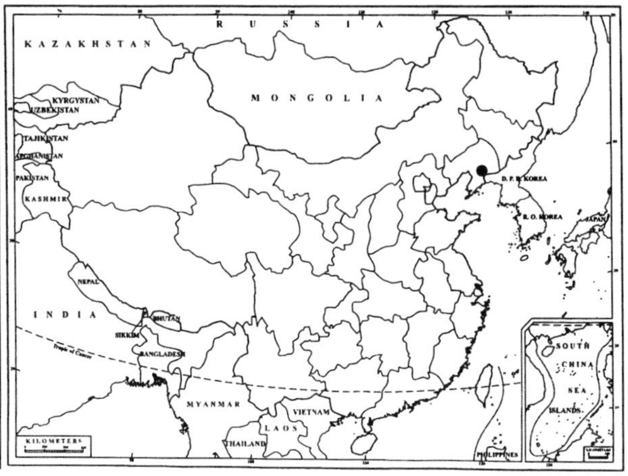

1. *Heterocladium angustifolium*

1. Plants robust; stem leaves oblong at the base, gradually lanceolate toward the apex; leaf margins entire or papillose; leaf cells papillose at upper corner; costa smooth on the back.. 1. *H. paludosum*
1. Plants rather small; stem leaves cordate at the base, suddenly narrowed to a ciliate apex; basal leaf margin multiplicate; leaf cells centrally papillose; costa papillose or setulosely papillose on the dorsal surfaces................. .... 2. *H. sachalinense*

1. ***Helodium paludosum*** (Aust.) Broth. 狭叶沼羽藓 xia-ye zhao-yu xian
Nat. Pflanzenfam. I(3): 1009. 1908. *Elodium paludosum* Aust., Musc. Appalach. 301. 1870. *Hypnum paludosum* Sull., Musci. Allegh. 6. 1846, *hom. illeg.* Type. U.S.A. Ohio, Columbus.

Plants rather robust, up to 6 cm long, yellowish to yellowish brown, in lax patches. *Stems* creeping and spreading, regularly pinnately branched; branches ca. 5 mm long; central strand undifferentiated; paraphyllia densely growing on stems and branches, usually filamentous, irregularly branched. *Stem leaves* appressed when dry, erect-spreading when moist, broad and concave at base, oblong to ovate, gradually acuminate, ca. 1.0–1.5 mm long, slightly plicate and decurrent; median leaf cells narrowly hexagonal, ca. 15–30 µm × 5 µm, thick-walled, finely papillose near the upper corners; apical cells elongate-rhomboidal. *Autoicous*. *Perichaetial leaves* ovate-lanceolate to lanceolate at the base, slenderly lanceolate toward apex. *Sporophytes* not seen.

Illustrations: P.-C. Wu (ed.) 2001 (Pl. 108, figs. 13–23).
Habitat: usually growing in alpine forest ground or bogs.
Distribution: China, Japan, Russia (Siberia), and North America.

Chinese specimens examined: JILIN: An-tu Co., *Koponen 36574* (H, PE); Fenghuangshan (Mt.), *X.-W. Kong 2274* (PE). NEI MONGOL: Daxinganling, *P.-C. Chen & C. Gao 956* (IFSBH, PE), 52°12′ N, 120°48′ E, *X.-L. Bai 298* (MO, NY, PE). SICHUAN: Song-pan Co., *Redfearn 35458* (MO).

The distribution of this species in China is very narrow, with only a few specimens collected.

2. ***Helodium sachalinense*** (Lindb.) Broth. 东亚沼羽藓 dong-ya zhao-yu xian
Nat. Pflanzenfam. I(3): 1018. 1908. *Bryochenea sachalinensis* (Lindb.) C. Gao & G.-C. Chang, Bull. Bot. Res. Harbin 2 (4): 118. 1982. *Thuidium sachalinense* Lindb., Acta Soc. Sci. Fenn. 10: 244. 1872. Type. Russia: Sakhalin.
Tetracladium osadae Sak., Bot. Mag. (Tokyo) 55: 533. 1941.

Plants medium-sized, ca. 3 cm long, narrowed at base, yellowish green or green above, in interwoven patches. *Stems* creeping, irregularly pinnately branched; branches ca. 5 mm long; central strand present; paraphyllia densely growing on stems and branches, foliose or filamentous, irregularly branched. *Stem leaves* appressed on the stems when dry, erect-spreading when moist; branch leaves cordate, ovate or triangular at base, suddenly narrowed to a lanceolate apex, ca. 1.5–2.0 mm long; costa vanishing in the middle or above the middle portion of leaf, papillose on back; margins serrate, ciliate at base; leaf cells rhomboidal or rounded quadrate, ca. 15 µm × 8 µm, thick-walled, centrally unipapillose; branch leaves widely ovate, concave; costa reaching upper portion of leaf; cells of branch leaves rhomboidal or oval, ca. 15 µm × 8 µm. *Dioicous*. *Perichaetial leaves* serrate, setulosely or shortly ciliate at margins. *Setae* yellowish brown or reddish brown; capsules cylindrical, erect to horizontal, up to 4 mm long; peristome features similar to those of *H. paludosum*. *Spores* nearly rounded, ca. 15 µm in diameter, finely papillose.

Illustrations: Plate 462, figs. 1–12.
Habitat: on humus or rocks in alpine or subalpine forest regions; alt. 1600–1700 m.
Distribution: China, Korea, and Japan.

Chinese specimens examined: HEILONGJIANG: Yi-chun Co., *Forest. Inv. Group s.n.* (PE). JILIN: An-tu Co., *Koponen 36713, 36715, 36760, 36775* (H, PE); Jiu-tai City, Tumengling, *K.-C. Lang 2* (PE); Mt. Changbai, *C. Gao 22117* (MO), *K.-C. Lang 20* (PE). NEI MONGOL: Geng-he Co., *D.-Z. Huang 264, 347* (PE). XINJIANG: Bu-er-jin Co., *R.-C. Ching 3142* (PE).

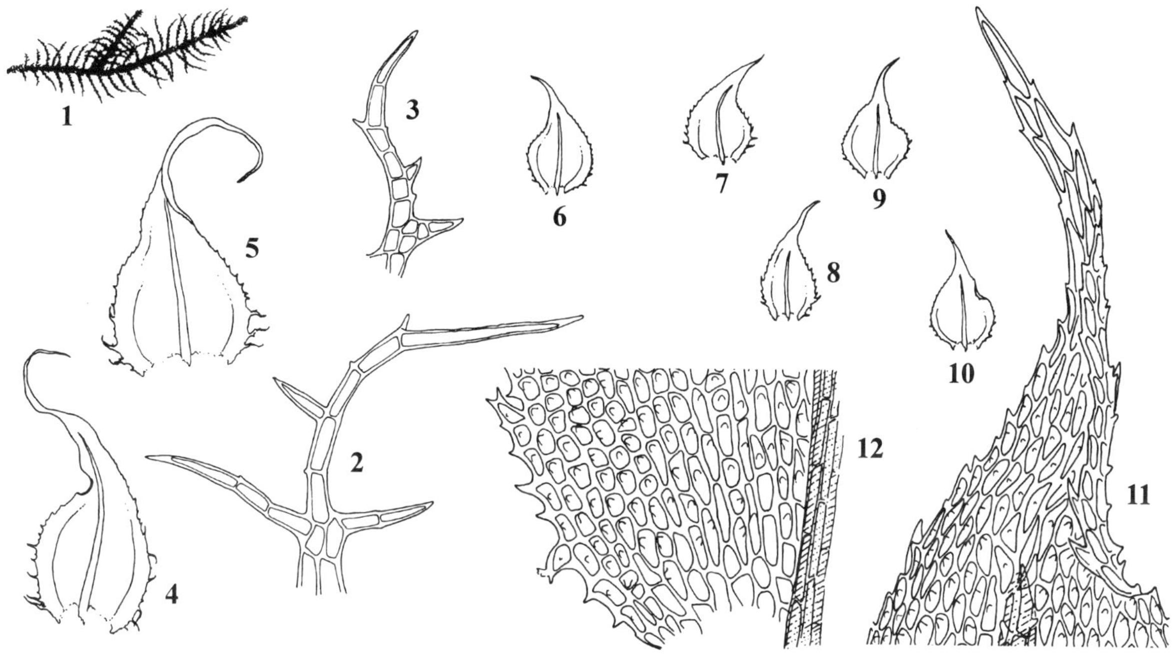

Plate 462. *Helodium sachalinense*. 1, habit (× 1). 2–3, paraphyllia (× 386). 4–5, stem leaves (× 38). 6–10, branch leaves (× 38). 11, apical leaf cells (× 386). 12, basal leaf cells (× 386). (1–12, drawn from *Koponen 36775*, H.)

9. ***Heterocladium*** Bruch & Schimp. in B. S. G. 异枝藓属 yi-zhi xian shu
Bryol. Fur. 5: 151. 1852. *Hypnum* subg. *Heterocladium* (Bruch. & Schimp.) Lesq., Man. Mosses N. America 317, 320. 1884.

Plants minute to medium-sized, usually dull green or yellowish green, in small mats. *Stems* creeping, irregularly branched to pinnately branched; paraphyllia few, slender or foliose; pseudoparaphyllia foliose. *Stem leaves* and branch leaves differentiated; stem leaves erect-spreading when moist or reflexed, leaf base cordately ovate, slightly decurrent, tapering gradually to a narrow point; leaf margins plane, serrulate; costa double, delicate, rarely inconspicuous or single; median cells nearly rectangular; basal cells small, quadrate; upper marginal cells shortly rhomboidal; cell walls smooth or minutely unipapillose to multipapillose; branch leaves rather small, widely ovate, with acute, obtuse or rounded apices, usually imbricate or secund. *Dioicous*. *Perichaetial leaves* pale, erect, obliquely spreading or squarrose above. *Setae* slender, smooth; capsules cylindrical, horizontal, pendent or recurved, with short neck, contracted under the mouth; annuli differentiated; opercula conical, with obtuse apices or beaks; exostome teeth narrowly lanceolate, lightly yellowish to yellowish brown; endostome nearly as long as the teeth, narrowly lanceolate, perforate; cilia well developed, usually 1–3. *Calyptrae* smooth.

There are about 10 species of *Heterocladium* in the world. Only one species is found in China. *Heterocladium papillosum* (Lindb.) Lindb. is treated in the genus *Pseudoleskeella.*

1. ***Heterocladium angustifolium*** (Dix.) R. Watanabe 狭叶异枝藓 xia-ye yi-zhi xian
J. Jap. Bot. 35: 261. 1960. *Rauia angustifolia* Dix., Rev. Bryol. Lichénol. 7: 111. 1934. Types. China: Liaoning (Ryonei), Mt. Koô (Fenghuang), on trees, *Kobayasi 3899b* (holotype BM), on dead trees, *Kobayasi 3900* (paratype BM).

Plants small, intricate, yellowish green, not glossy. *Stems* prostrate, 0.8–1.1 cm long, sparsely pinnately branched; branches ca. 4 mm long. *Stem leaves* small, appressed when dry, erect-spreading when moist, oblong-ovate-lanceolate or triangularly lanceolate, acuminate at apex, 0.40–0.45 mm × 0.10–0.15 mm; margins serrulate; costa inconspicuous; median leaf cells triangular to rhomboidal, 3–5 µm × 3–4 µm, thick-walled, smooth; apical cells oval, ca. 9 µm × 7 µm; branch leaves smaller, shortly ovate-lanceolate, ca. 0.15 mm × 0.10 mm. *Dioicous. Perichaetial leaves* ovate-lanceolate, sheathing at base, acuminate and falcate above, 1.5–2.0 mm × 0.5–0.6 mm; margins serrulate; costa single, slender; median cells oblong or rectangular, thick-walled; basal cells rectangular, smooth. *Setae* erect, reddish brown, ca. 9 mm long, smooth; capsules oblong-ovoid, symmetrical, brownish red, horizontal to nearly pendulous, neck short; opercula convex to shortly conic, reddish brown to brown; exothecial cells rhomboidal to hexagonal, thin-walled; annuli not differentiated; exostome teeth lanceolate, reddish brown, ca. 3 mm long, densely papillose above, cross-striolate below; endostome segments yellowish, narrowly lanceolate, nearly as long as the teeth, keeled; basal membrane ca. 1/2 the height of teeth; cilia short. *Calyptrae* cucullate, smooth. *Spores* pale yellowish, ca. 12 µm in diameter.

Illustrations: Watanabe 1972 (Pl. 37, 1–14).
Habitat: on base of trees or decaying wood.
Distribution: endemic to China.

Specimens examined: Reported from Liaoning (Watanabe 1972).

10. ***Leptocladium*** Broth. 薄羽藓属 bo-yu xian shu
Symb. Sin. 4: 97. pl. 3: 15. 1929.

Plants rather slender, rigid, yellowish green, in dense mats. *Stems* extensively creeping, densely covered with leaves and brownish rhizoids, pinnately branched; branches erect-spreading, simple, with obtuse apex. *Stem leaves* and branch leaves somewhat differentiated; stem leaves concave, imbricate, ovate or oblong-ovate, with a short or lanceolate apex, up to 0.5 mm long; leaf margins entire, erect; costa double, short, weak to inconspicuous; median leaf cells shortly linear, smooth or somewhat minutely papillose; basal cells loose; alar cells nearly quadrate, few and loose; branch leaves rather small, ovate, with a short apex. *Autoicous. Inner perichaetial leaves* narrowly lanceolate, with a slender apex; costa inconspicuous or absent. *Setae* slender, up to 1 cm long, reddish brown; capsules oblong-ovoid, brownish, nearly horizontal; opercula conical; exostome teeth widely lanceolate, densely striate; other features of peristome not seen.

This is a monotypic genus endemic to China. *Leptocladium* is very closely related to *Heterocladium* and *Leptopterigynandrum*; however, with an autoicous sexual condition and imbricate leaves, it is different from the latter two genera, which have a dioicous sexual condition and none imbricate leaves.

1. ***Leptocladium sinense*** Broth. 薄羽藓 bo-yu xian
Symb. Sin. 4: 97. pl. 3: 15. 1929. Type. China: Yunnan, Lu-djiang (Salwin), *Handel-Mazzetti 9324* (holotype H).

The description of this species is the same as that of the genus.

Illustrations: Plate 463, figs. 1–18.
Habitat: alt. 3800–4050 m.
Distribution: endemic to China.

Specimen examined: see the type citation above.

Brotherus (1929) discussed the generic relationships between *Leptocladium* and other genera within the Thuidiaceae; however, its systematic position was not well defined. Based on its pinnately branched stems, short, double or weak costae, and smooth leaf cells, P.-C. Chen et al. (1978) placed *Leptocladium* in the subfamily Heterocladioideae of the Thuidiaceae. Nevertheless, it is very close to the Leskeaceae systematically. Enroth (1997) placed this genus in the Amblystegiaceae.

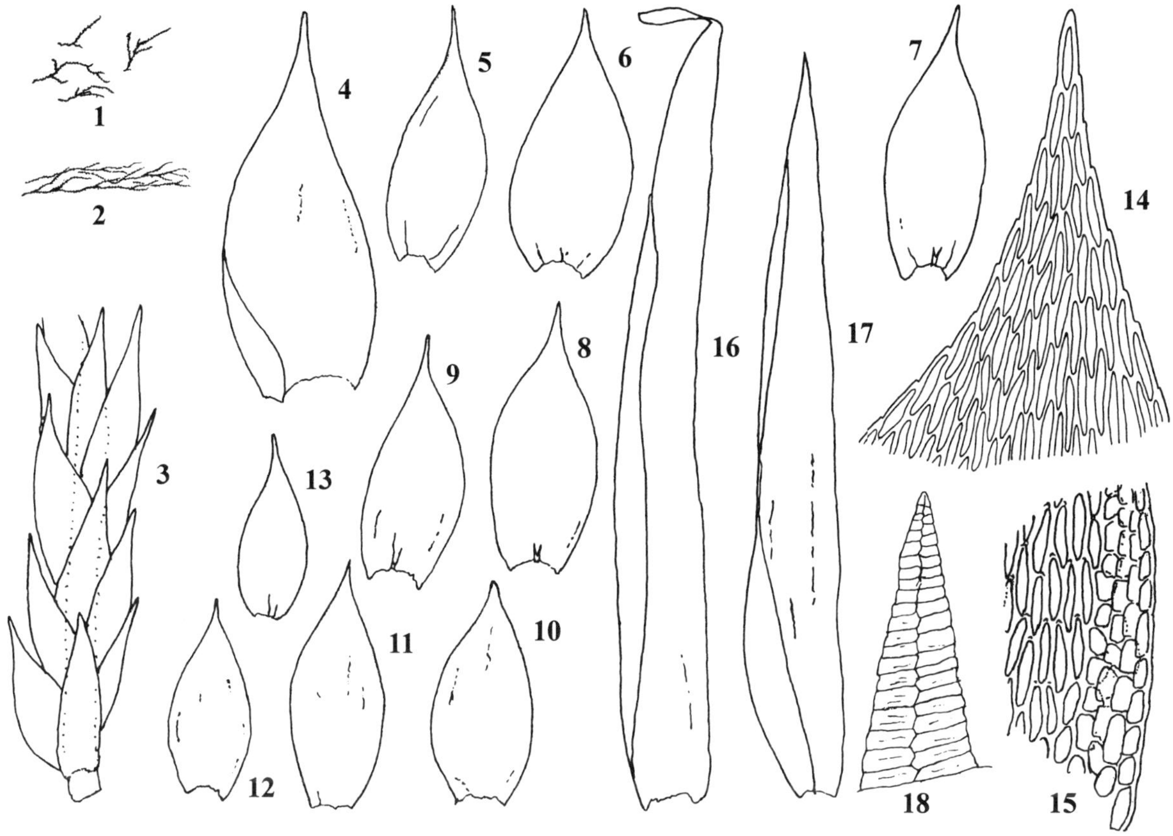

Plate 463. *Leptocladium sinense*. 1, loose habit (× 0.5). 2, tufted habit (× 0.5). 3, branch (× 50). 4, stem leaf (× 50). 5–11, branch leaves (× 50). 12–13, branchlet leaves (× 50). 14, apical stem leaf cells (× 315). 15, basal stem leaf cells (× 315). 16–17, perichaetial leaves (× 50). 18, peristome (× 200). (1–18, drawn from *Handel-Mazzetti 9324*, H.)

11. ***Leptopterigynandrum*** C. Müll. 叉羽藓属 cha-yu xian shu
Hedwigia 36: 114. 1897.

Plants rather slender, rigid when dry, dark yellowish green or brownish green, in dense or loose mats. *Stem leaves* and branch leaves not obviously differentiated, densely imbricate when dry, mostly ovate, oblong-ovate or widely ovate, lanceolate toward apex, often secund; margins plane, entire; costa double, short, weak or inconspicuous; leaf cells hexagonal, oblong-rhomboidal or oblong-oval; alar cells small, quadrate or round-quadrate, smooth. *Dioicous*. *Setae* slender, ca. 1.0 cm long; capsules erect, oblong-ovoid; opercula conical; peristome double, exostome teeth lanceolate, basal membrane low; endostome segments narrow; cilia absent. *Calyptrae* not seen.

Ten species are currently placed under this genus with 6 species better understood (Crosby et al. 2000). Most of the species are distributed in East Asia, mainly from the Himalayan regions. Two species occur in Southwestern South America. In China four species are known, with a distribution range in Hengduan Mts. and southwestern Xizang and Qinghai regions.

The study of *Leptopterigynandrum* in China was started by Brotherus (1924), who described two new species: *L. incurvatum* and *L. stricticaule*, from Yunnan and Sichuan. Later, Brotherus (1929) published another new species, *L. tenellum*, from Sichuan. P.-C. Chen et al. (1978) recognized all three species. X.-J. Li (editor-in-

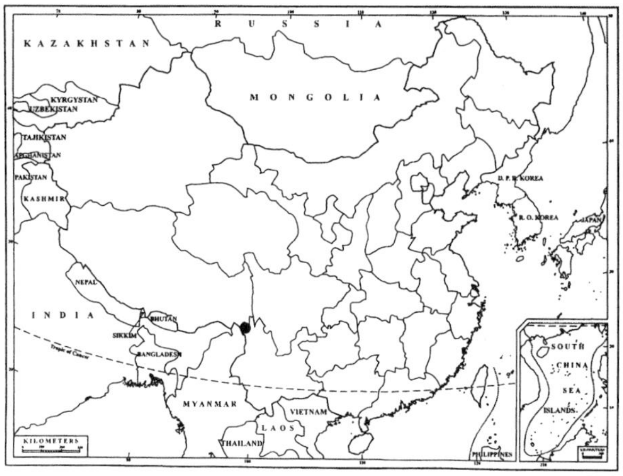

1. *Leptocladium sinense*

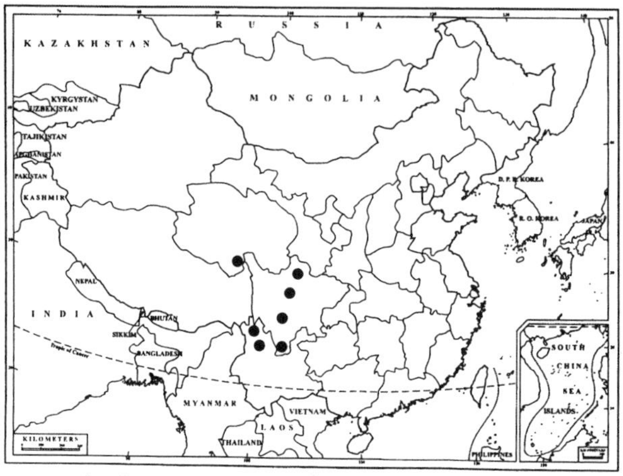

1. *Leptopterigynandrum austro-alpinum*

chief, 1985) treated four species in *Leptopterigynandrum* from Xizang. J.-C. Zhao (1993) reported *L. austro-alpinum* newly in East Tianshan of Xinjiang province. Recently B. C. Tan and Y. Jia (1997) found three species, including *L. incurvatum* Broth., *L. subintegrum* (Mitt.) Broth, and *L. austro-alpinum*. C. Müll. from Qinghai.

1. Plants loosely growing together; leaves oblong-ovate; costa weak or absent .... 4. *L. tenellum*
1. Plants densely growing together; leaves ovate or broadly ovate; costa short or weak 2
   2. Leaf apex mostly straight, never secund .... 3. *L. subintegrum*
   2. Leaf apex usually secund, rarely straight 3
3. Leaf cells rather small, hexagonal or nearly rectangular .... 1. *L. austro-alpinum*
3. Leaf cells rather large, rhomboidal or oblong-ovate .... 2. *L. incurvatum*

1. ***Leptopterigynandrum austro-alpinum*** C. Müll. 叉羽藓 cha-yu xian
Hedwigia 36: 114. 1897. Type. Argentina: April 1872, *Lorentz s.n.*
Leptopterigynandrum stricticaule Broth., Akad. Wiss. Wien Sitzungsber., Math.-Naturwiss. Kl. Abt. 1, 133: 577. 1924. Types. China: Sichuan (Setschwan), Yen-yuen Co., *Handel-Mazzetti 2338* (syntype H); Yunnan, Zhong-dien Co. (Chungtien), Piepun, *Handel-Mazzetti 4550, 4760* (syntypes H).

Plants minute, rather rigid, yellowish green, dull, in dense, small mats. *Stems* slender, up to 5 cm long, creeping, nearly pinnately branched, ascendant, 2–5 mm long, mostly branched once. *Leaves* imbricate when dry, widely ovate at base, nearly lanceolate or acuminate toward apex, strongly secund, ca. 1.4 mm long; basal leaf margins slightly revolute, entire or nearly entire; costa double, short and weak, rarely up to the middle of leaf; leaf cells hexagonal or nearly quadrate, marginal cells smaller, basal angle cells small, quadrate or rounded-ovate, opaque, smooth. *Sporophytes* not seen.

Illustrations: P.-C. Wu (ed.) 2001 (Pl. 79, figs. 1–11).
Habitat: usually on trees or calcareous rock surfaces; alt. 2900–4500 m.
Distribution: China, Mongolia, Russia (Siberia), North and Centeral America.

Chinese specimens examined: QINGHAI: Yu-shu Co., *B.-C. Tan 95-175, 95-202* (FH, MO). SICHUAN: between Yanyuan and Huaping, *Handel-Mazzetti 2338* (H); De-tuo Distr., *X.-J. Li 81-1518* (HKAS); Mao-wen Co., *Allen 7323H* (MO); Song-pan Co., *Allen 7077, 7086, 7286* (MO). YUNNAN: between Lijiang and Yulong Mt. *Handel-Mazzetti 8460* (H); Zhong-dian Co., *Handel-Mazzetti 8053* (H).

The leaves of this species are similar to those of *Leptopterigynandrum incurvatum* Broth., but leaf bases of *L. austro-alpinum* are rather wide and nearly ovate-triangular and leaf cells wider and shorter. B. C. Tan and Y. Jia (1997) synonymized *L. stricticaule* Broth. under the current species. They also considered the Himalayan *L. decolor* (Mitt.) Buck to be a likely synonym of this species.

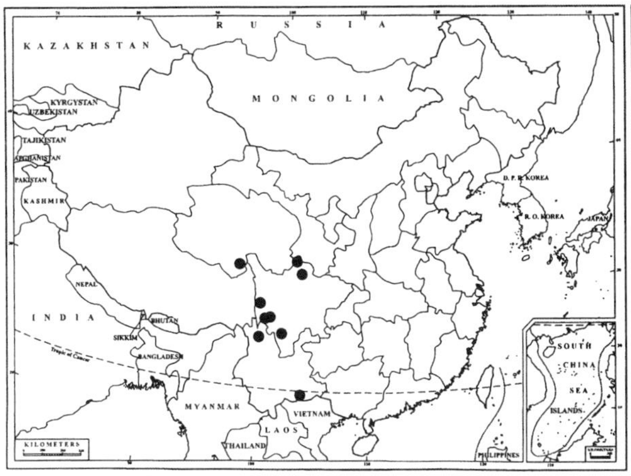

2. *Leptopterigynandrum incurvatum*

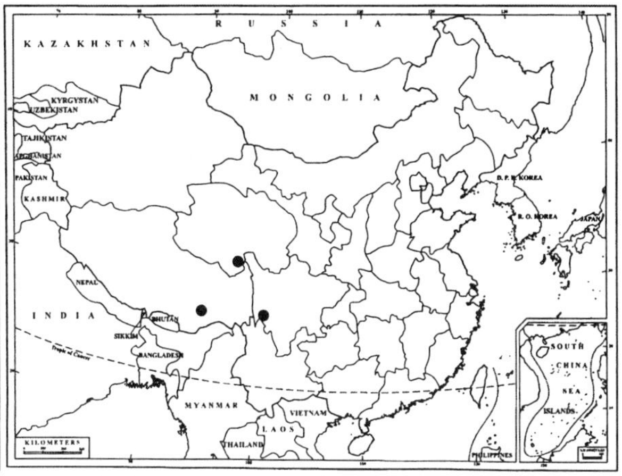

3. *Leptopterigynandrum subintegrum*

2. ***Leptopterigynandrum incurvatum*** Broth. 卷叶叉羽藓 juan-ye cha-yu xian
Akad. Wiss. Wien Sitzungsber., Math.-Naturwiss. Kl. Abt. 1, 133: 577. 1924. Type. China: Yunnan, Piepum (Piepen), *Handel-Mazzetti 4743* (holotype H).

Plants slender, stout, dark yellowish green in dense mats. *Stems* creeping, with dense rhizoids, erect-ascending above, pinnately branched; branches rather short, slightly circinate when dry. *Leaves* imbricate, ovate or

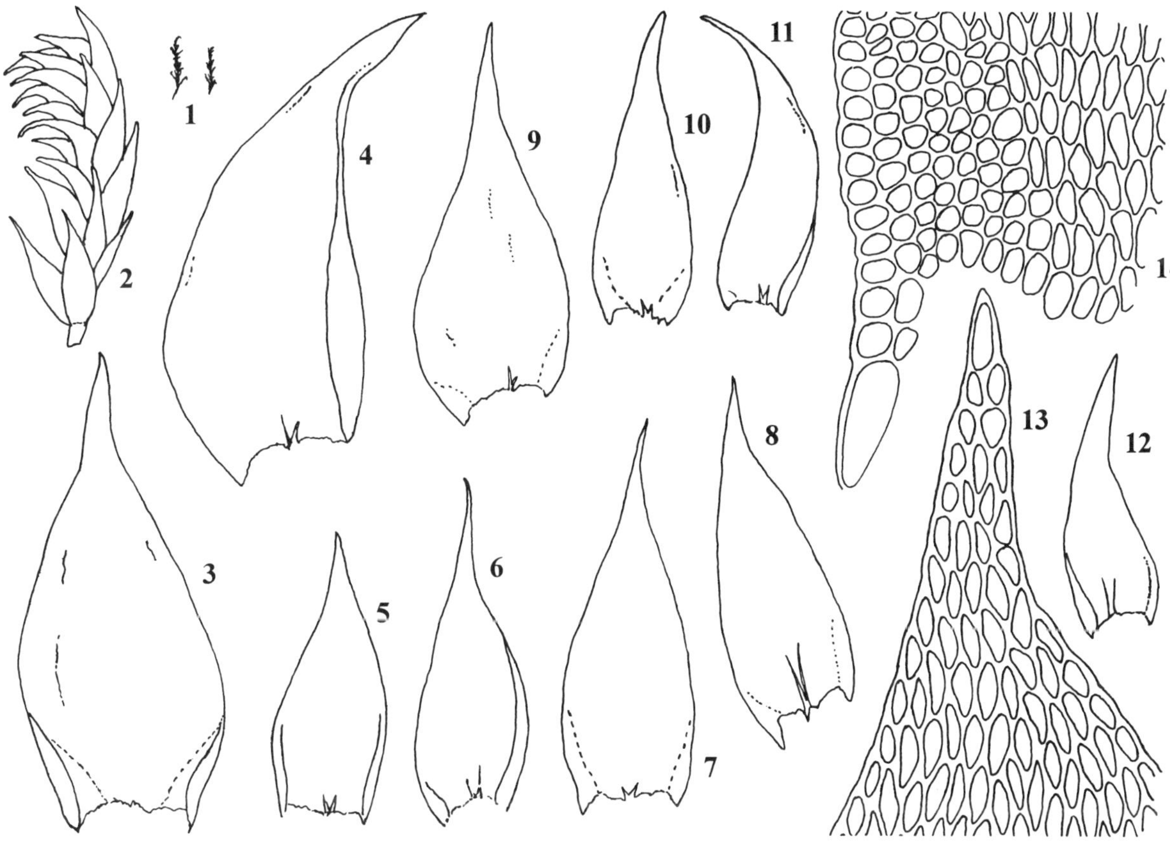

Plate 464. *Leptopterigynandrum incurvatum*. 1, habit (× 0.5). 2, branch (× 20). 3–4, stem leaves (× 50). 5–11, branch leaves (× 50). 12, branchlet leaf (× 50). 13, apical leaf cells (× 180). 14, basal leaf cells (× 180). (1–14, drawn from *Handel-Mazzetti 4743*, H.)

broadly ovate, gradually tapering to a lanceolate or narrowly lanceolate apex, flatly spreading, slightly secund, up to 1.4 mm long; margins plane, slightly recurved, entire or almost entire; costa double, short and weak; leaf cells rhomboidal, marginal cells smaller, opaque, rectangular or subrounded, thick-walled, smooth; branch leaves oblong-ovate, acuminate above. *Dioicous*. *Inner perichaetial leaves* erect, lanceolate; laminal cells linear, basal cells loosely rectangular or elliptic-hexagonal; costa inconspicuous. *Setae* slender, ca. 1.0 cm long, reddish; capsules elliptical, erect; exostome teeth lanceolate, smooth, loosely nodulose; cilia absent.

Illustrations: Plate 464, figs. 1–14.
Habitat: mostly on alpine rock surfaces or tree trunks; alt. 3000–4800 m.
Distribution: endemic to China.

Chinese specimens examined: QINGHAI: Yu-shu Co., *B.-C. Tan 95-182*, *95-185* (FH, MO). SICHUAN: Batang Co., *S. He 31614* (MO); Dao-cheng Co., *S. He 31682* (MO); from Muli Co. to Yongnin Co., *Handel-Mazzetti 7198* (H); Ma-er-kang Co., *Redfearn 34996* (MO); Nan-ping Co., *S. He 30201* (MO); Ruo-er-gai Co., *S. He 30702* (MO); Song-pan Co., *Allen 7301* (MO), *S. He 30554* (MO); Xiang-cheng Co., *S. He 31795* (MO). YUNNAN: Zhongdian Co., *Handel-Mazzetti 6992* (H).

The characteristics of this species are the leaves secund when dry and the oblong-oval or rhomboidal leaf cells that are larger than those of other closely related species.

3. ***Leptopterigynandrum subintegrum*** (Mitt.) Broth. 全缘叉羽藓 quan-yuan cha-yu xian
Nat. Pflanzenfam. (ed. 2), 11: 309. 1925. *Heterocladium subintegrum* Mitt., J. Proc. Linn. Soc., Bot., Suppl. 1:135. 1859. Type. India: northwestern Himalayas, temperate region, Simla, *T. Thomson s.n.* (holotype NY).

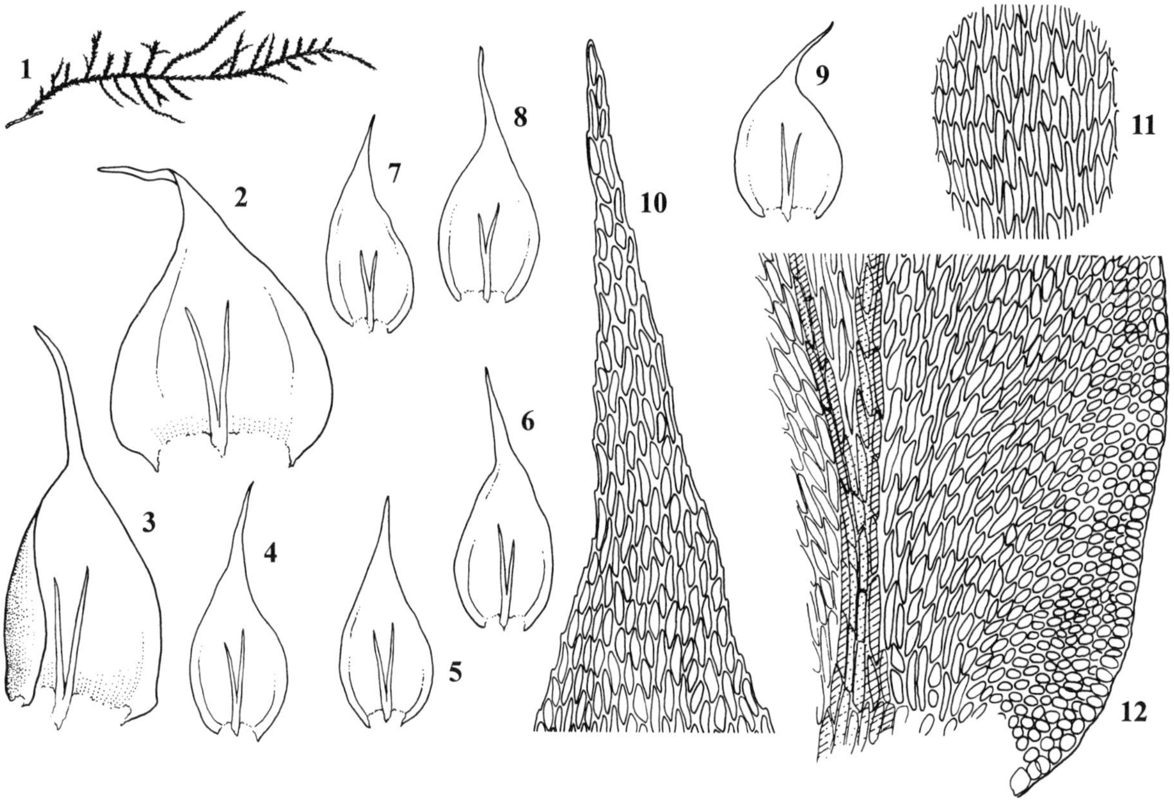

Plate 465. *Leptopterigynandrum subintegrum*. 1, habit (× 1). 2–3, stem leaves (× 38). 4–9, branch leaves (× 38). 10, apical leaf cells (× 260). 11, median leaf cells (× 260). 12, basal leaf cells (× 260). (1–12, drawn from *L.-S. Wang 814250*, HKAS.)

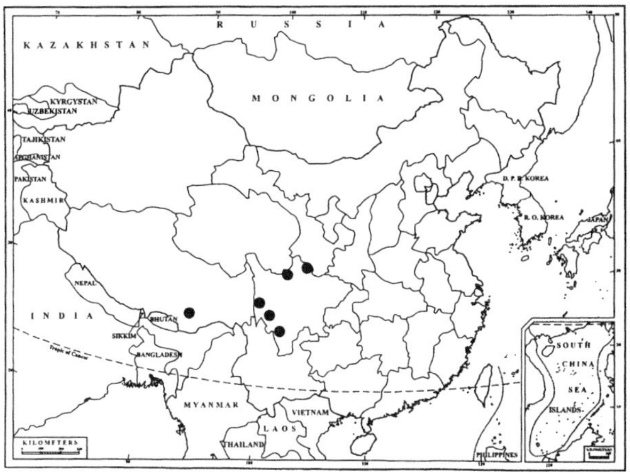

4. *Leptopterigynandrum tenellum*

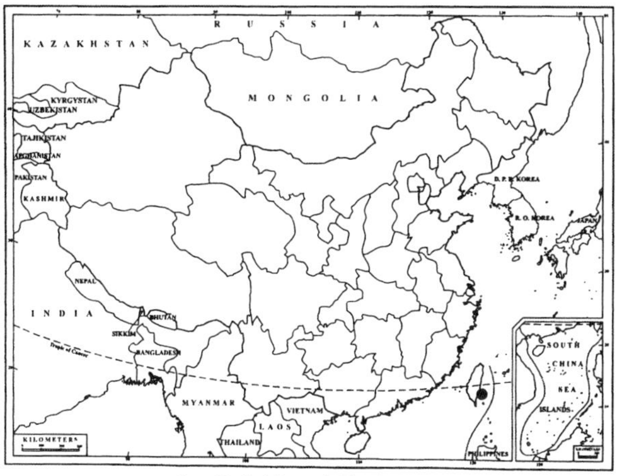

1. *Pelekium bifarium*

Plants slender, rigid, brownish green, in dense mats. *Stems* creeping, irregularly pinnately branched, with straight branches. *Stem leaves* and branch leaves identical, loosely imbricate when dry, ovate or ovate-triangular, tapering gradually to a narrow point; margins entire, revolute at the base; costa double, up to the middle of leaf; median leaf cells rhomboidal to rounded oval; alar cells rounded quadrate or oblong. *Sporophytes* unknown.

Illustrations: Plate 465, figs. 1–12.
Habitat: usually on the ground in alpine forest.
Distribution: endemic to China.

Chinese specimens examined: QINGHAI: Yu-shu Co., *B.-C. Tan 95-179* (FH). SICHUAN: Xiang-cheng Co., *L.-S. Wang 81-4250* (HKAS, PE). XIZANG: Mi-lin Co., *M.-Z. Wang s.n.* (PE).

The leaves of this species are mostly straight, not secund. This character is obviously different from that of other species in the genus.

4. ***Leptopterigynandrum tenellum*** Broth. 细叉羽藓 xi cha-yu xian
Symb. Sin. 4: 96. 4f. 1. 1929. Type. China: Sichuan (Setschwan), Mu-li Co., *Handel-Mazzetti 7365* (holotype H).

Plants delicate, yellowish green, in loose mats. *Stems* prostrate, up to 4–5 cm long, loosely pinnately branched, with abundant brownish rhizoids; branches up to 5 mm long, erect-spreading, densely foliate. *Leaves* erect spreading, slightly concave, widely ovate or ovate-lanceolate, narrow at apex, mostly secund, up to ca. 0.75 mm long; margins entire, narrowly revolute near base; costa lacking or short and double; leaf cells oblong-ovate, smooth; alar cells small and quadrate, thick-walled. *Dioicous*. *Sporophytes* not seen.

Illustrations: X.-J. Li (ed.) 1985 (Pl. 134, figs. 9–14).
Habitat: on bark of trees and rock surfaces in alpine forests; alt. 3600–3700 m.
Distribution: endemic to China.

Chinese specimens examined: SICHUAN: A-ba Co., *S. He 31041* (MO); Ba-tang Co., *S. He 31483* (MO); Dao-cheng Co., *S. He 31690* (MO); Nan-ping Co., *S. He 30169* (MO). XIZANG: Long-zi Co., *M. Zang 1278* (HKAS).

This species is characterized by its slender plants with distantly growing branches and by its leaf apices being narrow and plane.

12. ***Pelekium*** Mitt. 鹤嘴藓属 he-zui xian shu
J. Linn. Soc., Bot. 10: 176. 1868.

Plants slender, yellowish green or green, brownish when old, in loosely interwoven patches. *Stems* prostrate, up to 5 cm long, regularly bipinnately branched; central strand present; paraphyllia abundant, filamentous or foliose, simple, rarely branched. *Stem leaves* and branch leaves differentiated; stem leaves gradually forming lanceolate acumen from a widely ovate to triangularly ovate base, erect-spreading when moist, appressed when dry, up to 1 mm long; branch leaves irregularly arranged in 2 rows, concave, ovate or ovate-triangular; margins serrate; costa ending at the upper portion of leaves; leaf cells uni- to multipapillose. *Dioicous* or autoicous. *Perichaetial leaves* ovate-lanceolate or lanceolate, suddenly narrowed to a slender apex; margins serrate. *Setae* slender, yellowish brown or reddish brown; capsules cylindrical, horizontal or declined; opercula long beaked; exostome teeth 16, lanceolate, papillose in upper portion; endostome segments slender, finely papillose; basal membrane high or low; cilia usually 2. *Calyptrae* campanulate, hairy, lobed at base. *Spores* spherical.

Pelekium consists of three species, with their main distribution in tropical Asia. B.-Y. Yang and W.-C. Lee (1964) and Watanabe (1972) reported both *Pelekium bifarium* (Bosch & Sande Lac.) Fleisch. and *P. velatum* Mitt. from Taiwan. L. Zhang (1990) recently reported the former species from southern Yunnan. In the present study, two species are treated for China.

1. Dioicous; branch leaves usually arranged in 2 rows, erect-spreading when dry; leaf cells multipapillose .... 1. *P. bifarium*
1. Autoicous; branch leaves irregularly arranged, incurved or crisped when dry; leaf cell unipapillose .... 2. *P. velatum*

1. ***Pelekium bifarium*** (Bosch & Sande Lac.) Fleisch. 二列鹤嘴藓 er-lie he-zui xian
Musci Buitenzorg 4: 1513. 1923. *Thuidium bifarium* Bosch & Sande Lac., Bryol. Jav. 2: 123. 1865. Type. Indonesia: Sumatra (isotype BM).

Plants slender, green to yellowish green, brownish at base, in small lax patches. *Stems* procumbent, up to 5 cm long, regularly bipinnately branched; central strand present; paraphyllia densely covered on stems, usually not on branches. *Stem leaves* triangularly ovate, concave, acuminate to a piliform apex, appressed when dry, erect-spreading when moist, ca. 0.25 mm × 9 mm; margins entire; costa percurrent, hyaline above, smooth; median leaf cells rhomboidal to oblong, ca. 15 µm × 6 µm, thick-walled, unipapillose; branch leaves arranged in 2 rows, triangular to oblong-ovate at base, acuminate toward apex, ca. 1 mm long; margins serrate; costa slender, reaching 2/5 the leaf length, hyaline, multipapillose. *Sporophytes* unknown.

Illustrations: Watanabe 1972 (Pl. 25, figs. 18–36).
Habitat: usually on rocks in tropical and subtropical forests; alt. 580–1300 m.
Distribution: China, Indonesia (Java, Sumatra), Borneo, and the Philippines.

Chinese specimen examined: TAIWAN: Botel Tobago Isl, *Schwaege 104* (PE).

2. ***Pelekium velatum*** Mitt. 鹤嘴藓 he-zui xian
J. Linn. Soc., Bot. 10: 176. 1868. Type. Admiralty Is. (isotype BM).
Lorentzia longirostris Hampe, Nuovo Giorn. Bot. Ital., n. ser. 4: 288. 1872.
Pelekium fissicalyx C. Müll., Bot. Jahrb. Syst. 5: 87. 1883.
Pelekium lonchopodium C. Müll. in Geh., Biblioth. Bot. 13: 7. 1889.

Plants slender, usually yellowish green, in small patches. *Stems* prostrate, ca. 3 cm long, regularly pinnately branched; central strand present; branches about 3–7 mm long; paraphyllia abundant, filamentous, slightly branched. *Stem leaves* remote, appressed when dry, erect-patent when moist, abruptly narrowed from a triangularly ovate base to a lanceolate apex, ca. 0.6 mm long; leaf margins serrate; costa reaching leaf apex, hyaline, papillose on back; median leaf cells oblong-hexagonal, ca. 15 µm × 8 µm, thin-walled, unipapillose; branch leaves ovate to ovate-triangular, ca. 0.3 mm × 0.2 mm; costa vanishing at upper portion of leaf; leaf cells quadrate to hexagonal, unipapillose. *Autoicous. Perichaetial leaves* oblong or rounded ovate, suddenly narrowed to a piliform acumen. *Setae* slender, ca. 1.5 cm long, densely papillose; capsules symmetrical or asymmetrical, horizontal or pendulous, dark reddish brown; opercula conical, long beaked; exostome teeth yellowish brown, hyaline and minutely papillose above; endostome segments lightly yellow, as long as exostome teeth; basal membrane about 1/2 the length of endostome segments; cilia hyaline, usually 2. *Spores* ca. 15 µm in diameter, papillose.

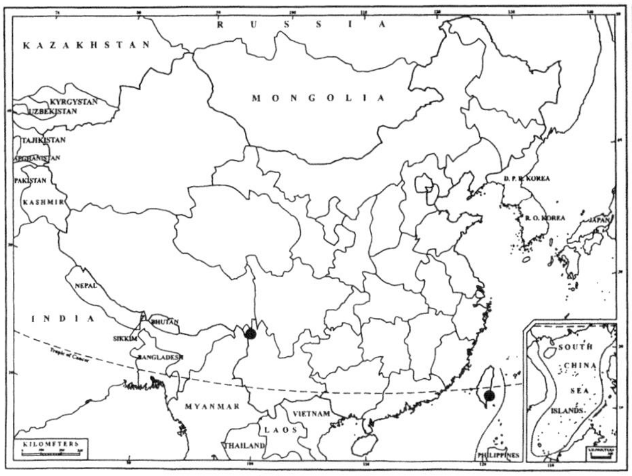

2. *Pelekium velatum*

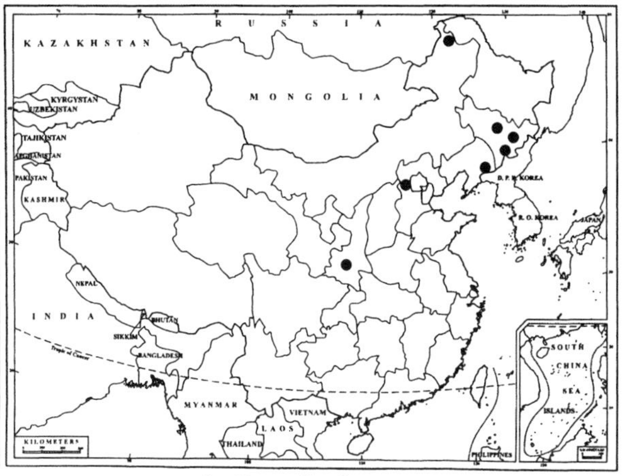

1. *Rauiella fujisana*

Illustrations: Watanabe 1972 (Pl. 25, figs. 1–17).
Habitat: mostly on shady and moist rocks.
Distribution: China, India, Thailand, Indonesia, the Philippines, New Guinea, and Pacific Islands.

Chinese specimens examined: TAIWAN: Botel Tobago Island, *Schwäge 120* (Noguchi herb., Watanabe 1972). YUNNAN: Gong-shan Co., *M.-Z. Wang 10608b* (PE).

13. ***Rauiella*** Reim. 硬羽藓属 ying-yu xian shu

Hedwigia 76: 287. 1937.

Plants rather small to medium-sized, up to 3 cm long, brownish green to yellowish green, not shining, rigid when dry, in loose interwoven mats. *Stems* creeping, spreading, regularly pinnately branched; apex of branches caudate; central strand present; paraphyllia numerous, filamentous or foliose, often forked, apical cell multipapillose. *Stem leaves* and branch leaves differentiated; stem leaves ovate or cordate at base, gradually or suddenly narrowed to a lanceolate apex; margins almost entire or slightly serrulate; costa rather thick, ending at upper portion of leaf; leaf cells oblong to rounded quadrate, thick-walled, multipapillose; branch leaves ovate or ovate-triangular, with a short acute apex. *Autoicous*. *Inner perichaetial leaves* lanceolate, widely ovate at base; margins serrate, but not ciliate; laminal cells linear, smooth; costa percurrent or excurrent. *Setae* slender, more than 1 cm long, reddish brown or yellowish brown, smooth; capsules erect-patent or horizontal, reddish brown, oblong-ovate to cylindrical; annuli differentiated; peristome double; exostome teeth narrowly lanceolate, yellowish brown, hyaline at apex, papillose; endostome segments yellowish, finely papillose; basal membrane high, at least 1/3 the length of endostome segments; cilia 2–3, hyaline, finely papillose. *Calyptrae* cucullate. *Spores* nearly spherical, finely papillose.

The characteristics of the genus *Rauiella* include: plants only pinnately branched, apices of stem leaves not papillose, paraphyllia filamentous to foliose, leaf cells multipapillose, and perichaetial leaves not ciliate. *Rauiella* consists of eight species in the world, and only one species is found in China.

1. ***Rauiella fujisana*** (Par.) Reim. 东亚硬羽藓 dong-ya ying-yu xian

Hedwigia 78: 278. 1937. *Thuidium fujisanum* Par., Index Bryol. 1281. 1898. *Thuidium cylindraceum* Mitt., Trans. Linn. Soc. London, Bot., 3: 190, *illeg. hom.* Type. Japan: Mt. Fuji, *Bisset s.n.* (isotype BM).
Rauia bandaiensis (Broth. & Par.) Broth., Nat. Pflanzenfam. I(3): 1005. 1907. *Thuidium bandaiense* Broth. & Par., Bull. Herb. Boiss., ser. 2, 2: 923. 1902.

Plants small, up to 3 cm long, green or dark green above, brownish green below, in thin interwoven patches. *Stems* creeping, regularly pinnately branched; branches ca. 5 mm long; central strand absent; paraphyllia dense, filamentous or foliose, often branched. *Stem leaves* appressed when dry, erect-patent when moist, triangular to triangularly ovate at base, less than 1 mm long, 0.5 mm wide, concave, gradually or suddenly narrowed to a short, lanceolate acumen; margins entire or serrate; costa extending to upper portion of leaf, spinosely papillose on the back; median leaf cells rectangular to nearly quadrate, ca. 12 µm × 9 µm, thick-walled, 2–6 papillae per cell, apical leaf cell linear, smooth; branch leaves ovate to triangularly ovate, gradually narrowed from a sheathing base to lanceolate apex; margins revolute, serrulate, papillose; costa slightly excurrent; median leaf cells linear, ca. 50 µm × 5 µm, smooth. *Setae* slender; capsules cylindrical, horizontal; peristome features the same as those described in the generic description.

Illustrations: Plate 466, figs. 1–16.
Habitat: on tree trunks, rarely on rotten logs or rock surfaces in temperate coniferous forests; alt. 1250 m.
Distribution: China, Korea, and Japan.

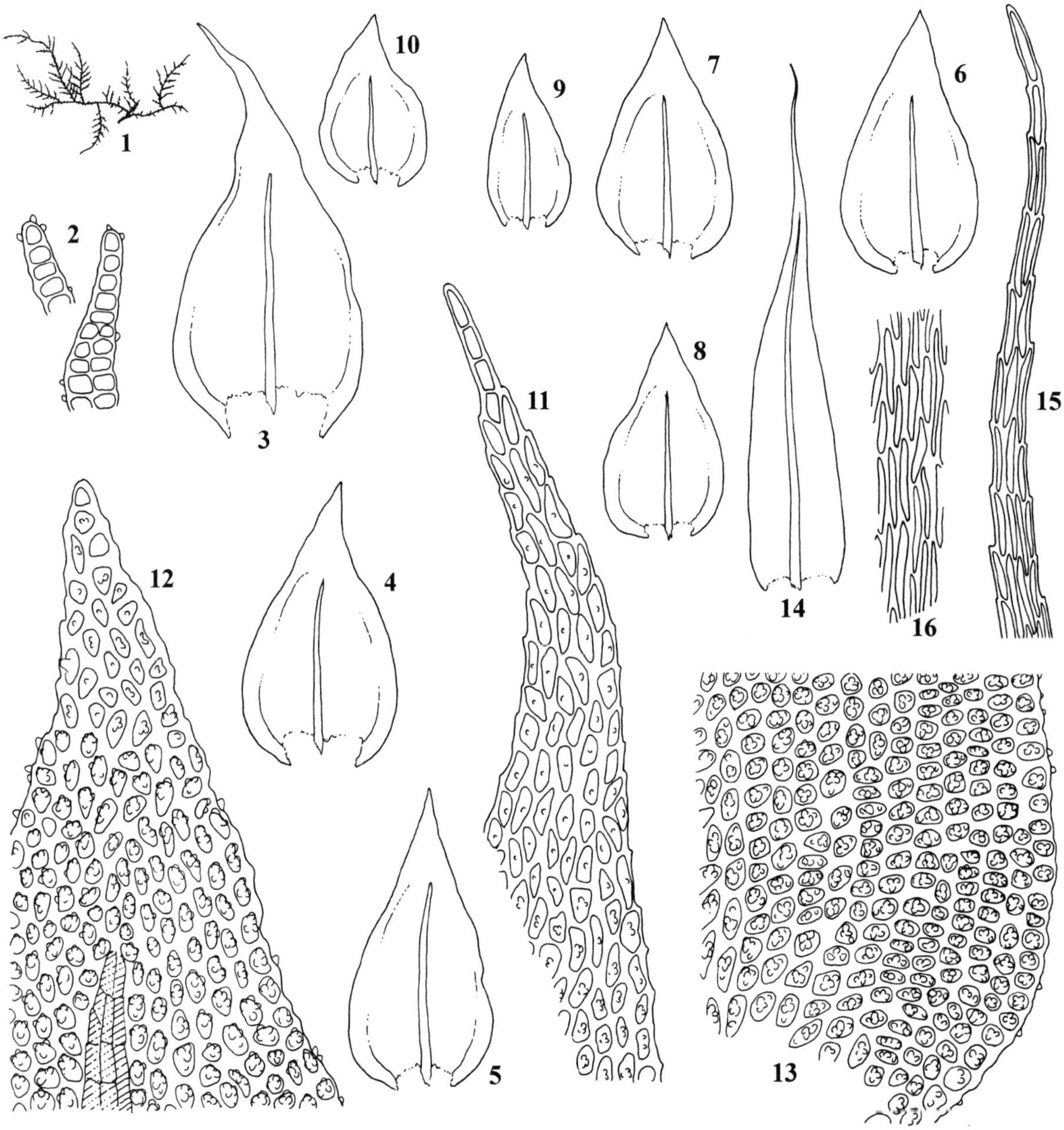

Plate 466. *Rauiella fujisana*. 1, habit (× 1). 2, paraphyllia (× 443). 3, stem leaf (× 90). 4–7, branch leaves (× 90). 8–10, branchlet leaves (× 90). 11, apical stem leaf cells (× 443). 12, apical branch leaf cells (× 443). 13, basal branch leaf cells (× 443). 14, perichaetial leaf (× 52). 15, apical perichaetial leaf cells (× 330). 16, median perichaetial leaf cells (× 330). (1–16, drawn from *Koponen 37101*, H.)

Chinese specimens examined: HEBEI: Xiao Wu-tai shan, *C.-Y. Yang 504b* (PE). JILIN: An-tu Co., *Koponen 37034* (G), *37101*, *37278* (H, PE); Jiao-he Co., *C. Gao 1392* (MO); Lin-jiang Co., *C. Gao 7739* (IFSBH). LIAONING: Kuan-dian Co., *Buck 23625*, *23754* (NY). NEI MONGOL: E-erguna zou qi, A-long Shan (Mt.), *X.-L. Bai 1021* (HIMC). SHAANXI: Hu Co., *M. Wang 400* (HKAS, PE, WUK).

14. ***Thuidium*** Bruch & Schimp. in B. S. G. 羽藓属 yu xian shu
Bryol. Eur. 5: 157. 1852.

Plants medium-sized to robust, green, yellowish green or brownish green, often in loosely interwoven mats. *Stems* procumbent to climbing, bi- to tripinnately branched; paraphyllia abundant on stems and branches, filamentous or foliose, often papillose; central strand present. *Stem leaves* and branch leaves differentiated or similar, but different in size; stem leaves ovate or ovate-cordate, usually with slender apex, narrowed at base, decurrent, multiplicate; leaf margins revolute, serrulate above; costa not extending to leaf apex, rarely excurrent; leaf cells isomorphic, mostly hexagonal or rounded to rounded-hexagonal, equally thick-walled, uni- or multipapillose. *Primary branch leaves* similar to the stem leaves, but smaller; secondary or young branch leaves minute, mostly ovate or oblong-ovate, concave; leaf margins erect; costa weak and short. *Autoicous* or dioicous. *Perichaetial leaves* lanceolate or ovate-lanceolate, with slender apices, sometimes leaf margins with long cilia; costa ending at leaf apex or slightly excurrent; leaf cells rectangular, smooth or papillose. *Setae* slender, smooth or densely papillose above; capsules horizontal or declining, ovoid or cylindrical, slightly curved, brownish, smooth; annuli 2–3 rows, revolute, deciduous; opercula acute to conical, obliquely beaked; stomata present; peristome double; exostome teeth yellowish or yellowish brown; basal membrane high, at least 1/3 the length of endostome segments; cilia 2–4, mostly nodulose, rarely rudimentary or lacking. *Calyptrae* cucullate or campanulate, smooth, rarely scabrous.

Thuidium is characterized by its regularly pinnately branched plants, its differentiated stem and branch leaves, its abundant paraphyllia on stems and branches, and its densely papillose leaf cells. *Thuidium* is the largest genus in the Thuidiaceae. Cardot and Thériot (1904) were probably the first authors dealing with the occurrences of two *Thuidium* species: *Thuidium japonicum* Dozy & Molk. and *T. pycnothallum* C. Müll. from China. Subsequent papers concerning the studies of the Chinese *Thuidium* were published by a number of authors (Brotherus 1925; Bartram 1935; P.-C. Chen & P.-C.. Wu 1965; P.-C. Chen et al. 1978; Dixon 1933; C. Gao 1977; X.-Y. Hu & P.-C. Wu 1991; Y. Jia et al. 1995; X.-J. Li (editor-in-chief) 1985; Redfearn & P.-C. Wu 1986; Reimers 1931; Potier de la Varde 1937; B.-C. Tan et al. 1994; B.-C. Tan & Y. Jia 1997; Watanabe 1964, 1980; P.-C. Wu et al. 1984, 1987; P.-C. Wu 1992; B.-S. Xu (editor-in-chief) 1989; C.-Y. Yang 1936; B.-Y. Yang & W.-C. Lee 1964; M.-X. Zhang 1978). Recently, Redfearn et al. (1996) enumerated 13 species and 2 varieties for China. In this study, we recognize 11 species. The occurrence of *Thuidium recognitum* needs confirmation. Species of *Thuidium* have been recorded throughout the country except Ningxia province. Most recently, Y.-M. Fang & Koponen (2001) published a new species, *Thuidium unipinnatum* Y.-M. Fang & Kop. and reported *T. kuripanum* (Dozy & Molk.) R. Watan. new to China. Unfortunately, we are not able to verify these species.

1. Stem leaves mostly ovate or widely ovate, with obtuse or shortly acuminate apices 2
1. Stem leaves usually triangular or cordate, with long, narrowly acuminate apices 5
    2. Leaf cells rhomboidal or oblong, unipapillose .... 11. *T. tamariscinum*
    2. Leaf cells oval or rhomboidal, multipapillose or stellately papillose, rarely unipapillose 3
3. Central strand undifferentiated; leaf cells stellately papillose .... 8. *T. pristocalyx*
3. Central strand present; leaf cells with 2–6 coarse papillae, rarely unipapillose 4
    4. Perichaetial leaves not ciliate .... 5. *T. glaucinoides*
    4. Perichaetial leaves long ciliate .... 9. *T. subglaucinum*
5. Plants medium-sized, usually not more than 5 cm long 6
5. Plants large, up to or more than 10 cm long 7
    6. Stem leaves ovate at base, tapering to acuminate above .... 4. *T. erosifolium*
    6. Stem leaves cordate at base, suddenly narrowed to a narrowly lanceolate apex .... 7. *T. plumulosum*
7. Costa of stem leaves ending below the apex .... 3. *T. delicatulum*
7. Costa of stem leaves percurrent to excurrent 8
    8. Stem leaves usually without or with short ciliate apices .... 1. *T. assimile*
    8. Stem leaves usually with long ciliate apices 9

9. Apices of stem leaves long piliferous, consisting of 6–10 uniserriate cells .... 2. *T. cymbifolium*
9. Apices of stem leaves shortly piliferous, consisting of 2 multiserriate cells 10
   10. Central strand of stems present; stem leaf bases triangular or triangularly ovate; perichaetial leaves ciliate .... 6. *T. kanedae*
   10. Central strand of stems absent; stem leaf bases cordate or cordate-ovate; perichaetial leaves without cilia...... .... 10. *T. submicropteris*

1. ***Thuidium assimile*** (Mitt.) Jaeg. 绿羽藓 lv-yu xian

Ber. Thätigk. St. Gallischen Naturwiss. Ges. 1876–77: 260. 1878. *Leskea assimilis Mitt.*, J. Proc. Linn. Soc., Bot., Suppl. 1: 133. 1859. Type. India: West Himalayas, Kumaon, *Strachey & Winterbottom s.n.*

Thuidium intermedium Philib., Rev. Bryol. 20: 33. 1893, *nom. illeg.*

Thuidium philibertii Limpr., Laubm. Deutschl. 2: 835. 1895. Synonymized by Touw (2001a).

Thuidium pycnothallum (C. Müll.) Par., Index Bryol. 1289. 1898. *Tamarisicella pycnothalla* C. Müll., Nuovo Giorn. Bot. Ital., n. ser. 3: 116. 1896. Syntypes. China: Shaanxi (Schen-si), Si-ku-tziu-san, Lao-iu-hur, July 1894, *Giraldi s.n.*; Huo-gia-ziez, Sept. 1894, *Giraldi s.n.*; Hua-tzo-pin, 20 June 1894, *Giraldi s.n.*; Khiu-Lin-san, June 1894, *Giraldi s.n.*; Zu-lu, Aug. 1894, *Giraldi s.n.*

Plants large, up to 7 cm long, yellowish green, brownish green when old, in large, interwoven mats. *Stems* creeping or ascending above, regularly pinnately branched; central strand present; paraphyllia densely growing on stems, foliose or filamentous, branched. *Stem leaves* ca. 1.8 mm long, triangularly ovate, plicate, gradually tapering to a narrow lanceolate and piliferous apex, mostly consisting of 2–5 cells, sometimes secund; margins more or less revolute, serrate; costa percurrent, papillose on back; median leaf cells rhomboidal or oblong, 5–20 µm long, thick-walled, unipapillose; apical cells hyaline, smooth; branch leaves concave, ovate to ovate-triangular, with a short apex, ca. 1 mm long, apical cells multipapillose. *Dioicous*. *Perigonial leaves* ovate-lanceolate, concave below. *Inner perichaetial leaves* narrowly lanceolate, up to 2 mm long, margins serrate, smooth, without hairs. *Setae* reddish brown, ca. 2 cm long, smooth; capsules about 3 mm long; peristome double; exostome teeth broadly lanceolate, reddish brown, papillose above; endostome segments as long as the exostome teeth, finely papillose; basal membrane about 1/2 the length of endostome segments; cilia 2–3. *Spores* 10–15 µm in diameter, finely papillose.

Illustrations: P.-C. Wu (ed.) 2001 (Pl. 102, figs. 9–17).
Habitat: on forest ground, rocks or tree trunks; alt. 2000–3600 m.
Distribution: China, Japan, Russia (Siberia), Europe, and North America.

Chinese specimens examined: GUIZHOU: Leigongshan (Mt.), *B.-C. Tan 91-1031* (FH). HEBEI: Dong-ling, *C.-Y. Yang 204* (BM). HENAN: Song Co., *J.-X. Luo 391* (PE). HUBEI: Mt. Shennongjia, *1980 Sino-Amer. Exped. No. 281B, 842* (NY). JILIN: Mt. Changbai, *C. Gao 22146* (NY), *B.-D. Xin 0275* (MO, PE). NEI MONGOL: E-erguna You Qi (Co.), *X.-L. Bai 194* (HIMC); Mt. Wala, 40°42′ N, 109°24′ E, *X.-L. Bai 185* (HIMC, NY). QINGHAI: Mengda, Tianchi, *H. D. Ding 21722* (PE); Yu-shu Co., *B.-C. Tan 95-1729* (FH, MO). SHAANXI: Bao-ji (Pao-ki) Co., *Hugh 360* (BM); Feng Co., *M.-X. Zhang 1364* (MO, PE); Fo-ping Co., *S.-E. Liu & P.-Q. Zhong 4410* (NY); Hu Co., *Z.-P. Wei 4809* (NY); Mt. Qin-ling, *P.-C. Chen et al. 25* (MO, PE), *Hugh 99* (BM); Mt. Tai-bai (Miao-wan), *Hugh 28* (BM); Xi-an (Seian) City, *Hugh 290* (BM). SHANDONG: Qing-dao City, *C. Gao & E.-Z. Bai 34572* (IFSBH, MO). SICHUAN: Mt. Omei, April 1982, *Ratcliffe s.n.* (BM); Nan-ping Co., *S.-Y. Zeng 105a* (HKAS); between Yanyuen Co. to Yonqnin, *Handel-Mazzetti 2948* (H); Xi-chang (Ningyüan) City, *Handel-Mazzetti 1514* (H). XINJIANG: He-jing Co., *S. He 951172* (MO); Urumqi Co., *S. He 951014* (MO). YUNNAN: An-ning Co., *Koponen 38028* (as *Thuidium cymbifolium*, NY); Lu-liang Co, Ma-ming-zhen (Ma-eul-chan), *Delavay s.n.* (BM, PC). ZHEJIANG: Tiantaishan (Mt.), *Faber 7* (H).

This species is characterized by its piliferous stem leaf apices that are usually 2–5 cells long and by having entire margins of inner perichaetial leaves.

2. ***Thuidium cymbifolium*** (Dozy & Molk.) Dozy & Molk. 大羽藓 da-yu xian

Bryol. Jav. 2: 115. 1867. *Hypnum cymbifolium* Dozy & Molk., Ann. Sci. Bot., ser. 3, 2: 306. 1844. Type. Indonesia: Java.

Thuidium japonicum Dozy & Molk. ex Sande Lac., Ann. Mus. Bot. Lugduno-Batavum 2: 297. 1865 [1866].

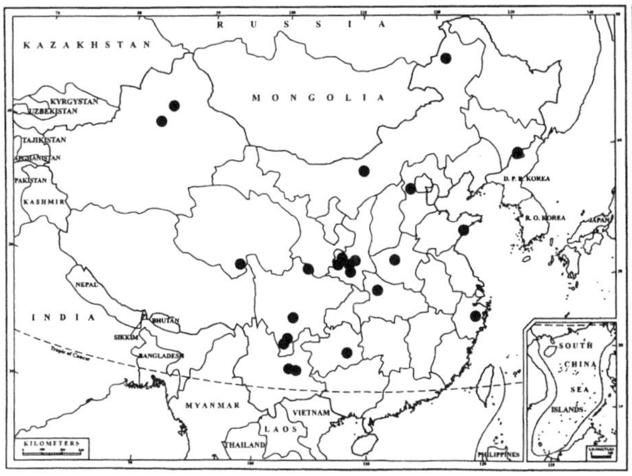

1. *Thuidium assimile*

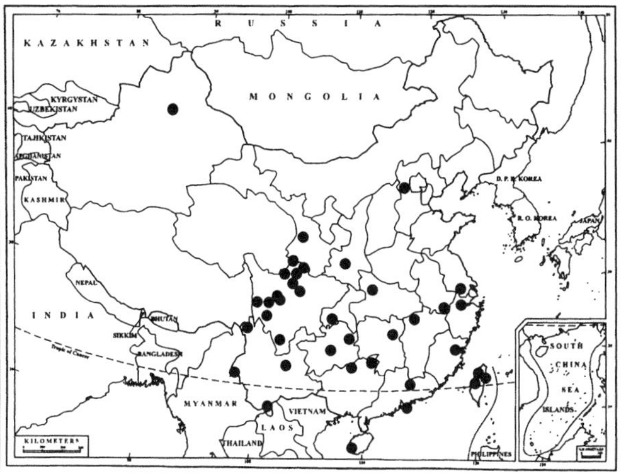

2. *Thuidium cymbifolium*

Plants robust, up to or more than 10 cm long, light green to dark green, yellowish brown when old, often in large, interwoven patches. *Stems* creeping, usually regularly bipinnately branched; branches about 1.5 cm long; central strand differentiated; paraphyllia dense on stems and branches, foliose or filamentous, apical cells multipapillose. *Stem leaves* loosely appressed when dry, erect-patent when moist, suddenly narrowed from a triangularly ovate base to a lanceolate to piliferous apex, usually 6–10 cells long; margins mostly revolute or slightly revolute, rarely plane, serrulate above; costa extending to the lanceolate apex, with papillae or paraphyllia on back; median leaf cells oval-rhomboidal to elliptic, 5–20 µm wide, setulosely unipapillose; branch leaves concave, ovate or oblong-ovate, with a short apex; costa up to 2/3 the leaf length. *Dioicous*. *Perichaetial leaves* widely ovate-lanceolate, margins entire. *Setae* yellowish brown to reddish brown, about 2 cm long; capsules cylindrical; peristome double; exostome teeth broadly lanceolate, reddish brown, papillose above; endostome segments as long as the exostome teeth, finely papillose; basal membrane about 1/2 the length of endostome segments; cilia 2–3. *Spores* ca. 20 µm in diameter, finely papillose.

Illustrations: P.-C. Wu (ed.) 2001 (Pl. 101, figs. 1–13).
Habitat: on shady moist rocks, humus, rotten logs or trunks in large patches.
Distribution: widely distributed in the world.

Chinese specimens examined: ANHUI: Huang-shan (Mt.), *P.-C. Chen et al. 6637* (PE), *7550* (MO, PE). CHONGQING: Nan-chuan Co., *P.-C. Wu 21155* (PE), *21018* (MO, PE). FUJIAN: Ning-de City, *J. Zhu 1112*, *1497* (HSNU). GANSU: *Potanin 682* (BM). GUANGDONG: Mt. Luo-fu (Loh Fau), *Merrill 11178* (BM). GUANGXI: Jiuwanshan (Mt.), *Y.-Jia 009* (PE), *M.-Z. Wang 46584* (PE), *Z.-Y. Li 89141* (PE); Xing-an Co., *S.-S. Hu 80403* (MO). GUIZHOU: Gui-yang City, *Boirieis 790* (PC); Jiang-kou Co., *B. C. Tan 91-1149* (FH, MO). HAINAN: Mt. Jianfongling, *P.-C. Chen et al. 358*, *779* (PE). HEBEI: Wei Co., *PE 2327* (MO). HONG KONG: *Ah Nin H17A* (BM). HUBEI: Mt. Shennongjia, *P.-C. Wu 381*, *462* (MO, PE). HUNAN: Chang-sha City, *Grubb 57* (BM). JIANGSU: Yi-xing Co., *P.-C. Wu & Y.-X. Jin 2148*, *2210* (MO, PE). JIANGXI: Wu-ning Co., *C.-M. Tan 97034* (MO). SHAANXI: Mt. Tui-kio-san, *Giraldi 2253* (as *Tamariscella pycnothalla*, BM, PC). SICHUAN: A-ba Co., *S. He 31054a* (MO); Ba-tang Co., *S. He 31633* (MO); Bao-xing (Pao-hsing) Co., *K.-L. Chu 3485* (BM); Dao-cheng Co., *S. He 31716* (MO); Ma-er-kang Co., *S. He 31113* (MO); Mt. Omei, *Raven et al. 76* (MO); Nan-ping Co., *S. He 30264* (MO), Ruo-er-gai Co., *S. He 30886* (MO); Song-pan Co., *S. He 30465b* (MO); Tian-quan Co., *S. He 32032* (MO); Wo-long Nature Reserve, *K.-Y. Lang 13152b* (PE), *J.-X. Luo 3234* (H, PE); Ya-jiang Co., *S. He 31979* (MO); Yan-yuan (Yenyuen) Co., *Handel-Mazzetti 3004* (BM). TAIWAN: Chia-yi Co., *Inoue 198* (BM, MO); Hsin-chu Co., *Shevock 18209* (MO); Hua-lien Co., *C.-I. Peng 83-76* (MO); Ilan Co., *C.-I. Peng 83-25* (MO); Miao-li Co., *Shevock 14337A* (MO); Nan-tou Co., *Shevock 14416* (MO); Tai-chung Co., *Shevock 18173* (MO). XINJIANG: Tian-shan Mts, *J.-C. Zhao 3446b* (HBNC). YUNNAN: An-ning Co., *Redfearn & M. Zang 33341* (MO); Kun-ming City, *Koponen 37452* (H, MO); Nu-jiang (Lu-djiang), *Handel-Mazzetti 9071* (BM); Rui-li Co., *X.-Y. Jia 910160* (MO); Xishuangbanna, *W.-X. Xu 6010* (PE). ZHEJIANG: Hang-zhou City, *Buck 23926* (NY); Lin-an Co., *P.-C. Chen 7918* (MO, PE); Mt. West Tian-mu, *R.-L. Hu 6* (MO).

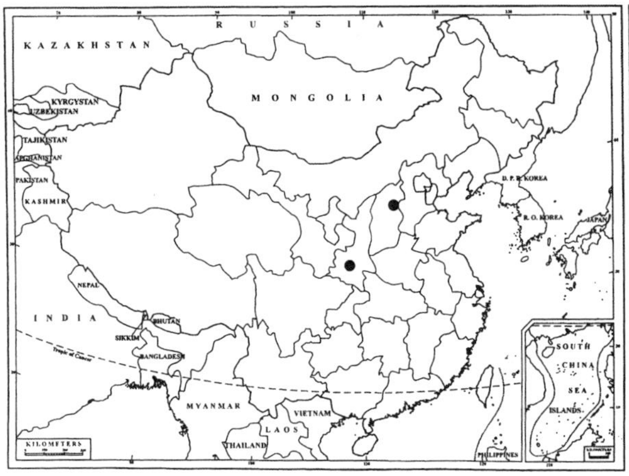

3. *Thuidium delicatulum*

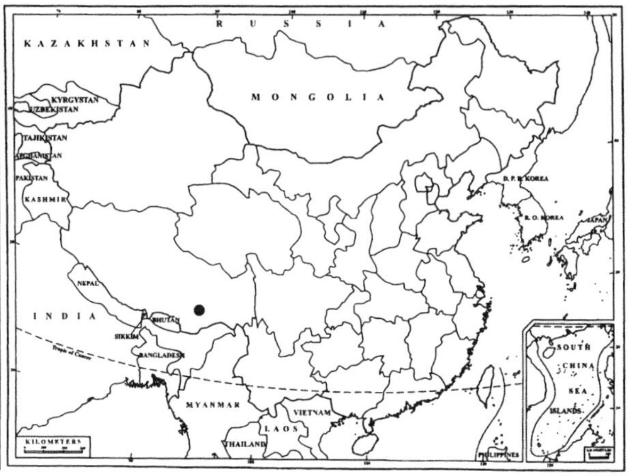

4. *Thuidium erosifolium*

3. ***Thuidium delicatulum*** (Hedw.) Schimp. 细枝羽藓 xi-zhi yu xian

Bryol. Eur. 5: 164 (fasc. 49–51. Monogr. 8). 1852. *Thuidium delicatulum* (Hedw.) Mitt., J. Linn. Soc. Bot. 12: 578. 1869, *hom. illeg. Hypnum delicatulum* Hedw., Sp. Musc. Frond. 250. 1801. Type. U.S.A.

Thuidium uliginosum Card., Bull. Soc. Bot. Genève, ser. 2, 3: 283. 1911.

Thuidium viride Mitt., Trans. Linn. Soc. London, Bot. 3: 188. 1891.

Plants large, up to 12 cm long, yellowish green to lightly brownish green, in large interwoven mats. *Stems* regularly tripinnately branched; branches 5–15 mm long; central strand present; paraphyllia dense on stems and branches, foliose or filamentous, branched and papillose. *Stem leaves* about 1 mm long, triangularly ovate, slightly plicate, concave, suddenly narrowed into a lanceolate acumen, often secund; margins serrate; costa reaching near leaf apex, multipapillose on back; median leaf cells oblong to rounded-rhomboidal, ca. 6–15 µm wide, thin-walled, unipapillose; branch leaves concave, usually ovate, with short apices; costa up to 4/5 the leaf length. *Dioicous*. *Perichaetial leaves* ovate-lanceolate, usually long ciliate at shoulder, rarely entire. *Setae* reddish brown, 3 cm long, smooth; capsules cylindrical, curved; exostome teeth reddish brown, grossly papillose above; endostome segments as long as the teeth, finely papillose above; lightly yellowish; basal membrane about 1/2 the height of segments; cilia 2–3, finely papillose. *Spores* about 20 µm in diameter, finely papillose.

Illustrations: P.-C. Wu (ed.) 2001 (Pl. 99, figs. 1–14).
Habitat: usually on humus or moist ground.
Distribution: China, Korea, Japan, Europe, and North, Central and South America.

Chinese specimens examined: SHAANXI: Mt. Tai-bai, *M.-X. Zhang 181* (XBGH). SHANXI: Shizui to Hengling, *M. Saito 71* (TNS).

4. ***Thuidium erosifolium*** S.-Y. Zeng 啮蚀叶羽藓 chi-shi-ye yu xian

Guihaia 10: 105. pl. 1. 1990. Type. China: Xizang, Lin-zhi Co., *Y.-G. Su 2362* (holotype HKAS; isotype PE).

Plants delicate, medium-sized, 3–4 cm long, yellowish brown, lightly brownish when old. *Stems* often slightly flexuous, irregularly pinnately branched; branches 0.5–1.0 cm long; paraphyllia usually consisting of 3 single cells. *Stem leaves* ovate-lanceolate at base, tapering to an acuminate apex; margins coarsely serrate, one cell long, more distinct near leaf apex; costa yellowish, stout, vanishing at leaf apex; median leaf cells hexagonal to polygonal, ca. 10 µm wide, thin-walled, centrally unipapillose; branch leaves ovate, with a shortly lanceolate apex, about 1/2 the length of stem leaves. *Sporophytes* not seen.

Illustrations: Plate 467, figs. 1–10.
Habitat: on tree trunks in forests.
Distribution: endemic to China.

Chinese specimens examined: See the type information cited above.

This species is characterized by its coarsely serrate leaf margins and thick, stout costa.

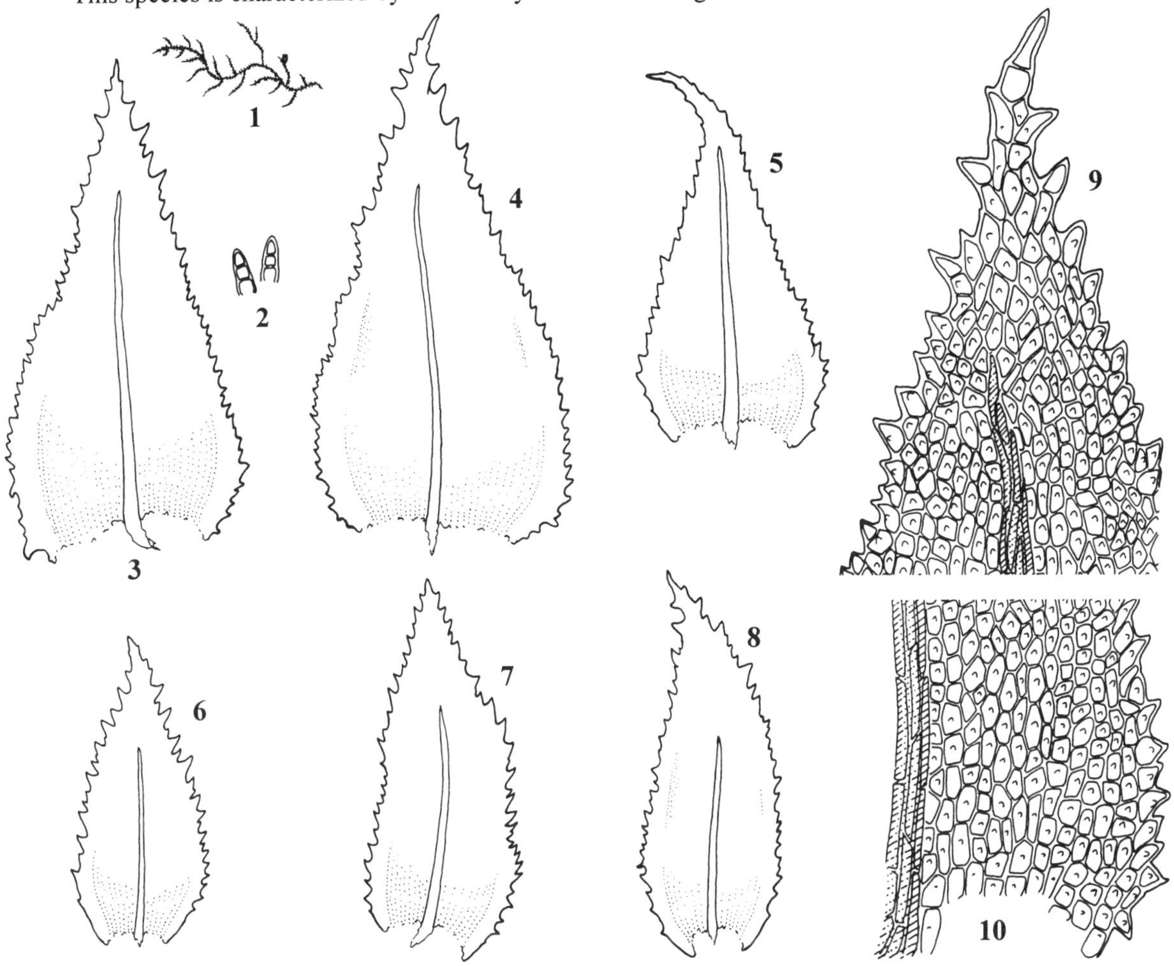

Plate 467. *Thuidium erosifolium*.1, habit (× 1). 2, paraphyllia (× 190). 3–4, stem leaves (× 190). 5, branch leaf (× 190). 6–8, branchlet leaves (× 190). 9, apical leaf cells (× 365). 10, basal leaf cells (× 365). (1–10, drawn from *Y.-G. Su 2362*, HKAS.)

5. ***Thuidium glaucinoides*** Broth. 拟灰羽藓 ni-hui-yu xian
Philipp. J. Sci. 3: 26. 1908. Type. The Philippines: Luzon.
Thuidium orientale Mitt. ex Dix., J. Bot. 51: 329. 1933.

Plants large, up to or more than 10 cm long, often lightly brownish green or yellowish green, in loose mats. *Stems* regularly bipinnately branched; branches ca. 5–10 mm long; central strand present; paraphyllia densely growing on stems or branches, foliose or filamentous, papillose. *Stem leaves* appressed when dry, erect-patent when moist, broadly ovate to ovate-triangular, with a short apex, ca. 0.8–1.0 mm long; margins serrate; costa well developed, reaching 3/4 the leaf length, setulosely papillose on back; median leaf cells oblong-ovate to elliptic, thick-walled, uni- to multipapillose, upper cells rounded hexagonal; branch leaves ovate to widely ovate, costa reaching 2/3 the leaf length. *Dioicous*. *Perichaetial leaves* lanceolate, piliferous; margins serrate, not ciliate. *Setae* reddish brown or yellowish brown, up to or more than 3 cm long; capsules oblong-ovate, horizontal; exostome teeth reddish brown, grossly papillose above; endostome segments as long as the teeth, finely papillose above; lightly yellowish; basal membrane about 1/2 the height of segments; cilia 2–3, finely papillose. *Spores* not seen.

Illustrations: P.-C. Wu (ed.) 2001 (Pl. 100, figs. 8–15).
Habitat: mostly on forest ground, rotten logs, under grasses or shady moist rocks; alt. ca. 1000 m.

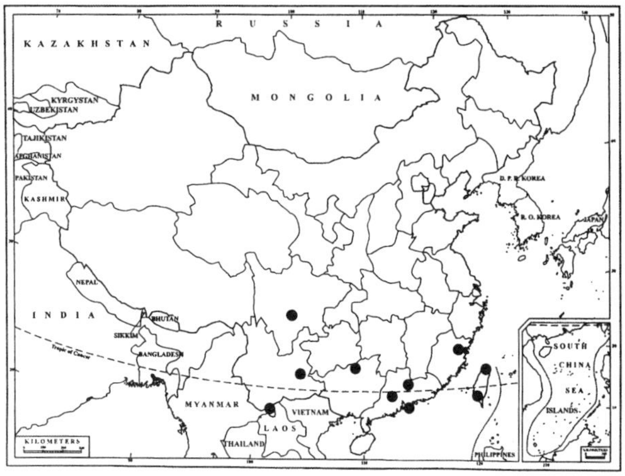

5. *Thuidium glaucinoides*

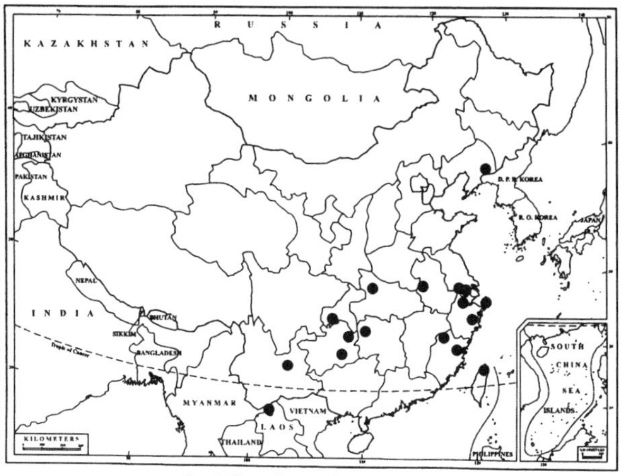

6. *Thuidium kanedae*

Distribution: China, Japan, Myanmar, Thailand, Vietnam, Malaysia, Indonesia, the Philippines, and Southern Pacific Islands.

Chinese specimens examined: FUJIAN: no exact locality, *H. H. Chung 290* (PE); Ning-de City, *J. Zhu 191, 1431* (HSNU). GUANGDONG: Mt. Ding-hu, *Redfearn et al. 34415* (MO, NY); Xin-feng Co., *Taam 424* (NY). GUANGXI: Jiuwanshan Mt, *Y. Jia 351, 369* (PE), *M.-Z. Wang 46538* (PE). HONG KONG: Xin Jie, Dapu, *Herklots 245* (CUHK, MO), *P. P. H. But & C.-H. Gao 27122* (CUHK). SICHUAN: Mt. Omei, April 1982, *Ratcliffe s.n.* (as *Thuidium delicatulum*, BM). TAIWAN: Kao-hsiung Co., *C.-C. Chuang 983* (MO, NY); Tai-pei Co., *C.-C. Chuang 5331* (NY); Tao-yuen Co., *M.-J. Lai 0276* (MO). YUNNAN: Lu-xi Co., *X.-Y. Jia 910059* (MO); Xishuangbanna, *W.-X. Xu 6015, 6094* (PE), *M.-Z. Wang 4419* (PE).

This species is characterized by its leaf cells with 1–3 papillae, its costa setulosely papillose on the back, and its perichaetial leaves without cilia. This species was reported from Guangdong as *Thuidium orientale*, which might be the first record of the species in China (P.-J. Lin et al. 1982).

6. ***Thuidium kanedae*** Sak. 短肋羽藓 duan-le yu xian
Bot. Mag. (Tokyo) 57: 345. 1943. Type. Japan: Kumamoto Pref., *Sakurai 11380*.
Thuidium toyamae Nog., J. Jap. Bot. 23: 115, f. 38. 1949.

Plants usually large, up to or more than 10 cm long, yellowish green to light green, brownish when old, in loosely interwoven mats. *Stems* regularly bi- to tripinnately branched; branches up to 1.5 cm long; central strand differentiated; paraphyllia densely on stems and branches, foliose or filamentous, branched, apex 2–4 cells long, papillose. *Stem leaves* appressed when dry, triangularly ovate to triangular, about 3 mm long, lanceolate toward piliferous apex, consisting of 3–10 cells; margins nearly plane to revolute, serrate above; costa stout, ending below leaf apex, setulosely papillose on back; median leaf cells rounded rhomboidal to oblong, 5–20 µm long, usually stellately papillose (2–4 papillae), rarely unipapillose; branch leaves concave, ovate to elliptic, with a short acute apex, ca. 1 mm long; margins serrate; costa reaching 2/3–4/5 the leaf length; with 2–4 papillae or single stellate papilla. *Dioicous. Perichaetial leaves* long ciliate above. *Setae* reddish brown to brown, up to or more than 3 cm long; capsules cylindrical, horizontal; exostome teeth broadly lanceolate, reddish brown, papillose above; endostome segments as long as the exostome teeth, finely papillose; basal membrane about 1/2 the length of endostome segments; cilia 2–3. *Spores* 15–20 µm in diameter, finely papillose.

Illustrations: Plate 468, figs. 1–8.
Habitat: usually on shady moist rocks, forest ground or fallen logs; alt. 780–1200 m.
Distribution: China, Korea, and Japan.

Chinese specimens examined: ANHUI: Jin-zhai Co., *Z.-M. Zeng T0228* (MO). CHONGQING: Nan-chuan Co., *M.-Z. Wang 860140* (MO, PE), *860435* (NY). FUJIAN: Ning-de City, *J. Zhu 1558* (HSNU); Mt. Wuyi, *C.-H. Zhang*

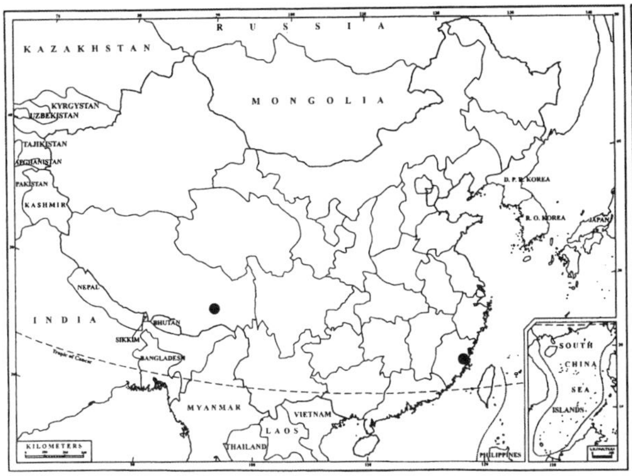

7. *Thuidium plumulosum*

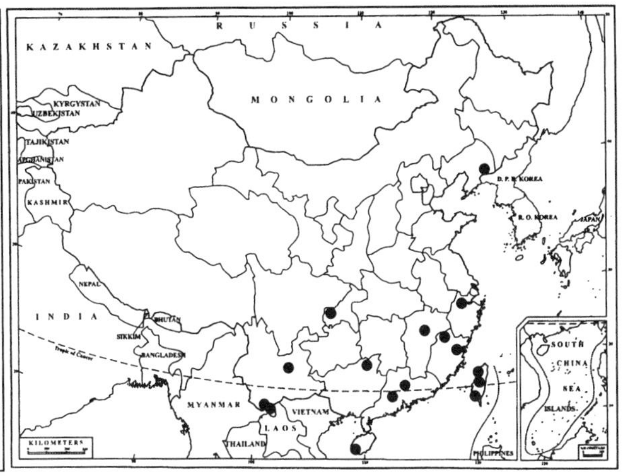

8. *Thuidium pristocalyx*

23-10, 23-26 (PE). GUIZHOU: Mt. Fan-jing, *S.-Y. Cheo 428* (NY); Mt. Lei-gong, *B.-C. Tan 91-826* (FH). HUBEI: Mt. Shennongjia, *P.-C. Wu 336* (MO). HUNAN: Bao-jing Co., *Z.-M. Zeng & R.-S. Tian 005* (PE). JIANGSU: Wu Co., *P.-C. Wu & Y.-X. Jin 1798* (MO, PE); Wu-xi Co., *P.-C. Wu & Y.-X. Jin 2141* (MO); Yi-xing Co., *P.-C. Wu & Y.-X. Jin 2045* (MO, PE). LIAONING: Kuan-dian Co., *Buck 23591* (NY). TAIWAN: Tai-pei Co., *C.-C. Chuang & Schofield 174* (NY). YUNNAN: Kun-ming City, *Handel-Mazzetti 69* (as *Thuidium cymbifolium*, BM); Meng-la Co., *Redfearn et al. 33744* (MO). ZHEJIANG: De-qing Co., *P.-C. Wu 1178, 1235* (MO, PE); Pu-duo Is., *P.-C. Wu 1348* (MO); West Tianmu shan (Mt.), *P.-C. Wu 1133* (PE); Ling-hai Co., *P.-C. Wu 711* (PE).

This species is very similar to *T. cymbifolium*, but differs in having stellately papillose leaf cells, which is a unique feature in *Thuidium*.

7. ***Thuidium plumulosum*** (Dozy & Molk.) Dozy & Molk. 毛尖羽藓 mao-jian yu xian

Bryol. Jav. 2: 118, 223. 1865. *Hypnum plumulosum* Dozy & Molk., Musc. Archip. Ind. 13. 1844. Type. Indonesia: Java.

Thuidium meyenianum (Hampe) Dozy & Molk., Bryol. Jav. 2: 121, 224. 1865. *Hypnum meyenianum* Hampe, Icon. Musc. 8. 1844.

Plants medium-sized, up to 6 cm long, slightly rigid, dark green, lightly brownish green when old, in loose flat patches. *Stems* usually bipinnately branched, in cross section outer cortical layer consisting of 3–4 layers of small, thick-walled cells, becoming larger cells, but rather thick-walled in the inner cortical layer; central strand consisting of 5–6 small cells; paraphyllia dense, filamentous, forked, papillose; branches ca. 0.5 cm long. *Stem leaves* remote, slightly plicate, with ovate-triangular or ovate-cordate bases, suddenly narrowed to piliferous apices, 1.0–1.5 mm long; costa vanishing below leaf apex; margins serrate; branch leaves ovate-triangular, younger leaves ovate to widely ovate, with narrowly obtuse apices, ca. 0.2–0.3 mm long; leaf cells rounded hexagonal, oblong, median cells 10–15 μm long, rather thick-walled, with single, sharp papilla on dorsal surfaces. *Dioicous*. *Inner perichaetial leaves* gradually narrowed from oblong bases to lanceolate, piliferous apices, margins long ciliate. *Setae* densely papillose, up to 3 cm long; capsules reddish brown, oblong ovoid; opercula shortly beaked; exostome teeth lightly yellowish to lightly red, densely transversely striate, hyaline at apex, papillose; endostome segments lightly yellowish; basal membrane ca. 1/2 the height of segments, papillose; cilia 3–4, nodulose. *Spores* nearly smooth, 12–15 μm in diameter.

Illustrations: Watanabe 1972 (Pl. 4, figs. 1–20, as *Thuidium meyenianum*).
Habitat: on tree trunks or rocks at low elevation in tropical and subtropical regions; alt. 400–500 m.
Distribution: China and Indonesia.

Chinese specimens examined: FUJIAN: Fozhou, *H.-H. Chung B142* (PE). XIZANG: Mo-tuo Co., *Y.-G. Su 2278-(1)* (HKAS, MO).

8. ***Thuidium pristocalyx*** (C. Müll.) Jaeg. 灰羽藓 hui-yu xian

Ber. Thätigk. St. Gallischen. Naturwiss. Ges. 1876–77: 257. 1878. *Hypnum pristocalyx* C. Müll., Bot. Zeitung (Berlin) 12: 573. 1854. Type. India: Nilghiri, *Schmid s.n.*

Thuidium glaucinum (Mitt.) Bosch & Sande Lac., Bryol. Jav. 2: 117. 1865. *Leskea glaucina* Mitt., J. Proc. Linn. Soc., Bot., Suppl. 1: 133. 1859.

Thuidium cochlearifolium Reim. & Sak., Bot. Jahrb. Syst. 64: 549. pl. 21 f. 5. 1931.

Plants large, up to 15 cm long, pale green or yellowish green, rarely dark green, in loosely interwoven patches. *Stems* usually regularly pinnately branched, reddish rhizoids in tufts at the apex or ventral surface of stems; central strand undifferentiated; paraphyllia sparse, foliose or filamentous, branched, sometimes lacking. *Stem leaves* appressed when dry, ovate to triangularly ovate, concave, 1–5 mm long; margins serrate; costa reaching 2/3 the leaf length, rarely forked above, usually smooth on the back, rarely papillose; median leaf cells oval to rhomboidal, ca. 5–20 μm long, thick-walled, stellately papillose; branch leaves ovate to widely ovate, about 1 mm long; leaf cells

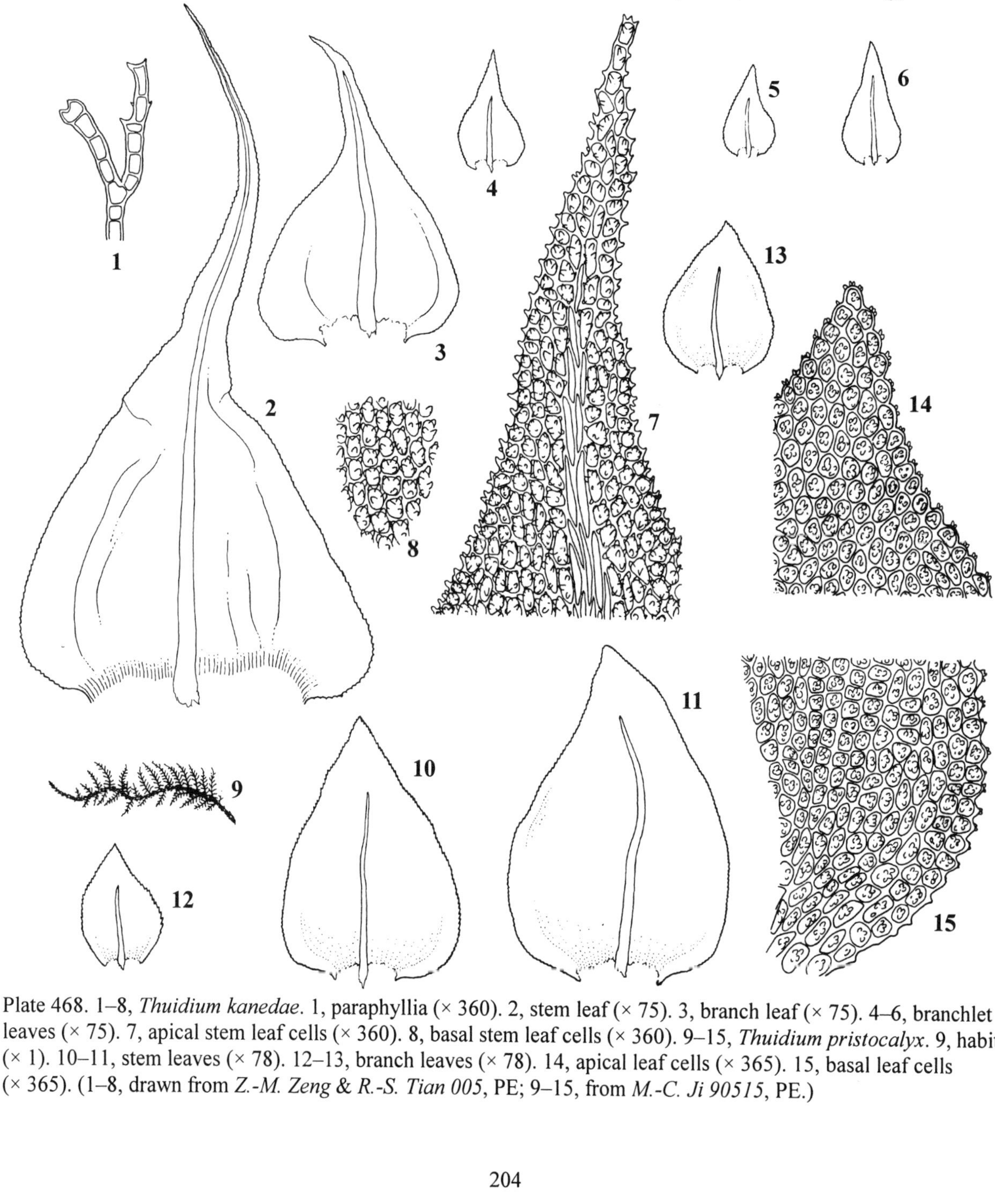

Plate 468. 1–8, *Thuidium kanedae*. 1, paraphyllia (× 360). 2, stem leaf (× 75). 3, branch leaf (× 75). 4–6, branchlet leaves (× 75). 7, apical stem leaf cells (× 360). 8, basal stem leaf cells (× 360). 9–15, *Thuidium pristocalyx*. 9, habit (× 1). 10–11, stem leaves (× 78). 12–13, branch leaves (× 78). 14, apical leaf cells (× 365). 15, basal leaf cells (× 365). (1–8, drawn from *Z.-M. Zeng & R.-S. Tian 005*, PE; 9–15, from *M.-C. Ji 90515*, PE.)

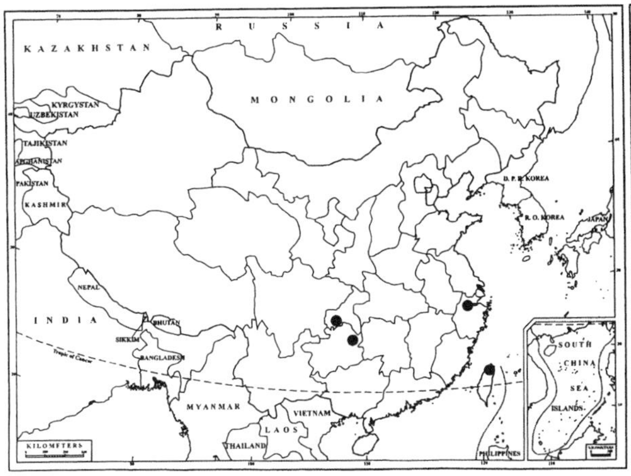

9. *Thuidium subglaucinum*

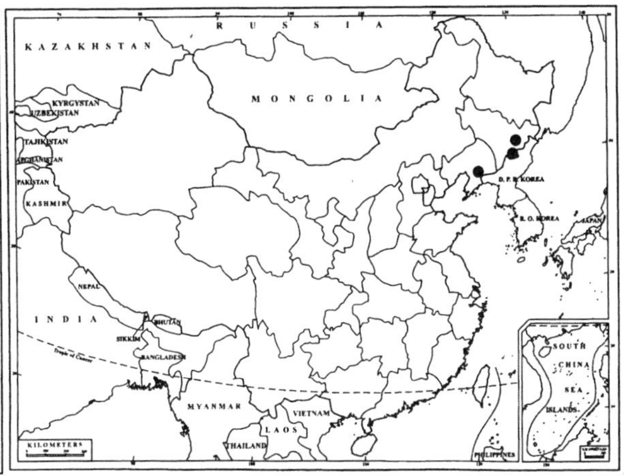

10. *Thuidium submicropteris*

similar to those stem leaves. *Dioicous*. *Perichaetial leaves* narrowly ovate-lanceolate, leaf margins serrate at the apex, entire at lower portion. *Setae* reddish brown or yellowish brown, up to 5 cm long; capsules long cylindrical; exostome teeth broadly lanceolate, reddish brown, papillose above; endostome segments as long as the exostome teeth, finely papillose; basal membrane about 1/2 the length of endostome segments; cilia 2–3. *Spores* ca. 15 µm in diameter, finely papillose.

Illustrations: Plate 468, figs. 9–15.
Habitat: on tree trunks, cliffs, rotten logs, humus or forest ground in low mountain regions; alt. 500–680 m.
Distribution: China, Korea, Japan, India, Sri Lanka, Malaysia, Indonesia, and the Philippines.

Chinese specimens examined: CHONGQING: Chongqing City, *P.-C. Chen 5128* (MO, PE). FUJIAN: Ning-de City, *J. Zhu 36*, *432*, *1121* (HSNU); Mt. Wuyi, *C.-H. Zhang 22-129* (PE). GUANGDONG: Mt. Dinghu, *H. Wu 85* (MO); Xin-feng (Sin-fung) Co., *Taam 1055* (NY). GUANGXI: Long-sheng Co., *J.-M. Jin 942-2* (MO). HAINAN: *P. Zeng 20* (NY); Le-dong Co., *Z.-H. Li 385183* (SYS, PE). JIANGXI: Yuntaishan (Mt.), *M.-C. Ji 90515* (JXAU, PE). LIAONING: Kuan-ding Co., *Buck 23770* (as *Thuidium glaucinum*, NY). TAIWAN: Gao-hsing City, *C.-C. Chuang 1063* (MO, PE, UBC); Miao-li Co., *Shevock 14362* (MO); Nan-tou Co., *C.-I. Peng 84-60* (MO); Tai-chung Co., *C.-C. Chuang 5942* (NY). YUNNAN: An-ning Co., *Redfearn 33300* (MO); Kun-ming Co., *Koponen 37496* (H, MO); Xishuangbanna, *W.-X. Xu 6002*, *6107*, *6158* (PE), *M.-Z. Wang 4375* (PE), *P.-C. Wu 21914*, *21983*, *21990* (PE), *Redfearn et al. 34282* (MO, NY). ZHEJIANG: Xitianmu Shan (Mt.), *B.-C. Tan 93-552* (FH).

9. ***Thuidium subglaucinum*** Card. 亚灰羽藓 ya hui-yu xian
Bull. Soc. Bot. Genève, ser. 2, 3: 283. 1911. Type. Korea: Quelpart, *Faurie 701*.

Plants large, but very delicate, ca. 10 cm long, brownish green to yellowish brown, in loosely interwoven patches. *Stems* ascending to nearly erect, irregularly to regularly bipinnately branched; central strand differentiated; paraphyllia numerous, foliose or filamentous, branched. *Stem leaves* involute when dry, widely ovate to ovate-triangular, with a short sharp apex, nearly 2 mm long; margins serrulate; costa reaching 3/4 the leaf length or ending below leaf apex, smooth or more or less papillose on the back; median leaf cells oval to oblong-oval, thick-walled, multipapillose; branch leaves ovate to widely ovate, reaching the middle of leaf length. *Dioicous*. *Perichaetial leaves* ovate-lanceolate, costa long excurrent, margins long ciliate on shoulder. *Setae* yellowish brown, ca. 2–3 cm long; capsules oblong-ovate; exostome teeth broadly lanceolate, reddish brown, papillose above; endostome segments as long as the exostome teeth, finely papillose; basal membrane about 1/2 the length of endostome segments; cilia 2–3. *Spores* 15–30 µm in diameter, finely papillose.

Illustrations: Watanabe 1972 (Pl. 11, figs. 1–18).
Habitat: on moist forest ground or shady wet rocks.
Distribution: China, Korea, and Japan.

Chinese specimens examined: CHONGQING: Nan-chuan Co., *X.-Y. Hu 0207* (MO, PE). GUIZHOU: Mt. Fanjing, *B.-C. Tan 91-1136* (FH). TAIWAN: Tai-pei Co., *S.-H. Lin s.n.* [Bryophytes of Taiwan Exsiccatae ser. 3, no. 148] (MO). ZHEJIANG: Hang-zhou City, *B.-C. Tan 93552* (NY).

This species is very close to *Thuidium glaucinum*, but differs in having ciliate perichaetial leaves and scatteredly papillose leaf cells. The latter species has entire perichaetial leaves and stellately papillose leaf cells.

10. ***Thuidium submicropteris*** Card. 短枝羽藓 duan-zhi yu xian

Beih. Bot. Centralbl. 17: 28. 1904. Type. Korea: Kan-Quen-to, *Faurie 64*.

Thuidium brevirameum Dix., Rev. Bryol. Lichénol. 7: 112. 1934. Type. China: Liaoning (Ryonei), Mt. Koô (Fenghuang), *Kobayashi 3891* (holotype BM); *Kobayasi 3899a, 3914* (paratypes BM).

Plants rather robust, up to 5 cm long, mostly brownish, green, in loosely interwoven mats. *Stems* creeping, regularly bipinnately branched; branches remote, ca. 0.5 cm long; central strand absent; paraphyllia sparse, foliose or filamentous, branched, unipapillose. *Stem leaves* incurved when dry, erect-patent when moist, abruptly narrowed from a cordate-ovate or cordate base to a hyaline, lanceolate apex, ca. 2 mm long; margins serrulate or entire; costa reaching middle or 2/3 the leaf length, papillose on the back; median leaf cells oval to oblong, 5–20 µm long, thick-walled, with a single stellate papilla; branch leaves concave, ovate to broadly ovate, 0.8 mm long, costa reaching upper portion of leaf. *Dioicous*. *Inner perichaetial leaves* ovate-lanceolate, margins entire, costa ending above the

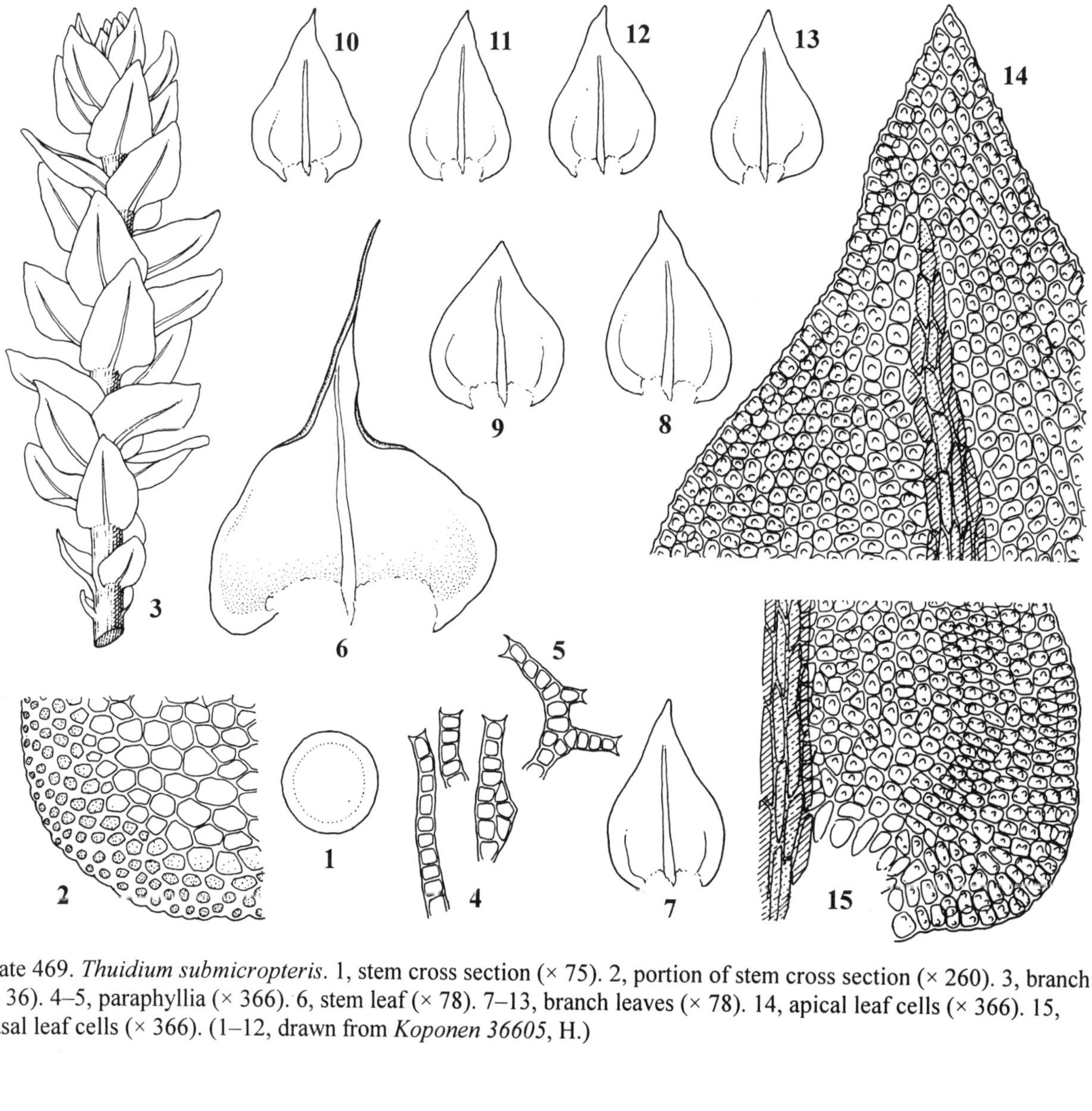

Plate 469. *Thuidium submicropteris*. 1, stem cross section (× 75). 2, portion of stem cross section (× 260). 3, branch (× 36). 4–5, paraphyllia (× 366). 6, stem leaf (× 78). 7–13, branch leaves (× 78). 14, apical leaf cells (× 366). 15, basal leaf cells (× 366). (1–12, drawn from *Koponen 36605*, H.)

middle of leaf. *Setae* yellowish brown to dark brown, 2 cm long; capsules ovoid; exostome teeth broadly lanceolate, reddish brown, papillose above; endostome segments as long as the exostome teeth, finely papillose; basal membrane about 1/2 the length of endostome segments; cilia 2–3. *Spores* 12–15 µm in diameter, finely papillose.

Illustrations: Plate 469, figs. 1–12.
Habitat: in temperate forest ground.
Distribution: China, Korea, and Japan.

Chinese specimens examined: JILIN: An-tu Co., *Koponen 36605* (as *Thuidium minutulum*, NY); Mt. Changbai, *T. Koponen 36603* (H, PE).

This species is very close to *Thuidium kanedae*, but the latter species has central strand and ciliate inner perichaetial leaves.

11. ***Thuidium tamariscinum*** (Hedw.) Bruch & Schimp. in B. S. G. 密枝羽藓 mi-zhi yu xian
Bryol Eur. 5: 163, 482. 1852. *Hypnum tamariscinum* Hedw., Sp. Musc. Frond. 261. 1801. Type. France.
Thuidium tamariscifolium Lindb., Not. Sällsk. Fauna Fl. Fenn. Förh. 13: 415. 1874, *hom. illeg.*

Plants large, up to 15 cm long, yellowish green or brownish green, in densely interwoven mats. *Stems* procumbent, regularly bi- to tripinnately branched; branches about 1 cm long; central strand present; paraphyllia numerous, foliose or filamentous. *Stem leaves* appressed when dry, erecto-patent when moist, widely triangular to triangularly ovate, gradually tapering into a short acumen, slightly plicate, about 1.0–1.5 mm × 0.5–1.0 mm; margins serrulate; costa stout, vanishing at upper portion of leaf, often papillose on the back; leaf cells rhomboidal to oblong, median leaf cells about 6–20 µm long, unipapillose; branch leaves ovate to widely ovate, concave shortly acute, about 0.50 mm × 0.75 mm long; costa vanishing below leaf apex. *Dioicous*. *Perichaetial leaves* larger, sheathing at base, suddenly narrowed to a long reflexous apex, slightly ciliate at shoulder, serrate above, costa excurrent. *Setae* slender, reddish brown, up to 2 cm long, smooth; capsules cylindrical, slightly asymmetrical, about 3 mm long, horizontal, dark reddish brown; opercula conical; exostome teeth reddish brown, grossly papillose above; endostome segments as long as the teeth, finely papillose above; lightly yellowish; basal membrane about 1/2 the height of segments; cilia 2–3, finely papillose. *Spores* ca. 20 µm in diameter.

Illustrations: Watanabe 1972 (Pl. 2, figs. 1–18).
Habitat: often growing with *Hylocomium splendens* and *Rhytidium rugosum* in subalpine forest regions.
Distribution: China, Japan, Russia (Far East), Europe, and Jamaica.

Chinese specimens examined: North China: without exact data. TAIWAN: Tai-pei Co., *M.-J. Lai 7330* (MO). Reported from Sichuan (Ma-er-kang Co., Nan-ping Co., and Song-pan Co.) by Y.-M. Fan & Koponen (2001).

This species is very close to *Thuidium delicatulum* (Hedw.) Mitt., but the latter species is distinguished by having papillosely piliferous leaf apices and densely papillose leaf cells.

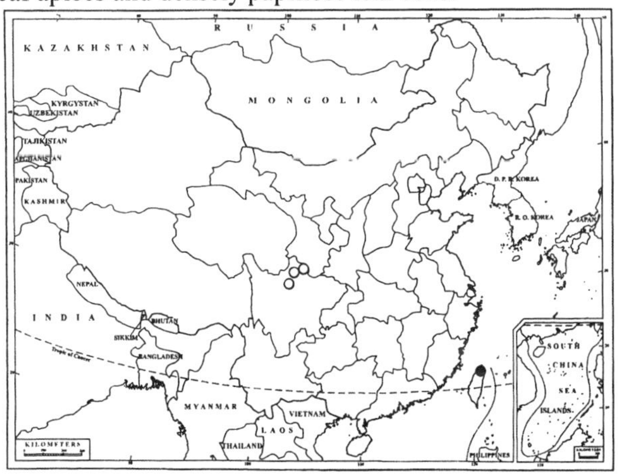

11. *Thuidium tamariscinum*

Literature Cited

Akiyama, H. 1990a. Taxonomic studies of mosses of Seram and Ambon (Moluccas, East Indonesia) collected by Indonesian-Japanese Botanical Expeditions II. J. Fac. Sci. Univ. Tokyo, Sect. 3, Bot. 14: 385–413.

Akiyama, H. 1990b. Taxonomic studies of mosses of Seram and Ambon (Moluccas, East Indonesia) collected by Indonesian-Japanese Botanical Expeditions II. Acta Phytotax. Geobot. 41: 129–142.

Akiyama, H. 1992. Taxonomic studies of mosses of Seram and Ambon (Moluccas, East Malesia) collected by Indonesia-Japanese Botanical Expeditions, V. Hypopterygiaceae. Acta Phytotax. Geobot. 43: 111–120.

Allen, B. H. 1987. A revision of *Leucomnium* (Leucomniaceae). Mem. New York Bot. Gard. 45: 661–677.

Ando, H. 1964. A revision of east-Asian species of *Homomallium*, a genus of Musci. I. Hikobia 4: 28–42.

Andrews, A. L. 1954. Taxonomic notes. XII. The families Rhytidiaceae and Hylocomiaceae. Bryologist 57: 251–261.

Bartram, E. B. 1935. Additions to the flora of China. Ann. Bryol. 8: 6–21.

Bartram, E. B. 1939. Mosses of the Philippines. Philipp. J. Sci. 68: 1–434.

Brotherus, V. F. 1906. Fragmenta ad floram bryologicam Asiae orientalis cognoscendam. II. Trudy Troickos.-Kjahtinsk. Otd. Priamursk. Otd. Imp. Russk. Geogr. Obšč. 8: 1–10.

Brotherus, V. F. 1907. Fabroniaceae–Leskeaceae, pp. 899-1008. *In* A. Engler & K. Prantl, "Die Natürlichen Pfanzenfamilien" I(3): v + 1246 pp. W. Engelmann, Leipzig.

Brotherus, V. F. 1924. Musci novi sinensis collecti a Dre. Henr. Handel-Mazzettii. II. Akad. Wiss. Wien, Math.-Naturwiss. Kl. Abt. 1, 133: 559–584.

Brotherus, V. F. 1925. Musci. *In* A. Engler & K. Prantl, "Die Natürlichen Pfanzenfamilien" ed. 2, 11: iv+542 pp. W. Engelmann, Leipzig.

Brotherus, V. F. 1928. Contributions à la flore bryologique du Cachemire. Ann. Bryol. 1: 28–46.

Brotherus, V. F. 1929. *In* H. Handel-Mazzetti, Symbolae Sinicae. Botanische Ergebnisse der Expedition der Akademie der Wissenschaften in Wien nach Südwest-China 1914/1918. IV. Musci. Julius Springer, Wien.

Buck, W. R. 1977. A taxonomic investigation of *Juratzkaea* Lor. and *Juratzkaeella*, gen. nov. Rev. Bryol. Lichénol. 43: 309–325.

Buck, W. R. 1987. Taxonomic and nomenclatural rearrangement in the Hookeriales with notes on West Indian taxa. Brittonia 39: 210–224.

Buck, W. R. & H. Crum. 1978. A re-interpretation of the Fabroniaceae with notes on selected genera. J. Hattori Bot. Lab. 44: 347–369.

Buck, W. R. & H. Crum. 1990. An evaluation of familial limits among the genera alligned with the Thuidiaceae and Leskeaceae. Contr. Univ. of Michigan Herb. 17: 55–69.

Buck, W. R. & D. H. Vitt. 1986. Suggestions for a new familial classification of pleurocarpous mosses. Taxon 35: 21–60.

Cardot, J. 1905. Mousses de l'ile Formosae. Beih. Bot. Centralbl. 19: 85–148.

Cardot, J. & I. Thériot. 1904. Mousses du Kouy-Tcheou (Chine), recoltes par E. Bodinier et determined par J. Cardot et I. Thériot. Bull. Acad. Int. Géogr. Bot. 13: 81–86.

Chen, P.-C. et al. 1963. Genera Muscorum Sinicorum, I. 304 pp. Science Press, Beijing.

Chen, P.-C. et al. 1978. Genera Muscorum Sinicorum, II. 331 pp. Science Press, Beijing.

Chen, P.-C. & P.-C. Wu. 1965. The preliminary study of the bryophytes of Mt. Hwangshan. Observations ad Florulam Hwangshanicum. pp. 1-59. Shanghai Science and Technology Press.

Chiang, T.-Y. & C.-M. Kuo. 1989. Notes on bryophytes of Taiwan (1–36). Taiwania 37: 74–156.

Crosby, M. R. 1974. Toward a revised classification of the Hookeriaceae. J. Hattori Bot. Lab. 38: 129–141.

Crosby, M. R., R. E. Magill, B. Allen & S. He. 2000. A checklist of the mosses. 320 pp. Missouri Botanical Garden, St. Louis.

Crum, H. A. & L. E. Anderson. 1981. Mosses of Eastern North America. 2 vols. Columbia Univ. Press, New York.

Dixon, H. N. 1933. Mosses of Hong Kong with other Chinese mosses. Hong Kong Naturalist, Suppl. 2: 1–31.

Dixon, H. N. 1934. Manchurian mosses. Rev. Bryol. Lichénol. 7: 105–116.

Dozy, F. & J. H. Molkenboer. 1861. Bryologia Javanica seu descriptio muscorum frondosorum Archipelagici Indici. II. Leiden.

Eakin, D. A. 1998. A taxonomic revision of the moss genus *Regmatodon*. Nova Hedwigia 67: 139–152.

Enroth, J. 1997. Taxonomic position of *Leptocladium* and new synonymy in Chinese Amblystegiaceae (Bryopsida). Ann. Bot. Fenn. 34: 47–49.

Fang, Y.-M. & T. Koponen. 2001. A revision of *Thuidium*, *Haplocladium*, and *Claopodium* (Musci, Thuidiaceae) in China. Bryobrothera 6: 1–81.
Fleischer, M. 1908. Musci Frond. Archip. Ind. Exsic. ser. 10, 451–500.
Gangulee, H. C. 1977. Mosses of Eastern India and adjacent regions. Fasc. 6, pp. 1464–1546. Published by the author, Calcutta.
Gangulee, H. C. 1978. Mosses of Eastern India and adjacent regions. Fasc. 7, pp. 1547–1652. Published by the author, Calcutta.
Gao, C. (ed). 1977. Flora Muscorum Chinae Boreali-Orientalis [Institute of Forestry and Pedalogy (ed.).] ix + 404 pp. Science Press, Beijing.
Gao, C. 1985. Theliaceae, Fabroniaceae, and Leskeaceae, pp. 294–314. *In* X.-J. Li (ed.), Bryoflora of Xizang, x + 581. Science Press, Beijing.
Gao, C. & G.-C. Zhang. 1982. Genera nova bryophytorum. Bull. Bot. Res., Harbin 2: 113–121.
Gao, C., G.-C. Zhang & T. Cao. 1981. Taxa nova bryophytarum Tibeticum. Acta Bot. Yunnan. 3: 389–399.
Gao, C., T. Cao & W.-H. Wang. 2001. Leskeaceae, pp. 121–156. *In* P.-C. Wu (ed.). Flora Bryophytarum Sinicorum. Vol. 6. Vi + 312 pp. Science Press, Beijing.
Granzow de la Cerda, I. 1989. Notes on five species of *Anomodon*, some with erroneous identity, including two new combinations. Bryologist 92: 381–386.
Granzow de la Cerda, I. 1997. Revision and phylogeny of *Anomodon* and *Herpetineuron* (Anomodontaceae, Musci). Contr. Univ. Michigan Herb. 21: 205–275.
He, S. & J. A. Snider. 2000. A taxonomic revision of *Symphyodon* (Musci: Symphyodontaceae). Bryologist 103: 52–81.
Herzog, T. 1925. Beiträge zur Bryophytenflora von Yunnan. Hedwigia 65: 147–168.
Herzog, T. & A. Noguchi. 1955. Beitrag zur Kenntnis der Bryophytenflora von Formosa und der Benachbarten Inseln Botel Tobago und Kwahwyoto. J. Hattori Bot. Lab. 14: 29–70.
Hong, R.-L. & R.-L. Hu. 1984. A study of mosses of Mt. Jiulong in Zhejiang. Res. For. Ecosys. 4: 207–241.
Hu, R.-L. & Y.-F. Wang. 1981. A survey on the bryophytes from the West Tian Mu Mountain in Zhejing Province. J. E. China Norm. Univ., Nat. Sci. Ed. 1: 85–104.
Hu, X.-Y. & P.-C. Wu. 1991. Study on the moss flora of Mt. Jinfu, Sichuan Province. Acta Phytotax. Sin. 29: 315–334.
Iwatsuki, Z. 1976. Bryological miscellanies. XXI. What is *Pseudoleskeopsis zippelii* (Dozy & Molk.) Broth.? J. Hattori Bot. Lab. 40: 139–143.
Iwatsuki, Z. 1991. Catalog of the Mosses of Japan. 182 pp. Hattori Botanical Laboratory, Nichinan.
Jia, Y., P.-C. Wu & J.-X. Luo. 1995. The mossflora of Mt. Jiuwan, Guangxi and its significance in dividing the boundary line between tropical and subtropical regions in China (2). Acta Phytotax. Sin. 33(5): 556–571.
Kindberg, N. C. 1897. Genera of European and North American Bryineae (Mosses). 410 pp. D. F. Bonniers, Göteborg.
Koponen, T. 1978. Modern taxonomical methods and the classification of mosses. Bryophyt. Biblioth. 13: 443–481.
Koponen, T., C. Gao, J.-X. Luo & I. Jarvinen. 1983. Bryophytes from Mt. Chang Bai, Jilin Province, Northeast China. Ann. Bot. Fennici 20: 215–232.
Kruijer, H. 1995. Systematiek van de Hypopterygiaceae. Buxbaumiella 38: 26–32.
Kruijer, H. 2002. Hypopterygiaceae of the World. Blumea, Suppl. 13: 1–388.
Lai, M.-J. 1976. The Hypopterygiaceae (Musci) of Taiwan. Taiwania 21: 150–158.
Lai, M.-J. & J.-R. Wang-Yang. 1976. Index bryoflorae Formosensis. Taiwania 21(2): 159–203.
Lawton, E. 1957. A revision of the genus *Lescuraea* in Europe and North America. Bull. Torrey Bot. Club 84: 281–307, 337–355.
Levier, E. 1906. Muscinee raccolte nello Schen-si (Cina) dal Rev. Giuseppe Giraldi. Nuovo Giorn. Bot. Ital., n.s. 13: 237–280.
Li, X.-J. (editor-in-chief). 1985. Bryoflora of Xizang. x + 581 pp. Science Press. Beijing.
Lin, P.-J., Z.-H. Yang & Z.-H. Li. 1982. A study of the bryophytes of Ding-Hu-Shan. Trop. Subtrop. Forest Ecosyst. 1: 58–76.
Lin, P.-J. & Z.-H. Li. 1985. Hypopterygiaceae, pp. 292–294. *In* X.-J. Li (editor-in-chief), Bryoflora of Xizang. x + 581 pp. Science Press, Beijing.
Lin, P.-J., T. Koponen & S. Piippo. 1992. Bryophyte flora of Jianfenling Mts., Hainan Island, China. Bryobrothera 1: 195–214.
Lin P.-J. & B. C. Tan. 1995. Contribution to the bryoflora of China (12): a taxonomic revision of Chinese Hookeriaceae (Musci). Harvard Pap. Bot. 7: 25–68.

Lin, Q.-W. 1989. The bryophytes of the Leigong Mountains Nature Reserve, pp. 342–354. *In* Z.-X. Zhou & M.-S. Yao (eds.), Scientific Survey of Leigong Nature Reserve. Guiyang.

Lin, Q.-W. & J. Zhou. 1991. Study on the bryophytes of Chishui *Alsophila* Natural Reserve, pp. 105–166. *In* Z.-X. Zhou (ed.), Scientific Survey of Chishui *Alsophila* Nature Reserve. Guiyang.

Lin, S.-H. & C.-K. Wang. 1977. Discovery of *Okamuraea brevipes* Broth. on *Eucommia ulmoides* Oliv., a medicinal plant from mainland China. Biol. Bull. Dept. Biol. Coll. Sci. Tunghai Univ. 46: 237–243.

Lou, J.-S. & P.-C. Wu. 1980. A preliminary report on the new bryophytes of Xizang (Tibet). Acta Phytotax. Sin. 18: 119–125.

Magill, R. E. (ed.). 1990. Glossarium polyglottum bryologiae: A multilingual glossary for Bryology. Monogr. Syst. Bot. Missouri Bot. Gard. 33: 1–297.

Mitten, W. 1859. Musci Indiae Orientalis. An enumeration of the mosses of the East Indies. J. Proc. Linn. Soc., Bot., Suppl. 1: 1–171.

Mohamed, H. & H. Robinson. 1991. A taxonomic revision of the moss families Hookeriaceae and Hypopterygiaceae in Malaya. Smithsonian Contr. Bot. 80: i–iv. 1–44.

Müller, C. 1896. Bryologia Provinciae Schen-si Sinensis. I. Nouvo Giorn. Bot. Ital., n.s. 3: 89–129.

Müller, C. 1897. Bryologia Provinciae Schen-si Sinensis. II. Nouvo Giorn. Bot. Ital., n.s. 4: 245–276.

Noguchi, A. 1951. Notulae Bryologicae II. J. Hattori Bot. Lab. 5: 40–42.

Noguchi, A. 1991. Illustrated Moss Flora of Japan. Part 4. (Hookeriaceae–Brachytheiaceae), 743–1012. Supplemented by Z. Iwatsuki & T. Yamaguchi. Hattori Botanical Laboratory, Nichinan.

Noguchi, A. & Z. Iwatsuki. 1972. Mosses of North Borneo. J. Hattori Bot. Lab. 36: 455–486.

Paris, E. G. 1909. Muscinées de l'Asie Orientale. 9. Rev. Bryol. 36: 8–13.

Potier de la Varde, R. 1937. Contribution à la flore bryologique de la Chine. Rev. Bryol. Lichénol. 10: 136–145.

Redfearn, P. L., Jr. & P.-C. Wu. 1986. Catalog of the mosses of China. Ann. Missouri Bot. Gard. 73: 177–208.

Redfearn, P. L., Jr., B. C. Tan & S. He. 1996. A newly updated and annotated checklist of Chinese mosses. J. Hattori Bot. Lab. 79: 163–357.

Reimers, H. 1931. Beiträge zur Moosflora Chinas I. Hedwigia 71: 1–77.

Sayre, G., C. E. B. Bonner & W. L. Culberson. 1964. The authorities for the epithets of mosses, hepatics and lichens. Bryologist 67: 113–135.

So, M.-L. 1995. Mosses and Liverworts of Hong Kong. xi + 162 pp. Heavenly People Depot, Hong Kong.

Stafleu, F. A. & R. S. Cowan. 1976–1988. Taxonomic literature. Second edition. Regnum Veg. 94, 98, 105, 110, 112, 115, 116.

Takaki, N. 1955. Researches on the Brachytheciaceae of Japan and its adjacent areas (I). J. Hattori Bot. Lab. 14: 1–28.

Tan, B. C. & Y. Jia. 1997. Mosses of Qinghai-Tibetan plateau, China. J. Hattori Bot. Lab. 82: 305–320.

Tan, B. C., Q.-W. Lin, M. R. Crosby & P.-C. Wu. 1994. A report on thc 1991 Sino-American Bryological Expedition to Guizhou Province, China: new and noteworthy additions to Chinese moss taxa. Bryologist 97: 127–137.

Tan, B. C., D. Boufford & T.-S. Ying. 1996. A contribution to the moss flora of Henan Province, China. Acta Bot. Yunnan. 18: 67–71.

Tan, B. & H. Robinson. 1990. A review of Philippine hookeriaceous taxa (Musci). Smithsonian Contr. Bot. 75: iv + 41 pp.

Taoda, H. 1977. Studies on the Fabroniaceae of Japan I. Hikobia 8: 46–58.

Thériot, I. 1907. Diagnoses d'espèces nouvelles. Monde Pl. Ser. 2, 9: 21–22.

Thériot, I. 1911. Diagnoses d'espèces et variétés nouvelles de mousses. 9. Bull. Acad. Int. Géogr. Bot. 19: 269–272.

Thériot, I. 1926. Reliquiae Coppeyanae. Bull. Soc. Sci. Nancy, sér. 4, 2(6): 717.

Thériot, I. 1929. Le genre *Pseudoleskeopsis*. Ann. Cryptog. Exot. 2: 5–22.

Touw, A. 2001a. A review of the Thuidiaceae (Musci) and a realignment of taxa traditionally accommodated in *Thuidium* sensu amplo [*Thuidium* Schimp., *Thuidiopsis* (Broth.) M. Fleisch., and *Pelkium* Mitt.], including *Aequatoriella gen. nov.* and *Indothuidium gen. nov.* J. Hattori Bot. Lab. 90: 167–209.

Touw, A. 2001b. A taxonomic revision of the Thuidiaceae (Musci) of tropical Asia, the western Pacific, and Hawaii. J. Hattori Bot. Lab. 91: 1–136.

Wang, C.-W. 1935. Bryological research and literature. Chin. Bot. Mag. 1(4): 415–426.

Watanabe, R. 1964. Thuidiaceae mosses collected by Dr. S. Nakanishi from Mt. Gokuan area in Formosa. J. Jap. Bot. 39: 273–282.

Watanabe, R. 1972. A revision of the family Thuidiaceae in Japan and adjacent areas. J. Hattori Bot. Lab. 36: 171–320.

Watanabe, R. 1980. A note on *Thuidium tibetanum* Salmon from Tibet, China. Misc. Bryol. Lichénol. 8: 145.
Watanabe, R. 1991. Notes on Thuidiaceae. J. Hattori Bot. Lab. 69: 37–47.
Wijk, R. van der, W. D. Margadant & P. A. Florschütz. 1962. Index Muscorum, vol. 2 (D–H). Regnum Veg. 26: 1–535.
Wijk, R. van der, W. D. Margadant & P. A. Florschütz. 1967. Index Muscorum, vol. 4 (P–S). Regnum Veg. 48: 1–604.
Wu, P.-C. 1992. The moss flora of Xishuangbanna, southern Yunnan, China. Trop. Bryol. 5: 27–33.
Wu, P.-C. (ed.). 2001. Flora Bryophytarum Sinicorum. Vol. 6. Vi + 312 pp. Science Press, Beijing.
Wu, P.-C., J.-S. Lou (J.-X. Luo) & M.-Z. Wang. 1984. Glossary of Terms and Names of Bryophytes. 124 + v. Science Press, Beijing.
Wu, P.-C., Y.-J. Ling & S.-L. Xie. 1987. A preliminary Study on the Bryophytes in Shanxi Province, North China. J. Shanxi Univ. 2: 88–92.
Wu, P.-C. & Y. Jia. 2000. A revision of Chinese Thuidiaceae (s.l., Musci). Acta Phytotax. Sin. 38: 256–265.
Xu, B.-S. (editor-in-chief). 1989. Cryptogamic Flora of the Yangtze Delta and Adjacent regions. Mosses and liverworts, p. 267–413. By Liu, Z.-L., Gao, C.-H. & D.-K. Li. B.-S. Science and Technology Publishers. Shanghai.
Yang, B.-Y. & W.-C. Lee. 1964. Bryophytic flora of Chi-You. Bot. Bull. Acad. Sin. 6: 181–195.
Yang, C.-Y. 1936. An enumeration of moss flora in Chihli Province. Sci. Rep. Nat. Tsing Hua Univ., B, Biol. Sci. 2: 111–135.
Zeng, S.-Y. 1991. A new species of the *Bryonoguchia* from Yunnan. Acta Bot. Yunnan. 13: 377–379.
Zhang, L. 1990. A study on the mosses of Menglun, Xishuangbanna, pp. 1–54. M.S. Thesis. Kunming Institute of Botany, Academica Sinica.
Zhang, M.-X. 1978. Flora Tsinglingensis. Tom III. Bryophyta (Par. 1). 329 pp. Science Press, Beijing.
Zhao, J.-C. 1993. A preliminary study on the bryophytes of eastern section of Tianshan Mountains, Northwest China. Chenia 1: 99–112.
Zhong, B.-G. & Y.-X. Xiong. 1990. A catalogue of Musci of Guizhou (II). J. Guizhou Norm. Univ. 14: 22–31.

Index to Latin Names

Names in **bold** indicate new nomenclature; in *italic*, new synonyms. Numbers in **bold** indicate pages containing descriptions for accepted taxa; in Roman, incidental mentions. Numbers followed by an asterisk (*) indicate pages on which illustrations appear.

Index to Chinese Names

This index includes Chinese names for families, genera, accepted species, and infraspecific taxa. The names are arranged according to the number of strokes in the first Chinese character of the names.

七 画 **(seven strokes)**

八 画 **(eight strokes)**

九 画 **(nine strokes)**

R
U
KAZAKSTAN
MONGOLIA
UZBEKISTAN
TADZHIKISTAN
KIRGHIZSTAN
TADZHIKISTAN
XINJIANG UYGUR ZIZHIQU
AFGHANISTAN
KASHMIR
PAKISTAN
NEI
QINGHAI
NINGXIA HUIZU ZIZHIQU
GANSU
XIZANG ZIZHIQU
NEPAL
INDIA
SICHUAN
SIKKIM
BHUTAN
Tropic of Cancer
INDIA
BANGLADESH
GUIZHOU
YUNNAN
ZIZHIQU
MYANMAR
VIET NAM
LAOS
THAILAND
THAILAND
SCALE 1:18000000
0 200 400 600